An Introduction to Galaxies and Cosmology

Compiled by a team of experts from The Open University and other UK universities, this revised edition has been designed for elementary university courses in astronomy and astrophysics. It starts with a detailed discussion of the structure and history of our own Galaxy, the Milky Way, and goes on to give a general introduction to normal and active galaxies including models for their formation and evolution. The latter half of the book provides an overview of the wide range of cosmological models and discusses the big bang and the expansion of the Universe. Written in an accessible style that avoids complex mathematics, and illustrated in colour throughout, this book is suitable for self-study and will appeal to amateur astronomers as well as undergraduate students. The book is also supported by a website hosting further teaching materials:
http://www.cambridge.org/galaxiesandcosmology

About the editors:

MARK H. JONES is Senior Lecturer and Staff Tutor in the Department of Physical Sciences at The Open University. He studied for his PhD in the X-ray Astronomy Group at the University of Leicester before carrying out post-doctoral research in space-based infrared astronomy at Queen Mary and Westfield College (University of London). His current research concentrates on the structure of the zodiacal cloud (interplanetary dust cloud). He is a Fellow of the Royal Astronomical Society and a Member of the Institute of Physics.

ROBERT J.A. LAMBOURNE is Professor of Educational Physics, Department of Physical Sciences at The Open University. He obtained his PhD at Queen Mary College (University of London) where he worked on the quantum theory of fundamental particles. His career in university teaching has spanned many areas of mathematics, physics and astronomy, with a particular emphasis on relativity and cosmology. Professor Lambourne is also a very active popularizer of science. He is a Fellow of the Royal Astronomical Society and a Fellow of the Institute of Physics. In 2002 he was awarded the Bragg Medal of the Institute of Physics in recognition of his contributions to physics education. He became a National Teaching Fellow in 2006 and was awarded the European Physical Society's Gero Thomas Commemorative Medal in 2014.

STEPHEN SERJEANT is Head of the Astronomy Discipline and Reader in Cosmology at The Open University. He studied for his doctorate at Oxford University, working on radio-loud quasars and on a gravitationally lensed hyperluminous galaxy. After spending time at Imperial College as a Research Fellow and as a Lecturer at the University of Kent, he joined The Open University in 2006. He currently works on star-forming galaxies and gravitational lenses. He co-leads the active galaxies science theme of the ATLAS Key Project on the Herschel Space Observatory and leads Herschel's legacy survey at the North Ecliptic Pole. Dr Serjeant is also the lead academic for the BBC science show *Bang Goes the Theory* and has consulted for *Stargazing Live* and *Light and Dark*. He is a Fellow of the Institute of Physics, a Fellow of the Royal Astronomical Society and a Fellow of the Higher Education Academy. He has written or co-written over 170 papers in academic journals.

An Introduction to
Galaxies and
Cosmology

Edited by Mark H. Jones, Robert J.A. Lambourne
and Stephen Serjeant

Authors:
Alan Cayless
Helen Fraser
Anthony W. Jones
Mark H. Jones
Robert J.A. Lambourne
Sean G. Ryan
Stephen Serjeant

Second Edition

Including material updated
from contributions by:
David J. Adams
Barrie W. Jones
Elizabeth Swinbank
Lesley I. Onuora
Andrew N. Taylor

The Open University   CAMBRIDGE
UNIVERSITY PRESS

PUBLISHED BY

CAMBRIDGE UNIVERSITY PRESS

University Printing House, Shaftesbury Road, Cambridge CB2 8BS, United Kingdom.

www.cambridge.org

Cambridge University Press is part of the University of Cambridge.

In association with THE OPEN UNIVERSITY

The Open University, Walton Hall, Milton Keynes MK7 6AA, United Kingdom.

www.open.ac.uk

First published 2003. Revised 2004, 2015.

Edited and designed by The Open University.

Typeset by The Open University.

Printed in the United Kingdom by Bell & Bain Ltd, Glasgow.

This publication forms part of the Open University module S282 *Astronomy*. Details of this and other Open University modules can be obtained from the Student Registration and Enquiry Service, The Open University, PO Box 197, Milton Keynes MK7 6BJ, United Kingdom (tel. +44 (0)845 300 60 90, email general-enquiries@open.ac.uk).

British Library Cataloguing in Publication data available on request.

Library of Congress Cataloguing in Publication data available on request.

Additional resources for this publication at www.cambridge.org/galaxiesandcosmology

ISBN 978 1 107 49261 5 paperback

3.1

CONTENTS

INTRODUCTION

To the unaided eye, the night sky offers a quite misleading impression of the distribution of matter in the Universe around us. We see stars in every direction, and we could be excused for jumping to the conclusion that the cosmos is simply filled with a more-or-less uniform distribution of stars. We now recognize this view to be incorrect: stars are, in fact, distributed in vast stellar systems called galaxies, and we live within one such system – the Milky Way Galaxy. The starting point of this book is to study our own Galaxy, before moving on to consider galaxies in general, and the special, but important, case of so-called *active galaxies*. Equally importantly, we shall also discuss the astronomical techniques and the scientific reasoning that has led to our current view of galaxies.

It took painstaking efforts by pioneering astronomers, most notably Edwin Hubble, to establish the existence of galaxies external to the Milky Way. However, once this leap had been made in the 1920s, the scene was set for an exploration of the Universe on ever larger scales. The picture that has emerged is that galaxies appear to be the 'building blocks' that make up the large-scale distribution of matter in the Universe. So, a question arises as to why the Universe is organized in this way – how, and why, do galaxies form? Although there are no simple answers to these questions, throughout this book we shall see how these problems are currently being tackled by astronomers.

Observations of galaxies play a key role in *cosmology* – the scientific study of the nature and evolution of the Universe as a whole. It was, again, Edwin Hubble who made the observations that revolutionized this field. His work led to the single most important result in cosmology: that the Universe is expanding – the distances between galaxies are increasing as time progresses. Once this idea is accepted, it is not difficult to deduce that, in the past, galaxies must have been closer together. This idea is taken to its logical extreme in the idea of the *big bang* model: at some finite time in the past – about 14 billion years ago – the separation between objects in the Universe would have been extremely small, and ever since this time the Universe has been expanding. Later in this book we shall explore the reasoning behind, and the implications of, this model in much more detail. However, since the idea of the big bang plays such a fundamental role in modern astronomy and cosmology it is worth here highlighting a few of the key features of this model:

- The Universe has a finite age, which is currently estimated to be 13.8 billion years.
- The Universe has been expanding since the very first instant of the big bang.
- The early stages of the evolution of the Universe were characterized by high temperatures and densities. Furthermore, at any given time, the density and temperature were highly, although not perfectly, uniform.
- Within the first few minutes of the big bang, nuclear reactions formed light nuclei. As a result, the fraction (by mass) of material in the Universe after these processes came to an end was about 76% hydrogen, 24% helium and a trace amount of lithium.

A surprising feature of modern astronomy is that much of cosmic history can be viewed directly. Light, and other forms of electromagnetic radiation, travels at the finite speed of 3×10^8 m s^{-1}. It takes about two million years for light to reach us from the nearby Andromeda Galaxy – so the images we take today actually show the appearance of this galaxy as it was two million years ago. Two million years is a very short time by cosmic standards, but the same principle applies to observations of much more distant galaxies. Current observations using the most sensitive telescopes can view galaxies as they appeared when the age of the Universe was just a few hundred million years. Such studies are now allowing astronomers to piece together a picture of the formation and evolution of galaxies over the history of the Universe. And this is not the limit of our vision – we can also detect radiation that gives us a view of the Universe as it appeared when it was only about 370 000 years old. Detailed studies of this *cosmic microwave background*, along with other observations, have heralded an era in which cosmologists can accurately determine the parameters of a model that describes our Universe. These advances have taken science to a position where we can probe the processes that occurred in the first few fractions of a second after the big bang.

In this book, we shall explore the implications of the latest observational techniques for astronomy and cosmology. We hope that, in doing so, we will share with you some of the sense of excitement that a scientific study of the Universe can bring.

Note: It is assumed that readers of this book already have an understanding of the fundamental aspects of stellar astronomy and the processes of stellar evolution. These topics are dealt with in detail in the companion volume to this book – *An Introduction to the Sun and Stars* (ISBN 978 1 107 49263 9).

CHAPTER 1
THE MILKY WAY – OUR GALAXY

1.1 Introduction

To the observer, the Milky Way is the faint band of diffuse light that arches across the night sky from horizon to horizon (Figure 1.1). This light comes mainly from a multitude of stars, although the unaided eye is unable to resolve these stars individually, hence the appearance of a 'band' of light. The stellar nature of the visible Milky Way was revealed about four hundred years ago when Galileo Galilei (1564–1642) made some of the earliest astronomical observations using a telescope.

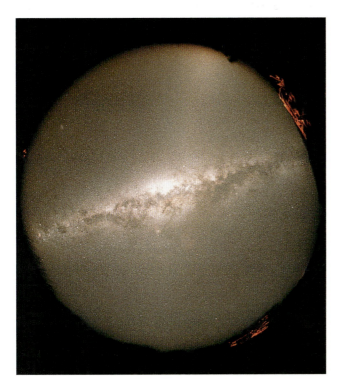

Figure 1.1 A photograph of one hemisphere of the night sky, showing the Milky Way stretching from horizon to horizon. The most prominent, central portion of the Milky Way is directly overhead in this image. The light comes mainly from the enormous number of stars that exist within our Galaxy; there are about 10^{11} in all, but many are too faint to be seen, or are obscured by the dust that is also part of the Galaxy. (D. di Cicco, Sky Publishing Corp.)

Although Galileo recognized the existence of huge numbers of stars in the Milky Way in around 1610, it was not until the 20th century that astronomers were able to deduce the distribution of those stars with any accuracy, and it is only over the past few decades that they have arrived at what is believed to be a true understanding of the nature of the Milky Way. We now know roughly how far the Milky Way extends in each direction in space, and that its light comes mostly from stars distributed in a flattened disc-like structure some 100 000 light-years across. We also know that, in addition to stars, the Milky Way contains gas and dust. Perhaps most astonishingly of all, the majority of astronomers have now become convinced that such familiar entities as stars, gas and dust account for no more than about 10% of the mass of the Milky Way. Most of the mass, a staggering 10^{12} times the mass of the Sun (i.e. $10^{12} M_\odot$, where $M_\odot \approx 2 \times 10^{30}$ kg), is now believed to be attributable to some kind of unidentified form of matter known, somewhat enigmatically, as 'dark matter'. The term **Milky Way** is now applied to this whole collection of entities – the system of stars, gas, dust, and dark matter – not just to the diffuse band of starlight resolved by Galileo.

Figure 1.2 An 'external' galaxy (NGC 6744) thought to be similar to the Milky Way. If this really represented the Milky Way, the Sun would be located about halfway between the centre and the edge of the flattened 'disc' of stars, gas and dust. This location, within a relatively thin disc, explains why we mainly see the Milky Way as a 'band' encircling the Earth. (S. Lee, C. Tinney and D. Malin/AAO)

Does anything exist beyond the boundary of the Milky Way? The answer is an emphatic 'yes'. At still greater distances, beyond the limits of the 100 000 light-year disc, astronomers have discovered other huge collections of stars, gas, dust and dark matter that, like the Milky Way, occupy relatively well-defined, and usually well-separated, volumes of space. These structures are called **galaxies**. Figure 1.2 shows one of these 'external' galaxies that is thought to be somewhat similar to the Milky Way in many respects. The Milky Way is simply our galaxy, the galaxy in which we have been born and have come of age. To emphasize this, the Milky Way is often referred to as 'the Galaxy', the capital 'G' distinguishing it from the billions of other galaxies in the observable Universe. In later chapters you will learn more about those other galaxies. Here in Chapter 1 we concentrate on the Milky Way.

In the sections that follow you will learn a great deal about the Milky Way. Section 1.2 provides a general overview of the Milky Way as a galaxy, including its structure, size and composition. Section 1.3 is devoted to the mass of the Milky Way. Sections 1.4 and 1.5 discuss some of the main structural components of the Galaxy in detail, and Section 1.6 considers how the Milky Way has changed with time. As you study these sections you will also gain insight into the process of astronomical science. You will see, in outline at least, how the nature of the Galaxy has been uncovered, how the disc of the Milky Way has been shown to be about 100 000 light-years across, how the Milky Way's mass has been roughly determined to be about $10^{12} M_\odot$ and how astronomers have determined that the Galaxy contains some stars that formed about 12×10^9 years ago. By examining the process of astronomy you will begin to understand how the making of careful observations, combined with the continuous review of their significance in the light of physical laws, enables findings and theories to be critically examined, refined and improved. Developing an understanding of this process is more important than learning any particular fact or figure.

1.2 An overview of the Milky Way

This section examines some of the general features of the Milky Way as a galaxy. It starts with an introduction to the main structural components of the Milky Way, and then goes on to examine the sizes of those components and their constituents. Detailed discussion of the mass of the Milky Way is deferred until Section 1.3.

1.2.1 The structure of the Milky Way

Since the Galaxy is a system of stars, it seems natural to consider its structure by mapping the locations of stars in space. While this approach seems simple enough, historically it was hampered by the fact that visible light is strongly scattered and absorbed by dust in the space between the stars. Some indication of this can be seen from Figure 1.1, in which the band of light from the Milky Way seems to be obscured by dark clouds. A solution to this observational problem is provided by observing in the near-infrared – at wavelengths of about 1–2 μm, just beyond the visible spectrum. Infrared light is subject to much less absorption and scattering (together termed *extinction*) by dust and this affords astronomers a much clearer

Figure 1.3 A map of the near-infrared emission from stars in the Galaxy based on star counts from the 2MASS survey. This map is centred on the centre of the Galaxy and extends about 180° along the Galactic plane. The diffuse objects in the lower right part of the map are satellite galaxies of the Milky Way called the Large and Small Magellanic Clouds. (2MASS/UMass/IPAC-Caltech/NASA/NSF)

view of the stars in the Galaxy. Figure 1.3 shows a remarkable map that was constructed from an all-sky near-infrared survey called 2MASS (2 Micron All Sky Survey). The map essentially shows the distribution of starlight across the sky. Two components of the Galaxy are immediately evident: a central concentration and a flattened disc.

Towards the centre of the Galaxy the density of stars increases, and the Milky Way appears to be 'thicker' there than further out. This broad, central region is called the **bulge.** Furthermore this bulge is elongated into a bar similar to the central regions of NCG 6744 (Figure 1.2). The Milky Way's bulge has a mass of around $10^{10} M_\odot$.

Most of the stars, including the Sun, occupy a flattened, disc-shaped volume. This is called the **Galactic disc**, or simply the **disc**, and has a mass of about $10^{11} M_\odot$. In addition to stars, the Galactic disc contains a substantial mass of gas and dust in the space between the stars. Furthermore interstellar gas extends, in what is called the gaseous disc, to a greater radius than the stellar disc. The mid-plane of the disc defines the **Galactic plane**, which plays an important part in providing a system of coordinates for defining positions in the Milky Way (see Box 1.1). The Sun is located very close to the Galactic plane, about halfway between the centre of the disc and its outer edge. From our vantage point, we see the disc edge-on from within. It is our location that causes most of the other stars in the Milky Way's disc to appear concentrated within a band looping around us – the source of the diffuse band of light that Galileo resolved into stars.

If the disc could be viewed face-on, we would see a spiral pattern due to the presence of bright features called **spiral arms**. (Such arms are an obvious feature of many galaxies; the evidence for their existence in the Milky Way is described in Section 1.4.) The overall spiral shape in Figure 1.2 is clear, but in detail the arms are fragmented and distorted. Although spiral arms are prominent, they stand out because they contain unusually hot, *luminous* stars, and not because they contain unduly large numbers of stars. The reasons why bright stars are concentrated in this way is discussed in Section 1.4. Spiral arms suggest that galaxies are rotating. Sure enough, the Sun and its neighbouring stars in the disc orbit the Galactic centre at speeds of about 200 km s^{-1}. However, as you will see later in this chapter, the stars and the pattern of spiral arms generally travel at *different* speeds.

BOX 1.1 GALACTIC COORDINATES

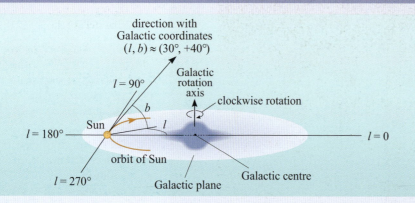

Figure 1.4 Galactic coordinates are centred on the Sun, and use the Galactic plane and the direction approximately towards the Galactic centre to define a frame of reference. A direction with $l \approx 30°$ and $b \approx +40°$ is shown.

Although astronomers use the equatorial coordinate system, based on right ascension (RA) and declination (often referred to as 'dec') to locate positions in the sky, when studying the Galaxy they often find it useful to use another coordinate system that reflects the Galaxy's symmetry about the Galactic plane as it is viewed from the Earth. In the system of **Galactic coordinates**, the direction of any object in the sky can be expressed in terms of its **Galactic latitude** (b) and **Galactic longitude** (l), both of which are angles normally expressed in degrees. Figure 1.4 shows how they are defined. The **Galactic equator** runs close to the mid-plane of the Milky Way's disc.

The origin of the coordinate system, the point $l = 0°$ and $b = 0°$, is defined to be *nearly* – but not precisely – in the direction of the Galactic centre, which is in the direction of the constellation Sagittarius. (The convention is to show the Galaxy with the North Galactic Pole upwards.) Galactic latitudes are measured north (b positive) or south (b negative) of the Galactic equator, so they range between $b = +90°$ and $b = -90°$. Galactic longitude ranges from $l = 0°$ (roughly towards the Galactic centre), eastwards through $l = 90°$ (roughly in the direction of Galactic rotation), and on to $l = 360°$.

While the majority of the stars are concentrated into the disc and the bulge, detailed observations reveal another component of the Galaxy. Surrounding the disc is a sparsely populated structural component called the **stellar halo**, or just the **halo**, the mass of which is only about $10^9 M_\odot$. Because the number of stars per unit volume is much lower in the halo than in the disc, the halo does not show up in the image in Figure 1.5. Although most stars in the halo are widely distributed, about 1% of halo stars are concentrated into dense **globular clusters** (Figure 1.6), usually containing 10^4 to 10^6 stars in a volume just 50 pc in diameter.

The halo is not flat like the disc, but rather has a spheroidal shape, with only slight flattening. For this reason, the halo (and often the bulge with it) is sometimes referred to as the **Galactic spheroid**, or simply the **spheroid**. (Geometrically, the term *spheroid* refers to the three-dimensional figure formed by rotating an ellipse about one of its axes). The name 'stellar halo' suggests there are only stars, not gas, in this component. In fact there is some gas, but present only at a very low density.

So far in our description of the structure of the Galaxy we have concentrated on the way in which stars are distributed. However, while stars are luminous and hence easily seen, they are only the visible tips of our Galactic iceberg. There is more to the Galaxy, *vastly* more, than meets the eye.

Figure 1.5 NGC 891, a spiral galaxy seen edge-on, observed in near-infrared light. (J. C. Barentine and G. A. Esquerdo, Kitt Peak, NOAO)

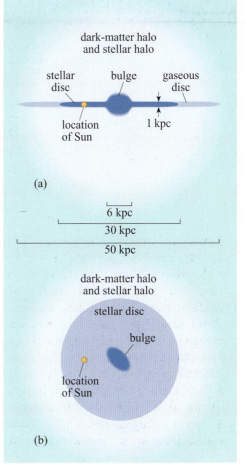

Figure 1.6 The globular cluster 47 Tuc. This globular cluster is one of the 150 or so globular clusters that are known to be associated with the Milky Way. (NASA/ESA)

Figure 1.7 (a) Edge-on and (b) face-on schematic views of the four major structural components of the Milky Way: the dark-matter halo, the disc, the stellar halo and the bulge. The sizes indicated in this figure are expressed in kiloparsecs (kpc), where 1 kpc ≈ 3260 light-years.

Rather surprisingly, the modern view of the Galaxy is that its largest and most massive component is a huge, roughly spherical cloud consisting of some kind of non-luminous matter. Since this non-luminous matter has never been directly observed at any wavelength, it is known as **dark matter**. The nature of dark matter is unknown at the time of writing. Its presence is revealed by the gravitational influence that it has on the more familiar forms of matter that can be directly observed. Despite knowing very little about dark matter, most astronomers have become convinced that the total mass of dark matter in the Milky Way is about ten times greater than the total mass of stars, and about 100 times greater than the total mass of gas and dust. Furthermore, this appears to be the case for other galaxies too; as you will see later, the nature, distribution and significance of dark matter is a recurring theme of great importance in the study of galaxies and cosmology.

The huge cloud of dark matter that is believed to be the main structural component of the Milky Way is usually referred to as the **dark-matter halo**. The mass of this component is so great that it is the gravity of the dark matter, rather than the gravity of all of the stars, that is primarily responsible for holding the Galaxy together. The gravitational influence of dark matter on luminous material allows astronomers to infer the shape of the dark-matter halo. They have concluded that the dark-matter halo, like the stellar halo, takes the form of a spheroid. Spheroids themselves come in a variety of shapes; there is some evidence that the dark-matter halo is an *oblate* spheroid, that is to say it resembles a sphere that has been flattened at its poles, although this is an area of ongoing debate.

Figure 1.7 is a schematic diagram of the major structural components of the Galaxy and their approximate sizes. This diagram should be compared with the face-on and edge-on images of galaxies in Figures 1.2 and 1.5, which are broadly similar to the Milky Way.

In summary, the Galaxy's major component is the *dark-matter halo*. Embedded within this is the *Galactic disc*, which is where most of the stars, gas and dust are found. The central region of the Galaxy is thicker than the rest of the disc, and is called the *bulge*. Surrounding the disc is the sparsely populated *stellar halo*. The disc contains bright *spiral arms*, and the bulge is elongated into a bar; the Milky Way is therefore a *barred spiral galaxy*.

■ What are the shapes and approximate masses of each of the four main structural components of the Galaxy?

☐ The dark-matter halo and the stellar halo are both slightly flattened (oblate) spheroids. The disc has the flattened circular form its name implies, and the central bulge is elongated into a bar. Very roughly, the stellar halo has a mass of $10^9 M_\odot$, the bulge mass is $10^{10} M_\odot$, the mass of the disc is $10^{11} M_\odot$ and the dark-matter halo has a mass of $10^{12} M_\odot$, although this last value is particularly uncertain.

1.2.2 The size of the Milky Way

A table of frequently used conversion factors and physical constants is provided in the Appendix.

Having introduced the main structural components of the Galaxy, we now examine their sizes. Although the size of the disc was given above as 100 000 light-years, distances in the Galaxy are usually measured in units of **parsecs** (pc) or **kiloparsecs** (kpc), where 1 kpc = 1000 pc. One parsec is equal to about 3.26 light-years, or 3.09×10^{16} m.

So, how big is the Galaxy? The answer depends greatly on which component you measure. The dark-matter halo is the most extensive component, but it is also the most difficult to measure since its presence is deduced only from its gravitational influence. The size of the dark-matter halo can be assessed from its effect on the motions of neighbouring galaxies. The Magellanic Clouds (Figure 1.3) are two small galaxies that are 50 to 60 kpc from the Milky Way, and the dark-matter halo apparently extends at least that far. So, in answer to the question: 'How big is the Galaxy?' you could cite the distance of the Magellanic Clouds as a lower limit on the radius of the dark-matter halo, implying a diameter of at least 100 to 120 kpc. Since the dark-matter halo cannot be observed very easily, you may prefer to consider a different question: 'How big is the disc of the Galaxy?' It turns out that even this more carefully posed question requires a cautious response. In short, the answer depends on which constituent of the disc you measure: stars or gas. The stellar disc of the Milky Way has a radius of at least 15 kpc. Observations indicate that the Sun is about 8.5 kpc from the centre of the Milky Way, which places it around halfway out in the stellar disc. This disc is around 1 kpc thick, which means some stars travel up to about 500 pc from the mid-plane of the disc.

■ If the radius of the Galactic disc is 15 kpc, then its diameter is 30 kpc. How many light-years (ly) is 30 kpc?

☐ 1 pc = 3.26 ly, so:
30 kpc = 30×10^3 pc $\times$ 3.26 ly pc^{-1} = 9.8×10^4 ly $\approx$ 100 000 ly.

In contrast to the stars, the gas and in particular the atomic hydrogen in the Galactic disc extends out to a radius of at least 25 kpc (although its density does fall considerably beyond 15 kpc). A clear example of the difference between the gaseous and stellar discs in a spiral galaxy is given by the barred spiral galaxy NGC 6744, which we observe almost face on, and which we believe is similar to the Milky Way (this is the galaxy pictured in Figure 1.2). Contours showing the atomic hydrogen gas density inferred from radio measurements of NGC 6744 are superimposed on the optical image in Figure 1.8. It has a gaseous disc that extends out to at least 1.5 times the radius of its stellar disc, maintaining the spiral structure seen in the visible image as it does so.

As the stellar halo has no substantial gaseous component, its size is given by the distribution of stars. However, defining the edge of this distribution is difficult, because the density of stars falls off gradually with distance from the centre of the Galaxy. For now we simply state that the stellar halo extends further than the disc, well beyond 20 kpc.

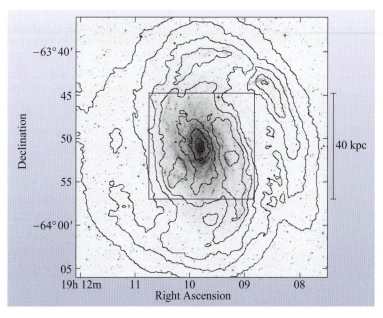

Figure 1.8 Contours of atomic hydrogen gas density in the barred spiral galaxy NGC 6744, based on radio observations, superimposed on an optical image from the Digitized Sky Survey. The central square corresponds to the image in Figure 1.2, and measures approximately 40 kpc on each side. Note that the gas extends well beyond the stellar disc, and that the spiral pattern is visible out to the edge of the gas disc. This galaxy is believed to be similar to the Milky Way. (Ryder *et al.*, 1999)

The bulge of the Galaxy takes the form of an elongated bar. The longest axis of this bar is in the Galactic plane and stretches out to about 3 kpc either side of the Galactic centre. The cross-sectional diameter of the bar is roughly 2 kpc. The mass of the bulge is much greater than that of the stellar halo, but its small size means that it has little relevance to any discussion of the overall size of the Galaxy.

Although approximate sizes for each of the Galaxy's main structural components have now been quoted, you may have noticed that the exact size of our Galaxy has still not been specified. It is always possible to define the size of the Galaxy as the size of *one* of the components, but such a definition would be rather arbitrary. It is more useful to retain a broad knowledge of the nature and scale of each of the structural components that make up our Galaxy.

If the Sun is 8.5 kpc from the Galactic centre and moving in a circular orbit at 200 km s^{-1}, how long will it take to travel once around the Galaxy? Express your answer in both SI units (seconds) and years. (Recall that the relationship between a body's speed, v, the distance travelled, d, and the time taken, t, is $v = d/t$, and that the circumference of a circle of radius R is $d = 2\pi R$.)

1.2.3 The major constituents of the Milky Way

The major structural components of the Milky Way – the dark-matter halo, disc, bulge and stellar halo – have now been introduced, and you have briefly encountered their constituents: dark matter, stars, gas, and dust. Now we look at these constituents more closely. We start with the dominant component, the dark matter.

Dark matter

Dark matter is detectable by its gravitational influence, but appears neither to emit nor absorb light nor any other form of electromagnetic radiation. The total mass of dark matter in the Milky Way seems to be about $10^{12} M_\odot$. However, at present we do not know the nature of this matter.

Everyday matter that we regularly encounter on Earth, and which constitutes the material in ordinary stars, is primarily made up of particles called **baryons**, the best known of which are protons and neutrons. Some of the dark matter may be dense, non-luminous, baryonic matter, but there is strong evidence (related to studies of the early Universe that are described in Chapter 6) indicating that much of the dark matter is non-baryonic. So, in addition to **baryonic dark matter**, which would be 'ordinary' matter that happened to be difficult to detect and about which we know very little, there is also a need for **non-baryonic dark matter**, about which we know even less.

Although the nature of non-baryonic dark matter is still a mystery, a number of proposals have been made regarding its possible composition. Most of these proposals assume that the non-baryonic dark matter consists of hypothetical fundamental particles of one kind or another. If any of these proposals is correct, then the dark-matter halo would simply be a vast cloud of these particles. Furthermore, since we are situated within this cloud, dark-matter particles should be detectable here on Earth. As we shall see in Chapter 8, experiments with this aim are underway or currently being developed.

Stars

There are about 10^{11} stars in the Galaxy, and, since the Sun's mass ($M_\odot \approx 2 \times 10^{30}$ kg) is typical, they have a combined mass of about $10^{11} M_\odot$, roughly one-tenth the total mass of dark matter. The vast majority of these stars occupy the disc.

Stars can differ from one another in their *mass*, their *age*, and their *chemical composition*; differences in these three fundamental parameters lead to differences in other properties, such as luminosity and temperature. The *spectral class* of a star is an important property that is closely related to its temperature. In order of decreasing surface temperature, the spectral classes are O, B, A, F, G, K, M. Many of the brightest stars in the Milky Way are large, bright, blue–white stars belonging to classes O and B, but by far the most common are small, faint, red stars belonging to class M. Stars form from large clouds of gas, and spend the greater part of their luminous lives as *main sequence stars* that are powered by the conversion of hydrogen into helium in their cores. The subsequent evolution of a star depends on its mass (the greater the mass, the shorter the life), but in most cases it includes the eventual enlargement of the star to form a *giant*, and the conversion of helium into a range of heavier elements.

Do the stars in each of the three Galactic components that contain stars – the stellar halo, disc and bulge – have the same range of masses, ages and compositions? Observations show that they do not. It turns out that the stellar halo and bulge contain much older stars than the disc, and that stars in the halo contain few elements heavier than helium. These differences led to categories of stars called **stellar populations** being defined. The differences between the populations tell us much of what we know about the origin and evolution of the Milky Way.

Before defining the three populations that are used in this book it is helpful to define a quantitative measure of a star's content of different chemical elements. Of particular interest is the degree to which a star is composed of elements heavier than helium – such elements are termed **metals** by astronomers. This can be quantified by the **metallicity**, Z, which is defined by the following equation.

$$Z = \frac{\text{the mass of elements heavier than helium in the object}}{\text{the mass of all elements in the object}} \qquad (1.1)$$

The metallicity of the Sun is $Z = 0.02$. That is, 2% of the Sun's mass comes from elements heavier than helium (while hydrogen and helium make mass contributions of 70% and 28% respectively). The reason why metallicity is such a useful quantity is that, with the exception of lithium, all metals are created by nucleosynthesis in stars. Roughly speaking, the metallicity of a star provides an indicator of the degree to which the material that makes up that star has passed through previous generations of stars in a process that is referred to as *cosmic recycling*.

Recent studies have indicated that the solar metallicity may be as low as 0.012.

Population I

Population I (often abbreviated to 'Pop. I') is associated with the disc, and includes many very young stars, some just a few million years old, but also includes some as old as 10^{10} yr. Their metallicities are mostly in the range $Z = 0.01$ to 0.04 (i.e. within a factor of two of the solar value) although some Pop. I stars with still lower metallicities exist. The stars of Pop. I move in essentially circular orbits (around the Galactic centre), which do not take them far above or below the plane of the Galaxy, and hence they are confined to the flat, circular structure that constitutes the Galactic disc.

Population II

Pop. II stars occupy the spheroid – the stellar halo and bulge – and turn out to be the oldest stars known, with ages exceeding 12×10^9 yr. Little or no interstellar gas is still associated with Pop. II stars, which is consistent with star formation in the spheroid ceasing long ago. Because this population is so old, only low-mass stars (which have long lifetimes) still shine as main sequence stars burning hydrogen in their cores. The more massive stars that formed at the same time as the surviving low-mass ones have already left the main sequence and are now red giants or white dwarfs.

For a long time it was thought that all Pop. II stars had much lower metallicities than do Pop. I stars, but it is now known that this applies only to the stellar halo, for which $Z < 0.002$, and where the lowest-metallicity stars detected so far have $Z \approx 1 \times 10^{-6}$. Some bulge stars, on the other hand, have the same metallicity as the Sun.

Unlike disc stars, Pop. II stars do not follow circular orbits, nor are they confined to the plane of the Galaxy. They move in eccentric orbits (see Figure 1.9), although still attracted to the Galactic centre, and may travel many kiloparsecs from the

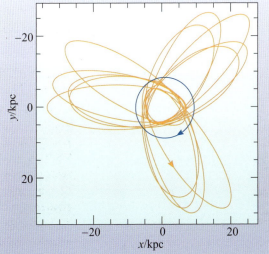

Figure 1.9 A face-on view of the Galaxy showing sample orbits for Pop. I (blue) and Pop. II (orange) stars. Shown is a clockwise, circular orbit for a Pop. I star 8.5 kpc from the Galactic centre, and an anticlockwise (retrograde), three-lobed orbit for a Pop. II star that takes it from 4 kpc to 35 kpc from the centre. This Pop. II star also travels up to 20 kpc above and below the disc, while the Pop. I star remains close to the Galactic plane.

Galactic plane. This is of course consistent with Pop. II stars belonging to the spheroid, briefly passing through the disc as they move from one side of the Galactic plane to the other. Such fleeting visitors to the Galactic disc are known as **high-velocity stars** because of their high speed relative to the Pop. I stars that belong to the disc. In contrast to the disc, there is almost no net rotation of the halo, so almost half of all halo stars travel in **retrograde** orbits (i.e. in the opposite sense to the more orderly disc stars, which all orbit in the clockwise direction as viewed from the North Galactic Pole).

Population III

The term Pop. III describes a theoretical population rather than one that has actually been observed. It encompasses stars that formed out of the unprocessed gas that would have been produced by the big bang. This material would have been almost entirely composed of hydrogen and helium. Even lithium, the next most abundant element in this gas, would only constitute one particle in a billion. Consequently the metallicity of these stars when they first formed ($Z \approx 10^{-9}$) would have been much lower than that of even the lowest metallicity stars of Pop. II ($Z \approx 1 \times 10^{-6}$). No Pop. III star has yet been observed, but theoretically they must have existed as the very first generation of stars in the Universe.

QUESTION 1.2

Stellar populations differ in age, metallicity and location, but sometimes stellar motion is used as an alternative criterion to location. Why is this reasonable?

Gas and dust

Most of the Milky Way's gas and dust lies in the disc, and is found within a vertical distance of 150 pc of the Galactic plane: it does not extend nearly so far from the mid-plane of the Galaxy as do the stars.

The gas is roughly 70% hydrogen and 28% helium (by mass). The remaining 2% is made up of the other elements. The hydrogen can exist in various forms depending on the density, temperature, and flux of ultraviolet (UV) radiation in each locality. In high-density, low-temperature (less than 100 K) environments with a low UV flux, the hydrogen is mostly in the form of **molecular hydrogen** (H_2). In environments where the temperature and/or the UV flux is high enough to free the hydrogen atom's single electron, there is a likelihood of finding **ionized hydrogen** (H^+, usually written HII and pronounced 'H-two'), particularly where the density is low enough to reduce the chance of the liberated electrons recombining with the positive ions. **Atomic hydrogen** (H, often written HI and pronounced 'H-one') occurs where conditions lie between the other two extremes. The total mass of gas in the Galaxy is estimated to be about 10% of the stellar mass.

- ■ What is the total mass of gas in the Galaxy, in solar masses?

- ☐ The mass of gas is 10% of the stellar mass of the Galaxy, and the latter is about $10^{11} M_\odot$, so the mass of gas is:
  $10\% \times 10^{11} M_\odot = (10/100) \times 10^{11} M_\odot = 10^{-1} \times 10^{11} M_\odot = 10^{10} M_\odot$

What astronomers call **dust** consists of tiny lumps of solid (condensed) compounds of carbon, oxygen, silicon and other metals. Most of the bulk of a dust grain comprises either carbonaceous (e.g. graphite) or silicate compounds. The outside of the grain is often surrounded by a coating, or *mantle*, of more volatile compounds such as water-ice (H_2O), ammonia (NH_3) and carbon monoxide (CO). Dust particles are typically 10^{-7} to 10^{-6} m (0.1 μm to 1 μm) across, close in size to smoke particles on Earth. The total mass of dust in the Galaxy is about 0.1% of the stellar mass.

> The term 'metals' is used here, and throughout this book, in its astronomical sense.

■ What is the total mass of dust in the Galaxy, in solar masses?

☐ The mass of dust is 0.1% of the stellar mass of the Galaxy, and the latter is $10^{11} M_{\odot}$, so the mass of dust is:
$$0.1\% \times 10^{11} M_{\odot} = (0.1/100) \times 10^{11} M_{\odot} = 10^{-3} \times 10^{11} M_{\odot} = 10^{8} M_{\odot}$$

The term **interstellar medium**, or **ISM**, is used to describe the gas and dust that occupies the space between the stars. On average the ISM contains about 10^6 particles per cubic metre, but the density and nature of the ISM varies greatly from one region to another, so this average does not have major significance. Almost half of the ISM (by mass) is contained in cool **dense clouds**, often called **molecular clouds** because they are rich in molecular hydrogen (H_2). These clouds occur with a wide range of masses, the most massive containing up to $10^7 M_{\odot}$ of gas and dust. Molecular clouds are usually many thousands of times denser than the average ISM, with typical diameters of 10 to 100 pc. However, these clouds account for only 1% or so of the volume of the ISM, despite contributing nearly 50% of its mass. Another contribution to the ISM comes from localized clouds (typically a few parsecs across) of hot, ionized gas known as **HII regions**. These regions account for a few per cent or so of the ISM's mass and volume. The Orion Nebula (see Figure 1.10) is one of the best known of these regions.

Much of the remaining volume of the ISM is occupied by an **intercloud medium** that may be hot or warm, depending on local conditions. Within the Galactic disc, the intercloud medium can be loosely regarded as a disc about 300 pc thick that is rich in atomic hydrogen (HI), and more or less uniformly distributed. However, this disc of intercloud medium, along with the thousands of individual clouds it contains, is embedded in a large body of hot intercloud medium that occupies at least part of the halo. There is still much uncertainty about the distribution of the intercloud medium, but the low density of this component of the ISM means that it contributes only a minor part of the ISM's mass, despite its large volume.

The ISM is intimately associated with star formation and stellar evolution: stars are born from cool dense molecular clouds, and they return matter to interstellar space in a variety of ways that gradually increase the proportion of heavy elements in the Galaxy. This process is known as *Galactic chemical enrichment*, and is discussed in greater detail in Section 1.6.1.

Figure 1.10 The Orion Nebula (M42), a prominent HII region in which the hydrogen is ionized by the UV radiation from a group of bright young stars. The Orion Nebula is about 6 pc in diameter and contains hundreds of solar masses of gas. The nebula forms a sort of blister on the face of a giant molecular cloud. This image combines observations taken from the Hubble Space Telescope and a ground-based telescope over a range of visible and near-infrared wavelengths. (NASA, ESA, M. Robberto (Space Telescope Institute/ESA) and the Hubble Space Telescope Orion Treasure Project Team)

We have now completed our survey of the main Galactic components and their constituents. We have seen that the Galaxy is made of dark matter, stars, gas and dust, and how it is structured as a dark-matter halo, disc, stellar halo and bulge. With this basic framework in place, we can now consider how astronomers have come to a detailed understanding of the Galaxy, starting with its single most important physical characteristic – its mass.

1.3 The mass of the Milky Way

In the preceding section the masses of various structural components of the Galaxy were simply stated. At the time, did you ask yourself how astronomers might know these values, or how uncertain they might be? You may be surprised to learn that reasonable estimates can be obtained from some quite simple calculations based on the influence of gravity on the motions of objects, although determining the mass of the entire Galaxy accurately is a major challenge.

In answering Question 1.1, you learned that the Sun takes roughly 260 million years to complete one orbit of the Galactic centre. To perform this calculation, you needed to know the speed at which the Sun travels in its orbit and its distance from the Galactic centre. The speed of the Sun was established in the 1920s, by Bertil Lindblad (1895–1965) and Jan Hendrik Oort (1900–1992), from the motion of Pop. I stars relative to Pop. II stars. As you saw in Section 1.2.3, Pop. II stars have little net rotation about the Galactic centre, and so provide a reference population relative to which the Sun's motion can be determined. Pop. I stars in our part of the Milky Way are streaming past the Pop. II stars at speeds that are typically 200 km s^{-1}. We will soon use this information to calculate the mass of the inner part of the Galaxy, but before doing so we introduce a graphical tool that plays an important part in the characterization and analysis of rotating systems.

A plot of speed against distance from the centre for the various parts of a rotating system is called a **rotation curve**. The rotation curve for a rigid wheel 3 m in diameter, making one revolution per second, is displayed in Figure 1.11a. As the figure shows, the speed (usually expressed in metres per second; m s^{-1}) of each part of the wheel increases in proportion to the distance from the centre. Note, however, that in the case of a rigid wheel the angular speed of each part about the centre – the angle that each part sweeps out per second, as viewed from the centre of the wheel – is the same. Finding that all parts of a body have the same angular speed is a characteristic of **rigid body rotation**.

Figure 1.11b shows the rotation curve of planets orbiting the Sun. Each planet takes a different length of time to complete one orbit, which means each travels at a different angular speed. This is characteristic of **differential rotation**, and is clearly different from rigid body rotation. As you can see, the rotation curve for such a system is *not* a straight line through the origin, so speed is *not* proportional to distance from the centre.

Figure 1.11c shows the rotation curve of the Milky Way. This has a different shape from Figure 1.11b, but it still shows that the speed is not proportional to distance from the centre, and therefore indicates differential rotation with a range of angular speeds. Figures 1.11b and 1.11c differ because there is no massive central body dominating the Milky Way, whereas the mass of the Sun dominates the Solar System.

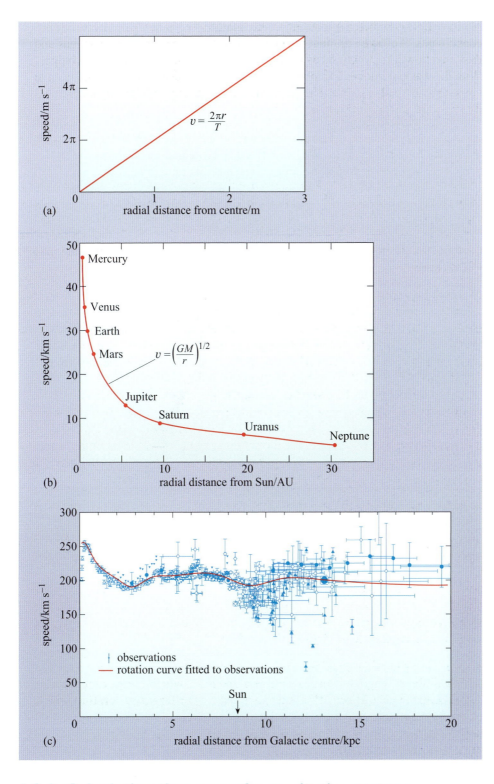

Figure 1.11 The rotation curves, showing speed v against radial distance r, of (a) a rigid wheel, 3 m in diameter, with a rotation period T of 1 s; (b) the planets of the Solar System; and (c) the Milky Way, based on Doppler-shift studies of gas clouds in the disc. Note that the rotation curve for the Galaxy is based on data that are subject to substantial observational uncertainties, and is therefore 'noisier' than the other two curves. This is because the observations are harder to make (especially for objects further from the Galactic centre than the Sun), and the analyses are plagued by additional complications, such as the non-uniform distribution of matter in the Galactic disc. ((c) Sofue *et al.*, 2009)

1.3.1 Calculating the mass of a gravitating system

The equations describing the rotation curve of a system governed by the gravitational attraction of a single central body relate the orbital speed and orbital radius to the mass of the central body. This is very important in astronomy, as it allows us to calculate the mass of a gravitating body from the motions of particles orbiting it. For an object in a circular orbit of radius r about a single, much more massive object of mass M, the equation describing the rotation curve (see Box 1.2) is

$$v = \left(\frac{GM}{r}\right)^{1/2}$$

The reciprocal of some quantity x is $1/x$. The symbol $\propto$ means 'is proportional to'. Note that $\sqrt{x} = x^{1/2}$

This equation indicates that the speed v is proportional to $1/r^{1/2}$, which means that the orbital speed falls as the radius increases, at a rate given by the reciprocal of the square root of the radius. We can write this in mathematical shorthand as $v \propto 1/r^{1/2}$. The curve plotted in Figure 1.11b has this functional form.

BOX 1.2 ROTATION CURVES FOR GRAVITATING SYSTEMS

According to Newton's second law of motion, the magnitude of the acceleration a_i of some body (labelled 'i') of mass m_i, due to a force of strength F, is

$$a_i = F/m_i \tag{1.2}$$

According to Newton's law of gravitation, the strength F_g of the gravitational force on each of the two point-like bodies of masses m and M, when their centres are separated by a distance r, is

$$F_g = GMm/r^2 \tag{1.3}$$

where G is the universal gravitational constant, 6.673×10^{-11} N m² kg⁻².

Substituting this expression for F_g into Equation 1.2 shows that because of the gravitational attraction of the body of mass M, the body of mass m will have an acceleration of magnitude

$$a_{g,m} = \frac{GMm/r^2}{m} = GM/r^2$$

Similarly, due to the attraction of the body of mass m, the body of mass M will have an acceleration of magnitude

$$a_{g,M} = \frac{GMm/r^2}{M} = Gm/r^2$$

In cases where M greatly exceeds m, written $M \gg m$, we can also write $GM/r^2 \gg Gm/r^2$, implying that $a_{g,m} \gg a_{g,M}$. That is, the acceleration of the more massive body has a magnitude, $a_{g,M}$, that is much smaller than the magnitude of the acceleration of the less massive body, $a_{g,m}$. Therefore the more massive body barely moves, and we can regard it as being the stationary centre of motion for the less massive body. One particularly simple form that this motion might take is for the less massive body to move around the more massive body at constant speed in a circular orbit. This kind of motion is known as uniform circular motion.

Any body moving in uniform circular motion at speed v about some centre at a distance r must be accelerating at all times. This acceleration is called its centripetal acceleration; it is always directed towards the centre of the motion and its magnitude is

$$a_{cen} = v^2/r$$

When this uniform circular motion is the result of the gravitational attraction between two bodies, the centripetal acceleration is provided by the gravitational acceleration, so for the less massive body in motion about the stationary, more massive one, we can write:

$$a_{cen} = a_{g,m}$$

so $\qquad v^2/r = GM/r^2$

and hence

$$v^2 = GM/r$$

This equation can be rearranged in two slightly different ways to give useful equations relating the orbital speed and the central mass of a two-body system:

$$v = (GM/r)^{1/2} \tag{1.4}$$

and

$$M = v^2 r/G \tag{1.5}$$

Note that the orbital speed of the less massive body does not depend on its mass. Also notice that we can deduce the mass of the central body from the speed and radial separation of an orbiting body, without knowing the mass of that orbiting body.

(a) Calculate the circumference of the Earth's orbit around the Sun. (The Earth is 150 million kilometres from the Sun, to three significant figures.) Give your answer in SI units.

(b) Calculate the speed at which the Earth orbits the Sun. Give your answer in SI units.

(c) Use the formula for the rotation curve to calculate the mass of the Sun from the orbital speed of the Earth. Give your answer in SI units. (For the universal gravitational constant, use the value $G = 6.673 \times 10^{-11}$ N m^2 kg^{-2}.)

(d) How many significant figures are there in your final answer, and why?

How does your answer to Question 1.3 compare with the modern value, $M_\odot = 1.9891 \times 10^{30}$ kg? Did you ever realize just how easy it would be to calculate the mass of the Sun? All you need to know is the speed and orbital radius of one small body in a circular orbit. You don't even need to know the smaller body's mass.

You could look up the orbital radii and periods of a few other planets to convince yourself that you get the same answer, but for now we have bigger fish to fry; let's calculate the mass of the inner part of the Galaxy!

Unlike the Solar System, the Milky Way does not have a single dominant mass at its centre. Rather, its constituents move under the gravitational influence of all the other constituents. This makes a detailed analysis very complicated, but it is still possible to make a simple estimate of the mass of the inner part of the Galaxy. The basis of this analysis is provided by the following result taken from Newtonian gravitational theory.

When the mass of a system is distributed in a spherically symmetrical manner about some central point, then the net gravitational force on a point-like object at some radius is due only to the mass *within* that radius. Furthermore, the net gravitational force is the same as if the mass inside that radius was all located at the centre.

In the case of the Milky Way, the distribution is not really spherically symmetrical, but mass outside a given radius has only a moderate effect, and a reasonable estimate of the mass can still be obtained using the method above. (Detailed calculations can be performed for more realistic mass distributions to confirm this.) If we adopt this procedure for the Galaxy, then the observed orbital speed at some radius r can be inserted into the rotation-curve equation to estimate the mass of the Galaxy enclosed within that radius from the Galactic centre. As this gives the mass within the radius r, rather than the total mass of the Galaxy, we denote that mass by $M(r)$.

(a) Following the technique used in Question 1.3c, calculate the mass of the Milky Way out to the distance of the Sun from the Galactic centre. Give your answer both in SI units and solar masses, assuming $M_\odot = 1.99 \times 10^{30}$ kg. (*Hint*: You know already that the Sun is 8.5 kpc from the Galactic centre, and that it moves at 200 km s^{-1} in a nearly circular orbit.)

(b) How many significant figures can you quote the result to, if you follow the usual mathematical rules? Is there any reason why you might deviate from this rule?

1.3.2 Using rotation curves

The answer to Question 1.4 shows that interior to the Sun's orbit at 8.5 kpc, the mass of the Galaxy is about $10^{11} M_\odot$. If we want to find the total mass of the Milky Way, we have to study its outskirts, where the orbiting material encloses virtually all the mass of the Galaxy. This is difficult, not least because it is difficult to determine exactly where the Galaxy ends. Even if we think there is not much more visible matter beyond a certain radius, we cannot be sure that we have found the 'edge' of any dark matter that is associated with the Galaxy.

Plotting a rotation curve can throw light on the question: 'Where does the Galaxy end?' We have just seen in Question 1.4 how it is possible to compute the enclosed mass at some radius in the Galaxy by knowing the orbital speed there. From an observed rotation curve, it is possible to compute the enclosed mass $M(r)$ at a whole range of radii, and doing so shows how $M(r)$ increases with radius, which therefore gives the distribution of mass. By seeing how the mass distribution is changing in the outermost measurable parts of a Galaxy, it is possible to have some idea of whether the mass distribution is tailing off near the last measurement.

The usual procedure for deducing $M(r)$ for a galaxy is to take an educated guess at the distribution of matter and then work out the rotation curve that such a distribution would produce. The initial guess is then adjusted until the modelled rotation curve agrees with the observed one. To see how this process works, in the following example and question we compute the rotation curves for some simple, assumed mass distributions, and then use these results to interpret what has been measured for the Milky Way, which is shown in Figure 1.11c.

EXAMPLE 1.1

Use the rotation-curve equation to help you to sketch a rotation curve for the following distribution of matter: $M(r) = M$, a constant, indicating a central mass only.

SOLUTION

A rotation curve is a plot of speed versus radius, so the more useful form of the rotation-curve equation (see Box 1.2) is $v(r) = (GM(r)/r)^{1/2}$ (Equation 1.4). To sketch the rotation curve, we need to know how v varies with r.

Since in the example $M(r)$ is a constant, M, the equation for the speed becomes

$$v(r) = (GM/r)^{1/2} = \text{const} \times 1/r^{1/2}$$

Note that we have put the constant parts of this expression into one term called 'const', and have kept the variable parts separate.

This equation shows that the rotation curve falls with increasing radius as $1/\sqrt{r}$, so your sketch of the rotation curve in this case should decrease in speed as radius increases, and it should flatten out towards large radius. One example of a system dominated by a central mass is the Solar System, so the rotation curve should match Figure 1.11b.

Historically, it was Johannes Kepler (1571–1630) who first recognized that a relationship of this form describes the motion of planets in the Solar System. This is the origin of the term **Keplerian orbit** that astronomers now use to refer to the motion of a body under the gravitational influence of a much more massive body.

QUESTION 1.5

Following the example presented above, use the rotation-curve equation to help you to sketch a rotation curve for each of the following distributions of matter.

(a) $M(r) = kr$ (for some constant of proportionality k)

(b) a uniform-density sphere, i.e. where

$$M(r) = \text{density} \times (\text{volume of sphere of radius } r) = \rho \times \frac{4}{3}\pi r^3$$

Figure 1.11c shows the measured rotation curve of the Milky Way. Some of the features of Figure 1.11c, such as the peak near the Galactic centre and the sharp dip that follows it, are more likely to be due to the inadequacy of the symmetry assumptions that underpin the analysis rather than real features of the rotation. However, the flatness of the rotation curve at large distances is thought to be real. It is flatness of this kind, extending well beyond the edge of the visible disc, which provides evidence for the presence of a substantial amount of non-luminous matter on the outskirts of the Milky Way – dark matter. If you compare all but the central kiloparsec of the Milky Way's rotation curve (Figure 1.11c) with your answers to Question 1.5, you will see that it is similar to the flat curve for $M(r) = kr$. Note that a mass distribution of the form $M(r) \propto r$ thins out with increasing radius, so a flat rotation curve does not imply constant density.

If most of the mass of the Galaxy were well within the largest measured radius, the rotation curve would be expected to decline quite rapidly with increasing radial distance, as in the case $M(r) = \text{constant}$ in the solution to Example 1.1. The answer to the question: 'What is the mass of the Galaxy?' really depends on the answer to another question: 'Where does the rotation curve turn down?'

Several independent investigations have failed to show any sign of a decline in the rotation curve of the Milky Way out to a radius of 20 kpc, indicating a substantial amount of matter at least out to that radius. This matter has not been directly observed at any wavelength and is therefore assumed to be some form of dark matter.

There are still many uncertainties about the distribution of mass in the Milky Way. Different assumptions about the radius of the Galaxy and the distribution of dark matter can easily provide estimates of the total Galactic mass that range from a conservative four times the mass of the stars, that is $4 \times 10^{11} M_\odot$, to a very substantial 60 times: $6 \times 10^{12} M_\odot$. The mass of the Galaxy can be assessed using the velocities of other objects besides disc gas, such as distant halo stars, globular clusters and nearby galaxies, but they too are based on specific assumptions and do not yet settle the question.

■ Could you calculate the mass of the wheel whose rotation curve is shown in Figure 1.11a, using the technique used in Question 1.4? Would it make any difference if most of the mass of the wheel were concentrated near the centre of rotation? Explain your answer.

□ You could not determine the mass of the wheel; the rotation curve contains no information on its mass because rotation of the wheel is governed by non-gravitational forces. (The forces are electrostatic, and act between the many neighbouring particles in the rigid structures of the wheel.) You could calculate the centripetal acceleration of various parts of the wheel, but those accelerations would not be related to its mass distribution. This answer would not be any different if the mass were concentrated near the axle.

1.4 The disc of the Milky Way

The disc is important to our understanding of the Galaxy for two reasons: it is the component to which most of the visible matter – stars, gas and dust – belongs, and it is the main site of ongoing star formation in the Milky Way. Many of the characteristic features of the disc that we can observe, such as the presence of young, high-metallicity stars, dense molecular clouds and HII regions, are directly connected with star formation, and play key roles in the process of cosmic recycling that we will discuss in Section 1.6.

1.4.1 Stellar clusters in the disc

Stars are the most luminous constituent of the Milky Way, and most stars reside in the Galactic disc. Usually stars do not form alone, but instead begin their lives in clusters or associations. This is a consequence of stars forming from dense molecular clouds (see Section 1.2.3) that contain enough gas to create large numbers of stars. **Open clusters** occupy regions of space, typically 2–3 pc across, where the density of stars is enhanced locally by a group of a few tens to a few hundred stars that formed at the same time. The cluster members are bound together by their mutual gravitational attraction. There are thousands of open clusters in the Galactic disc. Some are sufficiently prominent to be visible to the unaided eye; a notable example is Messier 7 in the constellation of Scorpius (Figure 1.12).

Because of their concentration close to the plane of the Galaxy, open clusters used to be called *galactic clusters*. However, this term is potentially confusing ('galaxy clusters', which is a term sometimes used to describe clusters of galaxies, are quite different) and its use is discouraged.

Figure 1.12 The open cluster Messier 7, also known as Ptolemy's cluster as it was first mentioned by the mathematician and astronomer Claudius Ptolemy as early as 130 AD. It is about 200 million years old. (ESO)

Open clusters are believed to have relatively short lives. With very few exceptions no individual open cluster is expected to survive for more than 10^9 years. The smallest open clusters live for even shorter periods, just a few million years, so they are mainly found in the spiral arms, since this is where star formation mainly takes place. Why do the clusters disrupt and dissipate in a time that is much shorter than the lives of many of the stars they contain? There are three processes that aid this. First, a cluster moving in the disc might be torn apart by a gravitational encounter with some other object such as a giant molecular cloud or another cluster. Secondly, gravitational interactions between a cluster's members can give one star enough energy to escape; given long enough, this process, called **evaporation**, would disperse all clusters. A third process that disrupts clusters is differential rotation, which was described in Section 1.3. The stars in an open cluster, while moving around one another, are also orbiting the Galactic centre and hence are subject to the effects of differential rotation that will stretch, distort and eventually destroy the cluster. The concentration of open clusters in the spiral arms must mean that the clusters have relatively short lives, otherwise motion relative to the arms would relocate them to various positions in the disc.

Stellar aggregates of another kind found in the disc, particularly within the spiral arms, are **OB associations**. OB associations have diameters of 100 pc or so, and densities not much greater than their general surroundings, but contain an unusually high proportion of O- and B-class stars (Figure 1.13). Many OB associations have a very young open cluster at their centre. The presence of numerous O and B stars, which have very short main sequence lifetimes, is a sure sign that OB associations are young, no more than a few million years old. About 70 OB associations are known.

1.4.2 The gaseous content of the disc

The gaseous interstellar medium (ISM) is intimately associated with stellar evolution. Stars form from cool dense clouds in the ISM, and at the ends of their lives they return matter to the ISM in a variety of ways that gradually increase the metallicity of the Galaxy. Gas and dust are therefore important constituents of the disc, and we examine their composition and distribution in this section. Included in the discussion are descriptions of how the gas and dust can be observed.

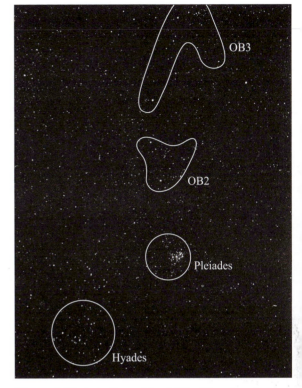

Figure 1.13 A portion of the Milky Way showing the Hyades and Pleiades open clusters, and OB associations in Taurus and Perseus.

Interstellar gas

As you saw earlier, the total mass of gas in the disc amounts to about 10% of the stellar mass. The gas forms a disc about 300 pc thick that is surrounded by hotter gas, which stretches out into the halo. Much of the gas is in the form of clouds, of which there are many thousands, with a range of temperatures and densities. The individual clouds probably take the form of sheets or filaments of gas rather than some idealized spherical shape. Regions of the ISM not occupied by clouds constitute the extensive but low-density *intercloud medium*. There the hydrogen exists mainly in the form of neutral atoms (HI), which are readily observed by their emission of radio waves at a wavelength of 21 cm (see Box 1.3).

Although we have said that the gas forms a disc with a thickness of 300 pc, this disc is not completely flat. Its mid-plane is flat out to a radius of 12 kpc from the Galactic centre, but at greater distances it is warped (tilted). This can be seen in the two parts of Figure 1.15, which show the gas distribution at 12 kpc and 16 kpc from the Galactic centre, as it would be seen from the Galactic centre rather than from the location of the Solar System. The images are built up from measurements of the hydrogen 21 cm emission line. The figure shows that at a distance of 12 kpc, the gas distribution is centred about the Galactic plane, indicating that the disc is flat at this distance from the Galactic centre. However, at 16 kpc the gas rises

BOX 1.3 THE 21 CENTIMETRE EMISSION LINE OF ATOMIC HYDROGEN

A major indicator of the distribution and line-of-sight velocity of neutral (atomic) hydrogen (HI) gas, not just in the Milky Way but other galaxies as well, is its emission line in the radio region of the spectrum at a wavelength of 21 cm. The origin of this **21 centimetre radiation** involves the relative *spins* of the electron and proton that constitute the hydrogen atom. These spins are illustrated in a classical (i.e. non-quantum) sense in Figure 1.14, where the proton and the electron are pictured as small spheres spinning at a fixed rate around axes through their centres. According to **quantum theory** the spin of an electron in a hydrogen atom is always either parallel to that of the proton, as in Figure 1.14a, or anti-parallel (i.e. opposed to it), as in Figure 1.14b. There is a small energy difference between the states shown in Figure 1.14, and it is the transition from the higher energy state (a) to the lower energy state (b) that gives rise to the 21 cm emission line.

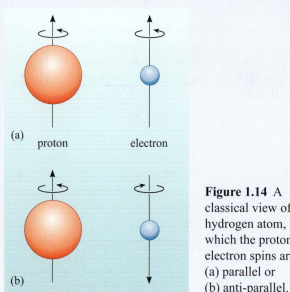

Figure 1.14 A classical view of the hydrogen atom, in which the proton and electron spins are (a) parallel or (b) anti-parallel.

■ What is the energy difference between these two states, in SI units (joules) and electronvolts?

□ The energy difference between the states is just the energy of the emitted photon. This is given by

$$\varepsilon = hf = hc/\lambda \tag{1.6}$$

Thus the energy difference between the two states associated with the 21 cm line is

$$\varepsilon = hc/\lambda$$

$$= 6.63 \times 10^{-34} \text{ J s} \times 3.00 \times 10^8 \text{ m s}^{-1}/0.21 \text{ m}$$

$$= 9.5 \times 10^{-25} \text{ J}$$

Since 1 eV = 1.60×10^{-19} J, the energy difference can also be expressed as

$$\varepsilon = 9.5 \times 10^{-25} \text{ J}/1.60 \times 10^{-19} \text{ J eV}^{-1}$$

$$= 5.9 \times 10^{-6} \text{ eV}$$

For us to observe this emission line, the hydrogen atoms must be in an environment where they can readily gain the energy required to raise them into the upper energy level at a reasonable rate compared with the rate at which they are reverting to the lower energy level by emitting radiation. One energy source is provided by collisions between hydrogen atoms as a result of their random thermal motion – this is an example of **collisional excitation**. For a reasonable proportion of such collisions to be sufficiently energetic, the average translational kinetic energy of an atom, e_k, must exceed the energy difference between levels, ε, such that $e_k \geq \varepsilon$. For thermal motion, the average translational kinetic energy is related to the temperature T of the gas particles via the equation $e_k = 3kT/2$, where k is the Boltzmann constant. Thus, by requiring $e_k \geq \varepsilon$, we get $3kT/2 \geq \varepsilon$ and hence $T \geq 2\varepsilon/3k$.

Putting in the value of ε, we get the requirement $T \geq 2 \times 9.5 \times 10^{-25} \text{ J}/(3 \times 1.38 \times 10^{-23} \text{ J K}^{-1}) = 0.046 \text{ K}$.

This condition is met everywhere in the ISM, so the 21 cm line is readily emitted wherever atomic hydrogen exists.

The radial velocity of a cloud of atomic hydrogen can be measured from the **Doppler shift** of the 21 cm emission line. In general, the radial velocity (i.e. the component of velocity along the line of sight) of an object that emits (or absorbs) radiation at a wavelength λ_{em} is given by

$$v_r = c(\lambda_{obs} - \lambda_{em})/\lambda_{em} \tag{1.7}$$

where λ_{obs} is the wavelength at which the radiation is observed and c is the speed of light. This relationship is valid provided that the radial velocity is much smaller than the speed of light; v_r must be less than about $0.1c$. Note also that the convention used in Equation 1.7, and throughout this book, is that an object moving *away* from the observer has a *positive* radial velocity.

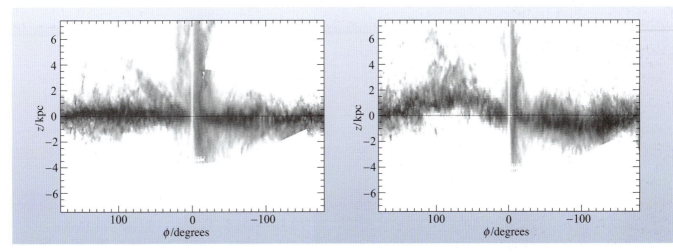

Figure 1.15 The Galactic HI density on cylindrical surfaces at (a) 12 kpc and (b) 16 kpc from the Galactic centre. Each map plots the distribution of gas at a vertical displacement, z, from the Galactic plane and an azimuthal angle, ϕ, measured in the Galactic plane. Each map is constructed as if viewed from the Galactic centre. The warping of the disc, which sets in around 16 kpc, is seen as a wave-like displacement of the gas from the equatorial plane, strongest around $\phi = \pm 90°$. (The vertical column of gas in the centre of each image is an artefact of the way the maps have been made.) (Binney and Merrifield, 1998; from data published in Burton, 1985; Hartmann and Burton, 1997; and Kerr *et al.*, 1986, courtesy of T. Voskes and B. Burton)

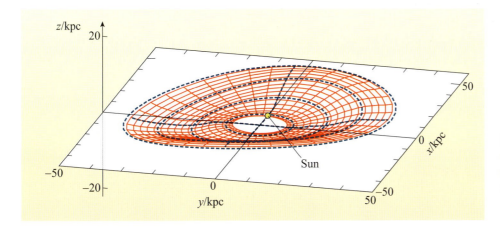

Figure 1.16 The warped mid-plane of the Galaxy as determined from measurements of HI. (Adapted from Kalberla and Kerp, 2009)

above and falls below the Galactic plane (see Figure 1.16). This demonstrates the presence of a warp, which can be interpreted as a tilt of the gaseous disc at distances around 16 kpc. The tilt is also present beyond 16 kpc. The gas reaches an altitude (z) of 1 kpc to 2 kpc above the plane in one azimuthal direction ($\phi = +90°$), and a similar distance below the plane in the opposite direction ($\phi = -90°$). Roughly one-quarter of galaxies have warped discs, and although possible causes of warping have been explored, such as tidal interactions with satellite galaxies, the phenomenon is not fully understood.

- Gas clouds cooler than about 100 K generally do not emit the 21 cm line; why not?

☐ At $T < 100$ K, hydrogen forms into H_2 molecules, so no atomic hydrogen remains.

Figure 1.17 Cool dense molecular clouds silhouetted against a bright HII region. This particular example is the Horsehead Nebula in Orion. (Anglo-Australian Telescope Board)

In the disc, about 50% of the mass of the hydrogen is molecular (H_2). A high molecular content is expected where: (i) the density is high, since this promotes the meeting of atoms; (ii) the temperatures are low (below 100 K) since this avoids the collisional disruption of molecules; (iii) the UV flux is low, since this avoids the UV-induced disruption of molecules. These are the conditions within the cool dense clouds that are often referred to as molecular clouds (see Figure 1.17). These clouds are found throughout the disc, but are particularly numerous between about 4 and 7 kpc from the Galactic centre. In dense clouds, not only is the hydrogen present mainly as H_2, but similarly all of the chemically reactive elements are predominantly combined into molecules. A particularly important molecule is carbon monoxide, CO, which is vital for detecting the presence of cold molecular gas (see Box 1.4). An example CO map showing the distribution of molecular gas around the Orion star-forming region is shown in Figure 1.18.

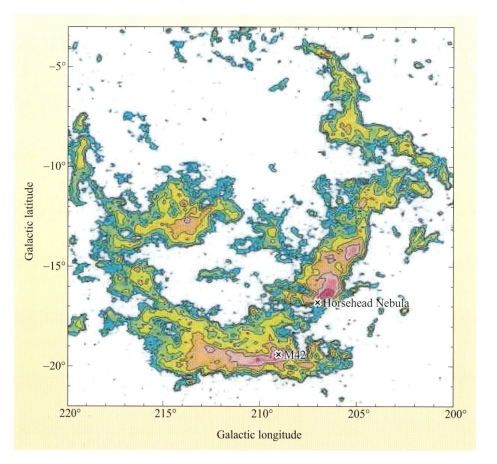

Figure 1.18 A map of the CO emission of the star-forming regions in Orion. The locations of the Orion Nebula (M42) and the Horsehead Nebula are shown. (Adapted from Wilson *et al.*, 2005)

BOX 1.4 CARBON MONOXIDE (CO) AS A TRACER OF MOLECULAR GAS

The energy associated with the rotation of a molecule is quantized, analogous to the way that the energy of an electron in an atom is quantized. This means that there are only certain rotational energies that a molecule can have, and changes in the rotational energy of molecules are accompanied by the absorption or emission of a photon. The differences between rotational energy states are very small, and so only low-energy photons are involved; the spectral lines for transitions between two rotational energy states are generally found at radio wavelengths.

However, molecular hydrogen (H_2), and other diatomic molecules that consist of identical atoms (such as C_2, O_2, and N_2) do not emit radiation from rotational energy transitions, for reasons connected with their symmetry. Hence the huge amounts of H_2 in the Galaxy are essentially undetectable. (H_2 does produce a weak infrared line and some ultraviolet spectral lines, but these are of very limited practical use.)

Most of the hydrogen in the ISM is in molecular rather than atomic form at temperatures below about 100 K. Consequently, the most abundant element in the Universe effectively becomes invisible at temperatures below 100 K.

However, the carbon monoxide (CO) molecule is composed of two dissimilar atoms, so its rotational transitions *can* be observed. This makes CO, which is reasonably abundant and believed to be distributed in the same way as H_2, an important tracer of cold molecular clouds in space. Carbon monoxide has strong radio emission lines at 1.3 and 2.6 mm that are used for this purpose.

Molecular clouds also contain some quite large molecules, such as ethanol (CH_3CH_2OH). Over 200 different molecules have been detected in dense clouds. In such an environment, only chemically unreactive elements, such as He and Ne, remain predominantly in atomic form.

The relatively small mass of ionized hydrogen (HII) in the ISM is contained in the intercloud media and, much more spectacularly, in HII regions (Figure 1.17). HII regions are frequently associated with dense clouds.

■ Why are hot HII regions often found in association with cool, dense clouds?

☐ New stars form within cool, dense clouds. Only very hot (and therefore massive) stars, particularly the short-lived but highly luminous main sequence stars of spectral classes O and B, can ionize hydrogen in their vicinity and thus produce HII regions. As massive stars are short-lived, they can be observed still in close association with the original dense clouds. Hence HII regions are found near the cool, dense clouds from which the O and B stars formed.

Interstellar dust

The nature of the soot-like dust grains that form a constituent of the interstellar medium was outlined in Section 1.2.3. The total mass of dust is about 0.1% of the stellar mass of the disc. Dust forms from atoms and molecules in the gaseous ISM that condense directly to solid particles; liquids do not form. The regions of the ISM that favour the formation of dust are those where the density is high and the temperature low. These conditions occur in the disc, where matter ejected from cool giants or supergiants in stellar winds moves away from the star and cools.

Dust grains emit a continuous spectrum of radiation that is similar to the black-body spectrum. The spectrum of dust may therefore be used to deduce its temperature. If dust particles are heated too much they **sublimate**, that is, they change directly from being solid to being gaseous without melting to form a liquid. The temperature at which sublimation occurs depends on the precise composition of the dust, but even the least volatile compounds (those that are most resistant to evaporation) sublimate at temperatures no greater than about 2000 K. In practice though, most interstellar dust grains are well below their sublimation temperature, nearer to 20 K than 2000 K, and are thus easily able to survive in the environment of space. The spectrum emitted by stars peaks in the visible part of the spectrum at wavelengths shorter than 2 μm, whereas the energy emitted by dust grains at a temperature of 20 K peaks around 200 μm (see Figure 1.19). At such a relatively

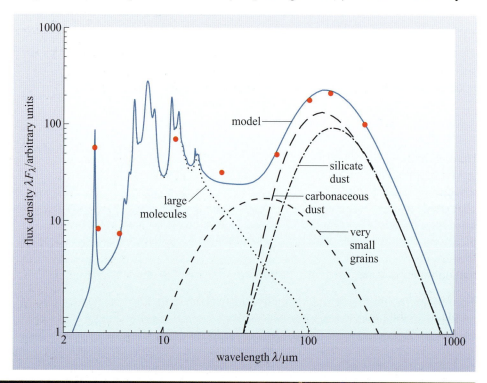

Figure 1.19 The infrared spectrum of radiation emitted by interstellar dust at a temperature of a few ×10 K. The red data points give the measured spectrum. The curves show expected contributions from dust grains, very small dust grains and large carbon-rich molecules. Note that the peak emission from this cold dust occurs at a wavelength $\lambda \sim 200$ μm (in the far-infrared). (Compiegne *et al.*, 2011)

Figure 1.20 A far-infrared image, made using observations from the Herschel Space Observatory, of part of the Orion giant molecular cloud complex. This image shows observations at three wavelengths ($\lambda = 70$ μm, in blue; $\lambda = 160$ μm, in green; $\lambda = 250$ μm, in red). Note the presence of the Horsehead Nebula on the far right-hand side of this image, which is observed in emission rather than absorption as in optical images (compare to Figure 1.17). (ESA/Herschel)

long wavelength, the peak of the dust emission is in the far-infrared region of the spectrum. The distribution of dust can therefore be deduced from far-infrared images, such as Figure 1.20.

1.4.3 A cross-section through the disc

Here we look in more detail at the way in which the distribution of stars and gas varies with distance perpendicular to the Galactic plane. We noted earlier (Section 1.2) that the density of stars declines with increasing distance from the mid-plane of the disc, and stated that 'the disc has a thickness of about 1 kpc'. In order to be more precise in defining a distribution that has no definite boundary, we need to make use of the mathematical entity known as the *exponential function*.

The vertical distribution of stars can be described by an exponential decay. More specifically, the **number density** of stars (i.e. the number of stars per unit volume), n, decreases with distance from the mid-plane, according to the equation:

The exponential function is explained in the Appendix.

$$n(z) = n_0 \, e^{-|z|/h} \qquad\qquad (1.8)$$

Here, n_0 is the number density of stars in the mid-plane; the positive quantity $|z|$ represents the distance above or below the mid-plane; h is an important distance parameter called the **scale height** of the disc, and it is this quantity that characterizes the 'thickness' of the disc. The quantity $n(z)$ is the number density of stars at a displacement z from the mid-plane, with $z > 0$ for points above the mid-plane and $z < 0$ for points below the mid-plane. The symbol $|z|$ is often read as 'the absolute value of z' or 'the modulus of z', since the modulus sign, $|\,|$, indicates that only the positive value of the enclosed quantity should be considered.

- ■ What is the ratio of the number density of stars at a distance h from the mid-plane, to the mid-plane number density?

- □ When $z = h$, Equation 1.8 becomes $n(h) = n_0 \, e^{-|h|/h} = n_0 \, e^{-1}$.

  You can evaluate e^{-1} using a calculator, or you can note that $e^{-1} = 1/e \approx 1/2.718$. Hence $n(h) \approx n_0 \times 1/2.718 \approx 0.37 n_0$, and hence $n(h)/n_0 \approx 0.37$.

That is, at a distance of one scale height above the mid-plane, the number density of stars is a factor $1/e$ or about 0.37 times the mid-plane number density.

The key idea here is that the scale height can be used to provide a quantitative measure of the vertical distribution of a component of the disc. Assigning scale heights to subpopulations of stars or classes of objects allows us to describe the vertical structure of the disc far more precisely than could be done by making crude statements about its 'thickness'. The scale height of a subpopulation doesn't say where that subpopulation ends, since there is no 'end', but it does indicate, in a precise way, the distance from the mid-plane at which the density of that subpopulation has significantly decreased. Below, we consider the scale height values for various constituents of the disc, and we comment on the significance of some of those values.

- ■ Can you state the scale height for the Sun? Justify your answer.

- □ The concept of a scale height has no meaning for a *single* object; it characterizes how the density of a *class* of objects varies, so it cannot be applied to individual members of the class.

Stellar components

Observations of the vertical distribution of stars reveal the existence of two components of the stellar disc – a **thin disc** that is concentrated towards the mid-plane of the Galaxy and a **thick disc** with a much greater scale height.

In detail, the scale height of the thick disc is approximately 1000 to 1300 pc. The scale heights of stars belonging to the thin disc are found to depend on their spectral classes. That is, thin-disc stars of different spectral types have different distributions above and below the mid-plane of the disc. In particular:

- Main sequence stars with spectral types G, K or M belonging to the thin disc (including the Sun) are distributed with a scale height of around 300 pc.

- The O and B stars belonging to the thin disc have scale heights of only 50 pc to 60 pc.

Now, we know that O and B stars are very young, whereas the G, K and M stars of the thin disc are Pop. I objects that span the whole age range of that population. It follows that the older stars of the thin disc are likely to be found further from the mid-plane than the more recently formed stars. This observed variation of scale height with age is thought to indicate an evolutionary process that operates within the disc. It is believed most stars form near the mid-plane, but once they have formed they are gradually scattered to greater heights by interactions with giant molecular clouds, which may be as massive as $10^7 M_\odot$.

Thick-disc stars are far less common than thin-disc stars in the mid-plane of the Galaxy. However, the relatively large value of its scale height indicates that the number density of thick-disc stars declines much more slowly than that of thin-disc stars as the displacement from the mid-plane increases. Thus, the thick-disc stars become relatively more important as distance from the mid-plane increases.

It is currently uncertain why the thin disc and the thick disc differ in this way; it could reflect different origins of the two subpopulations, or could be due to some event during the formation of the disc (such as a collision with a dwarf galaxy that added energy to the stars' orbits). This is one issue bearing on the origin and evolution of the Milky Way for which we do not yet have a satisfactory explanation.

The ISM

So far we have compared the scale heights of different classes of stars, but now we consider the vertical distribution of the gas and dust of the ISM. In this case we continue to represent the scale height by h, but rather than consider the number density of stars, n, we consider the mass density ρ of the ISM, measured in kg m^{-3} or some similar unit. Thus, the vertical distribution will be described by an equation of the kind $\rho(z) = \rho_0 \, e^{-|z|/h}$.

It is found that the ISM has a scale height around 150 pc. So, the gaseous component of the disc is more concentrated towards the Galactic plane than most G and K stars. The fact that the scale height of young O and B stars, 50 pc to 60 pc, is *less* than that of the gas from which they formed tells us about the star formation process. The rate at which stars are formed from the ISM is called the **star formation rate** (**SFR**) and may be measured in solar masses per year in any specified region (often the whole Galaxy). If the star formation rate were proportional to the local density of gas (that is, if SFR $\propto \rho$) then stars would be formed with the same height distribution as the gas. However, the O and B stars are more concentrated towards the mid-plane than is the gas. This indicates that the simple proportionality between SFR and ρ cannot be correct. Since young stars are more concentrated towards the mid-plane than is the ISM, it must be the case that star formation is more effective when the ISM density is higher. We have thus used the vertical distribution of gas and young stars in the disc to infer something about how the star formation rate depends on the density of gas.

This implies that if the SFR is related to ρ by a power law, i.e. $SFR \propto \rho^n$, then $n > 1$.

1.4.4 The spiral arms

So far we have examined the constituents of the disc, their distribution and composition. We have also seen that the Milky Way has a thin stellar disc, and in this way resembles the spiral galaxies we observe beyond our Galaxy. So it is not surprising that we find evidence for spiral structure in the Milky Way. In this section we look at the structure of the spiral arms and their possible origin.

Tracing spiral structure

As indicated in Section 1.2, it is believed that spiral arms stand out mainly because they contain concentrations of *bright* objects associated with recent star formation rather than being strong concentrations of *mass*. O and B stars are so short-lived (with lifetimes of only tens of millions of years) that, when you see them, they cannot be far from where they were born, so it is usual to regard the spiral arms as the main locations of star formation in the Milky Way.

- ■ Why would you expect HII regions to be associated with the spiral arms?

- ☐ HII regions need the presence of O and B stars to provide ionizing UV photons, and O and B stars are themselves associated with the spiral arms because they are short-lived stars that do not survive long enough to move very far away from their sites of formation.

At optical wavelengths, dust makes it difficult to see stars in the disc of the Milky Way if they are more than a few kiloparsecs away. Consequently, the spiral structure of the Galaxy is more easily mapped by using a combination of optical and radio 21 cm observations, since the latter are unaffected by dust and permit the detection of neutral hydrogen to much greater distances. A range of objects has been observed that appears to map out spiral structure. Such objects include dense molecular clouds, HII regions, open clusters, and OB associations. Objects that are used to map the locations of the spiral arms are called **spiral-arm tracers**.

Figure 1.21 (overleaf) indicates the distribution of bright HII regions and prominent clusters of young stars. They seem to trace out features that can be interpreted as the spiral arms of the Galaxy. The data in Figure 1.21 are shown overlaid on a schematic representation of the spiral arm structure of the Galaxy. Note the position of the Sun relative to the Galactic centre and the local spiral arms, named the **Sagittarius Arm** and the **Perseus Arm**. The Sun seems to be contained within a 'side spur' off one of these arms that is referred to as the **Orion Spur**. When looking at artists' conceptions of the Milky Way that include beautifully unbroken spiral arms, remember that scientific backing for such detailed views is almost entirely lacking.

The winding dilemma

We have discussed how the spiral arms can be traced using objects associated with recent star formation. However, in the remaining parts of this section, you will see that it is difficult to explain *why* star formation occurs along these reasonably well defined spiral tracks. It is also difficult to explain their persistence over long periods of time.

- How do we know that spiral arms last for long periods of time?

- ☐ Observations of the Milky Way do not tell us this. However, astronomers see many other spiral galaxies and are hence able to infer that spiral arms are not just short-lived structures.

In trying to account for the existence and persistence of spiral arms, one point must be kept clearly in mind: if the arms were composed of an unchanging set of stars, the differential rotation of the disc would cause the shape of the arms to alter with time, so an initially 'realistic' pattern of arms would soon cease to resemble any observed spiral arms. This is called the **winding dilemma**, and is explored further in Question 1.6.

QUESTION 1.6

Imagine that all disc stars have the same rotation speed of 200 km s^{-1} for all radial distances from 4 to 10 kpc from the Galactic centre, and suppose that a pattern of spiral arms like that shown in Figure 1.21 already existed in the Galaxy 4.5×10^9 years ago, when the Sun formed.

(a) How many orbits of the Galactic centre would the inner end of one of the arms, located 4 kpc from the Galactic centre, have completed since the formation of the Sun?

(b) How many orbits would the outer end of one of the arms, located 10 kpc from the Galactic centre, have completed in the same time?

(c) If the spiral arms had been made of the same stars throughout the lifetime of the Sun, what would they look like now? How well does your answer correspond to the actual appearance of spiral arms?

Even allowing for its oversimplifications, the answer to Question 1.6 indicates that there is no permanent population of 'spiral arm' stars. This suggests that the bulk of the stars that are currently in spiral arms must be moving relative to those arms, and will not remain in them for long.

An example of the passage of material through a spiral arm and consequent star formation is provided by M17. This is an HII region that is about 2 kpc from the Sun and lies within the Sagittarius Arm. Figure 1.22 shows an image of the region around M17 taken across several infrared wavelengths. M17 itself is seen as a bright star-forming region where gas is illuminated by the intense UV radiation of O and B stars. To the right of M17 is a series of dark filaments where the density in the molecular cloud is high enough to partially block infrared radiation. Star formation has actually started in these, so-called **infrared dark clouds**, as can be seen from the stars embedded in them, but it is at an early stage. It is thought that the highest-mass stars (O stars) are only formed once some star formation has begun and started to alter the environment of the cloud. There is a progression from the dark clouds to the HII region which corresponds to the million years or so that it takes for disc material to pass from the dark clouds to the HII region. Furthermore, to the left of M17, there is a region populated by bright O and B stars

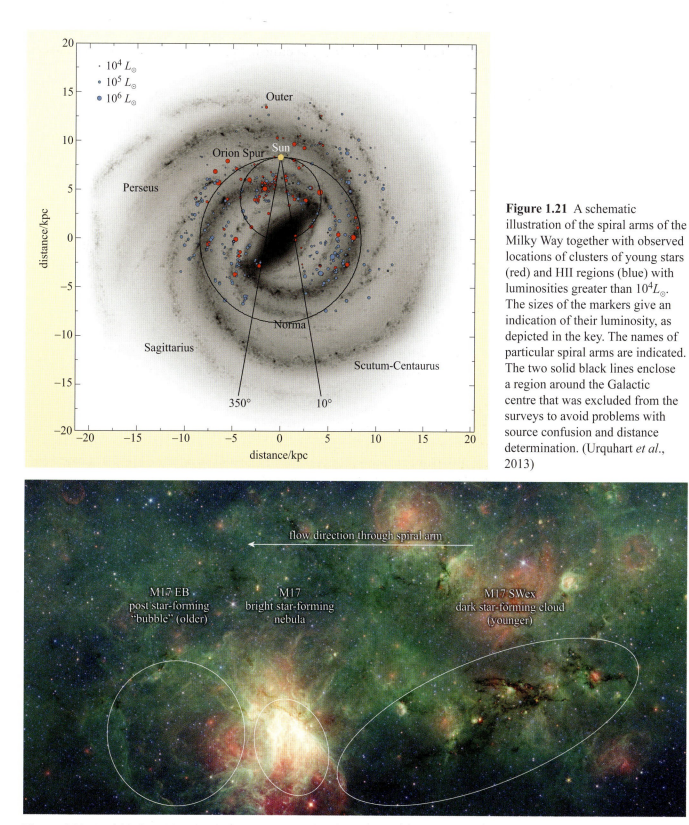

Figure 1.21 A schematic illustration of the spiral arms of the Milky Way together with observed locations of clusters of young stars (red) and HII regions (blue) with luminosities greater than $10^4 L_\odot$. The sizes of the markers give an indication of their luminosity, as depicted in the key. The names of particular spiral arms are indicated. The two solid black lines enclose a region around the Galactic centre that was excluded from the surveys to avoid problems with source confusion and distance determination. (Urquhart *et al.*, 2013)

Figure 1.22 A mid-infrared image obtained using the Spitzer Space Telescope of the region around the HII region M17. The image comprises three infrared wavelengths ($\lambda = 3.6$ μm, blue; $\lambda = 8$ μm, green; $\lambda = 24$ μm, red). Star formation is taking place in the dark infrared cloud labelled M17 SWex, but this is at an earlier stage than the HII region, which in turn is at an earlier stage than the OB association M17 EB. (NASA / JPL-Caltech / M. Povich (Penn State Univ.))

(an OB association) in a bubble of ionized gas – this is the expected fate of an HII region as the molecular and atomic gas is cleared away by the stellar radiation field. So the OB association to the left of M17 also reflects the motion of disc material through the spiral arm.

Spiral arm formation

How is the persistence of spiral arms reconciled with the differential motion of the stars and clouds that trace them? Various answers have been proposed to this question. One idea is that the spiral arms represent the current location of 'waves' of star formation that are travelling through the disc. Such waves need not participate in the differential rotation of the disc any more than sound waves travelling through air have to travel at the speed of the wind. If the waves formed a spiral pattern, as might be natural in a differentially rotating disc, then a persistent pattern of spiral arms might be an expected feature of the Galaxy.

An explanation along these lines, called **density wave theory**, was developed by C. C. Lin and Frank H. Shu in the 1960s. They treated the disc as an approximately smooth, axially symmetric distribution of matter in a state of steady differential rotation, but they assumed that such a disc would naturally develop regions in which the density was enhanced relative to the surrounding material. Lin and Shu argued that certain spiral patterns of density enhancement were especially favoured and could become self-perpetuating. Long-lived patterns of density enhancement of this kind are called **spiral density waves**.

A simple spiral density wave is expected to rotate about the Galactic centre, but to maintain its shape as it does so. That is, the pattern rotates rigidly, despite the fact that the material from which it is made rotates *differentially*. In fact, across most of the disc, the density wave moves more slowly than the matter in the disc. Only towards the outer edge of the pattern does the rotation speed of the density wave equal that of the disc. The radius at which this occurs is called the **co-rotation radius**. This means that throughout most of the disc, stars and gas approach the slowly moving density wave from behind, pass through it, and then move ahead of the wave's leading edge (this is the situation shown in Figure 1.22 – the flow direction is from behind the direction of rotation of the spiral arm). Gas is compressed when it enters the density wave. For a giant molecular cloud on the verge of forming stars, this increase in density might be enough to trigger star formation and thus account for the presence of the young objects that trace the arm.

Spiral arm formation can also be explored using numerical simulations. However, rather than the formation of a permanent spiral arm pattern as predicted by density wave theory, such simulations often show the development of fragmented arms (Figure 1.23) that are rather transient features – they come and go over timescales of hundreds of millions of years. The fragmented nature of such spiral arms is actually in reasonably good agreement with the appearance of many spiral galaxies that are believed to be similar to the Milky Way (Figure 1.2). The fact that arms may be transient is a much harder problem to address observationally – no single galaxy can be observed on such a timescale, so it is difficult to know if spiral arms are permanent or merely slowly changing features of the galaxy.

Although numerical simulation is starting to provide insights into how spiral arms may be formed, astronomers are still a long way from a full understanding of this process. So, despite the great importance of spiral arms in the Milky Way's disc, their origin should still be regarded as rather uncertain at present.

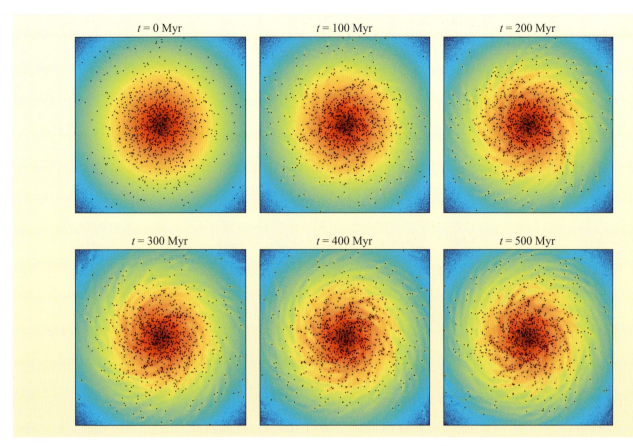

Figure 1.23 An example of a numerical simulation of a self-gravitating disc in a halo with the same mass as the Milky Way. This simulation has 100 million stars with 1000 giant molecular clouds (black dots) and shows the evolution of the disc over a time interval of 500 million years, starting from a uniform distribution. Each panel shows a simulation region 30 kpc in extent. After a simulation time of about 100 million years, spiral features become evident although their pattern changes with time. (D'Onghia *et al.*, 2013)

QUESTION 1.7

Assuming that the co-rotation radius of the Milky Way is at 15 kpc from the Galactic centre, and using the Galactic rotation curve given in Figure 1.11c, draw the rotation curve of a spiral density wave and estimate the speed at which the Sun would approach such a wave.

1.5 The stellar halo and bulge of the Milky Way

We now turn our attention from the disc – which is a region of ongoing star formation – to the stellar halo and the bulge. These two regions, which together form the Galactic spheroid, are locations where star formation has long since ceased, and where the oldest stars in the Galaxy are found. It is by studying these regions that we might hope to find clues about the early evolution of the Galaxy. With this in mind, we now proceed to examine these components of our Galaxy.

The stellar halo consists mainly of very old Population II stars which have low metallicities and which move in roughly elliptical orbits that are often highly inclined to the Galactic plane (Figure 1.9). Stars following such orbits plunge through the disc from time to time, but the spaces between stars, even in the disc, are so great that collisions are highly improbable.

■ Which spectral classes of main sequence star would you expect to be common in the stellar halo? What other kinds of star are likely to be abundant there?

☐ Only long-lived stars should be common. Apart from old main sequence stars of spectral classes K and M, there should be other, more highly evolved, stars such as red giants, horizontal branch stars, and asymptotic giant branch (AGB) stars. Even more highly evolved objects, such as white dwarfs, might also be expected.

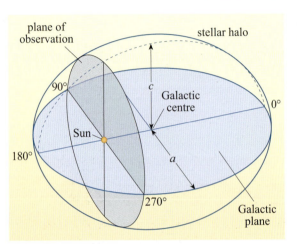

Figure 1.24 The shape of the stellar halo is determined by counting stars in a plane of observation cutting vertically through the Galaxy at the location of the Sun. Counts made at various Galactic latitudes determine the ratio of axes c/a (or axial ratio) of the stellar halo.

Measurements indicate that the stellar halo is somewhat flatter than a perfect sphere. Shape determinations are based on counts of the number of halo stars in a plane surrounding the Sun that is perpendicular to the direction of the Galactic centre (see Figure 1.24). Stars in this particular plane at a given distance from the Sun will all be at a common distance from the Galactic centre, so counts of such stars can be used to assess the shape of a section through the stellar halo. Such measurements indicate that the ratio of the axes a and c, shown in Figure 1.24, is $c/a \approx 0.8$, making the stellar halo an oblate spheroid.

1.5.1 Globular clusters

Globular clusters were introduced in Section 1.2.1, where it was stated that they typically contain 10^4 to 10^6 stars in a spherical region up to about 50 pc across. They are relatively prominent, and can easily be picked out and studied in unobscured regions of the sky. There are probably between 150 and 200 globular clusters in the Milky Way, about two-thirds of which are associated with the halo. About 1% of halo stars are found in these objects.

The stars in a globular cluster are most densely packed toward the centre. Because of this central concentration, the central regions of globular clusters are often overexposed in photographs. The number density of stars at the centre of a globular cluster is typically 10^4 pc^{-3}. This is approximately 10^5 times higher than the number density in the solar neighbourhood ($\sim 0.1 \text{ pc}^{-3}$), but it still leaves plenty of space between individual stars.

QUESTION 1.8

(a) What is the average separation of stars in the centre of a globular cluster? Give your answer in pc. (*Hint*: consider the average volume occupied by a single star.)

(b) How does this compare to the distance between the Sun and the stellar system of alpha Centauri, which is 1.3 pc away?

The distribution of globular clusters

Historically, globular clusters played a key role in developing our view of the size and shape of the Galaxy. Globular clusters are not distributed uniformly around the sky, but are concentrated in the direction of the constellation of Sagittarius. This was first recognized by Harlow Shapley, and is shown quite clearly in Figure 1.25a, which is a map of the positions of halo globular clusters on the sky. In 1917, on the basis of his studies of globular clusters, Shapley asserted that the centre of the Milky Way was in the direction of the constellation of Sagittarius. Realizing that globular clusters were more numerous near the Galactic centre, he correctly reasoned that the centre of their distribution indicated the centre of the Galaxy.

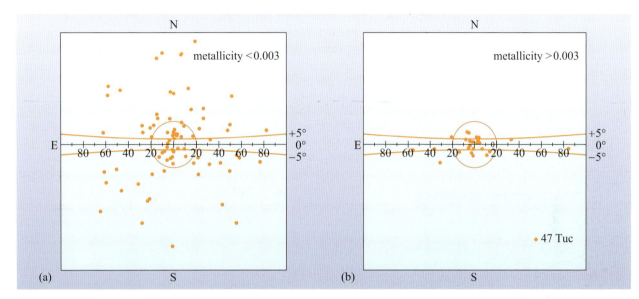

Figure 1.25 The locations of globular clusters on maps of the sky centred on the Galactic centre. The horizontal line is the Galactic plane, and the shallow curves mark Galactic latitudes $\pm 5°$. The circle has a radius of $20°$, which corresponds to ≈ 3 kpc at the distance of the Galactic centre where the globular cluster distribution is centred. Map (a) shows globular clusters with very low metallicity $Z < 0.003$ (the halo globular clusters), while map (b) shows clusters with $Z > 0.003$ (the disc globular clusters). The latter are much more concentrated towards the Galactic centre and the Galactic plane. (Zinn, 1985)

HARLOW SHAPLEY (1885–1972)

Harlow Shapley (Figure 1.26), the son of a farmer, was born in Missouri, USA. After a limited education he began work as a crime reporter on a small Kansas newspaper. Further study led to the University of Missouri, where he planned to take a degree in journalism. However the School of Journalism had not yet opened, so rather than waste a year he took an astronomy course and thereby began one of the most distinguished astronomical careers of the 20th century. He gained his PhD from Princeton in 1913, was a staff member at the Mount Wilson Observatory until 1921, and was then appointed Director of the Harvard College Observatory where he remained until 1952. Shapley began his studies of Cepheid variables in globular clusters while at Mount Wilson. Using the Cepheids as distance indicators (see Chapter 2) he found that the known globular clusters were concentrated about a location in the direction of Sagittarius that he identified as the centre of the Milky Way. Shapley overestimated the diameter of the Milky Way's disc by a factor of three, but he did comprehend its structure, and he recognised the off-centre location of the Sun. In a long and illustrious career, Shapley made many contributions to the study of the Milky Way and other galaxies.

Figure 1.26 Harlow Shapley. (Bachrach Portrait Studios)

Because globular clusters are so conspicuous, and hence can be recognized at great distances, they provide a means of determining the size of the stellar halo. Their distances can be derived from the brightness of their stars. Observations show that although there are many globular clusters within 20 kpc of the centre of the Galaxy, and a few beyond 37 kpc, there are none in between. Some astronomers treat the break at 20 kpc as a rough indication of the outer edge of the stellar halo, while others regard the more distant ones as indicating just how very extensive the stellar halo is! Unfortunately, the rather low number of globular clusters in the Galaxy – there are probably no more than about 200 – prevents any improvement in estimates based on globular clusters alone. In order to probe the outer part of the stellar halo, it is necessary to use the more numerous, but harder to recognize, non-cluster stars. Such objects are discussed in the next section.

Not all the globular clusters actually belong to the halo. Globular clusters with metallicity $Z > 0.003$ used to be regarded as halo objects, but are now recognized as belonging to the disc. They account for approximately one-third of all globular clusters in the Galaxy. Figure 1.25b shows the location of these relatively high-metallicity clusters. As you can see, they are mostly confined to the Galactic plane, as expected.

Globular cluster ages

As has already been noted, the halo harbours the oldest known stars. Those that are in globular clusters are particularly important because of the relative ease with which their ages can be determined. The globular clusters of the halo formed early in the evolution of the Milky Way, probably even before the Galaxy was a well-defined entity, so they can teach us something about the formation of the Milky Way. In addition, these oldest globular clusters provide a lower limit for the age of the Universe. Obviously, the Universe must be older than the oldest globular clusters it contains. There have been times when the observationally determined globular cluster ages have exceeded the age of the Universe estimated from some cosmological arguments. Such conflicts have been a source of great controversy, and their resolution is a significant sign of progress.

The age of a globular cluster is determined by analysing its Hertzsprung–Russell or H–R diagram. In order to see how this is done, we need to consider how the H–R diagram is drawn. Box 1.5 provides a description of the different forms of the H–R diagram that we shall refer to.

Figure 1.27 shows the H–R diagram of the globular cluster 47 Tuc, which is broadly representative of most globular clusters. It is presumed that the stars comprising a cluster all formed at the same time and had the same composition. When it was still young, all the cluster's stars would have belonged to the main sequence and the cluster's H–R diagram would have looked something like that of NGC 2362 in Figure 1.28a. As the cluster aged, the more massive stars of high luminosity would have left the main sequence and evolved to become supergiants. The point on the cluster's H–R diagram corresponding to stars that are just reaching the end of their time on the main sequence is called

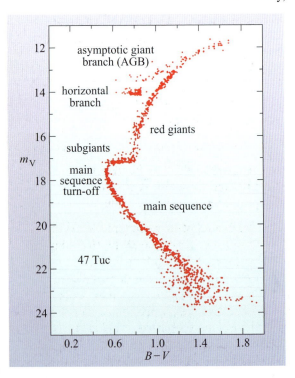

Figure 1.27 The H–R diagram of the globular cluster 47 Tuc. Some of the major features of the H–R diagram have been labelled. (Hesser *et al.*, 1987)

the **main sequence turn-off.** This is clearly seen in the cluster H–R diagram of Praesepe (Figure 1.28b) – the main sequence turn-off in this case is at $B - V \approx 0.2$, $M_V \approx 1$. With the further passage of time, the main sequence would have become progressively depopulated of massive stars.

BOX 1.5 H–R DIAGRAMS

Theoretical models of stars generally provide calculated surface temperatures and luminosities. The theoretical **H–R diagram** of a set of stars is therefore a plot of luminosity versus (surface) temperature. Such a plot is sometimes called a *temperature–luminosity diagram*. However, the temperatures and luminosities of real stars cannot be measured directly; they must be inferred from quantities that can be measured. It is quite common to plot the H–R diagram of real stars using the observed properties that correspond most closely to temperature and luminosity, which are *colour* and *brightness*.

■ Why does colour correspond to temperature?

□ Stars behave similarly to black bodies, and the continuous spectra emitted by black bodies peak at shorter wavelengths the hotter they are, so hotter stars look bluer.

Instead of just saying that a star is blue or red, astronomers express colour in a quantitative way by comparing the amount of energy received in one part of the spectrum with the amount of energy in another. One common way of measuring colour is to compare the energy received in the blue (B) part of the spectrum with the energy received in the green part. Since the green part of the spectrum is where the human eye has maximum sensitivity, this part of the spectrum is called the visual (V) band. From the ratio of the energy between the blue and visual bands, astronomers define a value they call the **colour index**, $B - V$, (pronounced 'B minus V'). For hot, white stars, $B - V \sim 0.0$, whereas for cool, red stars, $B - V \sim 1.0$. (You need not be concerned here about the details of how exactly the $B - V$ value is calculated from the observations.)

The brightness of a star is usually expressed using a magnitude scale, which describes its brightness relative to other stars. A difference of 2.5 magnitudes between two stars corresponds to a factor of 10 in brightness. The magnitude scale is logarithmic, so a difference of 5 magnitudes corresponds to a factor of $10 \times 10 = 100$ times in brightness, and a difference of 10 magnitudes corresponds to a factor of 10^4 times in

brightness. (It could be considered as a *faintness* scale, as stars whose magnitudes are at the positive end of the magnitude range are the faintest ones, while the brightest stars are at the negative end!) Magnitude determinations are often restricted to a specified part of the spectrum, such as the visual (V) band. Within such a specified band, a star has two magnitudes that are of interest. The first of these is the easily determined **apparent visual magnitude** (m_V), which directly compares the apparent brightness of stars, even though those stars may be at very different distances. The other kind of magnitude is the **absolute visual magnitude** (M_V), which compares the brightness that the stars would have if they were all at the same distance. It is this latter quantity that is most directly related to their intrinsic luminosity, but it is also the harder to determine accurately. The two kinds of magnitude are defined in such a way that their values would be equal for a star that was 10 pc away.

The precise relationship between the apparent and absolute visual magnitudes of a star at a distance d is given by

$$M_V = m_V - 5 \log_{10} (d/\text{pc}) + 5 \qquad (1.9)$$

where the distance is expressed in parsecs. This equation can be rearranged as

$$m_V - M_V = 5 \log_{10} (d/\text{pc}) - 5 \qquad (1.10)$$

The quantity ($m_V - M_V$) is called the **distance modulus**, and is widely used in astronomy as a quantity that provides a measure of distance.

In contrast to the theoretical temperature–luminosity diagram, the observational version of the H–R diagram plots a colour index on the horizontal axis and a magnitude scale on the vertical axis. Sometimes a diagram of this type is called a *colour–magnitude diagram*. Stars map out similar patterns in both the theoretical and observed versions of the diagram, as the physical information captured in the H–R diagram is essentially the same even in these different forms, and we use the name H–R diagram to refer to all the different forms.

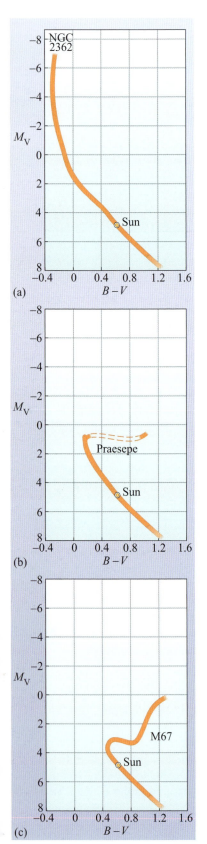

(a)

(b)

(c)

Figure 1.28 The H–R diagrams of three clusters of stars. (a) When the cluster is only a few million years old, as in the case of NGC 2362, essentially all its stars belong to the main sequence. (b) After a hundred million years, the age of the Praesepe cluster, all the high-mass stars will have left the main sequence. (c) After a few billion years, the age of M67, only low-mass main sequence stars remain. The surviving intermediate-mass stars will have evolved into red giants. All of these clusters are open clusters. As in the case of globular clusters, it is believed that all the stars in a given open cluster were formed at the same time from material of uniform composition. (Based on Arp, 1958)

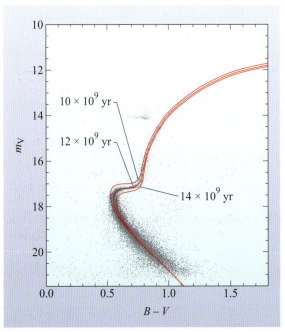

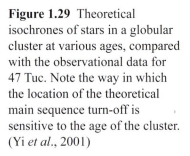

Figure 1.29 Theoretical isochrones of stars in a globular cluster at various ages, compared with the observational data for 47 Tuc. Note the way in which the location of the theoretical main sequence turn-off is sensitive to the age of the cluster. (Yi *et al.*, 2001)

The H–R diagram of a middle-aged cluster would have looked something like that of M67 (Figure 1.28c) – which begins to resemble the old-cluster diagram, Figure 1.27. The age of an individual star cluster can be gauged by making a calculation of the time required for the H–R diagram to evolve into the observed form.

The process of making a theoretical estimate of the time required for a cluster's H–R diagram to acquire a specific form involves the use of computer programs to model the evolution of the stars that make up the observed H–R diagram. As a result of such calculations, it is possible to plot the theoretical positions on an H–R diagram of stars in a cluster for a particular age of that cluster. Such a plot defines a curve on the H–R diagram that is called an **isochrone**. The most age-sensitive feature of a set of isochrones of differing ages is the location of the main sequence turn-off, so it is this that is used to date the clusters. Figure 1.29 shows a comparison between the theoretical isochrones and observations for one particular cluster (47 Tuc). As you can see, the age indicated in this case is about 12×10^9 years, although it should be noted that this value is dependent on the stellar evolution models used to calculate the isochrones. In general, different assumptions made in modelling will result in systematic differences between calculated isochrones, and there remains some uncertainty in using this technique to determine absolute ages.

Because of difficulties in finding precise *absolute* ages, there has been more progress with the question of the *relative* ages of different globular clusters. Studies indicate that some of the halo clusters are several billion years older than others. If this is correct, it indicates that the formation of the stellar halo was not, as was once thought, the result of the rapid collapse of a galaxy-sized gas cloud over just a few tens of million years. Either the collapse was a much more gradual process, or the stellar halo didn't form in a collapse at all, but rather by the coalescence of many different clouds. Under this second scheme, globular clusters formed either in these clouds or in collisions between them. In either case, the range of globular cluster ages indicates the timescale of the halo formation process.

1.5.2 RR Lyrae stars

The small number of globular clusters limits their usefulness as probes of the vast stellar halo. Fortunately, non-cluster stars belonging to the halo greatly outnumber globular clusters, and these stars can be used to probe the halo further. Members of a particular class of variable stars, known as **RR Lyrae** stars, are especially valuable for this purpose. These stars are named after the first star of their class to be studied, the star designated 'RR' in the constellation Lyra.

Low-metallicity, low-mass stars that have already been through the main sequence and red giant stages of evolution, and have become hot enough in their cores to burn helium, are called '**horizontal branch**' stars (see Figure 1.27). This name was given because all such stars have similar luminosities irrespective of their effective temperature, and they therefore form an almost horizontal band in the H–R diagram. RR Lyrae stars are a subset of the horizontal branch stars, and therefore have known absolute magnitudes.

Objects that, as a class, have known absolute magnitudes, are called '**standard candles**' because their known 'standard' light output allows the distance of any member of the class to be determined from a measurement of its apparent magnitude. (Uses for other standard candles in gauging the distances to other galaxies are discussed in Chapter 2). Horizontal branch stars meet this criterion, which makes RR Lyrae stars excellent standard candles, *provided* they can be recognized. How do you recognize an RR Lyrae star? They are the subset of horizontal branch stars falling in a region of the H–R diagram within which the outer layers of stars are unstable and pulsate. This region of the H–R diagram is called the **instability strip**. It is the pulsational properties of RR Lyrae variables that make them recognizable.

■ Horizontal branch stars that do not lie within the instability strip also exist. Could these be used easily as standard candles too?

□ The absolute magnitudes of non-variable horizontal branch stars are known, so they could in principle be used as standard candles. However, there is a practical difficulty: when you observe a non-variable star, it is not usually easy to tell whether it lies on the horizontal branch unless you already know its distance, in which case you don't need a standard candle. It is the distinctive variability of the RR Lyraes that identifies beyond doubt that they lie on the horizontal branch.

The known absolute magnitudes of RR Lyraes, and the relative ease with which they can be recognized, makes them useful for studying the distribution of matter in the halo.

QUESTION 1.9

The equation that relates the absolute and apparent magnitudes of an object to its distance d (in parsecs) is (Equation 1.10)

$$m_V - M_V = 5 \log_{10}(d/\text{pc}) - 5$$

The absolute magnitudes of RR Lyrae variables are $M_V \approx +0.5$. Assuming a well-equipped telescope could detect stars down to an apparent magnitude $m_V \approx +20.5$, to what distance could they be seen?

As RR Lyrae stars can be seen out to distances of tens of kiloparsecs, they are very useful probes of the stellar halo. They show that the number density of stars in the halo (i.e. the number per unit volume) falls off with distance from the Galactic centre, r, roughly in proportion to $1/r^3$, at least out to about 30 kpc. This is compatible with the decrease in number density seen for the globular clusters. As the globular clusters and RR Lyrae stars share the same spatial distribution, age, and metal content, it is fair to assume they are part of the same population.

We saw above that globular clusters provided the basis of one of the first reliable measurements of the direction and distance to the Galactic centre. RR Lyrae stars are also valuable for this purpose. Astronomers can observe RR Lyrae stars in the direction of the Galactic centre, and calculate the distance of each one they see. By noting the distance at which the number density of RR Lyrae stars reaches a maximum, astronomers have inferred a distance to the Galactic centre of 8.7 ± 0.6 kpc. This is clearly consistent with the value 8.5 kpc that is often adopted.

1.5.3 The Galactic bulge

We have seen how the globular clusters and RR Lyrae stars occupy the relatively sparsely populated spheroid, and have provided much information about the structure, size and age of the Galaxy, but there is an equally old component in the Galaxy that has a much greater density. This is the bulge, which occupies the central few kiloparsecs of the Galaxy.

Our view of the bulge is heavily obscured by dust, particularly the central region, which is the meeting place of the densest parts of the halo and the disc. However, it can be observed at infrared wavelengths (this was shown in Figure 1.3) where it manifests itself as a concentration of brightness, and a thickening around the Galactic centre. The equatorial radius of the bulge is about 3 kpc.

Observations at a wavelength of 2.4 μm reveal diffuse starlight from an old population of stars towards the Galactic centre. They show an enhancement to one side of the centre, suggesting an asymmetry in the distribution of stars around the central region. This asymmetry can be interpreted as evidence of a bar-like distribution of stars, as indicated by the innermost central contour in Figure 1.30.

Further information is provided by studying stellar motion. Unlike the situation in the disc, where stars travel in almost circular orbits around the Galactic centre, the

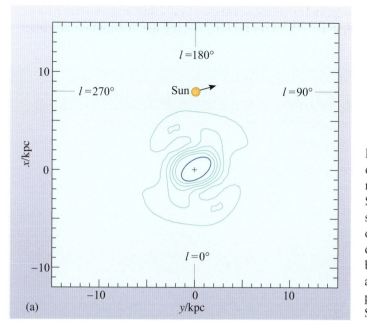

(a)

Figure 1.30 Schematic view of the size and orientation of a small central bar approximately 3 kpc long, based on numerous observations. The location and motion of the Sun are also shown. The innermost contour shows the bar shape and orientation suggested by 2.4 μm observations of diffuse starlight. Contours further out, based on the density of asymptotic giant branch (AGB) stars, show the bar extending out to a ring-like structure with a radius of approximately 5 kpc, which might be the densest, innermost portions of spiral arms. (Based on data from Blitz and Spergel, 1991 and Weinberg, 1992)

motions of stars in the bulge appear to be more diverse, with speeds in its outer regions around 100 km s^{-1}. Studies of the bulges of other galaxies, combined with information about the Milky Way, have assessed whether the bulges of galaxies are oblate spheroids that are symmetric about the galaxy's rotation axis or, whether they have three axes of different length. For the Milky Way, the evidence points to the bulge being a triaxial bar.

Once the existence of a bar is accepted, it prompts the question: 'Why does the bar exist?' Computer models of the motions of stars in galactic discs have shown that bars are very natural features. It turns out that discs without bars are difficult to produce, because a smooth disc dominated by rotational rather than random motion is unstable to the formation of a bar. That is, asymmetries in a disc can rapidly (within a few rotations of the galaxy) produce a bar structure; this behaviour of discs is called the **bar instability**. This suggests that there is no need to explain where the Galactic bar came from; the difficulty may be to explain why non-barred spirals do *not* have one.

1.5.4 The central black hole

We saw in Section 1.3 that it is possible to measure the mass of an object by its gravitational influence on other bodies that orbit it. We performed calculations for the mass of the Sun (for which we obtained the mass $M_\odot = 2 \times 10^{30}$ kg), and for the inner part of the Galaxy out to the radius of the Sun (for which we obtained $1 \times 10^{11} M_\odot$). In this section we examine another mass estimate, that of the compact object that has been found at the very centre of the Milky Way. Our findings have implications not only for the conditions in the Galactic centre, but also for the formation of our Galaxy and for many other galaxies – a topic that is discussed in subsequent chapters.

One of the first signs that the Galactic centre is a particularly interesting location in the Galaxy was the discovery of an intense, unresolved radio source known as **Sagittarius A*** (pronounced 'Sagittarius A star' and often written Sgr A*); an object that is thought to mark the precise centre of the Milky Way. Within 0.04 pc (1 arcsec) of Sgr A* lies a star cluster (Figure 1.31) whose members include hot, massive, and therefore young stars, whose age appears to be only about 10^7 yr. This suggests that either a burst of star formation occurred that recently, or that collisions and mergers of stars have occurred in the dense stellar environment.

It is possible to estimate the mass enclosed within a certain volume by studying the motion of material around that volume, as we did in Section 1.3, provided the observed motion is the result of gravitational forces alone. In the case of the Galactic centre, long-term observations have been performed to measure the positions of stars in this cluster. Rather remarkably, the motion of these stars is such that over an interval of about 20 years, their orbits have been determined as

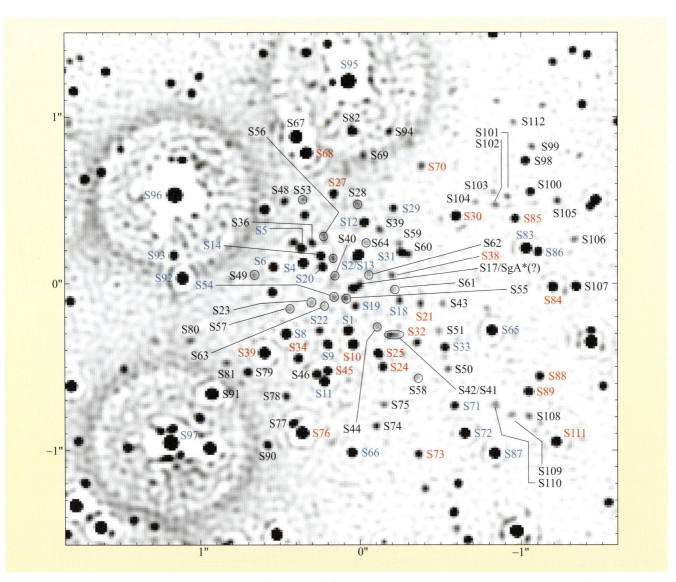

Figure 1.31 The star cluster surrounding the radio source Sgr A*. The spectral types of the stars are indicated by the colour of their labels (blue, types O–F; red, G–M; black, uncertain). While the position of Sgr A* is indicated, the identification of one of these infrared sources with the source itself is rather uncertain. This high resolution near infrared (H-band, $\lambda = 1.6$ μm) image was obtained using the VLT (Very Large Telescope). Note that the ring patterns around the brighter stars are artefacts. (Gillessen *et al.*, 2009a)

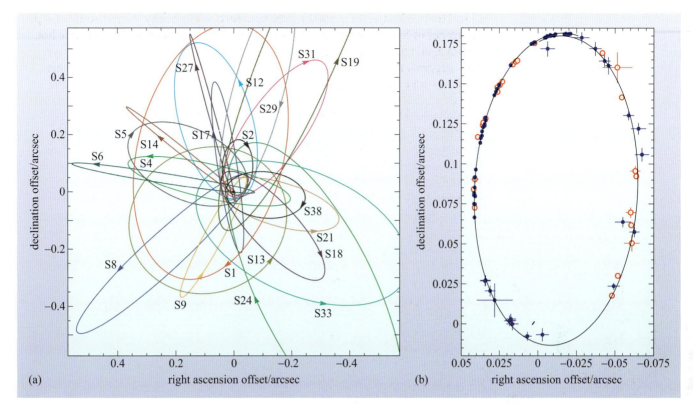

(a)

(b)

Figure 1.32 (a) The orbits of stars in the cluster centred on Sgr A* as determined from long-term monitoring in the near-infrared. Sgr A* lies at the position (0.,0.) in this map. The labels correspond to the stars shown in Figure 1.31, but note the change of scale. (b) The orbit of the star S2 (in Figure 1.31) as determined from near-infrared observations from 1992 to 2009. The star has an orbital period of 15.9 years and has now been observed over more than a full orbital cycle. Sgr A* lies at the position (0.,0.) in this map. The red and blue measurements display the results from two independent research groups ((a) Gillessen *et al.*, 2009a; (b) Gillessen *et al.*, 2009b)

illustrated in Figure 1.32a. Of particular note is the star labelled S2 in Figure 1.31, which is in a highly elliptical orbit around Sgr A* with an orbital period of 15.9 years. This star has now completed more than one full orbit (Figure 1.32b) in the time that it has been observed. Analysis of the orbital motion of this star, and others that have short orbital periods, allows the mass of Sgr A* to be determined. The current best measurement for this mass is $(4.3 \pm 0.4) \times 10^6 M_\odot$.

A highly probable explanation is that the enclosed mass takes the form of a *massive black hole* with a mass of $4.3 \times 10^6 M_\odot$ or so. Radio measurements of Sgr A* place an upper limit on its size, indicating that its radius cannot be more than a few times the distance from the Earth to the Sun. Although only an upper limit, this is already sufficient to confirm that the central object is very compact as well as enormously massive.

What are the implications for the formation of the Galaxy of finding a massive black hole in its centre? You will see in Chapter 3 that black holes probably exist at the centres of many other galaxies, and that in some galaxies the black hole gives rise to a wealth of other observational features that are not seen in our Galaxy.

We have now completed our survey of the major components of the Galaxy, and have seen hints of the processes that link them together. The final section of this chapter brings these ideas together to look at the evolution of the Galaxy.

1.6 The evolution of the Milky Way

In this final section of Chapter 1 we start to consider how the Milky Way has changed with time, and this naturally leads to the question of how the Galaxy was formed. In attempting to use observations of the Milky Way to illuminate its origin, we have to wind back the clock of star formation to see what the Galaxy would have been like at earlier times. We can hope for some success in doing this, because the Galaxy contains stars spanning a wide range of ages, and we can observe their different characteristics. Nevertheless, there is a limit to how much this process can reveal, since some information concerning the formation of the Galaxy has been erased, or at least obscured, by subsequent events. Consequently, the observations must be united with theoretical models that work forwards in time rather than backwards. Some models, such as the model of cosmic recycling that is elaborated on below, consider the way that material in the ISM is processed through stars, chemically enriched and injected back into the ISM again. Other models, which are discussed in subsequent chapters, look at the way dark matter and normal matter interact to form the pool of gas from which the first galactic stars formed. Yet another series of models looks at the way structure emerged in the early Universe, and how this allowed matter to form the kind of aggregates that developed into the groups and clusters of galaxies that surround us. As you will see later, understanding the formation and evolution of the Milky Way involves every part of this story, and every chapter in this book … and more. For that reason, our objectives in this section are rather modest. Rather than trying to present a full account of the origin and evolution of the Milky Way, we content ourselves with surveying some relevant observations, raising some difficult questions, and preparing the way for wider-ranging discussions of galaxy formation in later chapters.

1.6.1 The evolution of the interstellar medium

We saw in Section 1.2.3 that the stellar populations of the Galaxy have a range of ages, metallicities and locations. To understand why the different populations have different ranges of metallicity we have to consider in more detail where the metals (in the astronomical sense of the word) come from. As noted earlier, the only metal produced in the big bang was lithium; all others result directly or indirectly from nuclear reactions occurring in stars. The metallicity of a main sequence star corresponds to the metallicity of the ISM at the time the star formed. The fact that each stellar population contains a range of metallicities, and that the younger stars in a population tend to have higher metallicities, therefore suggests the operation of processes that progressively enrich the ISM by increasing its metallicity.

The ISM is a key element of the evolutionary picture, because it is the birthplace of stars and it is the repository for material ejected from stars towards the end of their lives. Dense molecular clouds form from the low-density ISM, and these give rise to a new generation of stars in a star cluster. The stars go on to produce metals. The metals may be ejected back into the ISM at the end of a star's life, or they may be locked away in a dense stellar remnant such as a white dwarf, a neutron star or a black hole. Metals that are ejected enrich the low-density ISM, and allow the cycle of enrichment to continue.

The return of matter from stars to the ISM occurs in several ways, via:

- stellar winds from cool giants/supergiants;
- the ejection of planetary nebulae, which are the envelopes of intermediate-mass stars – with initial masses less than about $11M_\odot$ – shed when the stars evolve to become white dwarfs;
- supernova ejecta containing most or all of a star's mass, some of which is enriched with metals.

By whichever route it happens, the transfer of chemically enriched material from a star to the ISM enriches the ISM and ensures that the next generation of main sequence stars to form in that region will have higher metallicity than its predecessor. Thus the chemical evolution of the Galaxy is a cyclic process involving star formation, element production within stars (nucleosynthesis), and the return of chemically enriched material to the ISM where it can form more stars. This process is sometimes called **cosmic recycling**, and is central to our understanding of the differences between Pop. I and Pop. II stars in terms of the way the Galaxy has evolved since its formation.

Although cosmic recycling will have taken place in each of the Galactic components (disc, bulge and stellar halo) that contain stars, it probably did not proceed at the same rate in each of them, nor does it necessarily continue at a significant rate in each of them today. The presence of various amounts of metals in stars of different ages allows astronomers to deduce the history of cosmic recycling in the various stellar populations, which in turn allows them to trace the star-formation history of the Galaxy.

A supernova produces an energetic shock wave that heats and ionizes a bubble of material in the ISM. Where large numbers of supernovae are occurring, such as in very active star-forming regions, these bubbles overlap, producing a very large cavity of hot, ionized gas called a **superbubble**. This hot material is restricted from expanding in the plane of the disc due to the pressure of the surrounding ISM, but it can expand vertically due to the much lower pressure of gas in the stellar halo. ISM material heated by supernovae may therefore enter the stellar halo via so-called **chimneys**, which are passages that open up where the hot, expanding material breaks out of the disc. Eventually, the hot material cools and probably returns in fragments to the disc. This mechanism has been termed a **Galactic fountain**, because of the motion of material squirted away from the disc and brought back again by gravity. It has been estimated that roughly $10M_\odot$ of material per year passes around this cycle.

Although the stellar halo now contains very little gas (almost none compared to the disc) there nevertheless exists a tenuous body of hot halo gas that is sometimes referred to as the **gaseous corona**. Absorption lines in the UV spectra of distant stars reveal the presence of this hot gas, which may be sustained by the Galactic fountain.

Is there any evidence for a returning flow of cooler gas? Possibly. Observations at wavelengths near 21 cm reveal the presence of clouds of atomic hydrogen, moving rapidly relative to the Sun. Many of these clouds have a significant velocity component towards the Sun. They are referred to as **high-velocity clouds**. It is difficult to determine the distance of these clouds, but they are known to be further away than most disc stars. They are often presumed to be in the Galactic halo, in which case they could represent the returning flow of gas ejected by the Galactic fountain. However, it is also possible that at least some of them are much further away than that. The distances of some clouds have been estimated at 400 kpc to 1000 kpc, comparable to the 750 kpc distance of the spiral galaxy M31. This would make them intergalactic rather than merely interstellar. Their origin is still a mystery.

■ How could the speed of such clouds, towards or away from the Sun, be determined?

☐ By looking for a shift in the wavelength (or frequency) of the 21 cm line, and assuming that it is a Doppler effect due to the motion of the cloud.

In addition to the return of material to the ISM, there is also the possibility of new matter entering the enrichment cycle, due to the infall of gas and dust from outside the Galaxy. This would come from the *intergalactic* medium, which consists of low density gas that is found between galaxies. Although little is known about such material in the neighbourhood of the Milky Way, the high-velocity clouds may be examples of such material.

Figure 1.33 provides a schematic overview of the key processes involved in cosmic recycling in the Galaxy. Once the various sources and sinks of gas are accounted for, what is the net rate of gain or loss of ISM gas today? The rate at which matter is entering the ISM is probably between $0.4M_\odot$ and $3M_\odot$ per year, of which no more than about $1.4M_\odot$ is infall from the intergalactic medium. The present rate at which mass is leaving the ISM, to form new stars, is probably somewhere between about $3M_\odot$ and $10M_\odot$ per year, and so it is likely that the rate of loss by the ISM is still exceeding its rate of gain.

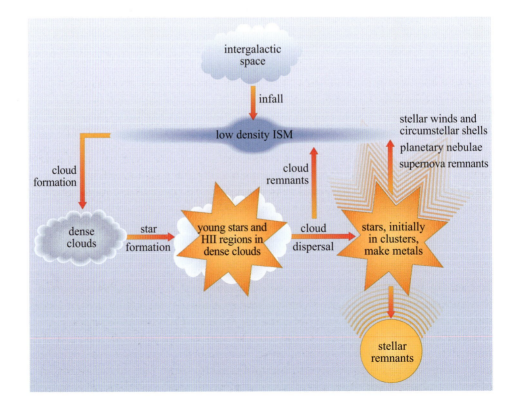

Figure 1.33 The evolution of the interstellar medium includes cycles, sources (infall from the intergalactic medium) and sinks (losses to stellar remnants).

QUESTION 1.10

By considering the processes in Figure 1.33, describe, in a couple of sentences, the amount and composition of the ISM that will exist in the Galaxy a *long* time in the future.

1.6.2 The evolution of the stellar populations

The stellar content of the Milky Way evolves in a number of ways as a result of cosmic recycling. One way is the change in the average chemical composition of stars that has taken place over time. We have already seen that within the Galaxy the youngest disc stars are also those with the highest metallicity. This makes good sense in terms of the picture of cosmic recycling. A corresponding tendency for old stars to have low metallicities was noted when we discussed the halo. Recall that the halo contains some of the oldest stars in the Galaxy and these have the lowest metallicities (down to $Z \sim 10^{-6}$ in some cases). This correlation between age and metallicity is known as the **age–metallicity relation**.

However, the correlation between age and metallicity is imperfect. The majority of bulge stars are probably as old as the stellar halo, but many of them have the same metallicity as the Sun. This combination of great age and high metallicity suggests that star formation proceeded very rapidly in the bulge, so that a large degree of cosmic recycling was achieved very quickly. This is consistent with the relatively high density of the bulge compared with the halo, since the star-formation rate, and hence the metal-enrichment rate, would be expected to be greater in a region of greater density.

Limitations such as these on the age–metallicity relation provide a useful reminder that in a complicated environment like the Milky Way, star formation is unlikely to have a simple evolutionary history.

QUESTION 1.11

Why would you expect surviving Pop. II main sequence stars to have lower metallicity than Pop. I main sequence stars?

QUESTION 1.12

Some people think that old stars have been undergoing nucleosynthesis for a long time, so when we observe Pop. II main sequence stars they should exhibit higher metallicity than the younger Pop. I stars because of the accumulated products of nucleosynthesis. Explain why this view is wrong.

1.6.3 New arrivals

To conclude our discussion of the evolution of the Galaxy, we note that some of its stars are recent arrivals. Astronomers have speculated for several decades that galaxies would on occasions collide, and that these events would lead to some remarkable shapes for the affected galaxies, as well as triggering new bursts of star formation as gas clouds are compressed in the collisions. (Stars, in contrast to gas clouds, are very small and almost never collide with one another, so the evolution of gas and stars differ greatly in collisions.) There are many examples of galaxies that appear to have collided recently; examples of such events are discussed in Chapter 2. What astronomers didn't realize until 1994 was that the Milky Way is involved in a collision *right now*.

What was discovered in 1994 is that a small galaxy, subsequently named the **Sagittarius dwarf galaxy**, is colliding with our Galaxy (Figure 1.34) and being shredded in the process. This collision is taking place on the far side of the Galaxy, which is why it went unobserved for so long. The collision is not a rapid event – the Sagittarius dwarf galaxy has been under the gravitational influence of the Milky Way for billions of years, and as it has interacted gravitationally with the

Figure 1.34 The collision of the Sagittarius dwarf galaxy with the Milky Way. The stars of the Sagittarius dwarf are very faint, so the dwarf galaxy is shown here as a red outline only. Astronomers have worked out that the Sagittarius dwarf is currently located on the far side of the Galaxy (at a distance of about 25 kpc from the Sun), almost in line with the Galactic centre (compare with Figure 1.1), and close to the southern side of the disc. (R. Ibata, R. Wyse and R. Sword/NASA)

Figure 1.35 (a) A map of the tidal streams that are attributed to the interaction of the Sagittarius dwarf galaxy with the Milky Way. Note that this map projection uses equatorial rather than galactic coordinates, and that the Galactic plane is masked out. The map shows the density of stars as measured by 2MASS (black and white) and the Sloan Digital Sky Survey (colour). (b) A schematic guide to the features in the map. Note the tidal streams from (RA = 50°, Dec. =+20°) to (300°, −30°) and (230°, −10°) to (120°, +20°) that wrap around the Galaxy, and the Sagittarius dwarf galaxy located at (284°, −30°). ((a) Adapted from Belokurov *et al.*, 2006 and Majewski *et al.*, 2003)

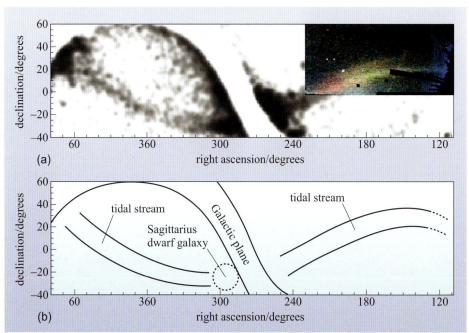

Galaxy, tidal effects have pulled off streams of stars. These tidal streams have been mapped crossing the stellar halo (Figure 1.35) and essentially wrap around the Galaxy. Furthermore, the paths of these streams are sensitive to the distribution of dark matter in the halo, thus offering another way to probe the mass distribution of the Galaxy. For our purposes however, the significance of these faint streams is that they illustrate that galaxies do not exist in isolation and that interactions between galaxies are common. This is a theme that we shall return to as we move on to consider the formation and evolution of galaxies in general.

1.7 Summary of Chapter I

We conclude this chapter with a summary table of the major structural components of the Milky Way (Table 1.1), followed by a revision of key points.

Table 1.1 The major structural components of the Milky Way.

Component	Shape	Dimensions	Main forms of baryonic matter		Mass/$M_\odot$	Motions
			Stellar	Gaseous		
dark-matter halo	oblate spheroid?	>50 kpc?	?	?	~10^{12}	?
disc	flat disc	radius ~15 kpc				
thin disc	spiral arms	thickness ~1 kpc	Pop. I Z ~0.005 to 0.04; O, B, stars in spiral arms	dense and diffuse clouds; intercloud medium, HII regions	stars ~10^{11} gas ~10^{10} dust ~10^{8}	circular differential rotation; confined to plane of disc scale height < 300 pc
thick disc		thickness ~2 kpc	Old Pop. I stars	little or no gas	stars ~10^{10}	almost circular, scale height ~1 kpc
spheroid						
stellar halo	oblate spheroid c/a = 0.8	radius >20 kpc	Pop. II Z <0.002	very little gas; high-velocity clouds?	stars ~10^{9} gas negligible	elliptical orbits, often highly inclined to Galactic plane
bulge	triaxial ellipsoid (bar)	radius 3 kpc	Z ~0.02 Massive black hole at centre		stars + gas ~10^{10}	
hot corona				tenuous hot gas		

Overview of the Milky Way

- The Milky Way – our Galaxy – is a barred spiral galaxy with four major structural components: a dark-matter halo that is only detected gravitationally, a disc, a stellar halo and a central bulge. The total mass is ~$10^{12}M_\odot$.

- The nature of the dark matter is unclear, but it may account for 90% of the total mass.

- The directly detectable matter consists mainly of stars (~90% by mass), gas (~10%) and dust (~0.1%).

- The disc is about 30 kpc in diameter and 1 kpc thick. The stellar halo is roughly spherical; its diameter is difficult to determine but estimates of more than 40 kpc are common. The nuclear bulge is a triaxial bar extending out to about 3 kpc from the centre.

- The stars of the Milky Way may be divided into a number of populations, each of which predominates in a particular region of the Galaxy. The very youngest stars are found mainly in the spiral arms. Population I stars reside in the disc. The oldest known stars, of Population II, are found mainly as individual stars of the stellar halo, and in globular clusters.

- The disc is in a state of differential rotation, with stars in the vicinity of the Sun taking about 2×10^8 yr to make a complete orbit of the Galactic centre.

The mass of the Milky Way

- The Galactic rotation curve allows the enclosed mass of the Galaxy to be measured. Interior to the Sun's orbit, the mass is approximately $10^{11} M_\odot$. Estimates for the total mass range from $4 \times 10^{11} M_\odot$ to $6 \times 10^{12} M_\odot$.

The disc

- There are about 10^{11} stars in all, with a total mass $\sim 10^{11} M_\odot$.

- Most stars and gas are approximately 70% hydrogen, 28% helium, and 2% heavier elements (metals) by mass.

- Hydrogen occurs in the form of molecules (H_2), atoms (HI) or ions (HII), according to local conditions. Molecular hydrogen is difficult to detect, however, so carbon monoxide (CO) is used as a tracer of H_2.

- Dust consists of μm-sized solid compounds, especially carbonaceous materials (such as graphite) and silicates with icy mantles, and accounts for about 1% of the ISM by mass.

- The Sun is one of the Pop. I stars, located about 8.5 kpc from the centre of the Galaxy, close to the mid-plane of the disc.

- The spiral arms are sites of active star formation. Attempts to trace the arms make use of young, short-lived objects in the disc such as bright HII regions, young open clusters, OB associations, dense clouds and clouds of neutral hydrogen gas.

- The spiral arm pattern might be caused by density waves – relatively slow-moving regions of density enhancement that rotate 'rigidly' around the Galactic centre. The compression of dense molecular clouds as they enter these regions might trigger the birth of the short-lived features that trace the spiral arms.

The stellar halo and bulge

- The stellar halo is roughly spherical but with polar flattening, resulting in an oblate spheroidal shape. Its radius is more than 20 kpc, although its density falls off with distance.

- The main constituents of the stellar halo are old, low-metallicity stars of Pop. II. The total mass of these stars is about $10^9 M_\odot$.

- About 1% of the halo stars are contained in globular clusters. These stars have ages up to about 12×10^9 years.

- The stellar halo includes a corona of tenuous, hot ($\approx 10^6$ K) gas, thought to be heated by Galactic fountains that are powered by supernovae. The high-velocity clouds of atomic hydrogen may be located in the stellar halo or may belong to intergalactic space.

- The bulge seems to have the form of a triaxial bar extending out to 3 kpc from the Galactic centre. Its outer regions rotate at about 100 km s^{-1} and its total mass is around $10^{10} M_\odot$.

- The bulge mainly consists of old stars of the same age as the stellar halo, although their metallicities seem similar to that of the Sun.

- Near the Galactic centre, the compact radio source Sagittarius A* lies at the heart of the Milky Way. From the motions of stars orbiting close to the Galactic centre, Sagittarius A* is believed to be a black hole of mass $4.3 \times 10^6 M_\odot$.

The evolution of the Milky Way

- A process of Galactic chemical enrichment is at work, driven by cosmic recycling, in which some of the gas removed from the ISM to form stars is returned to the ISM by stellar winds, planetary nebulae, and supernovae, enriched in the heavy elements (metals) that are formed within stars.

- Observations indicate an increase in the metallicity of newly formed stars with time, but this increase has proceeded at different rates in different parts of the Galaxy.

- The composition of the ISM has also evolved over time, although it is still predominantly hydrogen and helium.

- The Sagittarius dwarf galaxy is currently merging with the Galaxy and contributing new stars to the halo.

Questions

QUESTION 1.13

Why do high-velocity stars have lower metallicity than the Sun?

QUESTION 1.14

The Sun is 8.5 kpc from the Galactic centre, and is thought to be 4.5×10^9 years old. Use these data, and the rotation speed you can estimate from Figure 1.11c, to calculate the number of times the Sun has orbited the Galactic centre.

QUESTION 1.15

Assuming that optical views in the disc of the Milky Way are limited to a range of 5 kpc, estimate the fraction of the volume of the stellar disc that can be surveyed optically. What is the main cause of this limitation? (Assume that the thickness of the stellar disc is about 1 kpc.)

QUESTION 1.16

If the Sun was born in association with other stars, and originally formed part of an open cluster (which is not certain), why can we no longer see any evidence of that cluster?

QUESTION 1.17

How does density wave theory solve the winding dilemma?

QUESTION 1.18

Make a list of the kinds of astronomical objects that can be used in attempts to trace the spiral arms of the Milky Way.

QUESTION 1.19

What colour are the brightest stars in the Milky Way's globular clusters? Why?

QUESTION 1.20

'The Galaxy is not an unchanging body; rather, it continues to evolve.' Summarize the evidence for this assertion.

CHAPTER 2
NORMAL GALAXIES

2.1 Introduction

Although the Milky Way is vast in comparison with the scale of the Earth or our Solar System, it is actually very small in comparison to the visible Universe. We can justifiably ask: 'What lies beyond our own Galaxy?' The answer, we now know, is other galaxies – thousands of millions of them. These other galaxies that lie beyond the confines of the Milky Way can be broadly divided into two classes. The majority are called **normal galaxies**, which have a more or less unvarying luminosity that is roughly accounted for by the stars and interstellar matter that they contain. A minority of external galaxies – just over 2% – belong to a different class, that of **active galaxies**, characterized by an unusually high (and sometimes variable) luminosity that appears to be largely non-stellar in origin. This chapter focuses on normal galaxies; the more specialized subject of active galaxies is explored in Chapter 3.

The remainder of this chapter is divided into four major sections. The first deals with the classification of galaxies according to their shape and other readily apparent characteristics. This subject was pioneered by the great American astronomer Edwin P. Hubble (Figure 2.1) in the 1920s, and it is a version of Hubble's classification scheme that provides the basis for our discussion. The second section is concerned with how we determine a range of important physical properties of galaxies, including mass, luminosity and composition. The third section concentrates on the measurement of galactic distances. Distance determinations are difficult, but they are of great importance since they are critical to more complex considerations, including the extent to which galaxies gather together to form the clusters that will be discussed in Chapter 4. The fourth, and final, section concerns the origin and evolution of galaxies, a complex field of study that is widely regarded as one of the greatest challenges facing the current generation of astronomers and astrophysicists, and at the forefront of current extra-galactic astronomical research.

Note on terminology

Terms such as 'galactic distance' and 'galactic mass' have to be used with caution since some readers might interpret them as referring specifically to the Milky Way. In this text the term 'galactic' should be taken to refer to galaxies in general. However, when specific reference to the Milky Way is intended we follow the convention of Chapter 1 by spelling Galactic with a capital G. Thus, the term 'galactic centre' refers to the centre of any galaxy under discussion, whereas 'Galactic centre' refers specifically to the centre of the Milky Way.

2.2 The classification of galaxies

2.2.1 The Hubble classification

The meeting of the American National Academy of Sciences held at the Smithsonian Institute in Washington DC in April 1920 is widely regarded as a major event in the history of astronomy. This meeting was the scene of what has become known as 'Astronomy's Great Debate', in which Harlow Shapley (whose

EDWIN P. HUBBLE (1889–1953)

Figure 2.1 Edwin P. Hubble. (Caltech)

Hubble came from Marshfield, Missouri, USA, and originally intended to follow his father's profession by training as a lawyer. He studied law at the University of Chicago, and as a Rhodes Scholar at Oxford University. His career in law was short: at Chicago he had developed an interest in astronomy, and in 1914 he returned there to begin astronomical research. His career was interrupted by the First World War – he served in the US Army and was wounded in France. But in 1919 he took up a post at the Mount Wilson Observatory in California, and it was there, with his assistant Milton Humason, that he made his most significant contributions to astronomy. His most famous discovery, now called Hubble's law (described in Section 2.4 of this chapter), describes the simple relationship between the redshift of a galaxy and its distance. Hubble's law revolutionized the scientific view of the Universe, since it implied that the Universe is expanding and that it has a finite age.

In recognition of his outstanding contribution to astronomy, NASA and the European Space Agency named their major space-based optical observatory the Hubble Space Telescope in his honour.

work on the distribution of globular clusters was described in Chapter 1) and Heber D. Curtis (1872–1942) took opposing views on two fundamental issues: the determination of the size of the Milky Way, and the existence of external galaxies beyond the Milky Way. Shapley believed that the Milky Way was very large and occupied more or less the entire Universe, whereas Curtis favoured the idea that it was just one galaxy among many. By 1930 these two issues had largely been settled. Shapley's method of working out the size of the Milky Way had been widely accepted (although the value it provided was substantially revised in the 1950s) and the existence of external galaxies had been definitively established by Edwin Hubble. Indeed, so many external galaxies were then known that it had already become customary to divide them into a number of different classes, according to their **morphology**, that is, their 'visible' shape and structure.

Classification is a vital first step in many areas of science. A classification scheme is convenient; first, it groups galaxies together that seem to have something in common; but second, if galactic morphology reflects underlying physical properties or processes, then classification provides a meaningful basis to group together galaxies that are fundamentally similar.

Hubble first introduced a galaxy classification scheme in 1926, and in one way or another, this scheme is still widely employed. A modified version of this **Hubble classification** scheme is shown in Figure 2.2. The figure shows four major **Hubble classes** of galaxy: **elliptical**, **lenticular**, **spiral** and **irregular**. Both the lenticular and spiral classes are subdivided into **barred** and **unbarred** varieties. Finally the various classes and subclasses are divided into a number of **Hubble types**, denoted by a combination of letters and numbers.

Given that there are physical as well as morphological, differences between the various types and classes, the Hubble classification is of great importance in

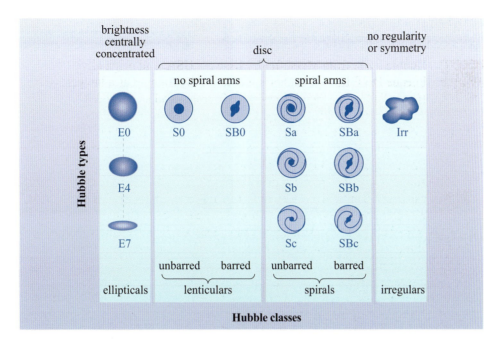

brightness
centrally
concentrated

disc

no regularity
or symmetry

no spiral arms

spiral arms

E0   S0   SB0   Sa   SBa   Irr

E4   Sb   SBb

E7   Sc   SBc

Hubble types

unbarred   barred   unbarred   barred

ellipticals   lenticulars   spirals   irregulars

Hubble classes

Figure 2.2 The Hubble classification scheme for galaxies.

astronomy. So how do we determine a classification? First, to place a galaxy into one of the four major Hubble classes, we consider four major questions:

- Is there any overall regularity or symmetry?
- Is the light concentrated in the centre?
- Is there a disc?
- Are there any spiral arms?

As discussed below, the answers to these questions puts a galaxy into one of the main Hubble classes. Other observational properties are then used to assign a galaxy to one of the types within its main class.

Elliptical galaxies

Elliptical galaxies are characterized by an overall elliptical outline when viewed in the sky, and a generally featureless appearance, combined with a light output that is highly concentrated in the galactic centre and decreases steadily with increasing distance from the centre. Galaxies in this class are divided into eight Hubble types that range from E0, for those that appear circular, to E7 for the most elongated (Figure 2.2). The whole number that follows the E in each type designation is determined by the relative size of the **semimajor axis**, a, and **semiminor axis**, b, of the observed ellipse (see Figure 2.3), and is obtained by multiplying the flattening factor $f = (a - b)/a$ by ten, and then rounding the result to the nearest whole number. The lack of elliptical galaxies with flattening factors greater than 0.7 probably indicates that such highly flattened ellipticals would be unstable.

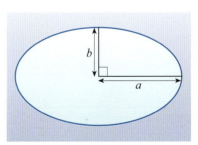

Figure 2.3 The semimajor axis, a, and the semiminor axis, b, of an ellipse.

Photographic images of various types of elliptical galaxy are shown in Figure 2.4 (overleaf). Note that elliptical galaxies are assigned a Hubble type according to their *visually observed* shape. The scheme takes no account of an elliptical galaxy's 'true' shape as a three-dimensional object. Nor does the appearance at non-visual wavelengths have any influence on the classification, even though it may differ from the visual appearance.

(a)   (b)   (c)

Figure 2.4 Elliptical galaxies of various Hubble types. (a) M89 or NGC 4552 (type E0), (b) M49 or NGC 4472 (type E4), (c) M5 or NGC 4621 (type E5). (*Note*: The prefix 'M' indicates that the galaxy is in the Messier catalogue that was published in 1784 by Charles Messier. 'NGC' stands for the New General Catalogue of Nebulae and Clusters (of stars), published in 1888 by Johan Dreyer. Most of the objects classified in the NGC are now known to be galaxies, but in 1888 the existence of galaxies beyond the Milky Way was unrecognized.) (NOAO)

Spiral galaxies

Spiral galaxies are characterized by a circular disc containing spiral arms, and a central nuclear bulge. The light is much more evenly distributed across the galaxy than in elliptical systems. Two examples of spiral galaxies are given in Figure 2.5 (M100 and M31). M100 is viewed almost face-on to its disc and the spiral arm pattern is clearly evident. M31 on the other hand is viewed at an oblique angle with the result that its disc appears elliptical. This elliptical appearance can be used to determine the angle of inclination between a line drawn at right angles to the disc and the line of sight from the Earth. This method allows spirals to be classified according to the shape they would have if viewed face-on (with a 0° angle of inclination).

(a)

(b)

Figure 2.5 Two examples of spiral galaxies: (a) M100, an SBc galaxy that is viewed almost face-on, and (b) the M31 galaxy in Andromeda (classified as type Sb). Note the presence of two smaller elliptical galaxies that are satellites of M31. ((a) ESO; (b) B. Miller)

Spiral galaxies are subdivided into two subclasses: S, for those without a central bar, or SB for those with a central bar. Each subclass is further divided into Hubble types a, b, c, where between a and c the winding spiral arms become progressively more open, and the central bulge smaller. This leads to six spiral types, Sa, Sb, Sc or SBa, SBb, SBc. Occasionally, galaxies cannot be unambiguously assigned to one of these classes, and are therefore described as a combination of the nearest two. For example, a galaxy that is intermediate between Sa and Sb is denoted Sab.

As you saw in Chapter 1, our Galaxy, the Milky Way has spiral arms and a bar, and these characteristics have led to it being classified as SBb or SBc (or SBbc – intermediate between SBb and SBc).

Lenticular galaxies

Lenticular galaxies are lens-shaped galaxies. Like spirals (and unlike ellipticals), they have a disc and a central bulge – but no spiral arms. The bulge may be quite large in comparison with the disc (more on average than in the case of the Milky Way) and is used to subclassify such galaxies. If the bulge is barred then the galaxy is classified SB0, otherwise S0. Lenticular galaxies are seemingly an intermediate class between ellipticals and spirals.

Photographic images of different types of lenticular and spiral galaxy are shown in Figure 2.6 (overleaf). Although Figure 2.2 shows spiral galaxies face-on, it is important to remember that when viewed from the Earth a spiral galaxy can be inclined relative to the observer, and that in many cases the spiral arms might be fragmented, and this is especially the case in Sa galaxies.

Irregular galaxies

As you would expect, anything that does not fit into the first three categories is classified as an irregular galaxy. These galaxies (class Irr) show little sign of symmetry or regularity. Some have bar-like structures and some show vague signs of spiral arms. Others are totally irregular. An example is the irregular galaxy IC 5152 shown in Figure 2.7a, and further examples of this class are the Large and Small Magellanic Clouds that are near neighbours of the Milky Way (Figures 2.7b and c).

Before leaving the topic of the Hubble scheme, it should be noted that classifying galaxies is not as unambiguous as it may seem from the above discussion and Figure 2.2. Some aspects are somewhat subjective, and different observers may classify the same object differently. In the case of galaxies with discs, it can be difficult to distinguish between lenticulars and various types of spiral when the angle of inclination is large and the galaxy is almost edge-on. One approach to countering the subjectivity of individuals involved in galaxy classification is to pool the classifications of hundreds, if not thousands of people – a principle that is put to good use in the online citizen science project called 'Galaxy Zoo'.

Another problem for morphological classification, and one that is harder to address, is that in optical images the appearance of a galaxy may be affected by such things as exposure time. For example, a short-exposure image of an elliptical galaxy will show only the bright central parts, while a longer exposure will reveal the outer parts as well and give a different impression of the overall shape.

Figure 2.6 Lenticular (S0, SB0) and spiral (S and SB) galaxies of various Hubble types. ((a), (c) Sandage and Bedke, 1994; (b) Sloan Digital Sky Survey; (d), (e), (f), (g) NOAO/AURA/ NSF; (h) David and Christine Smith/Adam Block/NOAO/AURA/ NSF)

(a) NGC 524 Hubble type S0

(b) NGC 936 Hubble type SB0

(c) NGC 7217 Hubble type Sab

(d) NGC 4650 Hubble type SBa

(e) M77 Hubble type Sb

(f) M91 Hubble type SBb

(g) M99 Hubble type Sc

(h) NGC 1073 Hubble type SBc

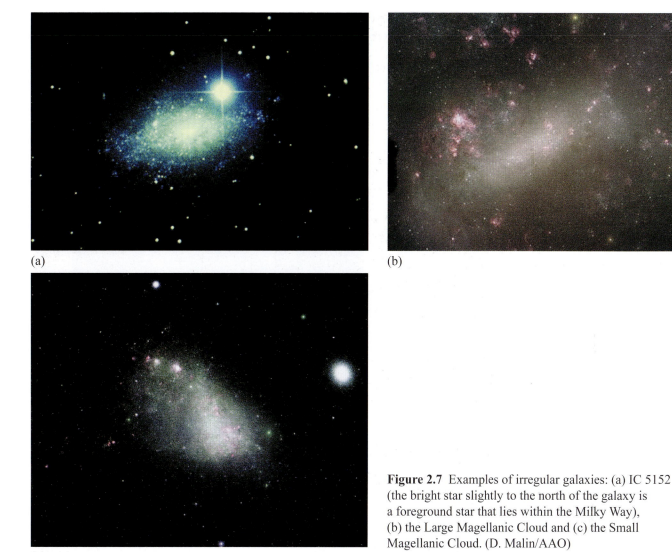

(a)

(b)

(c)

Figure 2.7 Examples of irregular galaxies: (a) IC 5152 (the bright star slightly to the north of the galaxy is a foreground star that lies within the Milky Way), (b) the Large Magellanic Cloud and (c) the Small Magellanic Cloud. (D. Malin/AAO)

Most people's mental picture of a galaxy is probably of a spiral, since these are the most photogenic and tend to predominate in illustrated books on astronomy. However, faint **dwarf elliptical** galaxies, with masses around $10^6 M_\odot$, such as the galaxy Leo I shown in Figure 2.8, are in fact the most common type of galaxy. Over 60% of galaxies are elliptical, fewer than 30% are spiral, and fewer than 15% are irregular. These figures are subject to some variation according to how the sample being surveyed is selected: there are relatively more E, S0 and SB0 galaxies in regions where galaxies are more densely clustered together, and in the most densely clustered regions S, SB and Irr galaxies are almost totally absent. In any survey it is important to account for bias: if only the most easily observable galaxies are surveyed (i.e. those with the greatest apparent brightness), then the faint dwarf ellipticals will be under-represented, and the biased sample obtained will give the impression that there are relatively fewer ellipticals than is actually the case. While mentioning statistics, it is also worth noting that about 60% of spirals and lenticulars are barred.

About 1% of galaxies cannot be assigned to any of the Hubble classes shown in Figure 2.2, even when their shape is known; some of these are briefly discussed in Section 2.2.3.

Figure 2.8 The dwarf elliptical galaxy Leo I lies at a distance of about 250 kpc from the Sun. Dwarf elliptical galaxies are very numerous, but their low luminosities make them difficult to detect. (D. Malin/AAO)

2.2.2 The physical characteristics of the Hubble classes

If you recall, we started on the journey of galactic classification by morphology (or visible shape) under the premise that if we grouped together galaxies that looked the same, we might expect that such galaxies would also share similar physical properties. In this section we explore the extent to which this premise holds true.

Originally Hubble (and many others) suggested that the classification system was directly related to galactic evolution, starting at elliptical galaxies and moving towards spiral and lenticular galaxies. However such a notion can be completely rejected – the different classes of galaxies do not represent an evolutionary sequence. As we shall see, the current belief is that the Hubble class of a galaxy is determined by its formation history. However, despite the lack of evolutionary progression through the Hubble classification, galaxies do have physical properties that show systematic variation with Hubble class or type.

One such property relates to rotation. The angular momentum of a galaxy depends on the rate at which the various parts of the galaxy rotate about the centre, and its degree of compactness. On the whole, the angular momentum per unit mass of elliptical and irregular galaxies is low, whereas it is relatively high for spirals and lenticulars. For spiral galaxies of a given mass, angular momentum increases through the types Sa, Sb, Sc (or SBa, SBb, SBc).

Another property correlated with Hubble class concerns the ratio of the mass of gas to the total mass of stars in a galaxy. More specifically, the ratio of the mass of gas that is in molecular or atomic form to the mass of stars, expressed as a percentage, varies considerably between Hubble classes. In the case of the Milky Way, this ratio is about 10% (see Chapter 1). For spiral types Sa to Sc (or SBa to SBc) this proportion is 5–15% (and increases in progressing from Sa to Sc or SBa to SBc), while the proportion is typically in the range 15–25% in irregulars. In contrast, many ellipticals and lenticulars have scarcely any molecular or atomic gas – the gas that is present in such galaxies is typically very hot (at temperatures exceeding 10^6 K) and is consequently highly ionized. The proportion of mass in the form of atomic or molecular gas is thus very low for ellipticals and lenticulars (1%, say). The presence or absence of molecular or atomic gas is related to a number of other important distinguishing features of the various Hubble classes.

Considering the variation of gas content with Hubble class, describe and explain any systematic variations you would expect to find between elliptical and spiral galaxies with regard to the following properties: presence of high-mass main sequence stars; proportion of stars in open clusters; prevalence of HII regions; abundance of Population I stars relative to Population II stars.

A number of other important galactic properties, such as mass, luminosity and diameter show wide variation in value from one galaxy to another, even within a given class, so cannot be simply correlated with Hubble class. However, the *range* of variation of each of these properties *can* be loosely correlated.

Generally, spiral galaxies have a fairly narrow range of mass, luminosity and diameter (masses, for example, are usually in the range from about $10^9 M_\odot$ to a few times $10^{12} M_\odot$), while masses of ellipticals vary from about $10^5 M_\odot$ to somewhat over $10^{13} M_\odot$. Irregular galaxies exhibit an intermediate range of properties (roughly 10^7–$10^{10} M_\odot$ in mass – even the largest irregulars are less than a quarter of the mass of the Milky Way).

Methods of diameter determination are described in more detail in the next section. However, it is probably already clear that in order to measure a galactic diameter, the shape of a galaxy must be well determined. An issue therefore arises with elliptical galaxies. Hubble types E0–E7 are based on the 'apparent' (i.e. projected, two-dimensional) shape rather than 'true' (three-dimensional) shape, which raises a number of questions about elliptical galaxies. Is it possible that all elliptical galaxies actually have the same three-dimensional shape, and that the apparent differences between the Hubble types from E0 to E7 are a result of viewing that single common shape in different projections?

Any three-dimensional shape that appears to be elliptical from all directions is called an **ellipsoid**. The shape of any particular ellipsoid is determined by the relative lengths of its three principal semi-axes, as indicated in Figure 2.9. If two of these semi-axes are of equal length, the ellipsoid will have a circular cross-section perpendicular to its third principal semi-axis. An ellipsoid of this kind is called a *spheroid*, and may be classified as an **oblate spheroid** or a **prolate spheroid** depending on whether the third principal semi-axis is shorter (former) or longer (latter) than the other two (see Figure 2.9). Ellipsoids with three unequal principal semi-axes do not appear circular from any direction, and are said to be **triaxial ellipsoids**.

Historically, it used to be assumed that every elliptical galaxy was an oblate spheroid (like Figure 2.9a), and that each elliptical galaxy rotated about the shortest axis so that the flattening was mainly due to this rotation – like the polar flattening of the Earth. Detailed research indicates that such assumptions are

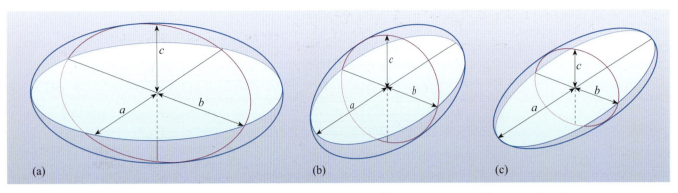

Figure 2.9 An ellipsoid is a three-dimensional body that appears to be elliptical from every direction. The shape of any particular ellipsoid can be specified by assigning lengths to the so-called principal semi-axes a, b and c shown in the diagram. (a) An oblate spheroid has $a = b > c$. (b) A prolate spheroid has $a > b = c$. (c) A triaxial ellipsoid has $a > b > c$.

not always justified; sometimes the observed line-of-sight velocities of stars in elliptical galaxies are simply not consistent with a spheroidal distribution and the rotations are often too slow to produce the observed flattening, particularly in the most luminous ellipticals. Moreover, some elliptical galaxies appear to be rotating about an axis other than the shortest axis, so their flattening is almost certainly *not* due to rotation.

Another peculiarity for some elliptical galaxies is that contours joining points of equal surface brightness do not have a common major axis – there appears to be some sort of 'twist' in the heart of the galaxy. In view of all these findings we must assume that at least some elliptical galaxies are triaxial, although the majority may be spheroidal.

2.2.3 The 'odd ones out': galaxies that don't fit the Hubble classification

It would be remiss to end this section without some reference to those galaxies that cannot be fitted into the Hubble classes shown in Figure 2.2, and to others that can only be accommodated with difficulty. Various amendments and extensions to the Hubble scheme have been proposed that reflect more accurately the true range of galactic shapes, but rather than discuss these in detail we shall just mention some of the limitations and shortcomings of the basic scheme.

Many galaxies have a more or less readily apparent Hubble type apart from the presence of some kind of abnormal feature. A case in point is the giant elliptical galaxy M87 (in the constellation of Virgo). A medium exposure image of this galaxy shows an apparently ordinary elliptical galaxy of type E0 or possibly E1, similar to the galaxy shown in Figure 2.4a. However, a short exposure that emphasizes the central regions of the galaxy (Figure 2.10a) shows an unusual feature – a 'jet' of material apparently spurting out from the core. Because of the presence of this jet, M87 is said to be a **peculiar galaxy** and is usually classified as having Hubble type E0p; the final 'p' indicating the presence of the peculiarity. (As you will see in Chapter 3, M87 is, in fact, an *active galaxy*; jets are seen in many active galaxies.) A jet is not the only feature that can make a galaxy peculiar.

Many peculiar galaxies appear to have been distorted in gravitational encounters with other galaxies: all interacting galaxies are peculiar. Some examples of interacting galaxies are given in Figure 2.10b–e. We shall see that interacting

(a)

Figure 2.10 Some peculiar galaxies: (a) M87, a peculiar elliptical galaxy with a jet emerging from its core, (b) interacting galaxies NGC 2207 and IC 2163, (c) part of a group of interacting galaxies called Stephan's Quintet, (d) interacting galaxies NGC 6872 and IC 4970, and (e) NGC 4676. ((a) J. Biretta/STScI/ Johns Hopkins University/NASA; (b) Hubble Heritage Team/STScI; (c) S. Gallagher (Pennsylvania State University); (d) ESO; (e) H. Ford/Johns Hopkins University/NASA)

(b)

(c)

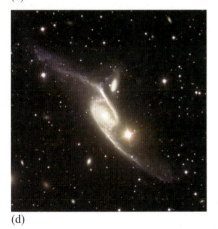

(d)

(e)

galaxies play an important part in current thinking about the way in which galaxies evolve. Finally, we should mention a **cD galaxy**. The name comes from an alternative astronomical classification of galaxy morphologies that is no longer widely used. The 'c' indicates a supergiant system, while the 'D' indicates that the galaxy has a large, diffuse envelope. An example of a cD galaxy is NGC 4874 (Figure 2.11, overleaf), a giant elliptical near the centre of a rich cluster of galaxies in the constellation of Coma. The densely packed centres of large clusters of galaxies often harbour cD galaxies, which are widely thought to result from the merger of several other galaxies. M87 is also sometimes given the alternative classification as a cD galaxy, and it is at the centre of a sub-clump of the Virgo cluster of galaxies. Detailed studies of some cD galaxies reveal the presence of several bright spots near the centre. These are sometimes interpreted as the nuclei of galaxies that have been absorbed as a result of merger events.

Figure 2.11 NGC 4874 (at the right of this image) is a giant elliptical galaxy that is classified as a cD galaxy. Note the large extent of the diffuse outer regions of this galaxy. This galaxy lies close to the centre of the Coma cluster of galaxies. (NASA/ESA)

QUESTION 2.2

Fill in the missing data in Table 2.1, which compares and contrasts various properties of three of the main classes of galaxy.

Table 2.1 A comparison of Hubble classes. (For use with Question 2.2)

Property	Ellipticals	Spirals	Irregulars
approximate proportion of all galaxies	$\geq 60\%$	$\leq 30\%$	$\leq 15\%$
mass of molecular and atomic gas as % of mass of stars		5–15%	
stellar populations			Populations I and II
approximate mass range			
approximate luminosity range	a few times $10^5 L_\odot$ to $\sim 10^{11} L_\odot$	$\sim 10^9 L_\odot$ to a few times $10^{11} L_\odot$	$\sim 10^7 L_\odot$ to $10^{10} L_\odot$
approximate diameter range[a]	$(0.01\text{–}5)\,d_{MW}$	$(0.02\text{–}1.5)\,d_{MW}$	$(0.05\text{–}0.25)\,d_{MW}$
angular momentum per unit mass			low

[a] d_{MW}, diameter of Milky Way.

2.3 The determination of the properties of galaxies

In Section 2.2 various assertions were made about the physical characteristics of the different types of galaxy in the Hubble classification scheme. Much of this information was summarized in Table 2.1, which indicates typical ranges for quantities such as luminosity, diameter, mass and composition without making any serious attempt to explain how such information was obtained. This section is concerned with just such determinations.

2.3.1 Luminosities and sizes of galaxies

With a sufficiently powerful telescope, most galaxies are seen as faint, extended objects with a brightness that varies from point to point. In studying the energy received from such objects, the quantity that is directly measured is the **apparent surface brightness** of the source. This quantity can be roughly thought of as the rate at which energy would reach a detector with a collecting area of 1 m^2 (i.e. the flux density) from a small region of angular area 1 arcsec^2 surrounding the point being observed (see Figure 2.12). Acceptable SI units of apparent surface brightness are W m^{-2} arcsec^{-2}. In practice, however, astronomers tend to express the observed surface brightness as an apparent magnitude and quote it in units of magnitudes arcsec^{-2}. Also, since such measurements are usually restricted to a particular range (or band) of wavelengths, measurements of apparent surface brightness are usually made in a specified waveband, such as the widely used V-band.

Since the apparent surface brightness of a galaxy tends to decline smoothly from its centre, images of galaxies are often displayed using *isophotal contours*. An **isophote** is simply a closed curve connecting points of equal apparent surface brightness, in much the same way that a closed contour line can be used to show points of equal height on a map. Figure 2.13 shows an example of such an image – in this case of the elliptical galaxy NGC 4278. Like all galaxies, this particular example has no sharp boundary or edge. There is no isophotal contour that marks the edge of the galaxy – the isophote with the greatest radius in this image simply represents the lowest level that can be seen above instrumental noise, and does not necessarily mark the physical boundary of the galaxy.

This extended nature of galactic images is a source of difficulty for those engaged in measuring galactic luminosities. Because galaxies 'fade out' it is hard to know where to stop measuring. However, galaxies of similar type (morphology and luminosity) usually have a similar **surface brightness profile**; that is to say, the way in which the surface brightness changes with distance from the centre of the galaxy is similar in all galaxies of a given type. By measuring the apparent surface brightness of a galaxy over its brighter parts it is therefore possible to calculate its total flux density by assuming that the surface brightness

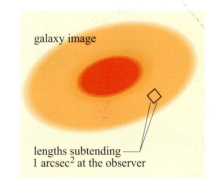

lengths subtending 1 arcsec^2 at the observer

Figure 2.12 The apparent surface brightness of a galaxy provides a measure of the flux originating in a particular angular area of 1 arcsec^2 surrounding the point being observed.

The V-band covers a range of wavelengths in the yellow–green part of the spectrum, the range over which the eye is most sensitive. The band is centred on a wavelength of 545 nm and has a bandwidth of 88 nm. Measuring the flux density in the V-band requires a greenish-coloured piece of glass (called a *filter*) to be placed between the sky and the camera, so that only V-band light makes it through to the camera.

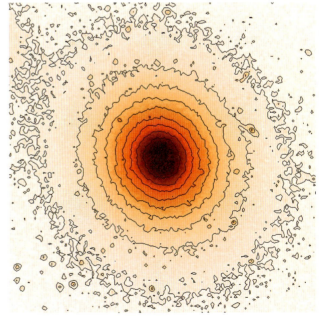

Figure 2.13 Isophotes of the giant elliptical galaxy NGC 4278. These isophotes correspond to measurements made in the red part of the visible spectrum. (Digitized Sky Survey/STScI)

over its unmeasured regions follows a standard surface brightness profile. Figure 2.14 shows an example of one of these surface brightness profiles for a lenticular galaxy, and how it corresponds well to the sum of model disc and bulge components.

Even when the total **flux density** from a galaxy is known, determining the corresponding luminosity is not entirely straightforward. For example, in the

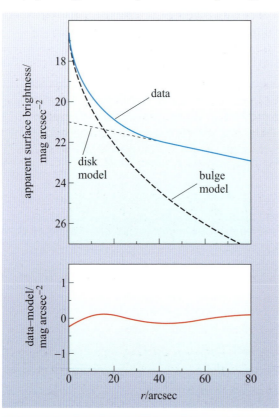

Figure 2.14 The top panel shows the apparent surface brightness as a function of angular distance (r) from the galactic centre for the S0 galaxy NGC 2911. The dashed lines show model contributions from a nuclear bulge and from a smooth disc, while the solid line shows the measured surface brightness profile. The lower panel shows the difference between the measured surface brightness profile and the model (disc and bulge): it can be seen that there is a very good agreement between the data and the model. (Adapted from Gadotti, 2008)

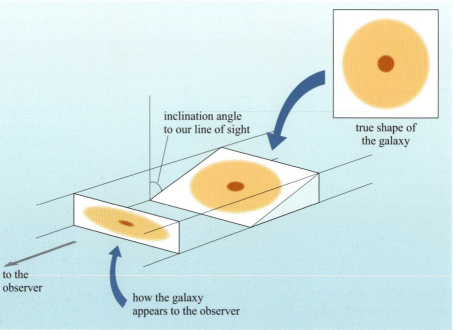

Figure 2.15 The observed shape of a galaxy arises from a combination of its true shape and its orientation relative to the observer. It is the projection of this shape onto the observed sky that influences our 'picture' of the galaxy, but to measure galactic properties such as total luminosity, we need to find the galaxy's 'true' shape.

case of spiral galaxies the orientation relative to the observer (see Figure 2.15) has an effect: because of its disc, a spiral galaxy will not radiate uniformly in all directions, and the orientation will also influence the extent to which radiation is scattered and absorbed by dust within the observed galaxy. However, the effects of orientation can be estimated from the ratio of major to minor axes in the observed galaxy (within a selected isophote). This ratio can then be used to compute a correction to the observed flux density that accounts for the effects of orientation. The 'corrected' flux density F, together with the distance d of the galaxy, then provides a value for the galactic luminosity L, after making due allowance for interstellar absorption within the Milky Way. Similarly, the angular diameter within a selected isophote can be used to estimate the angular diameter θ of the whole galaxy, which can in turn be used in conjunction with the distance of the galaxy to determine its linear diameter l. These procedures are subject to many uncertainties and assumptions, so they cannot possibly provide an absolutely precise value for any physical quantity. Rather, the aim is to approach the determination in a standardized way, so that comparisons between the luminosities and diameters of different galaxies are as meaningful as possible.

The results of luminosity and diameter determinations have already been summarized in Table 2.1. Basically, ellipticals are the most diverse class with luminosities that range from a few times $10^5 L_\odot$ for dwarf elliptical galaxies to about $10^{11} L_\odot$ for giant ellipticals (and cD galaxies). Spirals, on the other hand, occupy a narrower range, from about 10^9 to a few times $10^{11} L_\odot$. Irregulars occupy a wider range than do spirals, but are generally less luminous. Diameters follow a similar pattern, as Table 2.1 shows.

2.3.2 Masses of galaxies

You will recall from Chapter 1 that a great deal of uncertainty still surrounds the mass of the Milky Way, mainly due to the problem of assessing how much dark matter is associated with our Galaxy. Similar problems are associated with all galaxies. Here we briefly describe three methods for determining galactic masses. The key finding is that the masses of galaxies are typically about ten times greater than the mass of their stars. This is interpreted as strong evidence that the masses of all galaxies are dominated by dark matter.

Method 1: Rotation curves for spiral galaxies

The use of a *rotation curve* – a plot of rotational speed v against radial distance r from the galactic centre – to determine the mass of a spiral galaxy has already been described in Section 1.3. To recap: the rotation curve of a galaxy is measured observationally and compared with the theoretical rotation curve predicted by a model of the mass distribution in the galaxy, including its dark-matter halo. This theoretical mass distribution is then adjusted to find the best agreement between the observed and the predicted curves. The total mass in the final model then represents the estimated mass of the galaxy. The method suffers from various drawbacks, including the fact that (owing to assumptions about symmetry on which the method relies) it is rather unsuitable for spirals that have strongly pronounced bars, and because it gives only a lower limit to the galactic mass. The method also depends on knowing the radial distance r, which in turn requires that the distance of the galaxy is known. Nonetheless, it is relatively straightforward and has been used extensively.

The luminosity is given by $L = 4\pi d^2 F$ where d is the distance to the source, and F its flux density. The linear diameter is given by $l = d \times (\theta/\text{radians})$. Note that there are some subtleties to distance measurements in an expanding universe that you'll meet briefly later on.

■ Why does the method only give a lower limit on the mass? (This limitation was discussed in Chapter 1.)

☐ The method provides insight only into the mass within the largest value of r for which the rotation curve has been measured (see Section 1.2.2).

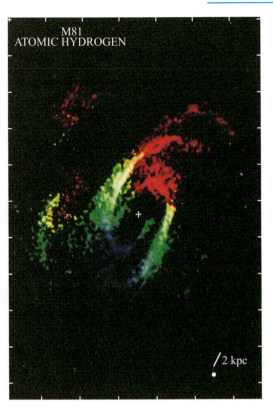

M81
ATOMIC HYDROGEN

/ 2 kpc

In order to ensure that observations of rotation curves extend out to the greatest possible values of r, it is usual to base them on Doppler shifts of the 21 cm radiation emitted by neutral hydrogen (Box 1.3). In nearby galaxies 21 cm emission can often be traced well beyond the optical limits of the disc. An example of the mapping of Doppler shifts is shown in a colour-coded form for the galaxy M81 in Figure 2.16. Of course, Doppler shift measurements only determine velocity components along the line of sight, so to derive a galaxy's rotation curve from the data shown in Figure 2.16 it is necessary to take into account the orientation of the galaxy relative to the observer. The rotation curve derived in this way for M81 and for several other galaxies is shown in Figure 2.17.

For more distant galaxies, the spatial resolution of the 21 cm measurements may be too low to map Doppler shifts at different locations in the disc. In this case a different approach can be adopted whereby a spatially unresolved measurement is made of the 21 cm line. Because this line is the sum of 21 cm emission from gas in all parts of the galaxy, it will be broadened due to the different Doppler shifts of the various contributions. While it is not possible to determine the rotation curve from such measurements, the maximum rotation speed in the disc can be found, and, from this value, a mass estimate of the galaxy can be obtained.

Figure 2.16 An image of the galaxy M81 that has been reconstructed from 21 cm radio observations. The intensity shows the strength of 21 cm emission, while the colour coding shows the extent to which the 21 cm line is Doppler shifted (red represents a high speed along the line of sight away from us, blue is a high speed towards Earth). The circle at the lower right of the map shows the angular resolution of the radio telescope used to make these observations. (Westerbork Synthesis Radio Telescope)

Irrespective of their limitations, 21 cm measurements, rotation curves and the related mass determinations have been of great value in modern astronomy. They have provided evidence of galactic mass distributions that extend well beyond the limits of visible discs, and have thus played an important part in giving dark matter its current significance.

Figure 2.17 Schematic representations of the rotation curves for some nearby galaxies. Note that each curve is marked with the catalogue designation and Hubble type of the galaxy to which it refers. (Mihalas and Binney, 1981)

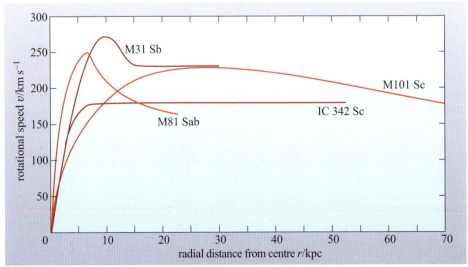

Method 2: Velocity dispersions for elliptical galaxies

We saw in Section 2.2 that elliptical galaxies have little, if any, rotation – so rotation curve methods will not help in determining the mass of an elliptical galaxy. Elliptical galaxies are often regarded as simple 'star piles' – swarms of stars that have long since settled down into galactic orbits under one another's gravitational influence. If this assumption of generally settled stellar motions within an elliptical galaxy is correct, then it is possible to predict the value of a quantity called the **velocity dispersion**, Δv, for a given elliptical galaxy. The velocity dispersion is a statistical quantity that provides a measure of the range of speeds of stars along a line of sight. In an elliptical galaxy of given shape, the velocity dispersion is expected to be proportional to the quantity $(M/R)^{1/2}$, where M is the mass of the galaxy and R is a scale length related to its size. This result, derived from what is known as the virial theorem (Box 2.1), may be applied to elliptical galaxies whose size R is known and for which Δv can be estimated from Doppler-shift measurements. This procedure seems to give reasonable values for M, although, once again, the mass obtained is only a lower limit since it is really that enclosed by the stellar orbits.

Box 2.1 THE VIRIAL THEOREM

The **virial theorem** is a very useful result that relates the total gravitational potential energy of a galaxy to the sum of the kinetic energies of all of the individual components that make up that galaxy. We won't prove this theorem here, but simply quote the relationship and show why such a result is reasonable.

Here let us consider an idealized galaxy that consists solely of stars (i.e. we will ignore any other components that might be present in a real galaxy). The total kinetic energy of the entire system (E_k) is simply the sum of the kinetic energies of all the stars that it contains.

Each star also has a gravitational potential energy, which arises from the gravitational interaction of that star with all the other stars in the galaxy. The gravitational potential energy of each star is a negative quantity because, by convention, the gravitational potential energy of a star is taken to be zero when it is so far from the galaxy that it is effectively free of its gravitational pull. Since any star would have to be given a *positive* amount of energy to enable it to attain this zero energy state, it must be the case that each of the stars in the assembled galaxy has a *negative* gravitational potential energy; that is, each star's gravitational potential energy is equal to −1 times the energy required to remove the star from the galaxy.

The total gravitational potential energy (E_g) of a galaxy can be defined in a similar way. By convention, the total gravitational potential energy is zero when all the

stars are so widely dispersed that each is effectively free of the gravitational influence of all the others. If you imagine a process in which stars are removed one by one from the galaxy, then it is easy to see that some positive amount of energy would be required to achieve the total disassembly that corresponds to the state of zero gravitational potential energy. The total gravitational potential energy of the assembled galaxy is then, the negative of this 'disassembly' energy.

The virial theorem states that when a galaxy exists in a stable state, such that it is neither contracting nor expanding, the total kinetic energy and the total gravitational potential energy are related by

$$E_k = -\frac{1}{2} E_g \qquad (2.1)$$

The proof of this statement would be a lengthy diversion here, but it is useful to note two features of this relationship. First, the negative sign is expected because the total kinetic energy E_k will be positive, while the total gravitational potential energy E_g must be negative. The second point is that the kinetic energy of the stars is less than the energy that would be required to completely disassemble the galaxy. This also seems reasonable: if the total kinetic energy were exceeded it could cause the stars to completely disperse and thus disrupt the galaxy.

The condition that the galaxy is in a stable state – neither contracting nor expanding – is vital. If this is not true then the virial theorem simply does not hold.

To illustrate this, consider a system of stars, distributed in such a way that they form a uniform spherical cloud. Furthermore, suppose that initially all the stars are stationary.

- ■ What is the total kinetic energy of this system?
- ☐ Since the stars are all stationary, the total kinetic energy is zero.

In this case the virial theorem does not hold, but this is, of course, a highly unstable situation. Released from rest, the system starts to collapse as each star accelerates under the gravitational influence of all the other stars. Thus the total kinetic energy of the stars increases. Because they are relatively small and widely separated, the stars are unlikely to collide with one another, although some will pass sufficiently close to others for their paths to be deflected. In this way kinetic and gravitational potential energy can be interchanged, although the total energy of the system remains constant throughout the collapse. The system as a whole contracts somewhat from its initial size, and this decreases the gravitational potential energy of the system (i.e. it becomes more negative). However, as the gravitational potential energy decreases there is a corresponding increase in the kinetic energy of the stars. Eventually, the system can be expected to settle into a state in which it is neither expanding nor contracting. When this happens the total energy is divided between kinetic energy and gravitational energy in just the way described by Equation 2.1. The virial theorem then applies to the system, which can be said to be in a **virialized** state.

It should be noted that the virial theorem applies to *any* system of bodies that interact solely by gravity, provided the system has become virialized. So the theorem should be expected to apply to spiral as well as elliptical galaxies, provided they have virialized. Likewise the virial theorem applies to any system consisting of bodies that interact solely by mutual gravitational interaction, such as globular clusters, and clusters of galaxies.

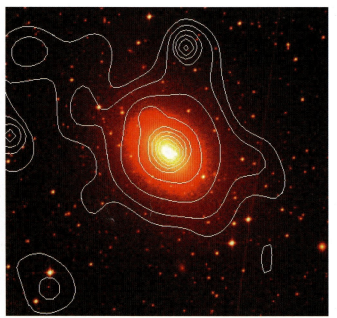

Figure 2.18 The elliptical galaxy NGC 3923. The optical image of the galaxy is shown overlaid with contours of X-ray emission, which extend further than the visible image. (Astrophysics and Space Research Group, University of Birmingham)

Method 3: X-ray halos of ellipticals

Some bright ellipticals have substantial halos of hot, diffuse gas with temperatures of several million kelvin. X-ray observations of such galaxies, such as shown in Figure 2.18, allow the extent, temperature and density of these gaseous halos to be determined. The mass measurement technique is based on the assumption that these hot, gaseous halos are gravitationally bound to their respective galaxies. Roughly speaking, a halo of a given extent and temperature implies that the galaxy has a certain mass: a very hot and extended halo would require the presence of a high-mass galaxy to prevent the gas from escaping. In practice, detailed X-ray measurements are compared with a theoretical model of the gaseous halo, and this allows the mass of the galaxy to be found. As you will see in Chapter 4, this method is similar in principle to a method that is used to measure the mass of clusters of galaxies.

2.3.3 The composition of galaxies

The answer to the question: 'What are galaxies made of?' should already be familiar to you. Broadly speaking, galaxies are made of dark matter, stars, and ordinary (baryonic) gas, together with various minority constituents such as dust. This section is concerned with the methods employed to determine the detailed compositions of individual galaxies. It aims to explain how we set about answering questions such as: 'How many stars are there in a galaxy?' or: 'What kind of stars predominate?' or: 'How much of the galaxy's mass is due to gas?'

Firstly, what fraction of the mass of a galaxy is in gas? In many galaxies, the majority of the gas is in the form of atomic hydrogen, and there is a relatively straightforward procedure for estimating the mass of atomic hydrogen in a galaxy, denoted M_H. The determination of M_H can simply be based on the flux density of 21 cm radiation received from atomic hydrogen in the galaxy. The mass fraction of gas in a galaxy is then quoted as M_H/M, where M is the mass contained in stars in that galaxy. Studies of the ratio M_H/M indicate a fairly systematic variation with Hubble class. Ellipticals have very little atomic hydrogen, with spirals making up a few per cent and irregulars 15–25%.

As observational techniques have advanced, it has become evident that certain galaxies are dominated by molecular hydrogen, H_2, particularly where star-formation is prevalent. As you saw in Box 1.4, molecular hydrogen gas can be traced by emission from the molecule CO.

At the other extreme in terms of gas temperature, we have already seen that some elliptical galaxies contain ionized hydrogen, which again cannot be measured from 21 cm line emission. Detailed X-ray observations can be used to measure the mass of ionized hydrogen in such galaxies. It appears that the proportion of mass of gas in ionized form in giant elliptical galaxies is similar to the proportion of mass of gas in atomic or molecular form in spiral galaxies.

Questions about the numbers and types of stars in a galaxy are interconnected and are best answered together. In principle, a galaxy might be expected to contain representatives of all the stellar types that were discussed in Chapter 1. However, star formation requires the presence of relatively cold gas – gas that is in atomic or molecular form. Since atomic or molecular gas is rare in elliptical galaxies, we might reasonably expect that there is little ongoing star formation in such systems. It is thus reasonable to suppose that those types of stars that belong to Population I will be exceptional in elliptical galaxies, but to what extent do observations support that idea? The absence of bright star-forming regions in elliptical galaxies is easily confirmed by direct observation, but such observations are not able to provide an accurate picture of the relative abundances of the various stellar types, since only the brightest sorts of star are visible individually in even the nearest galaxies. Another technique is needed to investigate the stellar content of a galaxy – one that can be based on the properties of a galaxy treated as a whole.

The main method used to investigate the stellar contents of galaxies is a technique known as **population synthesis**. This technique uses the **integrated spectrum** of most if not all of the galaxy. A major problem of stellar population synthesis is that cool, low-mass stars that contribute much of the total mass do not emit much visible light, so it is best to include infrared as well as visual observations.

The method of population synthesis starts by defining a number of different stellar categories – these are defined in terms of both stellar mass and metallicity. A plausible assumption is made about the relative numbers of stars in each category, and on this basis their contribution to the total mass and luminosity of the galaxy can be worked out. It is then possible to calculate the integrated stellar spectrum that would be expected from such a combination of stars at a given age. The result of this model spectrum is then compared with observations. The mix of stars and the age of the galaxy are adjusted and the process is repeated until there is a good match between the model and observed spectrum. Figure 2.19 shows how the spectrum of a galaxy is expected to evolve as it ages, given a starting point of a single burst of star formation.

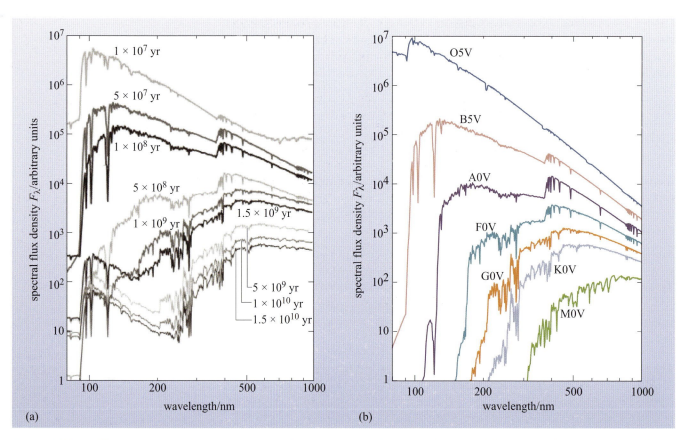

Figure 2.19 (a) The evolution of the spectrum of a galaxy using a population synthesis model in which all stars formed at the same time. The spectra correspond to the age of the galaxy after the initial star formation event. (b) Example spectra of some of the stars of different types that contribute to the luminous output of the model galaxy. At early times, the spectra in panel (a) are dominated by O and B stars in the UV, but as these stars are short-lived the UV light dies away quickly, and the older simulated spectra become dominated by the contribution from less massive stars. After about 1×10^9 years, the galaxy spectrum shows a brightening in the UV relative to the visible. This is due to a population of old stars that have been altered in binary star systems. Note the use of logarithmic scales for the spectra shown in (a) and (b). ((a) Han, Z. *et al.*, 2007; (b) model data as described in Castelli and Kurucz, 2004)

The relative importance of the various minor constituents can usually be deduced from observations made at a variety of wavelengths: for instance, sub-mm observations help to determine the dust content.

2.4 The determination of the distances of galaxies

2.4.1 Introducing distance determinations

Measuring the distances to external galaxies is a task of crucial importance in modern astronomy. There are three main reasons for this. First, if we want to determine properties of galaxies we often need to know their actual, rather than their apparent, sizes. If the distance d to a galaxy is known, then its angular size θ can be used in the formula $l = d \times (\theta/\text{radians})$ to find the actual size (provided θ is small, which in practice it always will be). Secondly, galactic distances are crucial to mapping the layout of the Universe. It is easy to identify the directions of observable galaxies and clusters, but only when their distances are known can their arrangement throughout space be fully determined. The third reason is that galactic distances hold the key to working out the age, evolution and ultimate fate of the

Universe – more will be said about these issues in Chapters 5–8. For the present it is sufficient to note that there are several different theoretical models of the Universe, and if galactic distances can be measured with sufficient accuracy then it may be possible to discount some of them.

So, the measurement of galactic distances is of great importance. But how is it done, and how reliable are the results? There are many methods of measuring galactic distances and new ones are being developed, or old ones revised and improved, all the time. We cannot hope to give an exhaustive account of this vast subject here, but we will outline the general principles and then examine a few methods in more detail, noting some of their limitations and shortcomings. Many of the methods for finding distances fall into a few broad categories.

Methods based on geometry

The basic idea behind the **geometrical methods** used to determine the distances to other galaxies is very simple: within an external galaxy, identify a feature of known linear diameter l, measure the angular diameter, θ, of that feature, then work out the feature's distance, d, using the formula

$$d = l/(\theta \text{ /radians})$$

The main shortcoming of this method is equally simple – there are few features of 'known linear diameter' in external galaxies, and even those features that do exist are unlikely to have accurately measurable angular diameters in any external galaxies apart from those that are very close to the Milky Way.

A good example of the geometric approach is the method used to make what is thought to be an accurate determination of the distance to the Large Magellanic Cloud (LMC), one of the nearest external galaxies. The LMC was the site of a supernova observed in February 1987. This event, known as SN 1987A, is shown in Figure 2.20. Three and a half years after the supernova occurred, electromagnetic radiation spreading outwards from the site of the explosion

(a) (b)

Figure 2.20 Supernova 1987A. (a) the progenitor to SN 1987A (arrowed) (b) the same region of sky after the supernova erupted. (D. Malin/AAO)

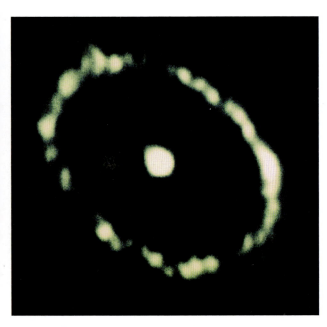

Figure 2.21 The gas ring observed three and a half years after the initial SN 197A explosion consists of material that was expelled by the progenitor prior to the explosion and was subsequently interacting with radiation from the supernova. (NASA/ESA)

encountered a ring of gas that had been ejected from the supernova's progenitor star thousands of years earlier. As the radiation met the ring it caused the various parts of the ring to brighten, leading to the effect shown in Figure 2.21. It appears from the figure that the supernova occurred at the centre of the ring, so it might be expected that each part of the ring would brighten at the same time. However, as seen from Earth, some parts of the ring were observed to brighten before others. The reason for this is easy to explain. The ring is inclined relative to the line of sight from Earth to the supernova, so some parts of the ring are closer to the Earth than others. Consequently there are time delays between the arrival of light signals that left different parts of the ring at the same time. Observations of the time delay between the brightening of the closest and furthest parts of the ring provide information that, together with other observations concerning the orientation of the ring, can be used to work out the ring's linear diameter. (Details of this calculation are indicated in Example 2.1.) Comparing this with the ring's observed angular diameter, the distance of the ring, and hence of the LMC itself, has been shown to be 52 ± 3 kpc. This result places the LMC on the outskirts of the Milky Way, in good agreement with independent determinations of its distance by other methods.

EXAMPLE 2.1

Find an expression for the diameter of the ring around SN 1987A in terms of the time delay (Δt) between the brightening of the nearest and furthest parts of the ring, and the ratio (b/a) between the semiminor and semimajor axes of the ring as it appears to an observer at the Earth. Assume that the ring is actually circular.

SOLUTION

First let's deal with geometry; the circular ring around SN 1987A (with diameter $2a$) appears to be elliptical because the ring is inclined with respect to our line of sight. This geometry is illustrated in Figure 2.22. When viewed from Earth (Figure 2.22a), the ring appears as an ellipse with semimajor and semiminor axes a and b, respectively.

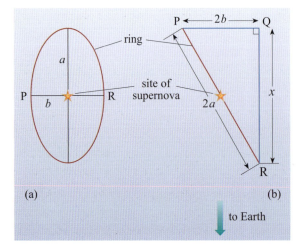

Figure 2.22 The geometry of the ring around SN 1987A. (a) A face-on view of the ring as seen from Earth. (b) A side view showing the relationship between points P and R on the ring.

A side view that cuts through the centre of the ring (Figure 2.22b) shows the relationship between a, b and the extra distance x ($= c \times \Delta t$) that light must travel if it is emitted from the far side of the ring (point P) compared with light that is emitted from the near side of the ring (point R).

The aim is to find an expression for the diameter of the ring ($2a$) in terms of quantities that can be measured in observations. One such quantity is x, which is related to the measured time delay ($c \times \Delta t$). Pythagoras' theorem can be applied to the right-angled triangle PQR to obtain an expression for x in terms of a and b,

$$x^2 = (2a)^2 - (2b)^2$$

The other quantity that can also be determined from observations is the ratio b/a – the ratio of the semiminor and semimajor axes of the image of the ring. So the aim of the algebraic manipulation is to obtain an expression for the diameter of the ring in terms of x and b/a.

Dividing both sides of the equation for x^2 by a^2 gives

$$\left(\frac{x}{a}\right)^2 = (2)^2 - \left(2\frac{b}{a}\right)^2 = (2)^2 \times \left(1 - \left(\frac{b}{a}\right)^2\right)$$

Taking the square root of both sides of this equation

$$\frac{x}{a} = 2\sqrt{\left(1 - \left(\frac{b}{a}\right)^2\right)}$$

This equation can be rearranged to give an expression for $2a$

$$2a = \frac{x}{\sqrt{\left(1 - \left(\frac{b}{a}\right)^2\right)}}$$

But $x = c \times \Delta t$ so the equation for the diameter ($2a$) can be written as

$$2a = \frac{c\Delta t}{\sqrt{\left(1 - \left(\frac{b}{a}\right)^2\right)}} \tag{2.2}$$

This expresses the linear diameter of the ring in terms of observable quantities (Δt and b/a), as required.

QUESTION 2.3

In the case of SN 1987A, light from the far side of the ring took 340 days longer to get to us than did light from the near side, and the angular diameter of the ring was measured to be 1.66 arcsec.

(a) Use Equation 2.2 to calculate the diameter of the ring around SN 1987A. (The ratio (b/a) can be estimated from Figure 2.21. *Hint*: as it is a ratio it has no units.)

(b) Using your answer from part (a), calculate the distance to SN 1987A. Express your answer in kpc.

Methods involving a 'standard candle'

This category covers a wider range of methods and is much more important than the previous 'geometric' category, but the basic idea is just as simple. A standard candle is an object of known luminosity embedded in the object whose distance is to be determined. In Chapter 1 you saw how RR Lyrae stars could be used as a standard candle for the purpose of determining the layout of the Milky Way's halo. In Section 2.4.1 we describe other astronomical standard candles that are used to determine galactic distances.

Once a standard candle has been identified, its distance is found by comparing the flux density, F, that it provides to observers on Earth, with its (known) luminosity L.

$$d = \sqrt{\frac{L}{4\pi F}} \tag{2.3}$$

Techniques that use this approach to measure distance are generically referred to as **standard candle methods**.

QUESTION 2.4

(a) Equation 2.3 expresses the relation between luminosity, flux density and distance in the absence of any absorption. What effect does absorption have on the observed flux density? Hence what is the effect of absorption on a distance estimate based on the observed flux density as given in Equation 2.3?

(b) In practice, it is not usually the flux density F over all wavelengths, but the flux density within a narrow wavelength range that is measured. Suppose that the V-band (i.e. the visual band) is used, and that in this band the observed flux density from a standard candle is F_V. What implication does this have for the use of that standard candle, and how should Equation 2.3 be modified to reflect this?

Standard candle methods have historically played an important role in galactic distance determinations and continue to do so. The two fundamental questions that limit every standard candle method are:

• Which astronomical sources are suitable to be standard candles?

• What is the luminosity of a selected standard candle?

Later in this section we will examine several examples of standard candle techniques and see how these issues are addressed.

The redshift method

This is a very important method of measuring distances to galaxies, and one that is quite distinct from geometrical or standard candle methods. The method is based on a correlation between the measured distances of galaxies and the observed spectral redshifts of those galaxies that was first recognized by Edwin Hubble in 1929. The details of this correlation and the precise definition of a quantity called the *redshift* (z) that characterizes the shifts of spectral lines will be treated in Section 2.4.6. However, it is worth noting here that, if the distance of a galaxy is represented by d and its redshift by z, then the correlation that forms the basis of Hubble's law is given by the simple proportionality

$$d \propto z$$

So, having deduced the existence of the redshift–distance relationship from the measured distances and redshifts of some galaxies, it is a relatively simple matter to use that relationship to infer the distances of other galaxies from measurements of their redshifts alone. The great advantage of this method is that the redshift z is easy to measure. The main disadvantage is that the redshift–distance correlation is only approximately true for any individual galaxy, so the value of d determined for a particular galaxy will only provide an approximate value for its distance. As you will see later, Hubble's discovery indicates that the Universe is expanding – one of the key observations in modern cosmology.

2.4.2 The distance ladder and its calibration

The major problem in distance measurement is that no single method spans the entire range of astronomical scales. Figure 2.23 lists some of the methods of measuring astronomical distances, and indicates the range of distances over which they have been used. (Note that the distance scale is logarithmic, and that the listing includes one method that is confined to the Milky Way.) Almost all methods of distance determination involve using distances found by one method to support another. The use of a chain of measurements, each relying on another, leads to the so-called **distance ladder** – a ladder in the sense that the accessibility of the upper steps depends on having appropriate steps, firmly in place, lower down.

Relating one step on the distance ladder to another involves a process of **calibration**, that is, the use of an established method of measurement to give absolute meaning to the relative measurements provided by some other method.

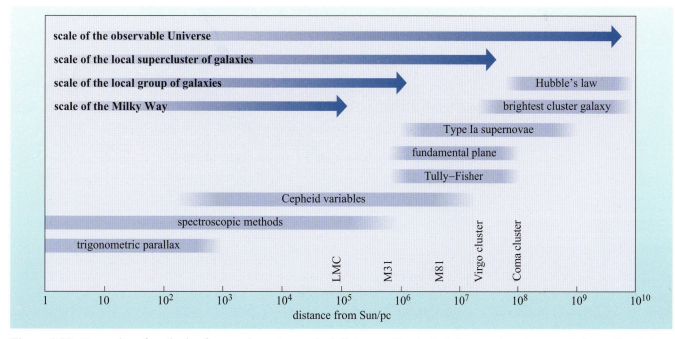

Figure 2.23 Examples of methods of measuring astronomical distances. Also included are various terms and size scales that will be explained later in this book.

1 Mpc (megaparsec) = 10^6 pc = 10^3 kpc.

For example, measuring galactic redshift is relatively simple, so determining the redshifts of several different galaxies is straightforward, and should give an indication of the *relative* distances of those galaxies. However, in order to use those redshifts to determine the *absolute* distance (in Mpc) of each of the observed galaxies it is necessary that the absolute distance to one or more of those galaxies should already have been determined using another established method of distance measurement, usually a well-checked standard candle method. By using the absolute distance measurements provided by one step on the distance ladder, the relative distances indicated by some other step can be checked, refined and calibrated so that they too can be interpreted as absolute distances.

The great advantage of this calibration process is that new methods of distance determination, which may extend our range of distance measurements, can be introduced and cross-checked against established methods that are limited to nearby objects. The disadvantage, however, is that any errors or uncertainties in the established methods of distance measurement are carried over to the newer methods when they are calibrated. The 1990s saw a significant improvement in the calibration of the distance scale, with the result that astronomers are now able to measure distances to remote galaxies (>100 Mpc, say) with an uncertainty of only about 15%. The following sections examine some of the methods of distance determination in more detail. This is by no means a comprehensive review of distance measurement methods, but it does highlight some of the more important techniques that have been applied in calibrating the distance ladder.

2.4.3 The Cepheid variable method

Cepheids are giant or supergiant stars with a variable light output that can change by as much as a factor of ten over a period that may be anything from about a day to about 100 days. The name 'Cepheid' is derived from δ Cephei, which was the first star of this class to be described. Cepheids have quite high average luminosities, so they are visible at large distances. A **light curve** showing the variation with time of the apparent visual magnitude of one particular Cepheid is given in Figure 2.24. The recognition that such stars could be used as standard candles was due to the collective efforts of three astronomers, Henrietta Leavitt (1868–1921), Ejnar Hertzsprung (1873–1967) and Harlow Shapley (Figure 1.26).

In 1907, while examining variables in one of our nearest neighbouring galaxies, the Small Magellanic Cloud (SMC), Leavitt discovered a correlation between the period (i.e. the time between successive peaks in the light curve) of a certain kind of variable star and its average apparent magnitude. Since all the stars in the SMC are at roughly the same distance from the Sun, it followed from Leavitt's discovery that there must also be a correlation between the period of such a variable and

Figure 2.24 The light curve of the variable star δ Cephei, which gives its name to the class of Cepheid variables. The period of such a star is the time between successive peaks of the light curve, and is about 5.4 days in this case. (Bok and Bok, 1974)

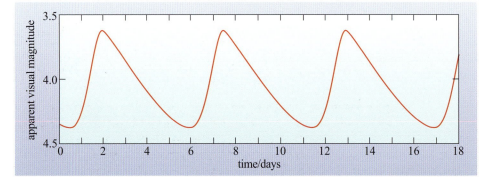

its average luminosity. The existence of this **period–luminosity relationship** implied that for any variable belonging to the same class as those studied by Leavitt, the average luminosity could be deduced from the (easily observed) period of variation. Of course, before the period–luminosity relationship could be used in this way it was first necessary to *calibrate* the relationship by using some established technique to determine the average luminosity of at least one variable star of the appropriate type. This was Hertzsprung's particular contribution – and it was no easy matter.

Hertzsprung's first achievement was recognizing that Leavitt's variables were in fact Cepheids. This insight reduced the calibration problem to that of measuring the average luminosity of a Cepheid, but the task was still not simple. There were many known Cepheids within the Milky Way, but none of them was close enough for its distance to be measured by the method of trigonometric **parallax**; consequently none of them had a reliably determined average luminosity. So Hertzsprung was forced to use more complicated and less reliable methods in order to obtain the average luminosity of a Cepheid. This he eventually did, although for various reasons, including a serious underestimate of the effects of interstellar absorption, his calibration was inaccurate and the resulting value for the distance of the SMC quite wrong.

Modern versions of the Cepheid period–luminosity relationship have, of course, been re-calibrated, and this method can now yield the distance to any given Cepheid with an uncertainty of just 15%. A recent version of the period–luminosity relationship is shown in Figure 2.25. There are three points about this that should be noted. First, despite its name, the period–luminosity relationship is usually presented as a relationship between period and average absolute magnitude (i.e. the average magnitude an object would have if viewed from a distance of 10 pc). Secondly, the absolute magnitude of a Cepheid will vary with time, but the average absolute magnitude will not, since it corresponds to the mean luminosity of the Cepheid over its full period of variation. Third, the magnitude used is usually restricted to one or other of the standard wavelength bands used by astronomers; in the case of Figure 2.25 the V (or visual) band has been chosen.

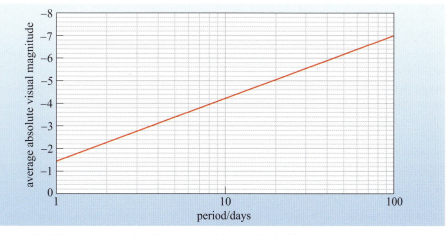

Figure 2.25 The period–luminosity relationship for Cepheid variables. Note that, in accordance with convention, the average luminosity is represented by an average absolute visual magnitude (which is calculated from the mean luminosity over a full period). The period–luminosity relation shown here is for Type I (classical) Cepheids. Type II Cepheids have lower luminosity than Type I Cepheids of comparable period.

QUESTION 2.5

A certain Cepheid is observed to have a period of 10 days. What is the average absolute visual magnitude (corresponding to the mean luminosity) of the Cepheid expected to be?

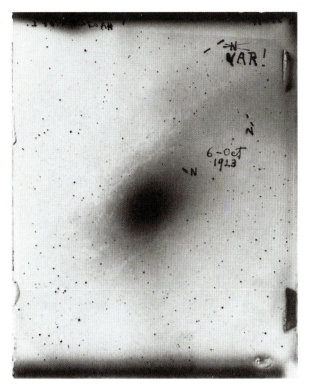

Figure 2.26 The photographic plate of M31 on which Hubble discovered a Cepheid. Note that the N (for nova) at the top right has been crossed out and VAR (for variable) written in its place. (Berendzen *et al.*, 1976)

The **Cepheid variable method** has played an important part in the history of astronomical distance measurements. It was used by Hubble to determine the distance to the spiral galaxy M31 (Figure 2.5b). This measurement, announced in 1924, convinced the majority of astronomers that there were galaxies external to the Milky Way and it thus ushered in the era of *extragalactic astronomy*. Figure 2.26 shows the photographic plate that Hubble was studying when he realized that what he had thought was a nova was in fact a Cepheid that held the key to M31's distance.

The sensitivity and high spatial resolution of the Hubble Space Telescope have allowed astronomers to use the Cepheid variable method to measure the distances of galaxies that are up to about 30 Mpc from the Milky Way. While this is not a very large distance in cosmic terms, it enables us to subsequently calibrate other distance measurement methods, which range further than this method. Consequently the Cepheid variable method is a vital step in building the cosmic distance ladder.

Although the period–luminosity relationship was discovered empirically, there is a good underlying physical explanation for it. The variation in luminosity of a Cepheid is thought to arise because the envelope of the star is pulsating (i.e. expanding and contracting regularly) with a period that depends on the mass of the star. The more massive the star, the larger and hotter (and thus more luminous) it is, and hence the period increases.

2.4.4 Type Ia supernova method

Figure 2.27 shows a supernova in the nearby external galaxy M82. As you can see, the supernova is very bright compared with the rest of its host galaxy and is easily discerned. Supernovae are classified as Type I or II mainly on the basis of their spectra (Box 2.2). Members of a subclass of Type I, **Type Ia supernovae**, appear to have approximately the same peak luminosity, and can therefore be used as standard candles. As you will see in Chapters 5 and 7, the use of Type Ia supernovae as standard candles is of immense importance to the study of the behaviour of the Universe on very large scales.

A supernova explosion occurs in one of two ways – either the rapid gravitational collapse of a stellar core nearing the end of its life, or by the sudden re-ignition of nuclear fusion within a white dwarf star. Either way, for a few weeks or maybe months, a supernova explosion can shine brighter than the rest of its entire host galaxy, and since certain types of supernova explosions appear to have similar peak luminosities, they can be used as standard candles, and consequently, for distance determination.

Figure 2.27 The discovery image of the Type Ia supernova 2014J in the relatively nearby galaxy M82. The location of the supernova is indicated. (UCL/University of London Observatory/S. Fossey, B. Cooke, G. Pollack, M. Wilde, T. Wright)

BOX 2.2 CLASSIFICATION OF SUPERNOVAE

Supernovae are classified into two broad groups depending on whether their spectra show hydrogen lines: Type I supernovae do not show hydrogen lines, whereas Type II supernovae do. Type II supernovae result from the core collapse of a supergiant star that has a significant amount of hydrogen in its envelope. This gives rise to a spectrum that contains hydrogen spectral lines.

The Type I class is subdivided according to the presence of silicon absorption lines in the spectrum. Those with silicon lines are classed as Type Ia, whereas those that lack silicon lines are classed as Type Ib or Type Ic. Very massive supergiants (with initial masses exceeding about $30M_\odot$) are thought to lose their hydrogen envelopes in intense stellar winds, and the core collapse of these stars gives rise to Type Ib and Ic supernovae.

Type Ia supernovae arise from a different mechanism to core collapse. It is believed that they occur in interacting binary star systems in which one star is losing mass by Roche lobe overflow. (Roche lobe overflow is a mechanism whereby an evolved star can lose matter to a companion.)

The star that gathers the overflowing material is a white dwarf that is close to the maximum mass that such a star can have. This mass is called the *Chandrasekhar limit* and has a value of about $1.4M_\odot$ (the exact value depends on the composition of the white dwarf). The white dwarf accretes material from its companion until it exceeds this mass limit, at which point the elements (usually carbon and oxygen) that make up the bulk of the white dwarf undergo rapid thermonuclear burning, following the re-ignition of nuclear fusion. This results in an energetic explosion that is seen as a supernova.

The idea that Type Ia supernovae could be used as standard candles on cosmological scales developed from observations of a handful of such objects whose distances were within the range of other distance determination techniques. This prompted intense study into the properties of such supernovae, and with some refinements, which we will return to in Chapter 7, it is now widely accepted that they are reliable standard candles. Detailed observational studies suggest that the peak luminosity of a typical Type Ia supernova is about $5.5 \times 10^9 L_\odot$. There is also a good physical reason to expect that Type Ia supernovae might have similar luminosities. As mentioned in Box 2.2, Type Ia supernovae are produced by the thermonuclear explosion of a white dwarf that is at the Chandrasekhar limit. Since this critical mass is similar for all white dwarfs, it is reasonable to suppose that they should all have similar luminosities.

The difficulty associated with observing Type Ia supernovae is that they are short-lived transient events, which from an observer's point of view, could happen at any time and at any location in the sky. By the 1990s automated telescopic systems were developed to conduct nightly imaging surveys covering reasonably large areas of the sky. These surveys flagged up, usually in a matter of hours after observation, any rapidly brightening source which could then be subject to spectroscopic follow up and long-term photometric monitoring.

■ Why would rapidly brightening sources identified from an imaging survey need to be investigated spectroscopically?

☐ To identify the source as a supernova, and in particular to confirm whether its spectrum confirms it as a Type Ia supernova.

QUESTION 2.6

Type Ia supernovae were first observed in three very nearby galaxies. List the items of information concerning these supernovae and their host galaxies that you would need in order to calibrate the Type Ia supernova method. Briefly explain how you would use the items in your list.

2.4.5 Galaxies as standard candles

The luminous output of an entire galaxy is potentially a very useful type of standard candle. However, as explained in Section 2.2, galactic luminosities vary greatly, so in order to use them as standard candles we must find subsets of particular groups of galaxies that have a common, reproducible luminosity. Some arguments in this method are based on simple statistics, others on the relationship between the kinematics of a galaxy (i.e. the motion of its constituent stars and/or gas) and its luminosity. In the remainder of this section we concentrate on the latter, and highlight the types of galaxies they are applicable to.

Spiral galaxies: The Tully–Fisher relation

In 1977 Brent Tully and Richard Fisher discovered that there is a correlation between the luminosity of a spiral galaxy and its maximum rotation speed (as indicated by the Doppler-broadened width of the 21 cm emission line in the galaxy's radio spectrum). This is known as the **Tully–Fisher relation**. The more massive (and, by assumption, more luminous) galaxies should be rotating more rapidly. (It was shown in Chapter 1 that orbital speed depends on the mass enclosed by the orbit.) The 21 cm line emitted by a massive spiral galaxy should therefore be expected to include contributions from hydrogen clouds travelling with a greater range of speeds along the line of sight (including a greater maximum speed), than the 21 cm line emitted by a less massive spiral. Consequently, the measured width of the 21 cm line should be greater for more luminous spiral galaxies. According to the Tully–Fisher relation, the luminosity of a spiral increases roughly in proportion to the fourth power of the maximum rotational speed V_{max}, so approximately

$$L \propto (V_{max})^4 \tag{2.4}$$

Tully and Fisher established and calibrated this relation using a sample of ten nearby spirals, the distances (and hence luminosities) of which had previously been determined using the Cepheid variable method. Provided the inclination of a spiral galaxy is accounted for (as well as any additional absorption or emission in a line of sight), this method is a reliable way to find unknown galactic distances to within ±15%.

Elliptical galaxies: The Faber–Jackson relation and the fundamental plane relation

For elliptical galaxies the analogue of the Tully–Fisher relation is called the Faber–Jackson relation (after Sandra M. Faber and R. E. Jackson, who discovered it in 1976). Elliptical galaxies show little rotation; the kinematics of their stars is dominated by random motions, whose velocity range is characterized by a quantity called the velocity dispersion Δv (see Section 2.3.2). The Faber–Jackson relation states that the luminosity of an elliptical galaxy is approximately related to the velocity dispersion by

$$L \propto (\Delta v)^4 \qquad\qquad (2.5)$$

In practice, it is hard to calibrate this relationship because it is almost impossible to determine the total flux from an elliptical galaxy (see Section 2.3.1). To make it easier, the relationship is usually calculated across the diameter of a specific isophote at which a certain level of surface brightness is reached.

Consequently, the Faber–Jackson relation is rather approximate; more usually astronomers use the so-called *fundamental plane relation*, which links velocity dispersion, the radius of an isophote and the surface brightness at that isophote. With this method galactic distances can also be measured to distant elliptical galaxies to ±15 %.

QUESTION 2.7

If two elliptical galaxies differ in their velocity dispersions by a factor of 1.2:

(a) What is the ratio of the luminosities of these galaxies? (Assume that the Faber–Jackson relation applies.)

(b) What is the ratio of their masses, if their radii are the same?

2.4.6 Hubble's law

Establishing Hubble's law

The terms 'red-shifted' and 'blue-shifted' refer respectively to displacements of identifiable spectral features, such as spectral lines, towards longer or shorter wavelengths than those at which they would normally be observed under standard laboratory conditions. One way that such shifting can arise is as a consequence of the Doppler effect, caused by the motion of the source of radiation relative to the observer. In such cases the amount by which a spectrum is shifted can be used to determine the speed at which the emitter is receding from, or approaching, the observer. However, spectra can be red-shifted or blue-shifted by other physical effects.

The spectra of distant galaxies are typically red-shifted. As you will see later, this is not simply the result of the Doppler effect, although they do arise from a change with time of the separation between us and the galaxies. A red-shifted spectrum corresponds to a separation that has increased over the time that light has been travelling to us. A schematic example of a red-shifted galactic spectrum is shown in Figure 2.28 (overleaf). In this case, which is typical of distant galaxies, the extent to which any of the spectral lines is red-shifted can be characterized by a single numerical value z that is said to be the **redshift** of the line.

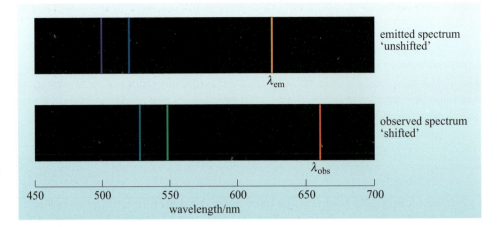

Figure 2.28 As a result of redshift, a spectral emission line emitted at a wavelength λ_{em} is seen by observers at a wavelength λ_{obs}.

To evaluate the redshift z in a specific case, all that is needed is a value for the observed wavelength (λ_{obs}) of a spectral line that has a known wavelength (λ_{em}) at the point of emission. The value of z can then be obtained from the definition

$$z = \frac{\lambda_{obs} - \lambda_{em}}{\lambda_{em}} = \frac{\lambda_{obs}}{\lambda_{em}} - 1 \qquad (2.6)$$

Generally speaking, all the spectral emission lines originating in a distant galaxy will be red-shifted to the same extent, so the redshift of any particular line in that galaxy's spectrum will also equate to the redshift of the galaxy itself. In the minority of cases where the spectrum of a galaxy is blue-shifted rather than a red-shifted, Equation 2.6 will give a negative value for z. Throughout this chapter we speak exclusively of redshifts, with the tacit understanding that a blue-shifted spectrum is characterized by a negative redshift.

■ Oxygen emits a spectral line at an 'unshifted' wavelength of 500.7 nm. Suppose that this line is observed in the spectrum of a galaxy at a wavelength of 596.0 nm. What is the redshift of the galaxy from which the line was emitted?

□ $z = \dfrac{(596.0 - 500.7) \text{ nm}}{500.7 \text{ nm}} = 0.190$

Hubble was one of the first astronomers to undertake the difficult task (in the 1920s) of measuring galactic redshift, and comparing his spectroscopic data with his distance determinations. In 1929, using a sample of just 24 galaxies he presented the first convincing evidence of a linear relationship between the redshifts and distances of galaxies. The basis of this relationship was introduced in Section 2.4.1; in modern notation the complete relationship is usually written as:

$$z = \frac{H_0}{c} d \qquad (2.7)$$

where c is the speed of light (3.00×10^8 m s^{-1}) and H_0 is a constant of proportionality known as the **Hubble constant**.

There are two important aspects of the redshift–distance relation that need to be emphasized. The first is that Equation 2.7 only applies over a relatively small range of redshifts – up to about 0.2. At higher redshifts the relationship between redshift and distance becomes more complicated. A more detailed discussion is provided in Chapter 5.

Secondly, even at low redshifts, the linear relationship between z and d is not a perfect one, as you can see from Figure 2.29. Nonetheless, Equation 2.7 does sum up the general trend and is one of the standard ways of expressing **Hubble's law**.

Hubble's law is usually interpreted as showing that the Universe as a whole is expanding (a more detailed discussion of this point is given in Chapter 5). Firstly, Equation 2.7 can be rewritten as

$$cz = H_0 d$$

We already stated that redshift is not attributable to the Doppler effect, although if the redshift is sufficiently small it is sometimes referred to as an *apparent* recession. In this situation, the quantity cz can be identified as the *apparent* speed of recession v of a galaxy. Equation 2.7 can then be expressed as

$$v = H_0 d \tag{2.8}$$

Thus, the speed of recession of a galaxy is proportional to its distance from us. This is just what would be expected in a Universe that was undergoing a uniform expansion. Every point would move away from every other point, and the speed of one point relative to another would be proportional to the distance between them. In such a uniformly expanding Universe, the Hubble constant would be very significant, since it would provide a measure of the rate of cosmic expansion. (We will see in Chapters 5–8 why this assumption doesn't hold at higher redshifts and therefore why this simple picture does not work.) It is important to note that although galaxies are generally seen to be moving away from the Earth, this does not imply that the Earth occupies any particularly special place in the Universe. In a uniformly expanding Universe *every* point moves away from *every* other point.

The distance–velocity relationship (Equation 2.8) is often quoted in books, and is sometimes confusingly also referred to as Hubble's law. In this book we adopt the convention that Hubble's law is a relationship between redshift and distance, as given in Equation 2.7.

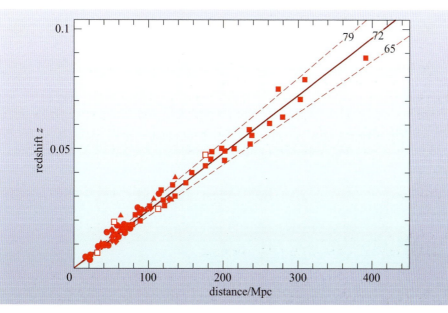

Figure 2.29 A plot of redshift against distance for some galaxies and clusters of galaxies. The different symbols represent different techniques that were used to measure the distances independently of the redshift. (We will discuss this further in Chapter 7.) The solid straight line shows the expected relationship between redshift and distance for a Hubble constant of 72 km s^{-1} Mpc^{-1}. The dashed lines show the expected relationship for H_0 values of 79 and 65 km s^{-1} Mpc^{-1} respectively. (Freedman *et al.*, 2001)

Using Hubble's law

Clearly, despite some deviations from the linear relationship, Hubble's law has great potential as a method of determining the distances of galaxies. Now that redshifts can be measured with relative ease, it would appear to be a simple matter to determine the distance of any galaxy if we know the value of the Hubble constant H_0. Evaluating H_0 is, of course, nothing other than the calibration problem for this method of determining distances. The determination of the Hubble constant has been one of the major problems in astronomy since the discovery of Hubble's law.

■ In what SI units would you express H_0 if you knew its value?

☐ z is a ratio of similar quantities (wavelengths) so it has no units. Thus, H_0 should have the same units as c/d. If we use SI units, then c (which is a speed) is measured in m s^{-1} and d (which is a distance) in m, with the consequence that H_0 could be measured in s^{-1}.

Given that astronomy is littered with unconventional units, you may not be too surprised to learn that astronomers tend not to express H_0 in terms of SI units. Instead, H_0 is usually given in terms of km s^{-1} Mpc^{-1}. Of course, since these units express a speed per unit distance, they are a valid alternative to the SI units of s^{-1}. It has already been mentioned that H_0 is an important quantity in cosmology, and for that reason we postpone a detailed discussion of how it has been established until Chapter 7, where it is considered along with other key quantities that describe properties of the Universe as a whole. For the present discussion it is sufficient to note that the value of H_0 is about 72 km s^{-1} Mpc^{-1} and that this value is believed to be correct within an experimental uncertainty of less than 10%.

QUESTION 2.8

Using a value for H_0 of 72 km s^{-1} Mpc^{-1} calculate the distance of a galaxy that has a redshift of $z = 0.048$. Check your calculation by referring to Figure 2.29.

QUESTION 2.9

If H_0 has the value 72 km s^{-1} Mpc^{-1}, what is its corresponding value in SI units (i.e. expressed in units of s^{-1})?

The origin of the scatter in the relationship between redshift and distance is an important consideration when using Hubble's law.

■ What would be the effect on the plot of redshift against distance if, in addition to the relationship described by Hubble's law, galaxies also had some random motion?

☐ Random motions would give rise to Doppler shifts that would result in positive and negative contributions to the total redshift. Thus the effect of random motions would be to introduce some scatter into the relationship between distance and redshift.

The random motion of individual galaxies, sometimes referred to as a **peculiar motion** in this context, seems to be the explanation for the scatter in the relationship between distance and redshift shown in Figure 2.29. Galaxies close to our own, where the effect of the Hubble expansion of the Universe is negligible, seem to have motions through space of a few hundred km s^{-1}. (In fact, these motions are often the result of galaxies being attracted, under the influence of gravity, towards over-dense regions of the Universe.) It is useful to know the typical distance at which the redshift due to the Hubble expansion is likely to exceed the redshift due to random motions.

QUESTION 2.10

It is assumed that galaxies have random velocities of typically 300 km s^{-1}.

(a) Calculate the typical redshift that would be expected for nearby galaxies (i.e. galaxies that are so close that the systematic redshifts predicted by Hubble's law can be ignored). Would this redshift necessarily be positive?

(b) At what distance does the redshift predicted by Hubble's law dominate over the spread in redshift calculated in part (a)? Assume that Hubble's law dominates when the Hubble redshift is a factor of ten greater than the typical redshift due to the random motion of galaxies, and that $H_0 = 72$ km s^{-1} Mpc^{-1}.

The answer to Question 2.10 indicates that Hubble's law is likely to be unreliable for measuring distances when the redshifts are less than 0.01 – this corresponds to a distance of about 40 Mpc (assuming $H_0 = 72$ km s^{-1} Mpc^{-1}).

Distances in an expanding universe

Finally, you need to be aware that the expanding Universe has a geometrical quirk that affects distance measurements. We won't describe this quirk in detail, and it only affects objects that are so distant that the Universe has expanded appreciably since the light left the object (so supernova SN 1987A is exempt, for example). The effect of this geometrical quirk is that distances measured in different ways end up with systematically different answers. For example, a distance measurement based on how big a galaxy appears on the sky will differ from a measurement to the same galaxy based on how bright it appears. The good news is that there are simple corrections to convert between distance measurements, and the corrections depend only on the redshift. For this reason, astronomers are sometimes careful to state what sort of distance is being referred to, in order to avoid ambiguity. This is why we're generally careful in this book to state (where relevant) how distances have been measured, but we're not otherwise concerned about how this geometrical quirk affects the measurements.

2.5 The formation and evolution of galaxies

Our view of the Universe exterior to our own galaxy is dominated by the presence of galaxies, so it is natural to ask how they form and how they change with time. There are two lines of attack in studying issues of galaxy formation and evolution. One approach is to use observations of the properties of galaxies and attempt to 'work backwards' to see what inferences can be drawn about how galaxies have evolved, and possibly even how they were formed. As we see later, the most distant observable galaxies are so far away that the light we see now was emitted when the Universe was less than 10% of its current age. Thus we are now in the remarkable position of being able to look back and see galaxies over a large fraction of cosmic history. Of course, we can only observe any one galaxy at one particular time, but by looking at changes in properties of populations of galaxies with redshift, astronomers can start to see the important features of galaxy evolution.

The alternative approach to 'looking backwards' is to consider how conditions in the early Universe are likely to have given rise to the structures we observe. This is possible because conditions in the early Universe were remarkably uniform and predictable. The seemingly complex structures of galaxies we observe in the present-day Universe have evolved from these simple conditions through a range of physical processes. As we shall see, the success of this approach depends on how well these physical processes can be modelled.

These two approaches are complementary. The aim of the theory of galaxy formation is to describe how galaxies arose from the conditions of the early Universe. Observations that look back from the present day over a significant fraction of the age of Universe provide the data against which such models must be tested.

In this section we begin with a brief introduction to the early Universe and then give an overview of the major theoretical ideas about how galaxies could have formed. We will introduce some cosmological concepts that are dealt with in more detail in Chapters 5–8. The observational approach is introduced by considering how isolated galaxies evolve, and then considering the role that interactions and mergers may have played in galactic evolution. The section concludes with a review of how deep surveys provide an insight into the formation and evolution of galaxies.

2.5.1 The evolving Universe

Any investigation of galaxy formation must be considered within the cosmological framework of the origin and evolution of the Universe as a whole. A detailed discussion of current ideas in cosmology is developed in Chapters 5–8 of this book. However in order to develop an understanding of galaxy formation, it is necessary here to introduce some key ideas about the history of the Universe. The most important concept is the widely accepted paradigm for describing the origin and evolution of the Universe called the *hot big bang* theory. The following points describe those features or consequences of the theory that are important in understanding galaxy formation and evolution.

- The Universe has a finite age, which is currently estimated to be about 13.8 billion years.

- The cosmic expansion that we see today, as implied by Hubble's law, has persisted since the earliest times, although the rate of expansion has not always had its current value. In particular, it is believed that the rate of cosmic expansion is currently accelerating.

- The physical conditions in the Universe at early times were characterized by extremely high temperatures and densities. However at any given instant, the density and temperature of matter throughout the early Universe were highly, but not perfectly, uniform.

- As the Universe expanded and cooled, protons and neutrons (i.e. particles of 'ordinary' baryonic matter) formed from more fundamental particles (quarks).

The number of protons exactly balances the number of electrons, which results in there being a net electric charge of zero on matter in the Universe.

- Within the first few minutes after the big bang, nuclear reactions resulted in the formation of helium nuclei. As a result of these reactions about 24% of the mass of baryonic matter was in the form of helium nuclei and about 76% in the form of hydrogen nuclei. Only very small traces of other light elements would have been produced.

- The matter that filled the early Universe also contained particles of non-baryonic matter. These account for the non-baryonic dark matter that now appears to be the dominant form of matter in galaxies.

- There has always been mutual gravitational attraction between all forms of matter (baryonic and non-baryonic) in the Universe.

- Despite its high degree of uniformity, the cosmic gas of baryonic and non-baryonic matter that filled the early Universe was subject to slight density fluctuations. In other words, there were regions of the early Universe where the actual density of matter departed slightly from the average.

An important aspect of observations in this framework is that light from galaxies at high redshift was emitted at earlier times in the history of the Universe. Although we have seen that redshift is related to distance by Hubble's law (for $z < 0.2$), it is the relationship between redshift and the **lookback time** – the time elapsed between the emission and observation of a galaxy – that is of more direct interest to astronomers studying the evolution of galaxies. Figure 2.30 shows how lookback time varies with redshift, assuming the currently well-accepted model for the expansion of the Universe. A galaxy with a redshift $z \approx 1$ corresponds to a lookback time of 8.0×10^9 years, while at $z \approx 10$, the lookback time is 13.5×10^9 years.

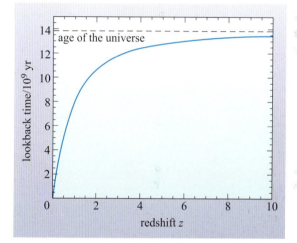

- ■ What is the age of the Universe at $z = 1$? If the fractional age at a given time is that time divided by the current age of the Universe, what is the fractional age of the Universe at $z = 1$? (Assume that the age of the Universe is 13.8×10^9 years.)

- ☐ Using Figure 2.30, at $z = 1$, the age of the Universe is
    $(13.8 \times 10^9 - 8.0 \times 10^9)$ years $= 5.8 \times 10^9$ years.
    The fractional age of the Universe is then
    $(5.8 \times 10^9$ years$)/(13.8 \times 10^9$ years$) = 0.42$ (or 42%).

Figure 2.30 The relationship between redshift z and lookback time for the currently accepted model of expansion of the Universe. Note that the age of the Universe in this model is 13.8 billion years.

So even at a rather modest redshift of 1, direct observations are possible that cover roughly half of the history of the Universe.

2.5.2 The origin of galaxies

The basic idea of galaxy formation is that the slight density fluctuations that were present in the early Universe have grown under the influence of gravity. As the Universe expanded the average density of the cosmic gas would have declined. However, against this background of a declining average density, some density enhancements of sufficient size, such as those shown in Figure 2.31, became more pronounced. Thanks to their own gravitational attraction, these density enhancements attracted matter from surrounding regions, increasing still further the lumpiness of the Universe. This process, known as **gravitational instability**,

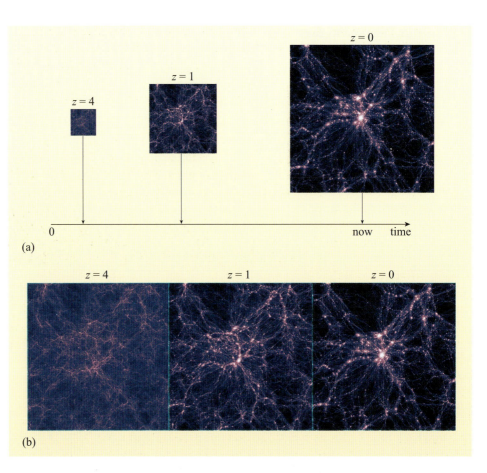

Figure 2.31 The effect of gravitational instabilities in a region of the expanding Universe dominated by dark matter. (a) The three panels illustrate how the simulated volume changes with time because of cosmological expansion. (b) The same panels as in (a) but now showing simulation volume at the same size so that the effects of gravitational collapse can be more clearly seen. Regions of enhanced density tend to grow along with the general cosmic expansion, but if sufficiently dense they may eventually defy the expansion and collapse. (Cosmological simulations from The Illustris Collaboration, 2014)

was therefore responsible for the production of localized regions in which clouds of cosmic gas and dark matter collapsed despite the general background of expansion. It is these collapsing clouds that are supposed to have been the 'seeds' of the galaxies and clusters of galaxies that surround us now.

So how did these collapsing clouds give rise to the galaxies that we see today? The simplest scenario for galaxy formation is that the collapse of a single over-dense region gives rise to a single galaxy. The mass contained in such a region would therefore correspond to the mass of the resulting galaxy. The mass would be mainly due to (non-baryonic) dark matter, but the region would also contain some baryonic matter, and this too would contribute to the galaxy. This baryonic matter would radiate away energy in ways that dark matter could not. Consequently, the baryonic matter settles into the centre of the dark-matter halo, where it forms the visible part of the galaxy. This type of formation process is often referred to as a **monolithic collapse** scenario. As we shall see later, this scenario has some attractive features, but at this point it is instructive to consider how the process of gravitational collapse is studied from a theoretical point of view, as this highlights other ways in which galaxies may have formed.

Since the way in which gravitational collapse proceeds depends on the distribution of mass, it is evident that the most dominant form of matter in the Universe – the dark matter – should play the key role in this process. The lack of knowledge about the nature of dark matter may seem an insurmountable problem, but astrophysicists have tackled this obstacle by making specific assumptions about the dynamical behaviour of the dark matter, and then investigating how a Universe that is dominated by this assumed form of dark matter would evolve. We shall return to discuss candidates for dark matter in Chapter 8, but we note here that it seems likely that the majority of the dark matter is in the form of fundamental particles that (at the time of writing) have yet to be detected. As you will see later, there are very good reasons for believing that the dark matter cannot consist of normal baryonic particles (protons, neutrons, etc.).

At this point we consider the two extremes for the dynamical behaviour of dark matter as it affects the growth of gravitational instabilities. In one, dark matter consists of slow-moving, massive particles. This kind of dark matter is referred to as **cold dark matter (CDM)**. The term 'cold' refers to the fact that the hypothetical dark matter particles have random speeds that are small compared with the speed of light, and this condition has applied throughout most of the history of the Universe. Computer-based simulations of the progress of gravitational collapse, in a Universe dominated by CDM, reveal that the first structures to form have masses of order $10^6 M_\odot$ – lower than those typically found for galaxies (typically $10^{11} M_\odot$) in the present-day Universe. As time progresses, larger-scale features develop by further collapse and by the merger of the lower mass structures that were formed previously. The overall picture is one in which proto-galactic fragments form early in the history of the Universe and many of the galaxies we see today are the result of mergers. This type of process is termed a **hierarchical scenario** or a **bottom-up scenario**, since galaxies are generally formed by the amalgamation of smaller entities.

The other extreme of behaviour is one in which the dark matter particles are rapidly moving, and goes by the name of **hot dark matter (HDM)**. The term 'hot' refers to the condition that the dark matter particles have speeds that are comparable to the speed of light. One effect of these high speeds is to wash out small-scale density fluctuations, leading to a distinctly different outcome from CDM models. A typical prediction of HDM models is that the first entities to form in the Universe would have much greater masses than individual galaxies. Structures with masses similar to present-day galaxies would form by the fragmentation of these larger entities. This type of process is called a **top-down scenario**. As we shall see in Chapter 4, there *is* structure on larger scales than individual galaxies – many galaxies exist within gravitationally bound clusters and within even more widely distributed large-scale structure. The major problem with HDM models is that in order to produce structures with masses that are typical of galaxies in the present-day Universe, they predict more structure on very large scales than is actually seen. Also, the notion that large-scale structure is formed before galaxies seems contrary to observation. Consequently, the HDM scenario seems to be ruled out by observation.

The currently favoured theory is called ΛCDM, which is a variant of the CDM scenario within a particular cosmological model – that is, a mathematical model that describes the expansion of the Universe. As you will see in Chapter 5, the expansion of the Universe is currently accelerating due to the influence of a property of space known as the *cosmological constant Λ* (a generalisation of this idea attributes the acceleration to a poorly-understood energy density of space termed *dark energy*). The 'Λ' of ΛCDM is taken to refer to expansion in which a cosmological constant (or dark energy) plays an important role.

In the ΛCDM model, structure in the Universe formed in a bottom-up scenario under the influence of CDM. In such a scenario it is expected that the first objects formed might be very high-mass stars, followed by structures on the scale of globular clusters. Galaxies would form by the merging of these smaller components.

Turning now to consider the morphology of galaxies within such scenarios, there are some general principles that are believed to hold. Simulations show that when a galaxy is formed by a merger of comparably-sized galaxies (sometimes called a **major merger**), then the morphology and the distribution of stellar velocities in the end product is typically as expected for an elliptical galaxy, even when the merging galaxies started off as spirals. If the two merging galaxies are very different in size (sometimes called a **minor merger**), a spiral disc in the larger galaxy can be preserved. For example, our own Milky Way is currently accreting several smaller galaxies, including the Large and Small Magellanic Clouds, and the Sagittarius dwarf galaxy (Section 1.6.3), but this has not disrupted the spiral disc of our galaxy. On the other hand, when the Milky Way eventually collides with the Andromeda Galaxy (M31) in five to ten billion years' time, the end result will likely be a giant elliptical galaxy.

The formation of spiral galaxies in hierarchical models is often thought to relate to the total angular momentum of the system. The amount of angular momentum per unit mass is observed to be higher in spiral galaxies than elliptical galaxies. To appreciate why this is so, it is useful to consider the monolithic collapse scenario in more detail. This scenario, in which a single cloud collapses to form a galaxy, provides a plausible account of the formation of spiral galaxies. If we consider an over-dense region that has acquired angular momentum from interactions with its neighbours, then collapse in the central parts would occur very rapidly to form a spheroidal distribution of stars that corresponds to the halo of the galaxy. However, in the outer parts of the collapsing region the orbits of gas clouds would intersect in the plane of rotation of the entire system. Collisions of such clouds would dissipate energy, leaving a thin rotating disc of gas.

This mechanism for forming a disc does not on its own explain why spiral galaxies have bulges, and why the bulges contain older stars than the spiral discs.

The likely solution to this problem is a scenario that combines elements of both the hierarchical growth and the monolithic collapse models. As in the hierarchical scenario, mergers produce elliptical galaxies of ever-increasing mass. However, there is also a long-term in-fall of gas from the environment surrounding these galaxies. If this gas has sufficient angular momentum, it forms discs around what were previously elliptical galaxies, thus transforming them into lenticular or spiral galaxies. Of course, at any time any two galaxies may undergo a merger, and thus (probably) form a new elliptical galaxy. An attractive feature of this scenario is that it provides a plausible explanation for the similarity of the elliptical galaxies and nuclear bulges of spiral galaxies – they are essentially the same types of structures, but in the case of a spiral a disc has grown around the nuclear bulge. The essence of this scheme is summed up in Figure 2.32, which shows a schematic history of a giant elliptical galaxy. Such a diagram, which illustrates how a galaxy is formed from the coalescence of many smaller galaxies, is termed a **merger tree**. At the time when this sequence starts – at the top of the tree – there are many low-mass elliptical galaxies in an environment that has a relatively high gas density. As time progresses, two types of process occur: there is a gradual in-fall of gas that forms discs, and there is a series of rapid merger events. The diagram indicates how the development of a disc depends on the time between merger events – a galaxy that has not undergone a merger for a long time is likely to develop a substantial disc.

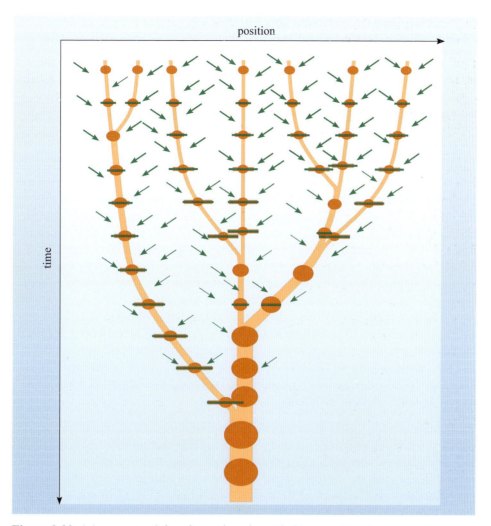

position

time

Figure 2.32 A 'merger tree' that shows the schematic history of the formation of a single giant elliptical galaxy by the merger of many smaller galaxies. The stellar content of galaxies is shown in dark orange. The steady accretion of neutral gas and of smaller galaxies in minor mergers is indicated in green. The result of every major merger event is an elliptical galaxy. The longer an individual galaxy goes without a major merger (those on the longest 'branches') the more substantial is the disc that develops by in-fall of intergalactic gas. Note also that as time passes, the density of intergalactic gas decreases, so discs grow more slowly at later times. (M. Merrifield (University of Nottingham))

When a major merger event occurs, an elliptical galaxy forms and all traces of the discs are wiped out. Furthermore, as time passes, the gas that originally surrounded these galaxies ends up within the galaxies. Thus the density of the intergalactic gas drops and so the rate at which discs form becomes slower with time. An attractive feature of this type of scenario is that galaxy formation starts to be seen as a continuing process rather than a one-off event that happened at a particular point in cosmic history.

So far, we have concentrated on the process of gravitational collapse – essentially, density enhancements grow by gravitational instability – and this behaviour can be modelled well in computer simulations. The greatest challenge in developing a theoretical model of galaxy formation is to incorporate the physical effects that arise once star formation starts in the Universe.

■ Apart from gravitational attraction, name three processes associated with massive stars that have an influence on the environment around these stars. What sort of effects do these processes have?

☐ Massive stars emit high-energy photons that are capable of ionizing any neutral gas nearby. Strong stellar winds from supergiants, and the supernovae that end the lives of some supergiants, have dynamical effects on any nearby material. Supernovae also result in a dramatic heating of the local interstellar medium as well as causing chemical enrichment.

The processes that occur on the scale of individual stars are believed to have an important effect on how galaxies form, but these effects are, as yet, rather poorly understood and difficult to incorporate into computer simulations. For instance, shock waves from early supernovae could compress gas clouds and trigger further star formation. On the other hand, the kinetic energy released by many supernova explosions could also have the effect of removing interstellar gas from protogalaxies, and this could inhibit further star formation. Such processes, which have a large-scale effect on star formation following some initial gravitational collapse, are referred to as **feedback**. In Chapter 3, you will meet another feedback mechanism, one that is related to the growth of supermassive black holes, much like the central black hole of the Milky Way in Section 1.5.4. Current modelling seems to indicate that the feedback associated with black hole growth plays a very important role in galaxy formation.

Predictions of what would happen in these kinds of complicated situations could, in principle, be made by a computer simulation that follows gas flows and individual stars within a galaxy. Such a level of detail is beyond the capabilities of even the most powerful computers available at present. However, significant progress has been made in recent years in developing simulations that accurately model the important physical processes by following sufficiently small mass elements that the morphology of simulated galaxies can be determined (Figure 2.33).

Simulations such as shown in Figure 2.33 are now starting to give astronomers confidence that many of the key physical processes in galaxy formation are reasonably well understood. Of particular importance in these simulations are the physical processes that determine whether the gas associated with particular galaxies can cool sufficiently to form the giant molecular clouds that are necessary for star formation.

Furthermore, such studies also offer insights into the merger process. Some of the key aspects of galactic mergers are expected to be as follows:

• A major merger between galaxies of similar masses tends to result in a galaxy with a morphology that is different from its progenitors. When one progenitor is much more massive than the other (a minor merger), the morphology of the resulting galaxy tends to resemble the more massive component.

• Discs are readily destroyed by major merger events. When a disc galaxy merges with a galaxy of much smaller mass (a minor merger), a likely result is that the disc will be thickened. (This effect was mentioned in Chapter 1 with respect to the formation of the Milky Way's thick disc.)

• The gas fraction of the progenitors has a large effect on the outcome of the merger. For gas-rich progenitors (so-called 'wet mergers') gas clouds will collide and be compressed, possibly leading to enhanced star formation. In gas-poor systems (so-called 'dry mergers'), stars will interpenetrate but no star formation will occur.

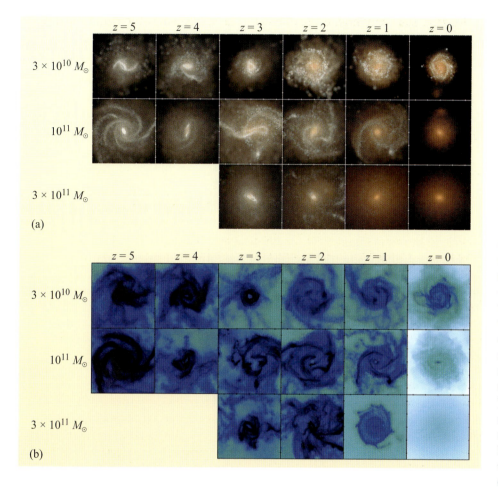

Figure 2.33 Examples of a galaxy formation from a numerical simulation for the growth of galaxies in a ΛCDM model. The simulation is based on gravitational collapse and various feedback processes which affect star formation. The examples show simulated high-mass galaxies at various redshifts. (a) The distribution of starlight. (b) The surface density of gas. Note that each simulated galaxy shown is an independent example (i.e. the diagram does not show how any particular galaxy changes with redshift). (The Illustris Collaboration, 2014)

While it is remarkable that modern simulations can reproduce structures that look like galaxies, a full test of such models is that they can reproduce, in a statistical sense, the populations of galaxies that we observe, not only in the local Universe, but also as we look back in time to higher and higher redshifts. In the remainder of this section we discuss how studies of the evolution of observable galaxies may allow astronomers to 'wind back the clock' and hence shed some light on the process of galaxy formation.

2.5.3 The evolution of isolated galaxies

The hierarchical scenario of galaxy formation suggests that interactions and mergers of structures smaller than galaxies must have occurred in the past. Hence the evolution of galaxies will depend both on those changes that are due to interactions and on those that are intrinsic to the galaxy itself. Since evolution of isolated galaxies and merger events are likely to have quite distinct observational consequences, we treat these two topics separately. Here we consider how an isolated galaxy might evolve with time, while interactions are the topic of the next section. Consider a hypothetical galaxy that has already formed and exists as a vast collection of stars, suffused by gas, dust, magnetic fields and radiation, all embedded in an overwhelming amount of dark matter. How would such a galaxy evolve? There are at least two important ways in which an isolated galaxy evolves. The study of each has become a specialized sub-field of galactic astronomy. Each is briefly described below.

Evolution of star formation rate

A key physical difference between elliptical and spiral galaxies is that the latter are sites of ongoing star formation. This qualitative difference in star formation rates is borne out by more detailed studies that measured the star formation rate (SFR) in different classes of galaxy. The SFR is usually quoted as the mass of stars formed per year. Figure 2.34 shows estimates of the star formation rates in ellipticals (and the bulge regions of spirals), and in the discs of spirals of types Sa, Sb and Sc. From this diagram it can be seen that the star formation rate in ellipticals (and bulges of spirals) was initially very high, but that it declined very rapidly. In fact, because star formation has effectively ended in elliptical galaxies, observational data concerning the star formation rate in such systems are very scarce.

The curve shown in Figure 2.34 for elliptical galaxies and bulges should be treated with a great deal of caution. It is clear that star formation has essentially stopped in such systems, but if elliptical galaxies are formed by mergers, then it is likely that many stars in such galaxies would have been formed prior to those merger events. Thus the age of an elliptical galaxy and the age of its stars may differ considerably.

In contrast to ellipticals, the star formation rate in the discs of spiral galaxies declines much more slowly after the formation of the galaxy. There is also a clear progression for different morphological types in that the star formation rate drops relatively quickly for type Sa galaxies, but is essentially constant for type Sc galaxies. Sb galaxies show intermediate evolution in star formation rate. Note that no curve is shown for irregular galaxies since the star formation rate in these systems varies considerably. However, it should be noted that at the present epoch many irregular galaxies do have high star formation rates, relative to their current total masses of stars.

Chemical evolution

The first stars, formed from the cosmic gas that filled the early Universe, should have consisted almost entirely of hydrogen and helium. In the terminology used in Chapter 1, these would have been stars belonging to Population III. As these stars burnt themselves out, some would have exploded in supernovae, enriching their surroundings with the products of stellar nucleosynthesis. This enriched material would have contributed to the formation of later generations of stars that would, consequently, have had a higher metallicity than the earliest stars. As this process continued, both the interstellar matter in a galaxy and the stars that were embedded in it would gradually change their chemical composition. The abundance of hydrogen would slowly decrease over time, whereas the abundances of helium and heavy elements would gradually increase.

- ■ Apart from supernovae, what other processes would you expect to contribute to the chemical enrichment of the interstellar medium?

- ☐ Apart from supernovae, all the other processes that cause stars to expel matter might be of relevance: stellar winds, the formation of planetary nebulae, and novae.

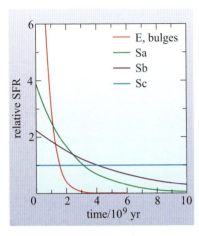

Figure 2.34 A schematic illustration of the star formation rate (SFR) as a function of age for galaxies of different types. Note that the time elapsed is the time since the event that formed the galaxy. The curve that represents the star formation rate in elliptical galaxies and the bulges of spiral galaxies is rather uncertain – the curve shown here should simply be taken as an indication that star formation in such systems took place as a rapid burst. (Kennicutt, 1998)

The study of chemical evolution is another highly technical field, particularly for spiral galaxies such as the Milky Way, where rates of star formation, and hence (presumably) of chemical enrichment, vary from one region to another. On the basis of chemical evidence it is possible to derive models for the changing rate of star formation, at least for the solar neighbourhood. These models depend on the extent to which matter from outside the Milky Way may have settled into the disc over time. This type of study may eventually help to establish whether it is likely that the discs of spiral galaxies developed from the gradual in-fall of material as outlined in Section 2.5.2.

2.5.4 The role of interactions and mergers in galaxy evolution

We have already seen that the favoured scenario for the formation of galaxies is one in which mergers and interactions play an important role, and here we look in detail at the observational consequences of such events. The numbers of galaxies that currently seem to be undergoing major interactions is rather low – only a few per cent of bright ($L > 10^{10} L_\odot$) galaxies. However, observational interest in these galaxies was heightened by the discovery that they tend to be sites of intense star formation.

Theoretical work on interactions began in the 1970s, when two Estonian brothers working in America, Alar and Juri Toomre, published the results of various simulations of **interacting galaxies**. The Toomres were soon followed by others, and a new field of research quickly developed, aided by the availability of ever larger computers. Some frames from one particular Toomre simulation are shown in Figure 2.35b, and the corresponding astronomical object – the so-called 'Antennae galaxies' (NGC 4038 and NGC 4039) – are shown in Figure 2.35a (overleaf). As you can see, the relevance of the simulation to the real world seems apparent and undeniable. Other peculiar systems have also been modelled with great success.

The widespread recognition that interactions might be of much more general importance followed the 1983 discovery, using the Infrared Astronomical Satellite (IRAS), that many galaxies shine more brightly at infrared wavelengths than they do at visible wavelengths. Enhanced infrared emission is frequently associated with interacting galaxies. It is believed that if the galaxies contain a relatively large fraction of cool gas, the merger can promote star formation. This gives rise to a population of high luminosity O and B type stars and hence HII regions. The ultraviolet radiation from the hot stars is absorbed by dust in the interstellar medium, which then re-radiates this energy in the infrared. A schematic illustration of this is shown in Figure 2.36. Galaxies that show evidence for high star formation rates (SFRs) compared to their current mass of stars are termed **starburst galaxies**, and many of these systems appear to be galaxies that are undergoing or have recently undergone some type of interaction or merger.

Figure 2.37 shows the results of an observation of the Antennae galaxies made using the Infrared Space Observatory (ISO). Contours of infrared emission (at 15 μm) are shown overlaid on a Hubble Space Telescope image of the central regions of the interacting pair (a larger-scale optical image of the galaxies, with their tails, is inset at the top right of the figure). One of the surprising features is that the peak of the infrared emission occurs at a different location from the maximum optical emission. In fact, about half of the entire luminosity of this system is radiated in the infrared from the region that appears dark on the optical image. This shows that not only is the UV light from young stars reprocessed by dust, but that the dust completely shrouds the intense star-forming regions.

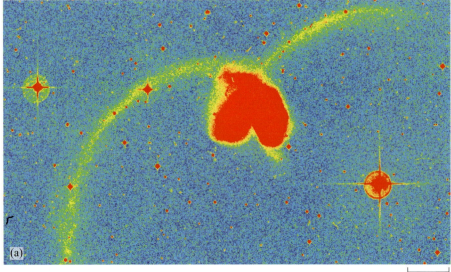

2 arcmin

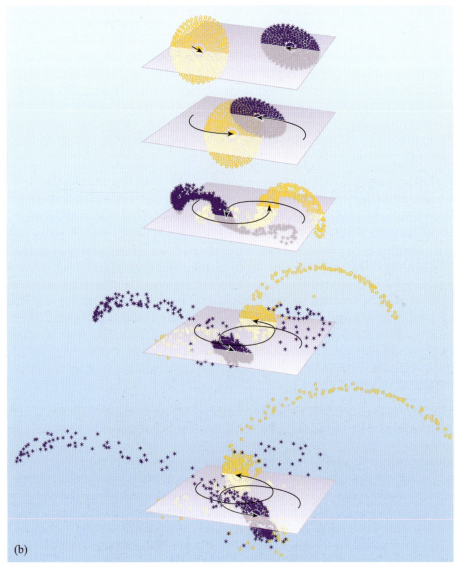

Figure 2.35 (a) The galaxies NGC 4038 and NGC 4039 are a colliding pair that go under the name of the Antennae galaxies. (b) A sequence of 'snapshots' from the pioneering numerical simulation carried out by Alar and Juri Toomre that successfully reproduces the main features of the Antennae galaxies. In the simulation, the colliding galaxies are represented by discs of stars, with each star interacting gravitationally with all the other stars in the simulated galaxies. The gravitational interaction of the two galaxies produces the 'tidal tails' of stars that account for the 'antennae' of NGC 4038 and NGC 4039. ((b) A. Toomre (MIT) and J. Toomre (University of Colorado))

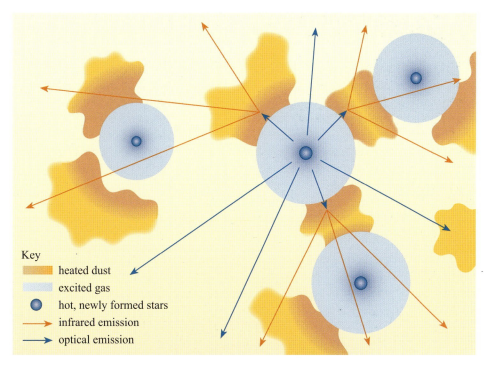

Key
- heated dust
- excited gas
- hot, newly formed stars
- → infrared emission
- → optical emission

Figure 2.36 A schematic diagram of HII regions and dust in a starburst galaxy. (For clarity, only the emission associated with one of the many hot stars in the HII region is shown.)

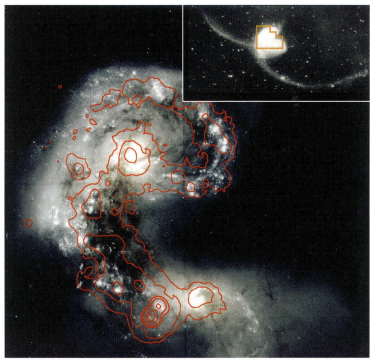

Figure 2.37 The Antennae galaxies, NGC 4038 and NGC 4039. The main image shows the infrared emission at 15 μm (as measured using the Infrared Space Observatory) as red contours that are overlaid on the optical image (obtained with the Hubble Space Telescope). The inset image at the top right is a large-scale optical image that shows the full extent of these interacting galaxies and their distinctive tails. The orange outline in the inset image shows the field of view of the Hubble Space Telescope that is shown in the main image. (Mirabel *et al*., 1998)

The reprocessing of stellar UV radiation by interstellar dust has dramatic consequences for the observation of galaxies in which intense bursts of star formation are going on. The dust absorbs UV and visible light from stars but it re-radiates all this energy as far-infrared radiation. Observations of interacting galaxies at the present time suggest that bursts of star formation may result in galaxies that appear to be very strong sources in the far-infrared, and may not be particularly strong sources in the UV and visible bands. As the sensitivity of far-infrared telescopes has improved, so more extreme examples of star-forming galaxies have been detected. The very brightest of these galaxies, which have luminosities in the far-infrared that exceed $10^{12}L_\odot$, are termed **ultra-luminous infrared galaxies**, or **ULIRGs**. A feature of ULIRGs in the local Universe is that their morphologies are almost always peculiar, which strongly suggests recent interaction. A key feature of these star-forming galaxies is the presence of cool gas in the progenitors; that is, they result from wet mergers.

It has already been noted in Section 2.5.2 that numerical simulations suggest the outcome of a major merger between two galaxies will be an elliptical galaxy, but to what extent is this borne out by observations? Of course, galactic mergers occur over a very long timescale, typically taking tens to hundreds of millions of years, so there is no possibility of observing an individual galaxy changing as a result of a collision. However there are galaxies that show evidence of collisions that took place in the past, and these provide key evidence to support the idea that ellipticals can be formed by mergers.

A galaxy that appears to have undergone a merger event in the past is NGC 7252. On the basis of a short exposure image it might be classified as an elliptical galaxy (Figure 2.38a). However a longer observation reveals a different picture: the low surface brightness parts of the galaxy show evidence of peculiar structure in this galaxy. In particular, there are long tails that are a tell-tale sign of a past collision (Figure 2.38b). Detailed studies of NGC 7252 reveal that the collision was between two spiral galaxies, and suggest that it is in the process of becoming an elliptical galaxy.

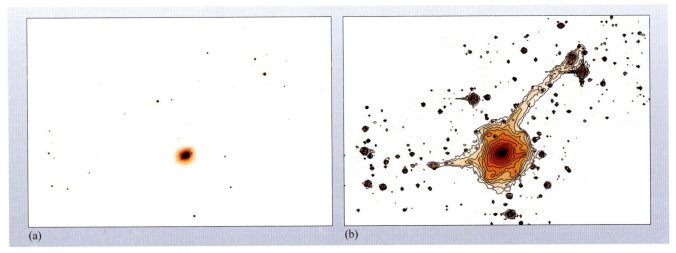

(a) (b)

Figure 2.38 The merger galaxy NGC 7252. Both images show the same field of view, but are sensitive to different levels of surface brightness. (a) Shows the brightest features of the galaxy (this is equivalent to taking a relatively short-exposure photograph). In this image the galaxy appears as an elliptical. (b) Shows features at very low surface brightnesses (this corresponds to taking a relatively long-exposure photograph). The features visible at low surface brightness show that the galaxy has undergone a merger event. (Data provided by NASA/IPAC Extragalactic Database from the observations of Hibbard *et al.*, 1994)

Historically, it was only morphologically peculiar galaxies that could easily be identified as systems that had undergone interactions and mergers. In the 1990s, however, developments in instrumentation opened up the possibility of detecting the aftermath of interactions in galaxies with apparently normal morphologies. The principle behind such studies lies in the fact that the stars from two interacting galaxies often retain distinctly different kinematic properties. In order to carry out such studies, a special type of instrument is used to map the spectrum at hundreds of locations over the face of a galaxy.

An example of a galaxy that has undergone a merger event, but shows no structural trace of interaction, is the galaxy NGC 4365. Figure 2.39a shows the surface brightness distribution of the galaxy, which appears to be a normal elliptical of type E3. By measuring the spectrum over the face of the galaxy a map can be formed that shows stellar velocities. This map is shown in Figure 2.39b; the blue areas represent regions in which stars are moving towards us and the red areas are where stars are moving away from us. In the outer part of the galaxy it appears as though the system is rotating around the long axis of the galaxy. However, in the inner parts of the galaxy the stars are rotating around the short axis. This strongly suggests that the galaxy was formed by the merger of two systems that had axes of rotation that were roughly perpendicular to one another. The map of spectra can also be used to study the stars that make up these two components, and it is found that the stars in both parts of the galaxy appear to have the same age, which is estimated to be in excess of 12×10^9 years. The inference drawn from this is that an ancient merger event caused a burst of star formation throughout the entire galaxy. While individual cases cannot prove that all elliptical galaxies formed by mergers, studies of this kind do lend support to the view that galaxy formation is the result of a hierarchical process.

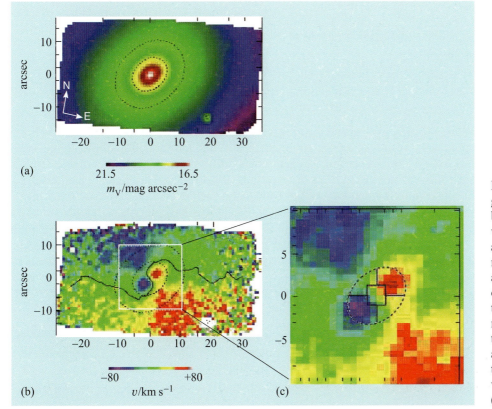

Figure 2.39 Observations of the galaxy NGC 4365. (a) The surface brightness distribution. (b) A velocity map of the same area as shown in (a); blue represents motion towards the observer and red is motion away from the observer. The colours indicate that the outer part of the galaxy is rotating around its long axis, while the inner region is rotating around an axis that is almost perpendicular to the long axis. (c) An enlarged view of the centre of the field of (b). (Davies *et al.*, 2001)

Finally, we note that there is evidence for mergers in which there is little gas available for star formation. Earlier we noted that cD galaxies tend to occur at the centres of massive clusters of galaxies. It is also the case that spiral galaxies are rare in such environments.

The large and very diffuse envelope that is characteristic of cD galaxies seems to be well explained by a process in which a massive elliptical grows by repeated mergers with lower-mass elliptical galaxies. This then, is an example of a 'dry merger' – one which has an effect on morphology, but with no associated star formation.

2.5.5 Observations of galaxy evolution by deep surveys

The most direct test of ideas of galaxy evolution is to make the most sensitive survey possible of a patch of sky well away from the plane of the Milky Way. If the field chosen is relatively free from any bright foreground stars or nearby galaxies then such a survey should detect galaxies over a range of redshifts. Sensitive surveys of this kind that include galaxies with large redshifts are known as **deep surveys**. Some of the best-known deep surveys have been conducted using the Hubble Space Telescope, starting with the Hubble Deep Field in the Northern Hemisphere (HDF-N), followed by the Ultra Deep Field (UDF) and the eXtreme Deep Field (XDF).

An image obtained for XDF is shown in Figure 2.40; it shows only some of the thousands of galaxies that were detected. Such surveys include many galaxies with redshifts between 1 and 3, and provide a sample of galaxies up to redshifts of about 8, and a candidate (i.e. not confirmed) galaxy at a redshift of 10.

■ What, approximately, was the age of the Universe when light was emitted from galaxies that are now observed with a redshift of 8?

☐ From Figure 2.30, at $z = 8$, the lookback time is 13.4×10^9 years. Hence the age of the Universe at $z = 8$ was $(13.8 - 13.4) \times 10^9$ years $= 4 \times 10^8$ years.

Thus deep surveys like the XDF provide a window back over cosmic history. The data from these surveys should provide information about various aspects of the way in which galaxies have evolved over the history of the Universe.

While the principle behind the observations is essentially simple – the observations are long exposures of seemingly blank fields – the interpretation of the image data is not at all straightforward. There are many difficulties in dealing with data from such surveys: not because the observations or instruments are flawed, but because the observations are at the limit of what can be achieved with current technology. One difficulty is that many of the observed sources have very low flux densities. As a result, many galaxies are too faint for their spectra to be measured, even using the largest available telescopes. This has meant that astronomers have had to develop techniques by which the redshift of a galaxy can be estimated from its broadband colours. A second major problem is that even though the Hubble Space Telescope has excellent angular resolution, many galaxies are only just resolved, so the task of determining their morphological class is very difficult.

Another aspect of deep surveys relates to the high redshifts of the observed galaxies. The relationship between the observed and emitted wavelengths and the redshift z of a galaxy is given by Equation 2.6. This can be rearranged to give

$$\lambda_{em} = \frac{\lambda_{obs}}{(1 + z)} \tag{2.9}$$

Figure 2.40 The XDF (eXtreme Deep Field) combines data from multiple deep surveys with the Hubble Space Telescope. This image shows a region with an area of about 4.7 arcmin^2, in which galaxies at redshift 8 can be observed and a candidate galaxy at $z \approx 10$ has been identified. (NASA/ESA, G. Illingworth, D. Magee, and P. Oesch, R. Bouwens, and the HUDF09 Team)

■ A galaxy with $z = 1$ is observed at a visual wavelength of 500 nm. At what wavelength was this light emitted? In what part of the spectrum is this?

□ Using Equation 2.9, the emitted wavelength was $\lambda_{em} = 500$ nm$/(1 + 1) = 250$ nm. This corresponds to the UV part of the electromagnetic spectrum.

Thus the observed visual images of many galaxies actually correspond to their appearance in the UV. This effect – that the observed image corresponds to emission at a quite different wavelength – is often referred to as **band-shifting**. UV emissions from galaxies tend to be dominated by the radiation from massive, short-lived stars. So basing the morphological classification of a high-redshift galaxy on a visual image could result in the galaxy being assigned to an inappropriate morphological class. On the other hand, the fact that UV emission can be observed directly (albeit shifted into the visible part of the electromagnetic spectrum) has benefits for determining star formation rates as we shall shortly see.

Despite the challenges posed by deep survey data, some results are beginning to be accepted by the wider astronomical community. Here we concentrate on two results from deep surveys.

Morphological change with redshift

If mergers do play an important role in the formation of the population of galaxies that we observe at the present time and, in particular, if massive ellipticals are formed by merger events, then we might expect that the relative numbers of elliptical, spiral and peculiar galaxies might vary over cosmic history. Most notably, the fraction of massive elliptical galaxies would be expected to increase with time as they are formed by the merger of galaxies of all types.

This is indeed what seems to be found. Deep surveys allow astronomers to select samples of galaxies according to redshift. Figure 2.41 shows the results of one such study, which combined data from a wide-area survey of local galaxies with two deep surveys made using the Hubble Space Telescope. It is clear that the fraction of ellipticals (combined here with lenticulars) has increased substantially between $z \approx 1$ and the present day. It can also be seen that the number of galaxies that are peculiar, in the sense that they are likely to be interacting, was much higher in the past. The picture that emerges is that at a lookback time of 8 billion years ($z = 1$), the galaxy population was quite different to the present day, and in a qualitative sense at least, the merger scenario offers some explanation for the change in population. However, many details remain to be understood about the morphological transformations of galaxies.

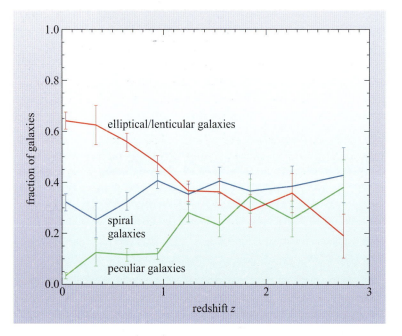

Figure 2.41 The observed change in the fraction of massive galaxies ($M > 10^{11} M_\odot$) by morphological type with redshift. (Buitrago *et al.*, 2013)

The star formation history of the Universe

Deep surveys provide a way of investigating how the star formation rate within galaxies has varied over the history of the Universe. We have already seen that bursts of star formation result in the creation of stars across a wide range of masses. The luminous output from a star-forming region is dominated by the emission from high-mass stars. Thus the UV luminosity of a galaxy may be used to provide a measure of its star formation rate (assuming one can accurately estimate the formation rate of all stars from measurements of the most massive ones). If a survey is made of star formation rates of galaxies as a function of their measured redshift, then it is possible, in principle, to determine how the rate of star formation has varied with cosmic history.

■ What is a potential problem in determining star formation rate from the UV luminosities in galaxies?

☐ The UV luminosity may not be a good tracer of the star formation rate. We know from studies of starburst galaxies in the local Universe that intense bursts of star formation are often shrouded in dust. Consequently, when observing regions where the star formation rate is high, much of the emerging radiation may be in the far-infrared.

Despite this problem, deep optical surveys, either from the Hubble Space Telescope or from large ground-based telescopes, provide a wealth of data on the UV emission from galaxies at high redshift. Consequently, a great deal of effort has gone into the careful analysis of such data, by correcting as well as possible for the effects of dust extinction. This has resulted in measurements of the variation of the cosmic star formation rate with redshift, up to $z \approx 8$.

An alternative approach would be to use the far-infrared emission of such galaxies, and this has been possible with infrared space telescopes such as Spitzer and Herschel. It can often be more difficult to make redshift measurements for galaxies discovered at far-infrared wavelengths, because the dust obscuration makes emission lines difficult to detect with optical telescopes. This situation is changing with sub-millimetre wave and radio telescopes, such as the Atacama Large Millimetre/sub-millimetre Array (ALMA), which can detect emission lines at wavelengths unaffected by dust extinction. In the meantime, the results of infrared surveys are in good agreement with the UV results where they overlap in redshift. The variation in cosmic star formation rate with redshift, as determined by both UV and far-infrared surveys, is shown in Figure 2.42.

Figure 2.42 shows that the star formation rate has varied dramatically over the history of the Universe. The maximum rate occurred at a redshift of about 2, corresponding to a lookback time of about 10 billion years. Since that time, the star formation rate has dropped by a factor of about 10 to the value we observe at the present time. Studies such as those illustrated in Figure 2.42 indicate that throughout the Universe, about 50% of stars were formed by a redshift of 1.3, and that 25% of stars were formed after $z \approx 0.7$.

This diagram can also be compared to population synthesis models of galaxies (Section 2.3.3), to see whether the star formation rate measurements agree with the stellar populations that would be expected within galaxies over cosmic history. This is an area of active research, but broadly speaking the results of such modelling show that the star formation history of the Universe is consistent with galaxies being formed by interactions and mergers over a long spread of time. One scenario that seems to be ruled out is the idea that galaxies were formed at one single time in the history of the Universe. Thus the results of studies of the star formation rate in the Universe also seem to support the hierarchical model of galaxy formation.

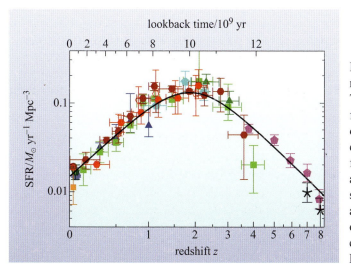

Figure 2.42 The star formation rate (SFR) as a function of redshift z (lower axis) and the lookback time (upper axis). The star formation rate is defined as the mass of stars that are formed per year in a volume that is currently 1 Mpc³. Because of the expansion of the Universe, a volume that currently encloses 1 Mpc³ would have been smaller in the past: the star formation rate shown here takes this change in volume into account. The different symbols indicate results from different surveys, but the red, deep red and orange data points are from analysis of infrared emission from galaxies whilst all the others are based on UV emission. The bars drawn through each data point indicate estimated uncertainties. (Madau and Dickinson, 2014)

2.6 Summary of Chapter 2

Morphology

- Mainly according to their shape, most galaxies can be assigned to one of four different classes: elliptical, lenticular, spiral or irregular.

- In the modified form of the Hubble classification scheme, shown in Figure 2.2, the spirals and lenticulars can be subdivided into barred and unbarred subclasses. The spirals can be further divided into a number of Hubble types. The ellipticals are also divided into a number of Hubble types.

- Irregular galaxies are generally chaotic and asymmetric, although some exhibit a bar and others show some traces of spiral structure.

- Over 60% of galaxies are elliptical, fewer than 30% are spiral, and fewer than 15% irregular. Dwarf ellipticals are the most common type.

Physical properties of morphological classes

- Elliptical galaxies are essentially ellipsoidal distributions of old (Population II) stars, almost devoid of cold gas and dust. Their three-dimensional shape is difficult to determine, but some at least appear to be triaxial ellipsoids with very little rotation. Some of the smaller ellipticals may be oblate spheroids.

- Lenticular galaxies appear to be an intermediate class between the most flattened of elliptical galaxies and the most tightly wound spirals. They show clear signs of a disc and a central bulge, but they have no spiral arms and little cold interstellar gas.

- Spiral galaxies have a disc, a central bulge and often a central bar. Within this class, spiral arms may be more or less tightly wound and the bulge may be more or less prominent in relation to the disc. (The Milky Way is a spiral galaxy and was traditionally described as being of Hubble type Sb or Sc, but it is now known to have a bar and is probably best described as type SBbc.)

- The largest normal galaxies are the cD galaxies – giant ellipticals that may have been formed in mergers and are often found close to the centres of clusters of galaxies.

The measurement of the physical properties of galaxies

- The surface brightness of galaxies varies from point to point. Continuous lines passing through points of equal surface brightness are called isophotes. As it is often difficult to determine the edge of a galaxy, observations of galaxies are often confined to the region within some specified isophote.

- Empirical relations obtained from observations of nearby galaxies are used to estimate quantities such as the luminosity and angular size of a distant galaxy, on the basis of its flux density within a given isophote.

- Galactic masses are generally hard to measure. However, the methods that may be used to determine them include the use of rotation curves for spirals, velocity dispersions and X-ray halos for ellipticals, and velocity dispersions for clusters of galaxies.

- The masses of galaxies are of the order of ten times greater than the estimated masses of the matter that can be attributed to stars and interstellar gas, indicating that they contain substantial amounts of dark matter.

- The stellar content of galaxies can be estimated through the process of population synthesis.

The distance scale

- Determining the distances of galaxies is of great importance in astronomy. Distance information can be crucial to the determination of other galactic properties, and it plays a vital part in investigations of the large-scale distribution of galaxies.

- There are many different methods of distance determination. Those applicable to galaxies include geometrical methods, standard candle methods and the redshift method based on Hubble's law.

- The various methods, taken together, form a galactic distance ladder in which the calibration process that most steps require is usually dependent on the accuracy of other steps lower down the ladder. This means that uncertainties tend to increase as the ladder is climbed.

- The distances of remote galaxies may be determined using Hubble's law, which relates distance to redshift. The calibration of this method (i.e. the process of determining the Hubble constant) is such that distances can be determined with an uncertainty of about 10%.

The formation and evolution of galaxies

- Galaxies are thought to have formed as a result of the gravitational instability of the expanding gas of baryonic and non-baryonic matter produced by the big bang.

- Galaxy formation can be studied theoretically using numerical techniques that extrapolate forward in time from the conditions of the early Universe. The major uncertainty in such an approach lies in incorporating the many processes that affect star formation. The most likely scenario for galaxy formation is of a bottom-up process in which cold dark matter played a vital role. A feature of this scenario is that interactions and mergers play an important role in galaxy formation.

- Interacting galaxies at the present epoch may be sites of intense star formation.

- Galaxy evolution can be studied observationally by performing deep surveys that sample a large fraction of cosmic history. Such studies can reveal how the distribution of galaxies between different morphological types has changed with time, and can give information about the evolution of the star formation rate in the Universe.

- The results of deep surveys are generally in accord with a hierarchical scenario of galaxy formation.

Questions

QUESTION 2.11

If the ellipse shown in Figure 2.3 represented the outline of an elliptical galaxy, what would be its Hubble type?

QUESTION 2.12

Figure 2.43 shows three different galaxies. On the basis of these images alone, what would you expect the Hubble types of the three galaxies to be?

(a)

(b)

Figure 2.43 Three optical images of galaxies (for use with Question 2.12). (NOAO)

(c)

QUESTION 2.13

Why does the fact that elliptical galaxies have elliptical outlines imply that their stars occupy an ellipsoidal volume of space?

QUESTION 2.14

List some of the shortcomings of standard candle methods for finding distances.

QUESTION 2.15

Sketch some typical isophotal contours of (a) an E4 galaxy, and (b) a face-on S0 galaxy.

QUESTION 2.16

Why is it not possible to use the rotation curve method to determine the masses of elliptical galaxies?

QUESTION 2.17

(a) Why does it make good sense to model the stellar population of the central bulge of M31 separately from that of its disc?

(b) If asked to carry out a population synthesis of an E2 galaxy, which of the stellar categories listed below would you expect to be well represented? Explain your answer.

(i)   Lower main sequence stars

(ii) Red giants

(iii) Upper main sequence stars

(iv) Cepheid variables

QUESTION 2.18

Figure 2.44 shows a sort of Hertzsprung–Russell (H–R) diagram that has been divided into cells corresponding to various ranges of temperature and luminosity. Within some of the cells, a block has been drawn. The size of each block is proportional to the number of stars in a certain galaxy that fall within the given range of temperature and luminosity. The three parts of Figure 2.44 (a, b and c) represent, in a random order, population models for various stages in the evolution of a galaxy that formed all its stars in one huge burst of star formation. Write down the correct chronological sequence of the diagrams and explain the effects that such evolution would have on the appearance of the galaxy.

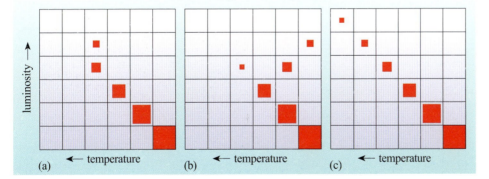

Figure 2.44 Galactic populations on a cellular H–R diagram. Note that these are not in chronological sequence.

CHAPTER 3
ACTIVE GALAXIES

3.1 Introduction

Even in images taken with the most modern equipment on a large telescope, it can be difficult to pick out the galaxies now known as 'active' from the other more normal galaxies. But if your telescope were equipped to examine the *spectra* of the galaxies, then the active galaxies would stand out. Normal galaxies contain stars that are generally similar to those in our own Galaxy; and spiral galaxies have additional similarities to the Milky Way in their gas and dust content. Active galaxies show extra emission of radiation, and this is most apparent from their spectra.

Active galaxies come in a variety of types, including Seyfert galaxies, quasars, radio galaxies and blazars. These types were discovered separately and at first seemed quite different, but they all have some form of spectral peculiarity. There is also evidence in each case that a very large amount of energy is being released in a region that is *tiny* compared with the size of the galaxy. It is usually found that the tiny source region can be traced to the nucleus of the galaxy, so the origin of the excess radiation is attributed to the **active galactic nucleus** or **AGN**. An active galaxy may be regarded as a normal galaxy *plus* an AGN with its attendant effects. As you will see, the different types of observed active galaxy can, to a large degree, be explained in terms of just two underlying *unified models* for AGN.

Note that the plural term 'active galactic nuclei' is also designated by the abbreviation AGN.

Galaxies that are clearly active seem to be quite rare in the nearby Universe, comprising only about 2% of the total population of galaxies. However, as we shall see, there is increasing evidence that most, if not all, massive galaxies must have passed through an active phase at some point.

The **engine** that powers the AGN, the tiny nucleus of the active galaxy, is an enigma. It has to be able to produce 10^{11} or more times the power of our own Sun, but it has to do this in a region little larger than the Solar System. To explain this remarkable phenomenon, a remarkable explanation is required. This has proved to be within the imaginative powers of astronomers, who have proposed that the engine consists of an *accreting supermassive black hole*, around which gravitational energy is converted into electromagnetic radiation.

In Section 3.2 you will learn how spectroscopy can be used to distinguish different kinds of galaxy and to measure their properties. Section 3.3 then introduces the four main classes of active galaxies and describes how they can be recognized. Section 3.4 examines the evidence for the existence of black holes at the centres of active galaxies, and in Section 3.5 you will study a simple model that attempts to explain the key characteristics of active galaxies in an illuminating way. Finally, in Section 3.6, we consider some of the outstanding questions about the origin and evolution of active galaxies. We begin by looking at the spectra of galaxies.

3.2 The spectra of galaxies

This section reviews what you have already encountered about the spectra of galaxies. The topic will be further developed to equip you to appreciate the spectra of active galaxies.

■ List the four main constituents of a galaxy.

☐ Dark matter, stars, gas and dust.

Congratulations if you remembered dark matter! But even though it is the main constituent of a galaxy, dark matter does not contribute to the spectrum of the galaxy so we need not consider it any further. The spectrum of a galaxy contains contributions from stars, gas and (sometimes) dust.

The spectrum of a *star* normally consists of a continuous thermal spectrum with absorption lines cut into it (Figure 3.1). As you probably know, it is possible to learn a lot about the star from a study of these absorption lines.

■ List what can be learned about a star from its absorption lines, briefly indicating the measurements that would need to be made on the lines.

☐ The strengths and widths of the absorption lines contain information about the star's chemical composition, surface temperature and luminosity. By looking for Doppler shifts in the lines you can measure radial velocity and, if the Doppler shifts are periodic in time, you can detect the binary nature of a star.

The *gas* in a galaxy is partly visible in the form of hot clouds known as HII regions, which you came across in Chapter 1. Such regions are usually only seen where there is ongoing star formation, and so are prominent in spiral and irregular galaxies. The optical spectrum of an HII region consists of just a few emission lines, as can be seen in Figure 3.2. HII regions can make a substantial contribution to the spectra of galaxies because they are very bright. The only other gaseous objects in a normal galaxy to emit at optical wavelengths are supernova remnants and planetary nebulae, and these are faint compared with HII regions.

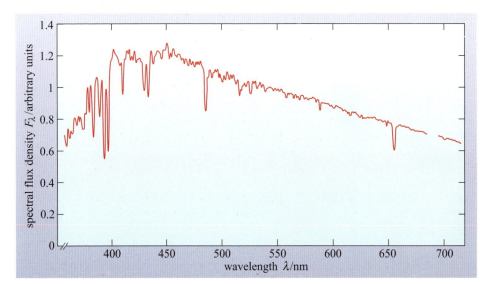

Figure 3.1 The optical spectrum of a star – in this case of spectral type F5 – shown as the spectral flux density, F_λ, plotted against wavelength. (From data described in Silva and Cornell, 1992)

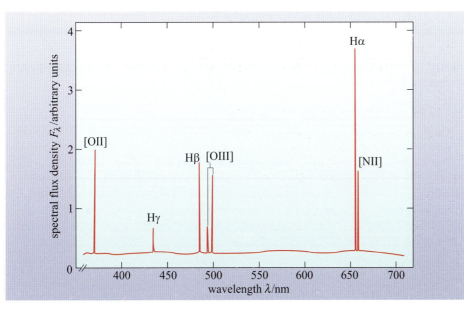

Figure 3.2 The schematic spectrum of a typical HII region, showing emission lines. HII denotes a singly ionized hydrogen atom, NII represents a singly ionized nitrogen atom, and OII and OIII denote singly and doubly ionized oxygen atoms. [NII], [OIII] and [OII] denote particular electronic transitions in these ions – the meaning of the square brackets is explained in Section 3.2.1. $H\alpha$, $H\beta$ and $H\gamma$ are the first three Balmer lines of hydrogen.

The *dust* component of a galaxy, being relatively cool, does not lead to any emission features in the optical spectrum of a galaxy. The main effect of dust at optical wavelengths is to absorb starlight. However, as you saw in Chapter 1, dust can emit strongly at far-infrared wavelengths (λ of about 100 µm).

As a rule, optical absorption lines result from stars, and optical emission lines result from hot gas.

The spectra of stars and HII regions extend far beyond the optical region. The Sun, for example, radiates throughout the ultraviolet, X-ray, infrared and radio regions of the electromagnetic spectrum. The majority of the Sun's radiation is concentrated into the optical part of its spectrum but, as you will shortly see, this is not the case for active galaxies, for which it is necessary to consider all the observed wavelength ranges. We shall call this the **broadband spectrum** to distinguish it from the narrower *optical* spectrum. You will recall that the word optical means visible wavelengths plus the near ultraviolet and near infrared wavelengths that can be observed from the ground, and extends from 300 to 900 nm. The optical spectrum is just one part of the broadband spectrum, albeit an important part. The spectrum of a normal galaxy is the composite spectrum of the stars and gas that make up the galaxy. Some of the absorption lines of the stars and some of the emission lines of the gas can be discerned in the galaxy's spectrum. As well as being able to work out the mix of stars that make up the galaxy, astronomers can measure the Doppler shifts of these spectral lines and so work out the motions within the galaxy as well as the speed of the galaxy through space.

In the case of active galaxies, the spectrum shows features *in addition* to those of normal galaxies, and it is from these features that the active nucleus of the galaxy can be detected.

3.2.1 Optical spectra

Normal galaxies

Normal galaxies are made up of stars and (in the case of spiral and irregular galaxies) gas and dust. Their spectra consist of the sum of the spectra of these components.

The optical spectra of normal stars are continuous spectra overlaid by absorption lines (Figure 3.1). There are two factors to consider when adding up the spectra of a number of stars to produce the spectrum of a galaxy. First, different types of star have different absorption lines in their spectra. When the spectra are added together, the absorption lines are 'diluted' because a line in the spectrum of one type of star may not appear in the spectra of other types. Second, Doppler shifts can affect all spectral lines. All lines from a galaxy share the redshift of the galaxy, but Doppler shifts can also arise from motions of objects within the galaxy. As a result, the absorption lines become broader but shallower. Box 3.1 explains how this Doppler broadening comes about.

HII regions in spiral and irregular galaxies (though not, of course, ellipticals) shine brightly and contribute significantly to the spectrum of the galaxy. The optical spectrum of an HII region consists mainly of emission lines, as in Figure 3.2. When the spectra of the HII regions and the stars of a galaxy are added together, the emission lines from the HII regions tend to remain as prominent features in the spectrum unless a line coincides with a stellar absorption line. Of course, the Doppler effect applies to spectral lines whether they are in emission or absorption, so emission lines too are broadened because of the motion of HII regions within a galaxy.

BOX 3.1 DOPPLER BROADENING

The Doppler effect causes wavelengths to be lengthened when the source is moving away from the observer (*red-shifted*) and shortened when the source is moving towards the observer (*blue-shifted*). Light from an astrophysical source is the sum of many photons emitted by individual atoms. Each of these atoms is in motion and so their photons will be seen as blue-shifted or red-shifted according to the relative speeds of the atom and the observer. For example, even though all hydrogen atoms emit Hα photons of precisely the same wavelength, an observer will see the photons arrive with a spread of wavelengths: the effect is to broaden the Hα spectral line – called **Doppler broadening**.

In general, if the emitting atoms are in motion with a range of speeds Δv along the line of sight to the observer (the *velocity dispersion*) then the Doppler broadening is given by

$$\Delta\lambda/\lambda \approx \Delta v/c \qquad (3.1)$$

where c is the speed of light, and λ is the central wavelength of the spectral line.

Why would the atoms be in motion? An obvious reason is that they are 'hot'. Atoms in a hot gas, for example,

will be moving randomly with a range of speeds related to the temperature of the gas. For a gas of atoms of mass m at a temperature T, the velocity dispersion is given by

$$\Delta v \approx \left(\frac{2kT}{m}\right)^{1/2} \qquad (3.2)$$

where k is the Boltzmann constant (1.38×10^{-23} J K^{-1}).

QUESTION 3.1

Calculate the velocity dispersion for hydrogen atoms in the solar photosphere (temperature $\sim 6 \times 10^3$ K). Then work out the width in nanometres of the Hα line (656.3 nm) due to thermal Doppler broadening.

It is very common for Doppler broadening to be expressed as a speed rather than $\Delta\lambda$ or even $\Delta\lambda/\lambda$. So astronomers would say that the width of the solar Hα line is about 10 km s^{-1}.

You can also see that thermal Doppler broadening depends on the mass of the atom so, for the same temperature, hydrogen lines will be wider than iron lines.

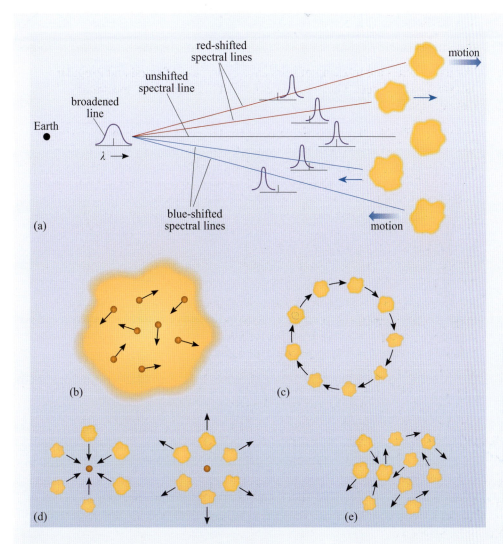

Figure 3.3 Doppler broadening arises when the source of a spectral line contains atoms moving at different speeds along the line of sight (a). This can be due to (b) thermal motion of atoms in a gas, (c) rotational motion of a galaxy, (d) inflow or outflow of gas from a centre, (e) chaotic motion in a gas cloud.

Thermal motion is not the only way in which a velocity dispersion can arise. Bulk movements of material can also broaden spectral lines.

■ What kinds of bulk motions could give rise to Doppler broadening?

□ For a line to be broadened, the emitting atoms must be moving at different speeds along the line of sight. This could occur where a gas cloud is rotating, where gas is flowing inwards or outwards from a centre, or where gas is in turbulent or chaotic motion.

So a galaxy rotating about its centre will produce a spectrum in which the lines are broadened. Normal galaxies have Δv values of between 100 and 300 km s^{-1}, which you can see is far higher than thermal motions in a hot gas such as the Sun's photosphere.

Whether the bulk motion is a rotation, an infall, an outflow, or just turbulence makes no difference; the net effect will be a broadened line whose width is proportional to the range of velocities present.

■ How might you distinguish thermal broadening in a spectrum from broadening due to bulk motions?

□ Thermal broadening depends on the mass of the individual emitting atoms (heavy atoms move more slowly) so lines from different elements will have different values of $\Delta\lambda/\lambda$. Broadening from bulk motion will affect all spectral lines equally, they will have the same value of $\Delta\lambda/\lambda$.

Doppler broadening applies equally to emission and absorption lines. The broadening is due to the motion of the emitting or absorbing atoms (Figure 3.3).

From the Galactic rotation curve in Figure 1.11c, estimate the broadening of lines from our own Galaxy in km s^{-1} if it were observed edge-on by an astronomer situated in a distant cluster of galaxies. (Assume that our Galaxy is not spatially resolved in such observations.)

The term 'forbidden line' arose from quantum theory. The permitted lines all obey a certain set of rules in that theory, whereas the forbidden lines break these rules.

One more feature of emission lines from HII regions needs to be mentioned, and that is the presence of so-called **forbidden lines**, as opposed to the others, which are called permitted lines. Most spectral lines that are seen astronomically can be produced in regions of either high or low gas density. Forbidden lines are produced only in regions of very low density; this is because the excited states responsible for their production are so long-lived that, at higher densities, the atom or ion is likely to be de-excited by collision with another particle before a photon can be emitted spontaneously. Such low densities cannot be achieved on Earth, which is why these lines are not observed in the laboratory. When they are observed astronomically, we can be sure that they have been produced in a region of extremely low density. They are prominent in the spectra of active galaxies and are denoted by square brackets [ ]. Strong forbidden lines seen in HII regions include [NII] at 655 nm and [OIII] at 501 nm (see Figure 3.2).

So what will the spectrum from a normal galaxy look like? It depends what kind of galaxy it is. The optical spectrum of an *elliptical galaxy* is a continuous spectrum with absorption lines. Such spectra are typically similar to the population synthesis model that you saw in Figure 2.19 for ages 5×10^9 years or greater. Sensitive observations of elliptical galaxies typically reveal the presence of many absorption lines, although these lines are somewhat broader and shallower than those seen in individual stellar spectra. There are usually no emission lines, because elliptical galaxies have no HII regions. The overall shape of the spectrum looks like that of a K-type (fairly cool) star because cool giant stars dominate the luminous output of the galaxy.

Figure 3.4 The optical spectrum of the normal spiral galaxy NGC 4750. It shows absorption lines and some emission lines. (Note that because of the Doppler shift caused by the motion of the galaxy, a particular spectral line is not necessarily at the same wavelength in all the figures in which it appears. Also note that this is a real, and not a schematic spectrum. Consequently this trace is more erratic than a schematic spectrum because of the presence of many faint absorption lines and the effect of instrumental noise.) (Kennicutt, 1992)

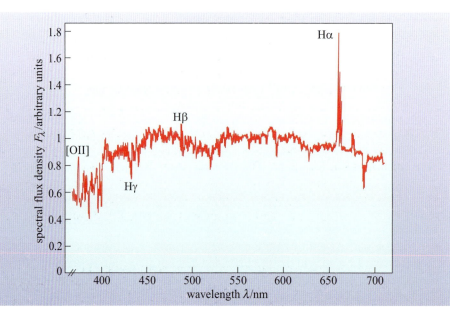

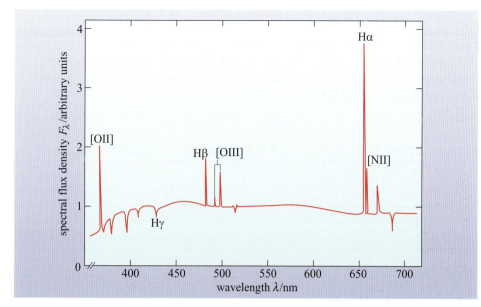

Figure 3.5 Spectrum of a mystery galaxy shown schematically. Note the strong emission lines, which have approximately the same width as those in normal spiral galaxies.

The optical spectrum of a *spiral galaxy* consists of the continuous spectrum from starlight with a few shallow absorption lines from stars, plus a few rather weak emission lines from the HII regions. Figure 3.4 shows an example. Note that the Hα line in this spectrum is a result of both absorption from stars and emission from HII regions.

■ Why has there been no mention of dust so far?

☐ Because we are only discussing optical spectra. Other than dimming the starlight, dust has no prominent emission or absorption lines in the optical region.

Before moving on to consider the spectrum of active galaxies, look at the spectrum in Figure 3.5.

■ How does the spectrum of the mystery galaxy in Figure 3.5 compare with those in Figures 3.2 and 3.4? How would you interpret the difference?

☐ The spectrum in Figure 3.5 shows very strong emission lines, similar to the spectrum of an HII region in Figure 3.2. Although the stellar absorption spectrum is present, the line spectrum is dominated by HII regions rather than stars. It looks like a galaxy with more HII regions than normal.

In fact, Figure 3.5 is the spectrum of a *starburst galaxy*. In Chapter 2 you saw that starburst galaxies are otherwise normal galaxies that are undergoing an intense episode of star formation. They contain many HII regions illuminated by hot, young stars, and the emission lines show up clearly in the optical spectrum. We mention starburst galaxies here because, as you will see, their spectra have a resemblance to active galaxies, and it is important to be able to distinguish them.

Active galaxies

Figure 3.6 (overleaf) shows an optical spectrum of an example active galaxy. It is immediately apparent that the emission lines are stronger and broader than in the spectrum of a normal galaxy shown in Figure 3.4. They are also broader than in the spectrum of the starburst galaxy in Figure 3.5. It is as if a component producing strong and *broad* emission lines had been added to the spectrum of Figure 3.5.

■ From what you have learned so far, what might be the nature of this component?

☐ The strong emission lines suggest that the galaxy contains hot gas similar to an HII region. The broad lines imply that the gas must be either extremely hot or in rapid motion.

Now answer Question 3.3, which will help you decide which of these two possibilities is the more likely.

QUESTION 3.3

Measure the wavelength and width of the Hβ line in Figure 3.6 (at half the height of the peak above the background) and so make a rough calculation of the velocity dispersion of the gas that gave rise to it. If the line widths are due to thermal Doppler broadening, estimate the temperature of the gas.

The answer to Question 3.3 is quite surprising. Not only is the implied temperature higher than the *core* temperatures of all but the most massive stars, it is also inconsistent with the process by which Hβ emission occurs, since at such temperatures any hydrogen would be completely ionized. In fact, the relative strengths of various emission lines can be used to estimate the temperature of the gas, and this is found to be only about 10^4 K. So the broadening cannot be thermal. The alternative explanation is *bulk* motions of several thousand kilometres per second. These are very large velocities indeed, and imply that large amounts of kinetic energy are tied up in the gas motions. We shall return to the nature of these motions later in this chapter.

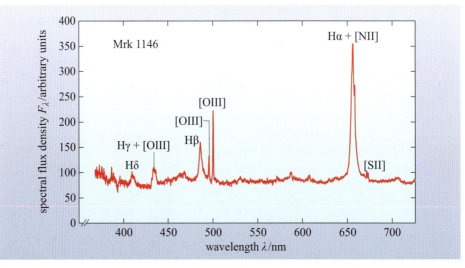

Figure 3.6 The optical spectrum (adjusted to zero redshift) of the active galaxy Markarian 1146. Note the strong and broad emission lines, especially the hydrogen lines Hα, Hβ and Hγ. The forbidden lines remain narrow ([OIII] at $\lambda = 436$ nm is almost coincident with Hγ). (Data from Sloan Digital Sky Survey, Data Release 6)

3.2.2 Broadband spectra

The broadband spectrum is the spectrum over all the observed wavelength ranges. To plot the broadband spectrum of any object it is necessary to choose logarithmic axes.

- ■ Why is it necessary to use logarithmic axes?
- □ Because both the spectral flux density, F_λ, and the wavelength vary by many powers of 10.

Figure 3.7 shows the broadband spectrum of the Sun: it has a strong peak at optical wavelengths with very small contributions at X-ray and radio wavelengths.

Normal galaxies

Figure 3.8 (overleaf) shows schematically the broadband spectrum of a normal spiral galaxy. It resembles that of the Sun, although the peak occurs at a slightly longer wavelength and there are relatively greater spectral flux densities at X-ray, infrared and radio wavelengths.

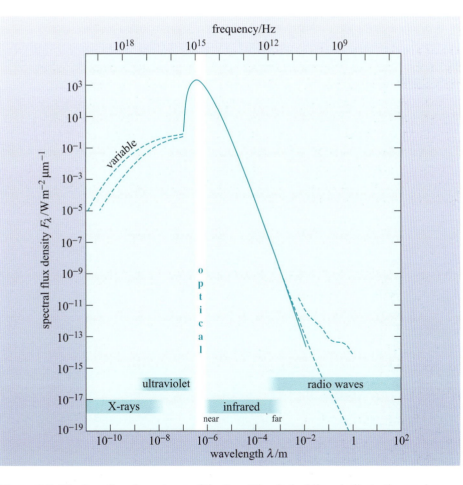

Figure 3.7 The broadband spectrum of the Sun. The dashed lines indicate the maximum and minimum in regions where the flux density varies. (Adapted from Nicolson, 1982)

■ List the objects in a normal galaxy that emit at (a) X-ray, (b) infrared and (c) radio wavelengths.

☐ (a) X-rays are emitted by X-ray binary stars, supernova remnants and the hot interstellar medium.

(b) Infrared radiation comes predominantly from cool stars, dust clouds, and dust surrounding HII regions.

(c) Radio waves are emitted by supernova remnants, atomic hydrogen and molecules such as CO.

From Figure 3.8 you would conclude that the spectrum peaks in the optical, but there is a subtlety in the definition of F_λ that needs to be addressed.

■ Look again at the broadband spectrum in Figure 3.8. Is this galaxy brighter in X-rays or in the far-infrared ($\lambda \sim 100$ μm)?

☐ The F_λ curve is higher in the X-ray region, so the galaxy appears to be brighter in X-rays than in the far-infrared (far-IR).

Obvious, isn't it? Well, appearances can be misleading. The spectral flux density F_λ is defined as the flux density received over a 1-μm bandwidth (see Box 3.2 overleaf). At far-IR and radio wavelengths that bandwidth is a tiny fraction of the spectrum. But at shorter wavelengths, 1 μm covers the entire X-ray, UV and visible

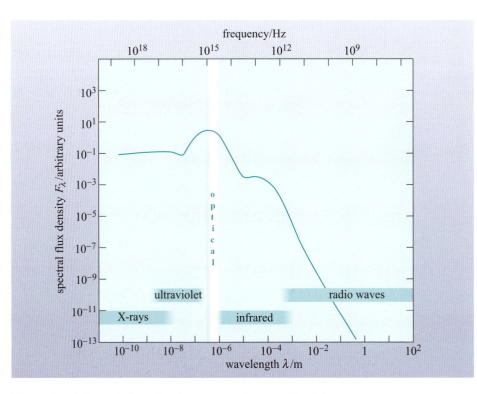

Figure 3.8 Schematic broadband spectrum of a normal spiral galaxy.

regions of the spectrum! So F_λ will under-represent the energy emitted by a galaxy in the far-IR (and radio wavelengths) and exaggerate the energy emitted in X-rays.

To correct this bias in F_λ spectra, astronomers often plot the quantity λF_λ instead. $\boldsymbol{\lambda F_\lambda}$, pronounced 'lambda eff lambda', (with units of W m^{-2}) is a useful quantity when we are comparing widely separated parts of a broadband spectrum. If the spectrum in its normal form of F_λ against λ is re-plotted in the form of λF_λ against λ (still on logarithmic axes), then the highest points of λF_λ will indicate the wavelength regions of maximum power received from the source. A broadband spectrum plotted in this way is known as a **spectral energy distribution** (or **SED**) because the height of the curve is a measure of the energy emitted at each point in the spectrum.

In Figure 3.9, λF_λ has been plotted against λ for the normal galaxy spectrum of Figure 3.8, and it can be clearly seen that this curve has a peak at optical wavelengths, confirming what was suspected. But it also shows that more energy is being radiated at far-IR wavelengths than in X-rays, the opposite of the impression given by Figure 3.8. From now on in this chapter broadband spectra will be plotted as SEDs with λF_λ against λ on logarithmic axes.

You may have found the concept of λF_λ difficult to grasp. If so, don't worry about the justification, but just accept that a λF_λ plot allows you to compare widely differing wavelengths fairly, whereas a conventional F_λ plot does not.

Mathematically, an important property of a λF_λ spectrum plotted is that a flat (horizontal) spectrum has an equal amount of energy per logarithmic interval.

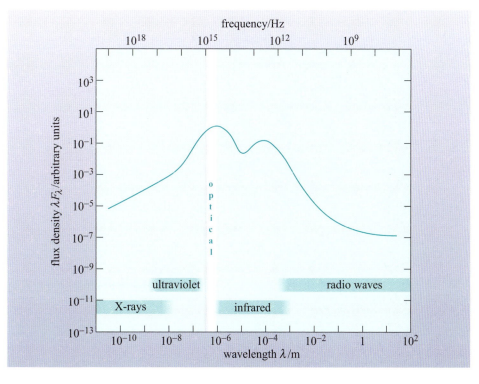

Figure 3.9 The spectral energy distribution (SED) of the galaxy in Figure 3.8. The vertical axis is now λF_λ instead of F_λ.

BOX 3.2 FLUX UNITS

Astronomers use several different units to measure the electromagnetic radiation received from an object.

Flux density, F, is the power received per square metre of telescope collecting area. It is measured in watts per square metre, $W\,m^{-2}$.

Spectral flux density is the flux density measured in a small range of bandwidth. As bandwidth can be expressed either in terms of wavelength (λ) or frequency (v) there are two kinds of spectral flux density in common use. F_λ is measured in watts per square metre per micrometre ($W\,m^{-2}\,\mu m^{-1}$) and F_v is measured in watts per square metre per hertz ($W\,m^{-2}\,Hz^{-1}$). The former is preferred by optical and infrared astronomers (who work in wavelengths) and the latter by radio astronomers (who work in frequencies). The special unit, the *jansky* (Jy), is given to a spectral flux density of $10^{-26}\,W\,m^{-2}\,Hz^{-1}$, in honour of the US engineer Karl Jansky (1905–1950) who made pioneering observations of the radio sky in the early 1930s.

Both flux density and spectral flux density are commonly (though inaccurately) referred to as *flux*.

Note that the symbol v (Greek letter 'nu') is commonly used to denote the frequency of electromagnetic radiation. In this book, the convention is to use f to denote frequency.

QUESTION 3.4

Astronomers observe two galaxies at the same distance. Both have broad, smooth spectra. Galaxy A is seen at optical wavelengths (around 500 nm), and yields a spectral flux density $F_\lambda = 10^{-29}\,W\,m^{-2}\,\mu m^{-1}$; it is not detected in the far infrared at around 100 μm (the upper limit to the measured flux density is $F_\lambda < 10^{-32}\,W\,m^{-2}\,\mu m^{-1}$). Galaxy B appears fainter in the optical and gives $F_\lambda = 10^{-30}\,W\,m^{-2}\,\mu m^{-1}$ around 500 nm, and the same value at around 100 μm. Which (on these data) is the more luminous galaxy?

Active galaxies

Figure 3.10 shows the spectral energy distribution of an active galaxy.

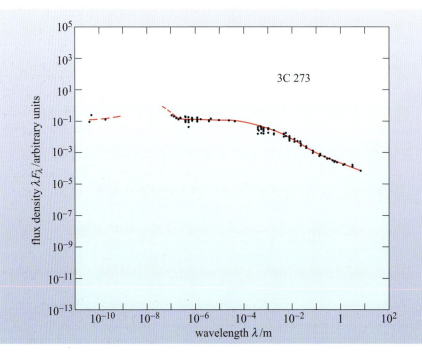

Figure 3.10 The spectral energy distribution of an active galaxy, the quasar 3C 273. The filled circles are measurements and the red curve shows the spectrum as determined from the data. (Data provided by NASA/ IPAC Extragalactic Database)

■ In broad terms, what is the major difference between the SED of the normal galaxy in Figure 3.9 and the SED of the active galaxy in Figure 3.10?

☐ Compared with the (unquantified) peak emission, the SED of the active galaxy is much flatter than that of the normal spiral galaxy. This indicates that there is relatively much more emission (by several orders of magnitude) at X-ray wavelengths and at radio wavelengths.

For the active galaxy (known from its catalogue number as 3C 273) the peak emission is in the X-ray and ultraviolet regions. Many other active galaxies are bright in this region and the feature is known as the 'big blue bump'. In some active galaxies, though not this one, the infrared emission is prominent. These galaxies emit a normal amount of starlight in the optical, so they must emit several times this amount of energy at infrared and other wavelengths – this is another feature that distinguishes active galaxies from normal galaxies. It means that we have to account for *several times* the total energy output of a normal galaxy, and possibly a great deal more. A normal galaxy contains 10^{10} to 10^{11} stars, so we need an even more powerful energy source for active galaxies.

The term **spectral excess** is used (rather loosely) to refer to the prominence of infrared or other wavelength regions in the broadband spectra of active galaxies. In particular, it is often used to indicate the presence of emission in a certain wavelength region that is over and above that which would be expected from the stellar content of a galaxy.

QUESTION 3.5

Now that you have some experience in interpreting the spectra of galaxies, look at the SED of the galaxy NGC 7714 in Figure 3.11. Describe as fully as you can what the diagram tells you about this galaxy. Can you guess what sort of galaxy it is?

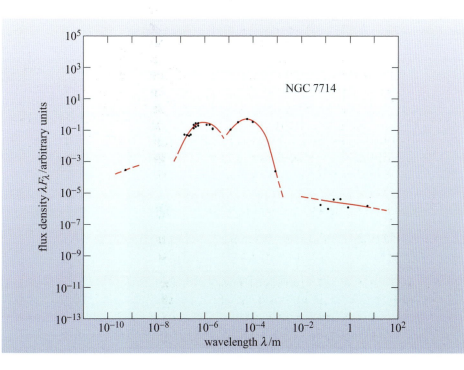

Figure 3.11 The spectral energy distribution of the galaxy NGC 7714 (for use with Question 3.5). (Data provided by NASA/IPAC Extragalactic Database)

3.3 Types of active galaxies

In this section you will learn about the observational characteristics of the four main classes of active galaxies: Seyfert galaxies, quasars, radio galaxies and blazars. This will set the scene for subsequent sections in which we will explore the physical processes that lie behind these different manifestations of active galaxies.

3.3.1 Seyfert galaxies

In 1943 the American astronomer Carl Seyfert (1911–1960, see Figure 3.14) drew attention to a handful of spiral galaxies that had unusually bright point-like nuclei. Figure 3.12 shows NGC 4051, one of the first **Seyfert galaxies** to be identified. Subsequently, it has been found that compared to normal galaxies, Seyfert galaxies show an excess of radiation in the far infrared and at other wavelengths. Even more remarkably, at some wavelengths, including the optical, this excess radiation is *variable*. Variability is discussed in detail in Section 3.4.1 − suffice it to say here that the variability implies that the emission from a Seyfert galaxy must come from a region that is *tiny* compared to the galaxy itself.

Spectra of the bright nuclei reveal that Seyferts can be classified into two types by the relative widths of their emission lines (Figure 3.13).

Type 1 Seyferts have two sets of emission lines (Figure 3.13a). The narrower set, which are made up largely of the forbidden lines discussed earlier, have widths of about 400 km s^{-1}. Despite this considerable width the region emitting these lines is known as the **narrow-line region**. The broader lines, consisting of permitted lines only, have widths up to 10 000 km s^{-1} and appear to originate from a denser region of gas known as the **broad-line region.** As noted above, forbidden lines are sensitive to the gas density in the emitting region. An analysis of which lines are present allows the densities of the gas in the broad- and narrow-line regions to

Figure 3.12 NGC 4051 is a member of a class of galaxies known as Seyfert galaxies. In this optical image (at a wavelength of around 440 nm) a false colour scheme has been used to show features across a wide range of surface brightness. Blue and green regions have a low surface brightness, whereas yellow, red and white regions are relatively bright. The intense emission from the point-like nucleus of the galaxy is clearly evident. NGC 4051 is relatively close – lying at a distance of about 17 Mpc from the Milky Way. The field-of-view of this image is 4.0 arcmin × 4.5 arcmin. (Data provided by NASA/IPAC Extragalactic Database from the observations of Tully *et al.*, 1996)

be determined. These two regions are also characteristic of other types of active galaxy. Type 2 Seyferts only show prominent narrow lines (Figure 3.13b). The broad lines are either absent or very weak in the optical spectra of type 2 Seyferts.

In fact, these two types are not as clear cut as they first seemed, since weak broad lines have now been found in Seyferts previously classed as type 2. Types 1 and 2 are better understood as extreme ends of a range of intermediate Seyfert types classified according to the relative strengths of their broad and narrow lines. In a Seyfert 1.5, for example, there are broad and narrow lines, but the broad lines are not as strong as those seen in type 1 Seyferts.

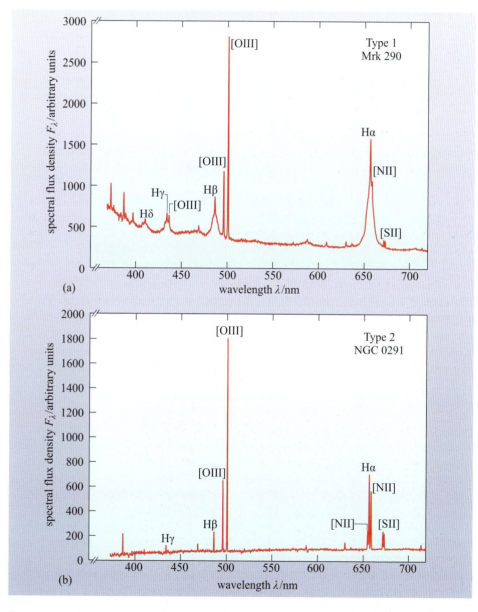

Figure 3.13 The optical spectra (adjusted to zero redshift) of two Seyfert galaxies. (a) Markarian 290, a type 1 Seyfert. (b) NGC 0291, a type 2 Seyfert. Note that the broad hydrogen lines (especially Hα and Hβ) visible in (a) appear narrower in (b). (Data from Sloan Digital Sky Survey, Data Release 6)

Figure 3.14 Carl Seyfert with the 24-inch telescope (that is now named in his honour) at the Dyer Observatory at Vanderbilt University. (B. Poteete)

Carl Seyfert (Figure 3.14) was born and grew up in Cleveland, Ohio. He entered Harvard with the intention of studying medicine, but became diverted from his career path after attending an inspirational lecture course in astronomy given by Bart Bok. Seyfert switched his attention to astronomy and remained at Harvard to carry out his doctoral research under the direction of Harlow Shapley (Figure 1.26). Following a post at Yerkes Observatory he was employed at Mount Wilson Observatory from 1940 to 1942. It was during this time at Mount Wilson that he carried out his observations into the type of galaxies that now carry his name. During the Second World War he managed to juggle several tasks: teaching navigation to the armed forces, carrying out classified research, and still finding time to partake in some astronomical research. He is notable for producing some of the first colour photographs of nebulae and stellar spectra – some of which were used in the Encyclopedia Britannica. After the war Seyfert gained a faculty position at Vanderbilt University in Nashville, Tennessee. He was the driving force behind the development of their observatory and was also an enthusiastic popularizer of science. He also found time to present local weather forecasts on television! He was tragically killed in a motor accident in 1960 at the age of 49. He died before the significance of Seyfert galaxies became fully apparent – the field of active galaxy research only became a key area of astronomy after the discovery of quasars in 1963.

3.3.2 Quasars

One of the most unexpected turns in the history of astronomy was the discovery of **quasars**. When first recognized, in 1963, quasars appeared at radio and optical wavelengths as faint, point-like objects with unusual optical emission spectra. The name comes from their alternative designations of 'quasi-stellar radio source' (QSR) or 'quasi-stellar object' (QSO), meaning that they resemble stars in their point-like appearance. Their spectra, however, are quite unlike those of stars. The emission lines turn out to be those of hydrogen and other elements that occur in astronomical sources, but they are significantly red-shifted.

Figure 3.15 shows the optical spectrum of 3C 273, which was the first quasar to be discovered (you have already seen its broadband spectrum in Figure 3.10). The redshift is 0.158, which corresponds to a distance of about 660 Mpc according to Hubble's law. Many other quasars are now known – a recent survey catalogues more than 100 000 – and the vast majority have even greater redshifts, the record (in 2014) is 7.1. All quasars must therefore be highly luminous to be seen by us at all.

The optical spectra of quasars are similar to those of Seyfert 1 galaxies, with prominent broad lines but rather weaker narrow lines. A composite spectrum for 700 quasars is shown in Figure 3.16. To form this spectrum, the individual quasar spectra were all corrected to remove the effect of redshift before being added together. Because many quasars have high redshifts, many of the features that are observed in the visible part of the spectrum correspond to emission features in the ultraviolet. In particular, the spectrum shows the **Lyman α** (Lyα) line that arises from the electronic transition in atomic hydrogen from the state $n = 2$ to $n = 1$. This line, which occurs at a wavelength of 121.6 nm, is clearly a very strong and broad line in quasar spectra.

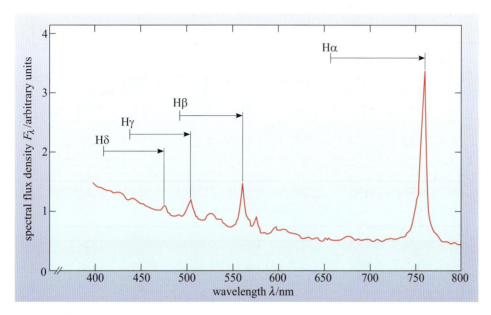

Figure 3.15 The optical spectrum of 3C 273, the first quasar to be discovered. The arrows show how the prominent hydrogen emission lines have been greatly red-shifted from their normal wavelengths. (Kaufmann, 1979)

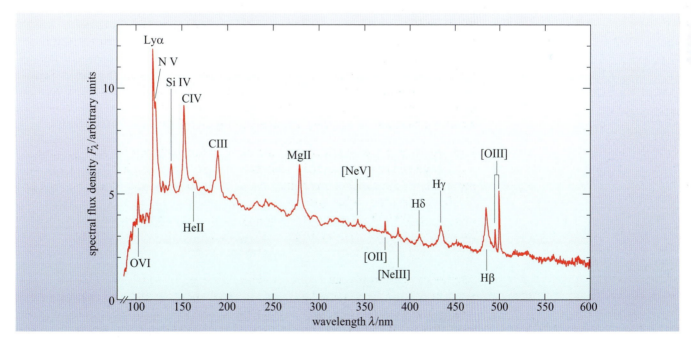

Figure 3.16 The mean optical spectrum of a sample of more than 700 quasars. The individual spectra were all corrected to remove the effect of redshift before the spectra were averaged. Note the broad emission lines. (Peterson, 1997, from data described in Francis *et al.*, 1991)

Quasars show spectral excesses in the infrared and at other wavelengths. About 10% of quasars are strong radio sources and are said to be *radio loud*. Some astronomers prefer to retain the older term QSO (quasi-stellar object) for *radio-quiet* quasars that are not strong sources of radio waves. The spectral energy distribution for a sample of radio-loud and a sample of radio-quiet quasars is shown in Figure 3.17. The big blue bump, hinted at in Figure 3.10, is particularly prominent here. Many quasars are also variable throughout the spectrum on timescales of months or even days.

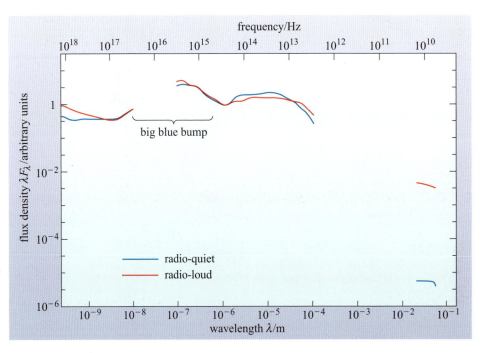

Figure 3.17 Mean SEDs for a sample of radio-quiet (blue line) and radio-loud (red line) quasars. The two curves are similar except at radio wavelengths. The 'big blue bump' is particularly prominent in this diagram. (Peterson, 1997, from data provided by Elvis *et al.*, 1994)

Detailed radio mapping shows that many of the radio-loud quasars have prominent *jets* that appear to be gushing material into space. In 3C 273 the jet is even visible on optical images (Figure 3.18).

Because quasars are so distant, it has been difficult to study the **host galaxies** that contain them. Recent work seems to show that there is no simple relationship between a quasar and the morphology of its host galaxy – while many quasar host galaxies are interacting or merging systems, there are also many host galaxies that appear to be normal ellipticals or spirals (Figure 3.19). It has also been found that the radio-loud quasars tend to be found in elliptical and interacting galaxies whereas the radio-quiet quasars (the QSOs) seem to be present in both elliptical and spiral host galaxies. It should be stressed however that the relationship between quasar host and radio emission is not clear cut, and that this is a topic of ongoing research.

Before their host galaxies were discovered in the 1980s quasars seemed much more puzzling than they do now. Indeed, for many years, there was a school of thought that supported the idea that quasars were not at such great distances as they are now thought to be, but were instead relatively close objects in which the redshift arose from some unknown physical process. The study of quasar host galaxies has all but dispelled this view and the modern picture of a quasar is of a remote, very luminous AGN buried in a galaxy of normal luminosity.

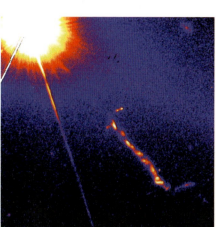

(a)

(b)

(c)

(d)

5 arcsec

Figure 3.18 Images of the nearest quasar, 3C 273. (a) An optical (V band) image shows a faint jet of material emerging from the star-like nucleus. The panels show the jet in more detail at (b) optical, (c) radio and (d) X-ray wavelengths. (Note that the different colours in panels (b), (c) and (d) represent different levels of intensity.) ((a), (b) Hubble Space Telescope; (c) MERLIN/Jodrell Bank Observatory; (d) Chandra X-ray Observatory)

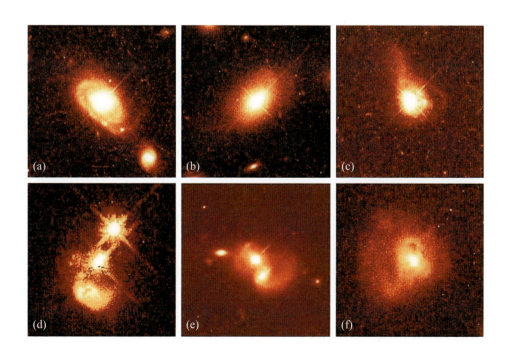

(a)

(b)

(c)

(d)

(e)

(f)

Figure 3.19 Examples of quasar host galaxies as observed at optical wavelengths with the Hubble Space Telescope. Quasars seem to occur in normal and interacting galaxies. The host galaxies shown here appear to be: (a) a normal spiral galaxy, (b) a normal elliptical galaxy, and (c) to (f) interacting or merging galaxies. (Note that the different colours represent different levels of intensity.) (J. Bahcall (Institute for Advanced Study, Princeton) and M. Disney (University of Wales, Cardiff))

3.3.3 Radio galaxies

Radio galaxies were discovered accidentally by wartime radar engineers in the 1940s, although it took another decade for them to be properly studied by the new science of radio astronomy. Radio galaxies dominate the sky at radio wavelengths. They show enormous regions of radio emission outside the visible extent of the host galaxy − usually these *radio lobes* occur in pairs.

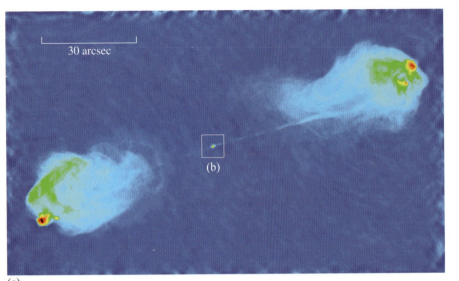

(a)

The first radio galaxy to be discovered, and the brightest on the sky, is called Cygnus A (Figure 3.20). Radio maps show the two characteristic lobes on either side of a compact nucleus. A narrow jet is apparent to the right of the nucleus and appears to be feeding energy out to the lobe. There is a hint of a similar jet on the left. Jets are a common feature of radio galaxies, especially at radio wavelengths. They trace the path by which material is being ejected from the AGN into the lobes.

Cygnus A is an example of the more powerful class of radio galaxy with a single narrow jet. The second jet is faint, or even absent, in many powerful radio galaxies; we will consider the reasons for this shortly. Note the relatively inconspicuous nucleus and the bright edge to the lobes, as if the jet is driving material ahead of it into the intergalactic medium.

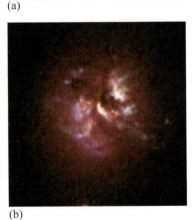

(b)

Figure 3.20 (a) The Cygnus A radio galaxy consists of two bright 'lobes' on either side of a compact nucleus. The lobe on the right is connected to the nucleus by a narrow jet. The white box shows the extent of (b), the host galaxy of Cygnus A. It is believed to be a giant elliptical galaxy with morphological peculiarities. The galaxy is at a distance of about 240 Mpc. This optical image combines observations made in the blue, visual (V) and near-infrared bands. ((a) Data provided by NASA/IPAC Extragalactic Database from the observations of Perley *et al.*, 1984; (b) R. Fosbury, ESO)

The jets of weaker radio galaxies spread out more and always come in pairs. These galaxies have bright nuclei, but the lobes are fainter and lack sharp edges. You can see an example in Figure 3.21. This is M84, a relatively nearby radio galaxy in the Virgo cluster of galaxies.

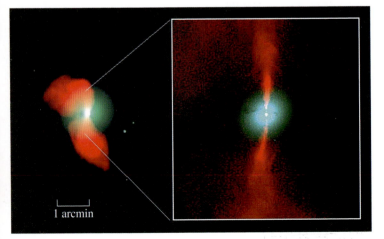

Figure 3.21 The radio galaxy M84. The radio emission is shown in red while the optical image of the galaxy is indicated in blue. The distance to M84 is about 18 Mpc. The inset shows an expanded view of the inner regions of the jets and the bright nucleus. (A. Bridle, NRAO)

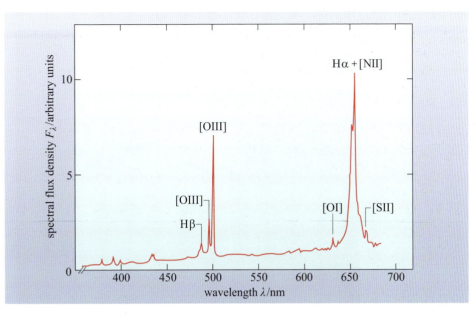

Figure 3.22 The optical spectrum of the nucleus of the radio galaxy 3C 445 (adjusted to zero redshift). (Osterbrock *et al.*, 1976)

Each radio galaxy has a point-like radio nucleus coincident with the nucleus of the host galaxy. It is this feature that is reminiscent of other classes of active galaxies and which is believed to be the seat of the activity. The nucleus shows many of the properties of other AGN, including emission lines, a broadband spectrum that is far wider than that of a normal galaxy, and variability.

The optical spectrum of the nucleus of a radio galaxy looks very much like that of any other AGN. Like Seyferts, radio galaxies can be classified into two types depending on whether broad lines are present (*broad-line radio galaxies*) or only narrow lines (*narrow-line radio galaxies*). Figure 3.22 shows an example of a spectrum of a broad-line radio galaxy.

Figure 3.23 (overleaf) shows maps of radio, optical and X-ray wavelengths of Centaurus A, which is the nearest radio galaxy to the Milky Way. The optical image (Figure 3.23b) shows that it is an elliptical galaxy with a dust lane bisecting it.

- Given that Centaurus A is an elliptical galaxy, does anything strike you as incongruous about Figure 3.23b?

- Elliptical galaxies are supposed to have negligible amounts of dust, so the thick dust lane seems very strange indeed!

The galaxy is obviously not a normal elliptical and this is a clue to the nature of radio galaxies. In fact, it is now thought that Centaurus A was formed by the collision of a spiral galaxy with a massive elliptical, the dust lane being the remains of the spiral's disc. We will come back to this interesting topic later in the chapter.

M87 (also known as Virgo A) is such a well-known radio galaxy that it must be mentioned at this point. In the optical region it, too, appears as a giant elliptical galaxy at the centre of the nearby Virgo cluster of galaxies. It seems that most radio galaxies are ellipticals. The single bright jet in the galaxy (Figure 3.24, overleaf) is reminiscent of the jet in the quasar 3C 273, shown in Figure 3.18.

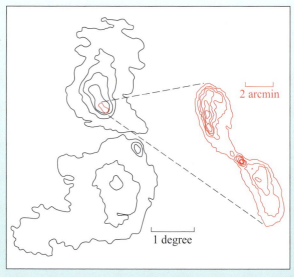

(a)

(b)

Figure 3.23 The Centaurus A radio galaxy.

(a) A radio map shows the lobes that extend over more than 9 degrees of the sky.

(b) An image at the visible wavelengths shows that the host galaxy is an elliptical galaxy with a dust lane bisecting it (the inner radio lobes are shown superimposed on this image).

(c) An X-ray image from the Chandra X-ray Observatory clearly shows the jet and the point-like nucleus in the inner parts of the galaxy.

((a) Seeds, 1998; (b) D. Malin/AAO; radio contours in red are data provided by NASA/IPAC Extragalactic Database from the observations of Condon *et al.*, 1996; (c) R. Kraft, SAO/NASA)

(c)

(a)

Figure 3.24 (a) Optical and (b) radio images of the giant elliptical galaxy M87 clearly show the presence of a 'one-sided' jet that extends from the active nucleus. Note that (a) and (b) are at the same scale. (NASA, NRAO and J. Biretta, STScI)

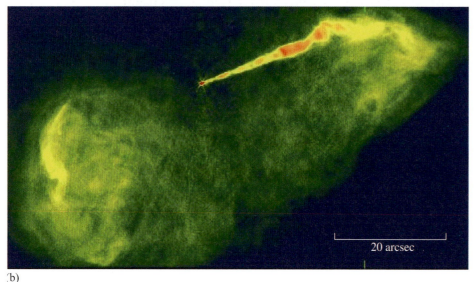

(b)

3.3.4 Blazars

Blazars appear star-like, as do quasars, but were only recognized as a distinct class of object in the 1970s. They are variable on timescales of days or less. All are strong and variable radio sources. There are two subclasses.

BL Lac objects are characterized by spectra in which emission lines are either absent or extremely weak. At first, they were mistaken for variable stars until their spectra were studied. (Their name derives from BL Lacertae, which is the variable-star designation originally given to the first object of this type to be studied.)

Over 1300 BL Lacs are known. Figure 3.25 shows three examples of a survey of BL Lac host galaxies that was conducted with the Hubble Space Telescope. In most cases the host galaxy appears to be elliptical and the stellar absorption lines help to confirm the redshift of the object.

Flat-Spectrum Radio Quasars (FSRQs) have similar properties to BL Lacs except that they have strong broad and narrow emission lines.

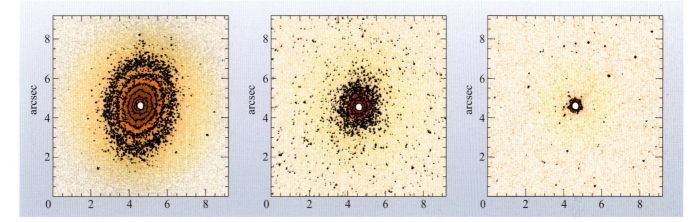

Figure 3.25 Examples of Hubble Space Telescope observations of BL Lac objects. This sequence shows the isophotes around three BL Lac objects: (left) 0548−322 − with a clearly imaged elliptical host galaxy, (middle) 1534+014 − which is resolved and can be shown to have isophotes that correspond to a normal elliptical galaxy, (right) 0820+255 − in which the host galaxy is unresolved. In all three cases the emission from the point-like AGN has been masked out. (Adapted from Scarpa *et al.*, 2000)

3.3.5 A 'non-active' class – the starburst galaxies

We end this section by drawing a distinction between the classes of active galaxy that are described above and the *starburst galaxies* mentioned earlier. As you have seen, starburst galaxies are essentially ordinary galaxies in which a massive burst of star formation has taken place. Their spectra show emission lines from their many HII regions and infrared emission from dust but, in the main, they do not show unusual activity in their nuclei.

Although it is clear that there are starburst galaxies that are not active galaxies, it does appear that some active galaxies are undergoing a burst of star formation. The link between these two phenomena when seen in the *same galaxy* is far from clear, but it is believed that the growth of supermassive black holes in active nuclei (of which more later) is somehow deeply connected to the creation of stars in the surrounding galaxy. As you will see, it is possible that both rapid star formation and activity in the galactic nucleus may be triggered by a galactic collision or merger. It is also possible for the active nucleus to drive out gas from the surrounding galaxy, starving the galaxy of the fuel for star formation. In this way, the active nucleus can in some ways regulate the formation of stars in the surrounding galaxy. The details are a subject of much active research, as you will see.

Take a few minutes to jot down as many differences that you can think of between normal galaxies and each of the four types of active galaxy. Are there any characteristics that all active galaxies have in common?

3.4 The central engine

From the previous section you will have discovered that one thing all active galaxies have in common is a compact nucleus, the AGN, which is the source of their activity. In this section you will study the two properties of AGN that make them so intriguing – their small size and high luminosity – and learn about the energy source at the heart of the AGN, the central engine.

3.4.1 The size of AGN

AGN appear point-like on optical images. It is instructive to work out how small a region these imaging observations indicate. Optical observations from the Earth suffer from 'seeing', the blurring of the image by atmospheric turbulence. The result is that star-like images are generally smeared by about 0.5 arcsec or more. A specialised technique, called adaptive optics (AO), can be used on ground-based telescopes to improve the resolution to about 0.05 arcsec or so. Similar resolutions can be obtained by observing from above the Earth's atmosphere with the Hubble Space Telescope. What does this mean in terms of the physical size of an AGN?

An arc second is 1/3600 of a degree and there are 57.3 degrees in a radian. So 0.05 arcsec corresponds to an angle of $0.05/(57.3 \times 3600)$ rad $= 2.4 \times 10^{-7}$ rad. For such a small angle, the linear diameter l of an object is related to its distance d by $l = d \times \theta$, where θ is its angular diameter in radians (Figure 3.26).

The nearest known AGN is NGC 4395, a Seyfert at a distance of 4.3 Mpc and it, too, is unresolvable. So its linear size l must be less than $(4.3 \times 10^6) \times (2.4 \times 10^{-7})$ pc $= 1.0$ pc. So, for a nearby AGN, we can place an upper limit of order 1 pc on its linear size. (For a more distant AGN, this upper limit is correspondingly larger.) A parsec is a tiny distance in galactic terms. Even the nearest star to the Sun is more than one parsec away, and our Galaxy is 30 kpc in diameter. So the point-like appearance of AGN tells us that they are *much* smaller than any galaxy.

A second approach to estimating the size of an AGN comes from their variability. The continuous spectra of most AGN vary appreciably in brightness over a timescale of about a year, and several vary over timescales as short as a few hours (about 10^4 s), especially at X-ray wavelengths (see Figure 3.27). This variability places a much tighter constraint on the size, as you will see.

To take an analogy, suppose you have a spherical paper lampshade surrounding an electric light bulb. When the lamp is turned on, the light from the bulb will travel at a speed c and will reach all parts of the lampshade at the same time, causing all parts to brighten simultaneously. To our eyes the lampshade appears to light up instantaneously, but that is only because the lampshade is so small.

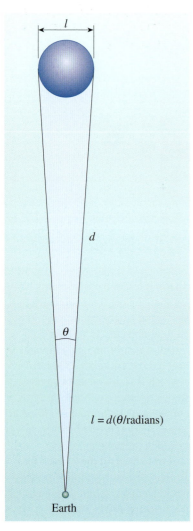

Figure 3.26 Schematic diagram to show how the linear size l of an AGN may be worked out from its angular size θ and distance d.

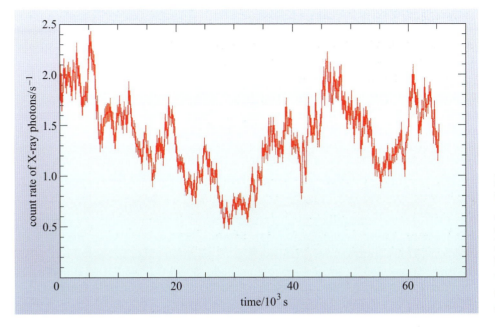

Figure 3.27 An example of X-ray variability, shown by the Seyfert galaxy MCG-6-30-15 during an observation made by the Chandra X-ray Observatory. The fastest fluctuations are spurious noise, but the variability over a few thousand seconds is a property of the AGN. (Lee *et al.*, 2002)

In fact, light arrives at your eyes from the nearest point of the lampshade a fraction of a second before it arrives from the furthest visible point (Figure 3.28). The time delay for the brightening, Δt, is given by

$$\Delta t = R/c \qquad (3.3)$$

where R is the radius of the lampshade.

Now imagine the shade to be much larger, perhaps the size of the Earth's orbit around the Sun, and the observer is far enough out in space that the shade appears as a point source of light.

■ What is Δt for a lampshade with the same radius as the Earth's orbit?

☐ $\Delta t = (1.5 \times 10^{11} \text{ m})/(3 \times 10^8 \text{ m s}^{-1}) = 500 \text{ s}$

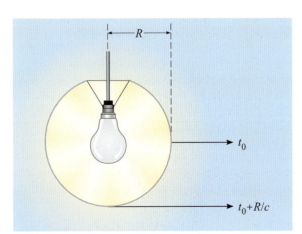

Figure 3.28 Light from the most distant visible point of a spherical lampshade will reach the observer a time R/c later than light from the near side. Fluctuations on timescales of less than R/c will not be observed.

So even if the lamp is switched on instantaneously, the observer will see the source take about eight minutes to brighten. Now suppose the bulb starts to flicker several times a second. What will an observer see? Even though the lampshade will flicker at the same rate as the bulb, it's clear that the flickering will have no effect on the *observed* brightness of the lampshade, since each flicker will take 500 seconds to spread across the lampshade and the flickers will be smeared out and mixed together. There is a limit to the rate at which a source (in this case the lampshade) can be seen to change in brightness and that limit is set by its size.

This argument may be inverted to state that if the observer sees a significant change in brightness of an unresolved source in a time Δt, then the radius of the source can be no larger than $R = c\Delta t$.

This kind of argument applies for any three-dimensional configuration where changes in brightness occur across a light-emitting surface. Of course, the argument is only approximate – real sources of radiation are unlikely to be perfectly represented by the idealized lampshade model that we have used here. The relationship between the maximum extent (R) of any source of radiation and its timescale for variability (Δt) is usually expressed as

$$R \sim c\Delta t \tag{3.3a}$$

(Where the symbol '~' is used to imply that the relationship is correct to within a factor of about ten.)

Returning now to the AGN, let us calculate the value of R for an AGN such as MCG-6-30-15. The timescale for variability that we shall use is the shortest time taken for the intensity of the source to double. By inspecting Figure 3.27 it can be seen that this timescale is about 10^4 s.

We have $R \sim c\Delta t$, so with $\Delta t = 1 \times 10^4$ s, we obtain $R \sim 3 \times 10^{12}$ m $= 1 \times 10^{-4}$ pc. This is a staggeringly small result – it is ten thousand times smaller than the upper limit we calculated from the size of AGN images – and corresponds to about 20 times the distance from the Sun to the Earth. The AGN would easily fit within our Solar System. The argument does not depend on the distance of the AGN. Hence the observed variability of AGN places the strongest constraint on their size.

One note of caution: the variability of AGN usually depends on the wavelength at which they are observed. Variations in X-rays, for example, tend to be faster than variations in infrared light. Does this imply that the size of an AGN depends on the wavelength? In a sense, yes, as we are seeing different radiation from different parts of the object. The X-rays seem to come from a much smaller region of the AGN than the infrared emission, so we must be careful when talking about 'the size' of an AGN.

QUESTION 3.7

An AGN at 50 Mpc appears smaller than 0.1 arcsec in an optical observation made by the Hubble Space Telescope, and shows variability on a timescale of one week. Calculate the upper limit placed on its size by (a) the angular diameter observation, and (b) the variability observation.

Other evidence also indicates the small size of AGN. Radio astronomers operate radio telescopes with dishes placed on different continents. This technique of *very long baseline interferometry* (VLBI) is able to resolve angular sizes one hundred or so times smaller than optical telescopes can. Even so, AGN remain unresolved.

3.4.2 The luminosity of AGN

It is instructive to express the luminosity of an AGN in terms of the luminosity of a galaxy like our own. The figure may then be converted into solar luminosities, if we adopt the figure of $2 \times 10^{10} L_\odot$ for the luminosity of our Galaxy.

Consider a Seyfert galaxy first. At optical wavelengths the point-like AGN is about as bright as the remainder of the galaxy, which radiates mainly at optical wavelengths. But the AGN also emits brightly in the ultraviolet and the infrared, radiating at least three times its optical luminosity. So one concludes that for a typical Seyfert, the AGN has at least four times the luminosity of the rest of the galaxy.

We have seen that a characteristic of a quasar is that its luminous output is dominated by emission from its AGN. However quasar host galaxies are not less luminous than normal galaxies, so the AGN of quasars must be *far* brighter than normal galaxies and must also be considerably more luminous than the AGN of Seyfert galaxies.

In the case of a radio galaxy, the AGN may not emit as much energy in the optical as Seyfert or quasar AGN, but an analysis of the mechanism by which the lobes shine shows that the power input into the lobes must exceed the luminosity of a normal galaxy by a large factor, and the AGN at the centre is the only plausible candidate for the source of all this energy.

A similar conclusion for AGN luminosity follows for blazars, which appear to be even more luminous than quasars. We examine why in Section 3.4.6 below.

QUESTION 3.8

Calculate the luminosity of an AGN that is at a distance of 200 Mpc, and appears as bright in the optical as a galaxy like our own at a distance of 100 Mpc. Assume that one-fifth of the energy from the AGN is at optical wavelengths.

One can conclude that AGN in general have luminosities of more than $2 \times 10^{10} L_\odot$ produced within what is, from an astronomical point of view, a tiny volume. The Sun's luminosity is about 4×10^{26} W, so a typical AGN has a luminosity of more than 8×10^{36} W. In fact, that's quite modest for an active galaxy, so for the purposes of this chapter we shall adopt a more representative value of 10^{38} W as the characteristic luminosity of an AGN.

You are now in a position to appreciate the basic problem in accounting for an AGN. It produces an *enormous* amount of power (luminosity) in what is astronomically speaking a *tiny* volume. This source of power is known as the engine. Current ideas about the workings of this engine are discussed in the next section.

3.4.3 A supermassive black hole

A **black hole** is a body so massive and so small that even electromagnetic radiation, such as visible light, cannot escape from it. It is its combination of small size and very strong gravitational field that makes it attractive as a key component of the engine that powers an AGN. You saw in Chapter 1 that there is good evidence of a black hole of mass $4.3 \times 10^6 M_\odot$ at the centre of the Milky Way. As you will see, it turns out that much more massive black holes are needed to explain AGN, and these are referred to as *supermassive black holes*.

A black hole, supermassive or otherwise, is such a bizarre concept that it is worth recapping. The material of which it is made is contained in a radius so small that the gravity at its 'surface', the so-called **event horizon**, causes the escape speed to exceed the speed of light. According to classical physics, any object that falls into it can never get out again. Even electromagnetic radiation cannot escape, which is why the hole is called 'black'. What goes on inside the black hole is academic – no-one can see. What might be seen is activity just outside the event horizon where the gravitational field is strong, but not so strong as to prevent the escape of electromagnetic radiation. It is this surrounding region that is of most interest to astronomers.

The radius of the event horizon is called the **Schwarzschild radius** and is the distance at which the escape speed is just equal to the speed of light. It is given by

$$R_S = 2GM/c^2 \tag{3.4}$$

Let us now calculate the maximum mass M of a black hole that is small enough to fit inside an AGN. In Section 3.4.1 you found that an AGN that varies on a timescale of one day must have a radius less than 3×10^{12} m. We shall see later that all of the emission from the AGN must come from a region that is outside of the Schwarzschild radius and that the size of this emission region is a few times bigger than R_S. Consequently, for this approximate calculation we shall adopt a size for R_S that is a factor of ten smaller than the size we calculated above for the emission region, i.e. $R_S = 3 \times 10^{11}$ m. Then, from Equation 3.4,

$$M = R_S \times c^2/2G$$
$$= (3 \times 10^{11} \text{ m}) \times (3.0 \times 10^8 \text{ m s}^{-1})^2/(2 \times 6.67 \times 10^{-11} \text{ N m}^2 \text{ kg}^{-2})$$
$$= 2 \times 10^{38} \text{ kg}$$

which is equivalent to

$$(2.0 \times 10^{38} \text{ kg})/(2.0 \times 10^{30} \text{ kg})M_\odot = 1 \times 10^8 M_\odot$$

This result shows that it is clearly possible to fit a black hole with an enormous mass within an AGN, although it does *not* prove that the central black hole has to be this massive. We will shortly see that there is a different argument that does show that mass of any black hole at the centre of an AGN must be about $10^8 M_\odot$. This is usually adopted as the 'standard' black hole mass in an AGN. It is some 10^7 times greater than the masses of the black holes inferred to exist in some binary stars that emit X-rays. Hence, the name **supermassive black hole** has been adopted.

3.4.4 An accretion disc

What will happen to matter that comes near a black hole? Consider a gas cloud moving to one side of the black hole, such as cloud A in Figure 3.29. The hole's gravity will accelerate the gas cloud towards it. The cloud will reach its maximum speed when it is at its closest approach to the black hole, but will slow down again as it moves away; it will then move away to a distance as great as the distance from which it started. Thus far nothing is new; the gas cloud will behave exactly as it would if it came near some other gravitationally attracting object, such as a Sun-like star.

Now, let us extend the argument to a number of gas clouds being accelerated towards the black hole from different directions in space. This time, as the gas clouds get to their closest approach they will collide with each other, thus losing some of the kinetic energy they had gained as they fell towards the hole. Therefore some, but not all, of the clouds of gas will have slowed to a speed at which they cannot retreat, so they will go into an orbit around the hole. Further collisions amongst the gas clouds will tend to make their orbits circular, and the direction of rotation will be decided by the initial rotation direction of the majority of the gas clouds. The effect of the collisions will be to heat up the gas clouds; the kinetic energy they have lost will have been converted into thermal energy within each cloud, and so the cloud temperature will rise.

So far, we can envisage a group of warm gas clouds in a circular orbit about the black hole. But the clouds of gas are of a finite size and, because they move in a Keplerian orbit, the inner parts of the gas clouds will orbit faster than the outer parts. A form of friction (*viscosity*) will act between neighbouring clouds at different radii and they will lose energy in the form of heat. The consequence of this is that the inner parts of the gas clouds will fall inwards to even smaller orbits. This process will continue until a complete **accretion disc** is formed around the black hole (Figure 3.30). The accretion disc acts to remove angular momentum from most of the gas in the disc so that if you look at the path of a small part of one gas cloud, you can see that it will spiral inwards. Since angular momentum is a conserved quantity the accretion disc does not actually diminish the total angular momentum of the system – it simply redistributes it such that most gas in the disc will move inwards. This process occurs only for a *viscous* gas – planets in the Solar System do not show any tendency to spiral in to the Sun because interplanetary gas is very rarefied. The viscosity causes the gas to heat up further, the thermal energy coming from the gravitational energy that was converted into kinetic energy as the gas fell towards the hole. The heating effect will be large for objects with a large gravitational field and so we might expect that accretion discs around black holes will reach high temperatures and become luminous sources of electromagnetic radiation.

The gradual spiralling-in of gas through the accretion disc comes to an abrupt end at a distance of a few (up to about five) Schwarzschild radii from the centre of the black hole. At this point the infalling material begins to fall rapidly and quickly passes through the Schwarzschild radius and into the black hole. Note that the accretion disc is located *outside* the event horizon, where the heat can be radiated away as electromagnetic radiation. The accretion model is of such interest because an accretion disc around a massive black hole can radiate away a vast amount of energy, very much more than a star or a cluster of stars. It is this radiated energy that is believed to constitute the power of an AGN.

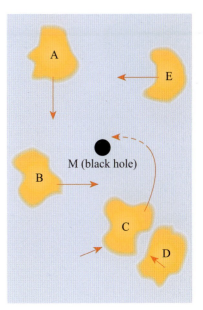

Figure 3.29 Schematic diagram of discrete gas clouds falling towards a black hole. Clouds C and D are shown colliding. This will allow the clouds to become trapped in an orbit around the black hole.

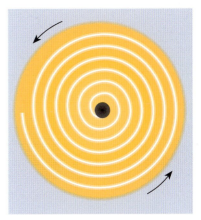

Figure 3.30 A rotating accretion disc; the line shows the spiral infall of one particle.

You may be wondering how large the accretion disc is; after all, the accretion disc as well as the black hole has to fit inside the AGN. The accretion disc gets hotter and therefore brighter towards its inner edge. The brightest, and hence innermost part is what matters. Since this is at only a few times the Schwarzschild radius, there is no problem of size.

■ Estimate the extent of the brightest part of the accretion disc for a black hole of mass $10^8 M_\odot$. How does this compare with the radii of planetary orbits in the Solar System?

□ From Section 3.4.3 we know that the Schwarzschild radius is about 3×10^{11} m, which is twice the radius of the Earth's orbit or 2 AU. The brightest part of the accretion disc could then extend to about five times this distance or about 10 AU, which is about the radius of Saturn's orbit.

3.4.5 Accretion power

Calculations based on the above accretion disc hypothesis show that if a mass m falls into the black hole, then the amount of energy it can radiate before it finally disappears is about $0.1mc^2$, or about 10% of its rest energy. Other than matter–antimatter annihilation, this is the most efficient process for converting mass into energy ever conceived. A comparable figure for the nuclear fusion of hydrogen in stars is only 0.7% of the rest energy of the four hydrogen nuclei that form the helium nucleus.

QUESTION 3.9

How much energy could be obtained from 1 kg of hydrogen (a) if it were to undergo nuclear fusion in the interior of a star, (b) if it were to spiral into a black hole? Would you expect to get more energy if it were to chemically burn in an oxygen atmosphere?

Now let us apply the idea of an accreting massive black hole to explain the luminosity of an AGN. We have to explain an object of small size and large luminosity. The Schwarzschild radius of a black hole is very small, and the part of the accretion disc that radiates most of the energy will be only a few times this size. The luminosity will depend on the rate at which matter falls in. Suppose that the matter is falling in at the rate Q (with units of kg s^{-1}), this is known as the **mass accretion rate**. We can now work out the value of Q to produce a luminosity L by writing

$$L = 0.1Qc^2 \quad \text{or} \quad Q = L/(0.1c^2) \tag{3.5}$$

Using the values $L = 10^{38}$ W and $c = 3 \times 10^8$ m s^{-1}, we get $Q = 10^{22}$ kg s^{-1}. Converting this into solar masses per year using $1M_\odot \approx 2 \times 10^{30}$ kg and 1 year $\approx 3 \times 10^7$ s, we get $Q \approx 0.2M_\odot$ per year. Is there a large enough supply of matter for a fraction of a solar mass to be accreted every year? Most astronomers think that the answer is yes, and that even higher accretion rates are plausible – after all our own Galaxy has 10% of its baryonic mass in gaseous form, so there is at least $10^{10}M_\odot$ of gas available.

■ Does this estimate of the accretion rate require a supermassive black hole, or will any black hole such as one of $5M_\odot$ do?

☐ The mass of the black hole does not enter into the above calculation. So on this basis a $5M_\odot$ black hole would seem to be sufficient.

Moreover, the mass calculated in Section 3.4.3 is an upper limit. So, why is a *supermassive* black hole needed? To see why, we ask: is there any limit to the power L that can be radiated by an accretion disc around a black hole, or can one conceive of an ever-increasing value of L if there is enough matter to increase Q?

There *is* a limit to the amount of power that can be produced, and it is called the **Eddington limit**. As the black hole accretes faster and faster, the luminosity L will go up in proportion, that is to say the accretion disc will get brighter and hotter. Light and other forms of electromagnetic radiation exert a pressure, called **radiation pressure**, on any material they encounter. (This pressure is difficult to observe on Earth because it is difficult to find a bright-enough light source.) Around an accreting black hole with a luminosity of 10^{38} W, the radiation will be so intense that it will exert a large outward pressure on the infalling material. If the force on the gas due to radiation pressure exactly counteracts the gravitational force, accretion will cease. This process acts to regulate the luminosity of an accreting black hole.

To work out the Eddington limit, it is necessary to balance radiation pressure against the effects of the black hole's gravity. Consider an atom of gas near the outer edge of the accretion disc. The force on it due to radiation pressure is proportional to L, whereas the gravitational force is proportional to the mass M of the black hole (assuming the mass of the accretion disc to be negligible). A balance is achieved when $L_E = $ constant $\times M$, where L_E is the Eddington limit. Full calculations give

$$L_E/W = (1.3 \times 10^{31})M/M_\odot \qquad (3.6)$$

This is the upper limit of the luminosity of a black hole of mass M – the luminosity can be lower than L_E but not higher. The larger the mass M, the greater the value of L_E.

In fact, this is only a rough estimate. It assumes that the accreting material is ionized hydrogen (a good assumption) and that the hole is accreting uniformly from all directions (which is not a good assumption). The Eddington luminosity may be exceeded, for example, if accretion occurs primarily from one direction and the resulting radiation emerges in a different direction. Nonetheless, it is a useful approximation.

Putting $L = 10^{38}$ W into Equation 3.6, we find that $M = 7.7 \times 10^6 M_\odot$. So we see that we do need a *supermassive black hole* to account for the engine in an AGN. A mass of $10^8 M_\odot$ is assumed to be typical, although as you will see later, black hole masses ranging from a few $10^6 M_\odot$ to about $10^{10} M_\odot$ have been measured.

In summary, then, the Eddington limit means that the observed luminosity of quasars requires an *accreting supermassive black hole* with a mass of order $10^8 M_\odot$; the accretion rate is at least a significant fraction of a solar mass per year; and the Schwarzschild radius is about 3×10^{11} m.

3.4.6 Jets

You have seen that two kinds of active galaxies – quasars and radio galaxies – are often seen to possess narrow features called jets projecting up to several hundred kiloparsecs from their nuclei. If these are indeed streams of energetic particles flowing from the central engine, how do they fit with the accretion disc model? How could the jets be produced?

The answers to these questions are not fully resolved, but there are some aspects of the model of the central engine that probably play an important part in jet formation. A key idea is that the jets are probably aligned with the axis of rotation of the disc – since this is the only natural straight-line direction that is defined by the system. This much is accepted by most astrophysicists, but the question of how material that is initially spiralling in comes to be ejected along the rotation axis of the disc at relativistic speeds (i.e. speeds that are very close to the speed of light) is an unsolved problem. One mechanism that has been suggested requires that at distances very close to the black hole the accretion disc becomes thickened and forms a pair of opposed funnels aligned with the rotation axis, as illustrated in Figure 3.31. Within these funnels the intense radiation pressure causes the acceleration and ejection of matter along the rotation axis of the disc. Unfortunately, this model fails in that it cannot produce beams of ejected particles that are energetic enough to explain the observed properties of real jets. A more promising approach, but one which does not yet offer a full explanation, involves the interaction of a magnetic field with the accretion disc, leading to a field pattern that could potentially accelerate particles in directions perpendicular to the disc rotation axis.

If jets are ejected along the rotation axis of the disc, then why do quasars and the more powerful radio galaxies generally only appear to have a single jet? It seems improbable that the engine produces a jet on one side only, and it is thought that there are indeed two jets but only one is visible. In this model, two jets are emitted at highly relativistic speeds, and one of them is pointing in our direction and the other is pointing away. Due to an effect called *relativistic beaming*, the radiation from the jets is concentrated in the forward direction. The consequence of this is that if a jet is pointing even only very approximately towards us it will appear very much brighter than would a similar jet that is pointing in the opposite direction. (The special case of what happens when a jet is pointing directly at us will be considered in the next section.)

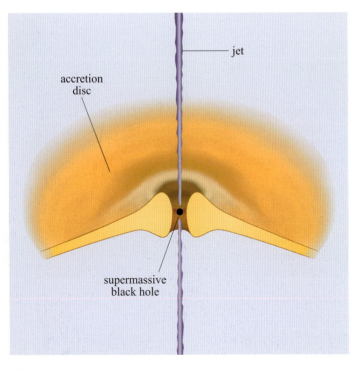

Figure 3.31 A scenario for the formation of jets in which the inner region of an accretion disc thickens to form two opposed funnels (for clarity, the accretion disc is cut-away to reveal the central black hole). The emission of radiation from the faces of the funnel leads to radiation pressure that acts to channel outflowing material into two relativistic particle beams called jets. Unfortunately this simple model cannot fully explain the observed properties of real jets.

accretion disc

jet

supermassive black hole

Estimate the accretion rate on to a black hole needed to account for the luminosity of a Seyfert nucleus that has twice the luminosity of our Galaxy. Express your answer in solar masses per year. What, other than the mass accretion rate, limits the luminosity?

3.5 Models of active galaxies

So far we have seen how the properties of the central engine of the AGN can be accounted for by an accreting supermassive black hole. Though there are many questions still to be resolved, this model does seem to be the best available explanation of what is going on in the heart of an AGN. But of course all AGN are not the same. We have identified four main classes and in this section we will attempt to construct models that reproduce the distinguishing features of these four classes.

Figure 3.32 shows the basic model that has been proposed for AGN. It is a very simple model, and does not account for all AGN phenomena, but it does give you a flavour of the kinds of ideas that astrophysicists are working with. You can see that the central engine (the supermassive black hole and its accretion disc) is surrounded by a cloud of gas and dust in the shape of a torus (a doughnut shape). The gap in the middle of the torus is occupied by clouds forming the broad-line region and both in turn are enveloped by clouds forming the narrow-line region.

We begin by looking at the accretion disc.

Figure 3.32 A generic model for an active galaxy. (a) The central engine is a supermassive black hole surrounded by an accretion disc with jets emerging perpendicular to the accretion disc. (b) The engine is surrounded by an obscuring torus of gas and dust. The broad-line region occupies the hole in the middle of the torus and the narrow-line region lies further out. (c) The entire AGN appears as a bright nucleus in an otherwise normal galaxy. Note that the jets extend to beyond the host galaxy and terminate in radio lobes.

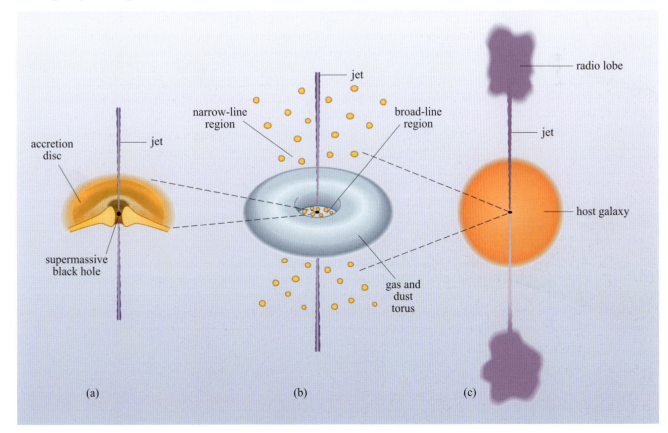

3.5.1 The accretion disc

As noted above, most of the emission from the accretion disc is expected to come from a region that has a radius of about five times that of the Schwarzschild radius – typically about 10 AU or so. Such a tiny region at an intergalactic distance is well below the spatial resolution of any type of telescope, but there are other observational signatures that can potentially probe this region. In particular, high energy emission lines in the X-ray part of the spectrum are believed to originate in the accretion disc. In the case of the Seyfert galaxy MCG-6-30-15, whose variability was illustrated in Figure 3.27, the X-ray spectrum shows an extremely broad emission line with a width of about 100 000 km s^{-1}. This is consistent with the motion expected in this part of the accretion disc itself. Furthermore, the line is greatly distorted as if it originated in the intense gravitational field near a black hole.

3.5.2 The obscuring torus

If an AGN consisted solely of the central engine, observers would see X-rays and ultraviolet radiation from the hot accretion disc (accounting for the 'the big blue bump' in Figure 3.17) and, apart from the jets, very little else. To account for the strong infrared emission from many AGN, the model includes a torus of gas and dust that surrounds the central engine.

The dust particles – condensed grains of carbonaceous and silicate material – will be heated by the radiation from the engine until they are warm enough to radiate energy at the same rate at which they it receive it. As dust will vaporize (or sublimate) at temperatures above 2000 K, the cloud must be cooler than this.

QUESTION 3.11

Assuming that dust grains radiate as black bodies, estimate the range of wavelengths that will be emitted from the torus.

Note: A black-body source at a temperature T has a characteristic spectrum in which the maximum value of spectral flux density (F_λ) occurs at a wavelength given by Wien's displacement law

$$(\lambda_{peak}/m) = \frac{2.90 \times 10^{-3}}{(T/K)}$$

So such a dust cloud will act to convert ultraviolet and X-ray emission from the engine into infrared radiation, with the shortest wavelengths coming from the hottest, inner parts of the cloud.

From a very simple dust cloud model, it is easy to understand why AGN so often emit most of their radiation in the infrared. Almost certainly, dust heated by the engine is observed in most AGN, although the dust may be more irregularly distributed than in our simple model, and the torus may have gaps in it. Some small amount of the infrared radiation will generally come from the engine itself, though, and in BL Lacs it is probable that most of the infrared radiation comes from the engine. The variability that was discussed in Section 3.4.1 applies to radiation from the engine at X-ray and optical wavelengths (and sometimes at radio wavelengths). The infrared emission from the torus is thought to vary much more slowly, as you would expect from the greater extent of the torus.

Note that this torus is *not* the same as the accretion disc surrounding the black hole, though it may well lie in the same plane and consist of material being drawn towards the engine.

It is possible to make a rough estimate of the inner radius of the torus by asking how far from the central engine the temperature will have fallen to 2000 K, the maximum temperature at which the grains can survive before being vaporized. This distance is called the **sublimation radius** for the dust.

If it is assumed that grains absorb all the radiation that falls on them, and they radiate energy away as black bodies, the distance r from an AGN of luminosity L, at which the dust has a temperature T is given by:

$$r \approx \left(\frac{L}{16\pi\sigma T^4} \right)^{1/2} \tag{3.7}$$

where σ is the Stefan–Boltzmann constant ($\sigma = 5.67 \times 10^{-8}$ W m^{-2} K^{-4}).

QUESTION 3.12

Calculate the dust sublimation radius, in metres and parsecs, for an AGN of luminosity 10^{38} W. (Assume that dust cannot exist above a temperature of 2000 K.)

For typical luminosities, the inner edge of the torus is three or four orders of magnitude (i.e. 1000 to 10 000 times) bigger than the emitting part of the accretion disc that is contained within the central engine in Figure 3.31. So, is there any direct evidence for such a structure? Remarkably, very high resolution radio observations have been used to measure the motion of molecular gas within about 1 pc of the central engine in several nearby active galaxies. An object that has been studied in this way is NGC 4258 (Figure 3.33, overleaf), which is a low luminosity Seyfert 2 galaxy. Molecular clouds close to the centre of the galaxy emit radio waves, which are naturally amplified in a similar way to the amplification of light in a laser – such sources are called *masers* (for microwave amplification by stimulated emission of radiation). The Doppler shift of these maser sources allows a rotation curve to be plotted (Figure 3.33d), which shows that the molecular clouds are in a thin disc and that they are moving in Keplerian orbits around a central mass. This then is direct evidence for material circulating in a region beyond the dust sublimation radius. While the observed disk is thinner than the dusty torus proposed in the standard AGN model, the observations are only sensitive to molecular gas, which may be concentrated into a thin disc within the torus. Furthermore, this rotation curve shows that the galaxy harbours a mass of $3.9 \times 10^7 M_\odot$ within a radius of 0.15 pc from its centre – an observation that provides very strong support for the idea that the central engine is indeed a supermassive black hole. Even more remarkably (and despite the host being a spiral galaxy) the source also has radio jets. These jets seem to be aligned along the rotation axis of the disc, and this lends support to the ideas of jet formation that were outlined in Section 3.4.6.

Figure 3.33 The Seyfert 2 galaxy NGC 4258. (a) A composite image showing optical continuum emission (shown in blue/green) and Hα emission (red). The Hα emission traces the spiral arms of the galaxy and two curved outflows from the central region of the galaxy. The latter features are more clearly visible in (b) a radio map at $\lambda = 20$ cm of the same region. (c) A map of the positions of radio measurements of the central regions of the galaxy. The black dot represents the location of the large central mass, believed to be a supermassive black hole. The red, blue and green spots on the schematically illustrated disc are the sources of maser emission (from water molecules). Also shown is a radio map at 22 GHz showing radio emission from the very central regions. Note that this emission is aligned in a direction roughly perpendicular to the disc. (d) The rotation curve as deduced from the maser emission in the disc, indicating the presence of a central mass of $3.9 \times 10^7 M_\odot$ within the central 0.1 pc of the galaxy. (NRAO/AUI and L. Greenhill (Harvard-Smithsonian Center for Astrophysics))

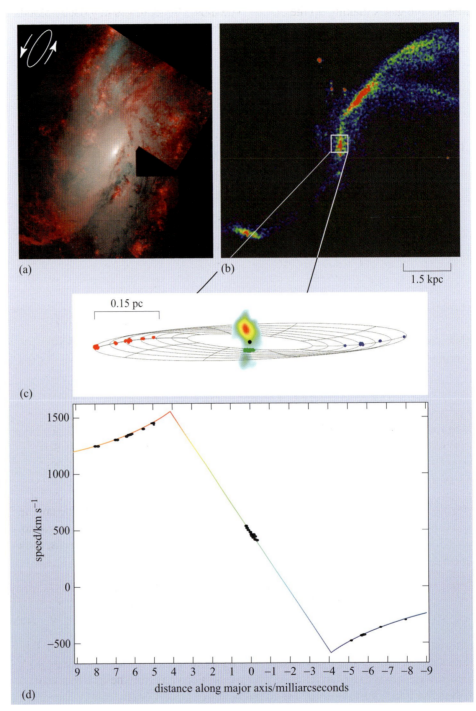

3.5.3 The broad- and narrow-line regions

In our model, the engine is surrounded by gas clouds (Figure 3.32). You have already seen how common these are in our own and other galaxies, so it is reasonable to expect them to be present in at least the spiral galaxies that contain AGN. If these gas clouds are illuminated by ultraviolet or X-rays from the engine they will absorb the ultraviolet or X-ray energy, and will emit the characteristic lines of the gases making up the clouds. The most abundant gas in galactic clouds is hydrogen, and, sure enough, the Hα and other lines of hydrogen appear strongly in the observed spectra of AGN.

What about other spectral lines that might be expected? Fortunately we get clues from objects in our own Galaxy, the HII regions, which consist of gas clouds illuminated by sources of ultraviolet radiation, albeit at a lower luminosity. These HII regions emit strong lines of nitrogen and oxygen, [NII] and [OIII], in the optical. Sure enough, the lines that appear in the optical spectra of AGN turn out to be just what you would expect from a gas of normal cosmic composition surrounding an AGN.

As you have discovered, there appear to be two kinds of line-emitting regions known as the broad-line region (BLR) and narrow-line region (NLR). If we interpret the spectra in terms of the density (inferred from the presence or absence of forbidden lines) and motion of gas clouds (inferred from line widths), then the BLR corresponds to dense fast-moving clouds and the NLR to low-density, more slowly moving clouds.

It is not possible to see the motion in great detail, but these motions are probably associated with the strong gravitational field surrounding an AGN. The orbital speed of a cloud will increase as the distance from the central black hole decreases. Thus the faster moving BLR clouds are assumed to be closer to the centre than the slower moving NLR clouds.

Broad-line region

In the model, the clouds of the broad-line region surround the central engine within the opening in the middle of the dust torus. The radius of the BLR is of the order of 10^{14} m, placing it well inside the torus. At this distance from the black hole orbital speeds are several thousand kilometres per second, which is consistent with the typical speed of 5000 km s^{-1} that is measured from Doppler broadening. The clouds are fully exposed to the intense radiation from the engine (remember that any dust will have vaporized in this region) and will be heated to a high temperature. It is difficult to measure the temperature of BLR clouds, but it appears to be of the order of 10^4 K.

It has been estimated that the BLR of a typical AGN will have about 10^{10} clouds covering about 10% of the sky as seen from the central engine. The total mass of gas is less than $10M_{\odot}$, so it is utterly negligible compared with the black hole itself.

As you will have noted from Section 3.3, broad lines are not seen in every AGN. The general belief among astronomers is that every AGN has a broad-line region, but in some cases our view of the BLR clouds is obscured by the dust torus, so broad lines do not appear in the spectrum.

In NGC 4151, a prominent type 1 Seyfert galaxy, the broad lines are observed to vary as well as the continuous spectrum. The line variations lag about 10 days behind associated variations in the continuous spectrum. The usual interpretation is that the variations commence in the engine, where the continuous spectrum originates, then take 10 days to 'light up' the broad-line region. Using similar reasoning as we adopted above to measure the size of AGN (Section 3.4.1), the broad-line region must be a distance r of about 10 light-days from the engine. Supposing that the broad lines are Doppler-broadened by rotation around the engine, then one has a picture of regions of gas moving at a speed v of about 7000 km s^{-1} around a central engine of mass M at a radius r. The value of M can now be calculated from v and r, in the same way that the mass of our Galaxy inside radius r was inferred in Chapter 1. Using Equation 1.5,

$$M = v^2 r/G \qquad (1.5)$$

with $r = 10$ light-days (3×10^{14} m), and $v = 7 \times 10^6$ m s^{-1}, and converting into solar masses, we obtain $M = 10^8 M_\odot$. This technique, which goes by the name of **reverberation mapping**, has been used to measure black hole masses in several nearby AGN, yielding masses from a few times $10^7 M_\odot$ to over $10^9 M_\odot$. This range is consistent with the value of M for an accreting black hole calculated from consideration of the Eddington limit.

Narrow-line region

The model places the narrow-line region much further out from the central engine where orbital speeds are lower; 200–900 km s^{-1} is typical for the NLR.

An important consequence of the NLR being outside the dust torus is that it is always in view, so narrow lines will be seen even if the broad-line emitting gas is obscured.

QUESTION 3.13

The narrow-line region is the most extensive part of the AGN and envelops all the other components. Like the other parts, it is illuminated by the central engine. Bearing in mind the geometry of the dust torus, describe what the NLR might look like if a spaceship could get close enough to see it. From which direction would the observers have the best view?

So the model predicts that the NLR, if we could see it, would have a distinctive shape. You might think that such observations would be impossible, considering the tiny size of an AGN. But the NLR is the outer part of the AGN and has no real boundary. In fact, several NLRs have been imaged by the Hubble Space Telescope and one example, for the Seyfert galaxy NGC 5252, is shown in Figure 3.34. The double wedge shape reveals where the gas is illuminated by radiation shining from the centre of the torus. In this case the emission extends several kiloparsecs from the AGN and is known as an *extended narrow-line region*. The extended region is simply interstellar gas ionized by the radiation from the engine. This observation, and others like it, provides supporting evidence for the geometry of the dust torus and the NLR.

So even if we cannot observe the inner structure of an AGN, the regions around the nucleus are tantalizingly consistent with the model.

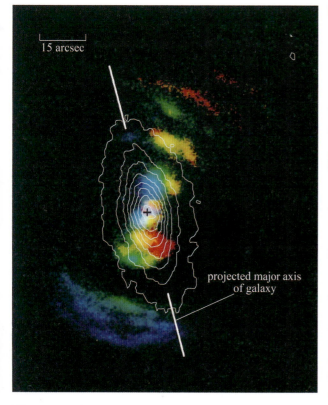

Figure 3.34 NGC 5252 is a type 2 Seyfert galaxy that is about 96 Mpc away. The white contours show the isophotes of the host galaxy (Hubble type S0). The coloured areas show emission from the extended narrow-line region: blue and red regions indicate emission from gas that is moving towards, or away from us, respectively (green and yellow regions have a low radial velocity). The emitting regions form two characteristic wedge shapes, or *ionization cones* that reveal where gas is illuminated by radiation escaping from the poles of the obscuring torus. (Morse *et al.*, 1998 with isophotal data from the Digitized Sky Survey/STScI)

3.5.4 Unified models

You are now familiar with the main components for building models of AGN: a central engine powered by an accreting supermassive black hole (with or without jets), clouds of dust, clouds of gas and accretion processes that can organize the gas and dust into a torus-shaped structure. Many attempts have been made to use these components to explain the different types of AGN. Two basic ideas underlie these models. First, all AGN are essentially the same and differ chiefly in the luminosity of the central engine, which in turn depends on the mass of the black hole and the mass accretion rate. Second, if the AGN contains a dust torus then the radiation observed will depend on the direction from which the AGN is viewed. Two possible schemes for such unified AGN models are shown in Figure 3.35. One is for radio-quiet AGN and the other is for radio-loud AGN.

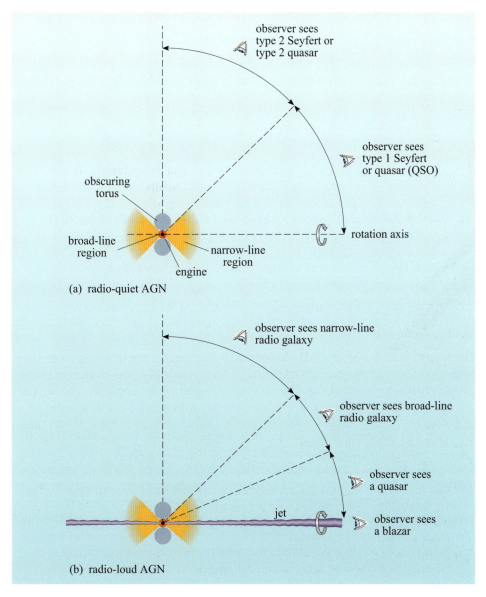

Figure 3.35 Two tentative unified models for AGN. (a) Radio-quiet AGN. (b) Radio-loud AGN. (The angles are approximate.)

Radio-quiet AGN

There has been a great deal of debate about whether there really are two different types of Seyfert or whether they can both be accounted for by the same model.

For example, suppose that you look at the model AGN in Figure 3.35a from a direction not too far from the rotation axis of the torus. You will see X-ray, UV (the 'big blue bump') and other radiation from the engine, broad lines from the broad-line region, narrow lines from the narrow-line region and infrared from the dust torus. You will observe the features associated with a type 1 Seyfert. If you look at the same model from a direction nearer to the plane of the torus, the X-rays and the broad optical emission lines will be hidden by the torus, and you will observe the features associated with a type 2 Seyfert. Observations strongly suggest that at least some type 2 Seyferts are essentially type 1 Seyferts seen from a different angle. This also accounts for the intermediate types, where the broad-line region is only partly in view.

■ The galaxy in Figure 3.34 is described as a type 2 Seyfert. Is this significant?

☐ Yes. In the unified model, type 2 Seyferts are seen from an angle close to the plane of the torus. This is the best viewing angle to see the shape of the NLR, as discussed in Question 3.13.

Does the same apply to other types of AGN? Radio-quiet quasars (QSOs) appear in many respects to be similar to type 1 Seyferts, showing both broad and narrow emission lines, but are much more luminous. There seems little doubt that Seyfert 1 galaxies and radio-quiet quasars differ primarily in luminosity.

It is also now known that there is a population of 'type 2' quasars that are high luminosity analogues of type 2 Seyferts. The high levels of dust obscuration in these sources mean that many of the usual tell-tale signatures of quasars are suppressed. However, as might be expected, such objects are luminous infrared sources, and in detail, their infrared spectra are distinct from the spectra of galaxies undergoing starburst events, and this allows type 2 quasars to be identified.

Radio-loud AGN

The second model (Figure 3.35b) is similar to the first, but now the engine is producing a pair of jets that will eventually end in a pair of lobes, as seen in radio galaxies and some quasars.

Looking at the model from the side, one expects to see narrow lines in the spectrum (but not broad lines) and two jets surrounded by extended lobes. This is a narrow-line radio galaxy. At an angle closer to the jet axis the broad-line region comes into view and a broad-line radio galaxy is seen. So far this is analogous to the two types of Seyfert, but now another effect comes into play. As you saw in Section 3.4.6, relativistic beaming will cause an approaching jet to be brighter than a receding jet, so as the angle decreases one jet will fade at the expense of the other and a radio galaxy with a single jet will now be visible (though there may well be two lobes). As the angle continues to decrease the intense source of radiation surrounding the black hole comes into view and the object appears as a quasar, with never more than one visible jet. Finally, a blazar is seen when the torus is face on to the observer who is looking straight down the jet. One distinguishing feature of the blazars is that the spectrum is dominated by a smooth continuous spectrum, which is what one would expect if the radiation is coming from the jet itself. Another feature of blazars is their rapid variability over a wide range of wavelengths, and this again is consistent with the idea of the emission arising from a jet.

Unification of the radio-loud sources is more contentious and this model is by no means the last word on the subject. It has been difficult to reconcile all the observed properties of the AGN with the model. For example, one test would be to examine whether the numbers of different kinds of AGN are consistent with what the model predicts.

■ Suppose that radio galaxies, radio-loud quasars and blazars were all the same kind of object but seen from different angles. From Figure 3.35b, which would you expect to be the most common? Which the least common?

□ Radio galaxies would be seen over the widest range of angles, so these would be the most common. Blazars, on the other hand, would only be seen over a narrow range of angles and would be relatively rare.

This simple approach is complicated by two things. First, AGN vary greatly in luminosity and distance, so the number observed is not necessarily a measure of how common they are. Powerful or nearby objects are more likely to show up in a survey than weak or distant objects. Second, AGN are visible over such large distances that the light from the more remote ones started on its journey when the Universe was considerably younger than it is today. The most distant quasars may no longer exist in the form in which they are observed. We shall return to that idea shortly.

At the moment the jury is still out, as they say, but astronomers are confident that even if the different kinds of radio-loud AGN are not identical siblings, they are at least close cousins.

Perhaps the most difficult question is why some AGN are radio-loud while most are radio-quiet. You have seen that radio-loud AGN tend to have hosts that are ellipticals. It was once thought that the presence of gas in spiral galaxies acted to suppress the emergence of jets from the engine, but that idea is no longer favoured. Current thinking relates the presence of jets to the angular momentum of the black hole, with only the faster spinning black holes able to produce jets. The novel element is that a high spin rate could be achieved not by accretion but by the merger of two massive black holes following the collision and merger of their host galaxies. As you saw in Chapter 2, there is other evidence that some giant elliptical galaxies are formed from mergers, so this seems a plausible, if yet unproven, explanation as to why the radio-loud sources are preferentially found in elliptical galaxies.

3.6 The origin and evolution of active galaxies

The model for AGN described in this chapter has been the subject of much heated debate and detailed observational tests since the idea of accretion onto supermassive black holes was first suggested as a mechanism to power active galaxies. It is now widely accepted by astronomers that such black holes do exist and that they provide an explanation for many of the observed properties of AGN. While detailed observation and modelling will continue in order to better understand the high energy environment associated with AGN, a major focus of inquiry is now on how the population of active galaxies has evolved over cosmic history and how this fits in to current thinking about the formation and evolution of galaxies in general.

3.6.1 Quasars over cosmic history

At the beginning of this chapter we asked whether active galaxies really are in a class of their own or whether most galaxies go through an active stage at some point in their lives. We can shed some light on this by looking for evidence that active galaxies evolve.

We can sensibly ask 'how long do AGN live?' As indicated earlier, we observe distant objects not as they are today, but as they were at the time their light was emitted. As electromagnetic radiation takes 3.2 million years to travel one megaparsec, even the relatively nearby quasar 3C 273 is seen as it was some 2.5 billion years ago, and those with the highest observed redshifts are seen at less than a billion years after the beginning of the Universe. So by studying remote quasars and comparing them with closer ones, it should be possible to see if they have changed over the lifetime of the Universe.

Astronomers have worked out the numbers of quasars in a given volume of space for different redshifts. When the expansion of the Universe is taken into account (by considering a comoving volume), the number density of type 1 quasars seems to have reached a maximum around a redshift of 2–3. This corresponds to a time of about 10 billion years ago, and the comoving density of quasars has been declining sharply ever since. Indeed, quasars were something like 10^3 times more common then than they are now. This suggests that the quasar phenomenon is short-lived, by cosmic standards. Where have they all gone?

- ■ Bearing in mind what you already know about quasars, what would you expect a 'dead' quasar to look like?

- ☐ As a quasar is believed to be an AGN within an otherwise normal galaxy, a dead quasar would look like a normal galaxy *without* an AGN.

- ■ How could you tell whether a normal galaxy once had a quasar inside it?

- ☐ Look in the nucleus! If the black hole model is correct, dead quasars will leave a supermassive black hole behind them.

So if quasars are indeed powered by supermassive black holes, it should be possible to find the 'relic' black holes in our local region of space, even where there are no obvious AGN. You saw in Chapter 1 that there is compelling evidence that our Galaxy, while not active, also harbours a black hole of mass $4.3 \times 10^6 M_\odot$

in its central region. Furthermore, it is thought that M31 (the Andromeda Galaxy), which is the nearest big spiral to the Milky Way, contains an object of about $1 \times 10^8 M_\odot$, and its small elliptical companion, M32, hides an object of $2 \times 10^6 M_\odot$. This prompts the question: do all galaxies harbour supermassive black holes? This has been the subject of much investigation, and the answer seems to be a qualified 'yes'. Supermassive black holes are present in the majority of galaxies that are dominated by a spheroidal component (i.e. ellipticals, lenticulars and some spirals). So it appears very plausible that active galaxies, rather than being distinct from normal galaxies, are simply a stage that many galaxies go through. A key question then, is whether this active phase is a natural consequence of the way in which normal galaxies evolve.

3.6.2 Active galaxies in the merger scenario

As discussed in Chapter 2, the widely accepted scenario for galaxy evolution is one in which interactions and mergers play an important role. We will follow through the implications of mergers for the provision of material to feed the black holes, but we will start by considering the fate of two supermassive black holes involved in a merger event.

■ If two galaxies, each containing a supermassive black hole, merge to form a large elliptical galaxy, what do you expect has to happen to these black holes?

☐ We would expect that after the merger of the galaxies, the newly formed galaxy would contain a single black hole, presumably formed by the merger of the two black holes from the original galaxies.

So if galaxies grow by mergers, then central black holes are also expected to merge. Is there any evidence for this scenario? Well, before the two black holes merge, they are likely to form a binary system, and although very rare, there are several known cases of galaxies, such as NGC 6240, which have two active nuclei. Evidence for coalescence events themselves is yet to be found, but the process by which two supermassive black holes come together is likely to be a source of, so-called gravitational waves. While gravitational waves have not yet (as of 2014) been detected directly, much research effort is currently going into the development of detectors that will act as gravitational wave observatories, with the potential to open up a new source of information about the distant universe.

An active galaxy is, of course, more than just a galaxy containing a supermassive black hole – there also needs to be accretion of material onto that black hole. Do mergers provide an explanation for the increased accretion rates needed to drive the active galaxy phase? Well, the main barrier to accretion to the centre of a galaxy is the angular momentum of that material. However, galaxy mergers tend to result in a system with low angular momentum per unit mass, such as happens when two spiral galaxies merge to form an elliptical. Numerical simulations indicate that some material, either in the form of stars or gas, can find its way to the central regions of the galaxy, and is potentially available to fuel the accretion disc around the black hole.

The higher rate of mergers and collisions in the past is consistent with the fact that the comoving density of type 1 quasars has been declining over the last 10 billion years. Work is on-going in studying black hole coalescence and the availability of fuel for the accretion process following merger events, but in general, it seems that the evolution of active galaxies fits in well to what is known about how galaxies evolve. However, it may also be the case that the active nucleus itself can play a role in influencing the evolution of its host galaxy, as we shall now see.

3.6.3 Do active galaxies play a role in galaxy evolution?

Many galaxies seem to have been through an active phase in which the accretion luminosity is high enough to affect the environment of the host galaxy. One way in which we know active galaxies are influencing their galactic homes is through outflows. The spectra of about 10% of type 1 quasars show broad *absorption* lines, which is interpreted as evidence for material being blown outwards from their nuclear regions. There is considerable interest in how such processes may influence star formation. This is an extremely complex process – AGN could, in principle, act to heat gas that then flows out of the galaxy much like the galactic fountains mentioned in Chapter 1. This would remove material that could have potentially gone into forming stars. Conversely, the bulk motion of heated gas against molecular clouds within the galaxy may result in compression of those clouds and enhance the star formation rate. Although there remains considerable uncertainty about the detailed mechanisms at play, it is well established that properties of black holes and their host galaxies are intimately linked. In particular, there is a relationship between the mass of the central black hole (M_{bh}) and the velocity dispersion of stars (σ) in the bulge of the host galaxy. This M_{bh}–σ relation is interpreted as strong evidence that supermassive black holes and host galaxies must have evolved together, and current research aims to understand the complexities of this co-evolution.

Note that in the context of AGN research, stellar velocity dispersion is commonly denoted by σ (Greek letter 'sigma') rather than Δv as used elsewhere in this book.

3.6.4 How, and when, did supermassive black holes form?

Given that the majority of massive galaxies contain a supermassive black hole, and that the evolution of the galaxy and its black hole seem to go hand-in-hand, it is of interest to ask how far back in cosmic history does this association go? In other words: which came first – the galaxy or its black hole? A key part of this problem is understanding how black holes form and grow. A mechanism that we know produces black holes is a core collapse supernova, which marks the end point of stellar evolution for a massive star. This process produces black holes with masses of order of magnitude $10 M_\odot$ – far smaller than the $10^8 M_\odot$ required to power a quasar. However, such 'seed' black holes seem to be the most obvious starting points for supermassive black holes. We can estimate how quickly a black hole can grow using similar arguments to those given earlier about the limiting accretion rate (Section 3.4.5). If a black hole is supplied with material at a rate that always matches the Eddington luminosity, and assuming that the accretion process radiates away 10% of the rest mass of the in-falling matter, then the time taken for a black hole to double in mass is 3.4×10^7 years.

QUESTION 3.14

Luminous quasars have been observed at a redshift $z \approx 7$, which corresponds to a time when the age of the Universe was about 750 million years.

(a) Write down an expression for the mass of the black hole, M_{bh} in terms of the mass of the seed black hole, M_{seed} and the number of times, n, that the mass has doubled in the intervening time.

(b) Calculate how many times the mass could double in the time available, stating any assumptions that you make.

(c) Is it possible for a $10M_\odot$ black hole to grow to become a supermassive black hole in this time?

Although the answer to Question 3.14 indicates that it would be just about possible for a supermassive black hole to have formed by $z \approx 7$, it would require that seed black holes were formed in the very early history of the Universe and that some of these black holes then grew at the fastest possible rate for hundreds of millions of years. It is currently not clear if this is a realistic scenario. Understanding this epoch, when galaxies and their black holes formed and grew together, is now a major goal of astrophysics. Progress in this field will require a new generation of observational facilities, especially in the X-ray part of the spectrum, which can probe galaxies and their central black holes at redshifts of 10 and higher.

3.7 Summary of Chapter 3

The spectra of galaxies

- The spectrum of a galaxy is the composite spectrum of the objects of which it is composed.

- The optical spectrum of a normal galaxy contains contributions from stars and HII regions. An elliptical galaxy has no HII regions and has an optical spectrum that looks somewhat like a stellar spectrum but with rather fainter absorption lines. A spiral galaxy has both stars and star-forming regions, and its optical spectrum is the composite of its stars and its HII regions (which show rather weak emission lines).

- The widths of spectral lines from a galaxy may be affected by Doppler broadening due either to thermal motion or to bulk motion of the emitting material.

- An active galaxy has an optical spectrum that is the composite of the spectrum of a normal galaxy and powerful additional radiation characterized by strong emission lines. The broadening comes from bulk motion of the emitting gas.

- A broadband spectrum comprises radiation from a galaxy over all wavelength ranges. To judge a broadband spectrum fairly, it is necessary to use a λF_λ plot on logarithmic axes, which is called a spectral energy distribution (SED).

- The SEDs of normal galaxies peak at optical wavelengths while the SEDs of active galaxies show emission of substantial amounts of energy across a wide range of wavelengths that cannot be attributed to emission from stars alone.

Types of active galaxy

- All active galaxies have a compact, energetic nucleus – an AGN.

- Seyfert galaxies are spiral galaxies with bright, point-like nuclei that vary in brightness. They show excesses at far infrared and other wavelengths, and have strong, broad emission lines.

- Quasars resemble very distant Seyfert galaxies with very luminous nuclei. They are variable. About 10% are strong radio sources thought to be powered by jets of material moving at speeds close to the speed of light.

- Radio galaxies are distinguished by having giant radio lobes fed by one or two jets. They have a compact nucleus like Seyfert galaxies. The compact nucleus is variable, and its emission lines may be broad or narrow.

- Blazars exhibit a continuous spectrum across a wide range of wavelengths and emission lines, when present, are broad and weak. They are variable on very rapid timescales.

The central engine

- An object that fluctuates in brightness on a timescale Δt can have a radius no greater than $R \sim c\Delta t$.

- The point-like nature of AGN and their rapid variability imply that the emitting region is smaller than the size of the Solar System.

- The central engine of a typical AGN is believed to contain a supermassive black hole of mass $\sim 10^8 M_\odot$ and Schwarzschild radius $\sim 3 \times 10^{11}$ m (2 AU).

- Infalling material is thought to form an accretion disc around the black hole, converting gravitational energy into thermal energy and radiation. A typical AGN luminosity of 10^{38} W can be accounted for by an accretion rate of $0.2 M_\odot$ per year.

- The maximum luminosity of an accreting black hole is given by the Eddington limit, at which the gravitational force on the infalling material is balanced by the radiation pressure of the emitted radiation.

- Jets are thought to be ejected perpendicular to the accretion disc.

Models of active galaxies

- The standard model of an AGN consists of an accreting supermassive black hole (the engine) surrounded by a broad-line region contained within a torus of infrared emitting dust and a narrow-line region.

- Unified models attempt to explain the range of AGN on the assumption that they differ only in luminosity and the angle at which they are viewed.

- One type of model attempts to unify radio-quiet AGN. Type 1 Seyferts and type 2 Seyferts differ only in the angle at which they are viewed. Radio-quiet quasars (QSOs) are similar to Seyferts but much more powerful. Evidence for this model is strong.

- Another set of models, in which the engine emits a pair of jets, attempts to unify radio-loud AGN. The observer sees a radio galaxy, a quasar or a blazar as the viewing angle moves from side-on to the jets to end-on.

- The difference between radio-loud and radio-quiet AGN may lie in the angular momentum of their black holes. The faster spinning holes may have arisen from mergers of black holes resulting from the collision of their host galaxies.

The origin and evolution of active galaxies

- Quasars were most abundant at redshifts of 2–3 and have been declining in number for the last 10 billion years.

- It is widely accepted that supermassive black holes are present in the majority of galaxies that are dominated by their spheroidal component. Only a small proportion of these black holes currently power luminous AGN – the remainder are dormant.

- It seems probable that AGN fade with time as the supply of accreting material is used up. Galactic collisions or mergers probably play an important role in allowing material to accrete onto a central supermassive black hole.

- Following the merger of two galaxies, their supermassive black holes are likely to coalesce. The observation of a small population of galaxies with two active nuclei supports this scenario.

- The seed black holes from which supermassive black holes grew must have formed at very early times in the history of the Universe.

Questions

QUESTION 3.15

Suppose that a galaxy has emission lines in its optical spectrum. A line of wavelength 654.3 nm is broadened by 2.0 nm. Estimate the velocity dispersion of the gas giving rise to the broadened spectral line. Is it likely to be a normal galaxy?

QUESTION 3.16

Calculate λF_λ flux densities in W m^{-2} in the radio, the far infrared and the X-ray regions, given the F_λ and λ values listed in Table 3.1. Which wavelength region dominates?

Table 3.1 For use with Question 3.16.

Region	λ	F_λ/W m^{-2} μm^{-1}	λF_λ/W m^{-2}
radio	10 cm	10^{-28}	
far-IR	100 μm	10^{-23}	
X-ray	10^{-10} m	10^{-20}	

QUESTION 3.17

Suppose that an unusual galaxy has broadband spectral flux densities F_λ at wavelengths 500 nm, 5 μm and 50 μm, of 10^{-27}, 10^{-28}, and 10^{-28} W m^{-2} μm^{-1}, respectively. By calculating λF_λ, comment on whether it is likely to be a normal or an active galaxy.

QUESTION 3.18

A particular galaxy has a large luminosity at X-ray wavelengths. One astronomer believes it to be a galaxy that happens to contain a large number of separate X-ray stars. Another astronomer believes that the X-rays indicate an active galaxy. How, by measuring the *spectrum* of the galaxy, could this question be resolved?

CHAPTER 4
THE SPATIAL DISTRIBUTION OF GALAXIES

4.1 Introduction

Having looked at the properties of individual galaxies – both normal and active – in some detail, it is now appropriate to consider how these galaxies are distributed in space.

Surveys of the region outside our own Milky Way show that there are galaxies all around us. Deep field images such as those taken by the Hubble Space Telescope (Figure 2.40) have revealed that galaxies are present in great numbers out to very large distances. As suggested in Chapter 2, these galaxies are *not* distributed uniformly; there is structure present on all but the very largest distance scales.

At the smaller end of this range of scales, a proportion of galaxies are found in *groups* or *clusters*, which consist of local concentrations of tens to thousands of galaxies. Clusters of galaxies typically are no more than a few megaparsecs in diameter, but are themselves organized into larger structures called *superclusters* that can extend for tens of megaparsecs. On even larger scales, the matter in the Universe seems to resemble a three-dimensional network in which regions of high galaxy density are connected by *filaments* and *sheets*. These structures surround large *voids* in which very few galaxies are found. This appears to describe the current distribution of galaxies up to the largest observable scales.

In this chapter we will discuss how astronomers have come to their present understanding of the way in which galaxies, and more generally, matter is distributed throughout space. This has been one of the great scientific endeavours of the past fifty years or so, and is a process that continues with long-term projects that aim to produce precise three-dimensional maps of vast regions of the Universe around us. Such endeavours are conducted with the ultimate aim of understanding how the observed structure has developed over the history of the Universe.

We start, however, by looking at the distribution of galaxies in our own neighbourhood.

4.2 The Local Group of galaxies

Our own Galaxy belongs to a modest concentration of galaxies known as the **Local Group**. The Milky Way itself has about a dozen satellite galaxies, including two particularly prominent ones: the Large and Small Magellanic clouds. These irregular galaxies are visible to the unaided eye as fuzzy patches in the southern sky (Figure 2.7b and c). (Note that images of many of the prominent members of the Local Group are given in Chapter 2.)

Slightly further away lies the largest of the Local Group galaxies – the Andromeda Galaxy or M31 (Figure 2.5b). This spiral galaxy (type Sb) is somewhat more massive than the Milky Way and, at a distance of 0.8 Mpc, is the most distant object visible to the unaided eye.

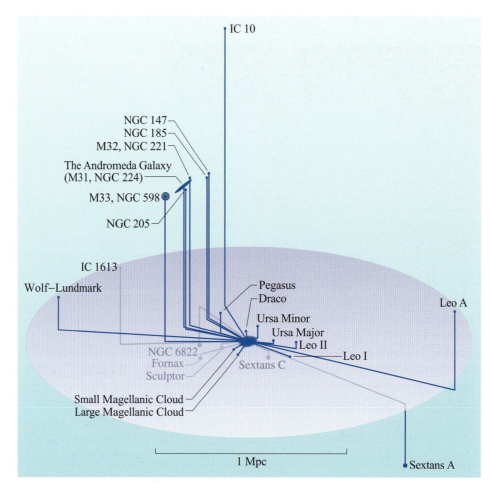

Figure 4.1 The main members of the Local Group of galaxies. Some are named after the constellations in whose directions they lie. In this diagram the Milky Way is located at the centre. (The disc separates the northern and southern halves of the celestial sphere.)

In total, the Local Group contains over 100 galaxies: Figure 4.1 shows how the main, i.e. brighter, members of the Local Group are distributed in space.

The Local Group occupies a volume of space approximately 2 Mpc across. This is to be compared with the 0.03 Mpc diameter of the Milky Way. The Local Group is almost certainly not a transitory bunching of galaxies, but is **gravitationally bound**. Each member moves in an orbit determined by the gravitational influence of the whole Local Group. Furthermore, in a bound system, galaxies cannot normally escape unless ejected as the result of a collision between clusters or other perturbations.

Most of the members of the Local Group are much less massive than the Milky Way. After M31 and the Milky Way itself, the next most massive members of the Local Group are the two Magellanic Clouds and the more distant spiral galaxy M33 (Figure 4.2) – each having a mass approximately an order of magnitude less than that of the Milky Way. Of the remaining galaxies, most are dwarf elliptical galaxies with masses of typically just a few per cent of that of the Milky Way. M32 (visible in Figure 2.5b as a small galaxy just below the centre of the image) is a small companion of M31 and is classified as type E2 since it is slightly elongated. Leo I (Figure 2.8) is a dwarf elliptical at a distance of about 0.25 Mpc.

Dwarf ellipticals may be common but because they are very faint they are difficult to detect. Because of this we are uncertain of the exact number of galaxies in the Local Group; we have already noted that the current count is around 100,

Figure 4.2 M33, a spiral galaxy in the Local Group. (D.Malin/IAC/RGO)

but it is possible that further dwarf ellipticals will continue to be discovered as astronomical techniques improve. This difficulty in detecting faint objects has implications for deep surveys – as we look at more distant groups or clusters, we see fewer of the less bright members.

Typically the term **group** is used to describe a concentration of a relatively small number of galaxies (a few to a few tens of bright galaxies), while **cluster** refers to similar structures containing greater numbers (up to thousands) of bright galaxies. The key physical characteristic of a group or cluster, however, is that they are gravitationally bound. Consequently, they are not just a collection of galaxies that happen to be located close to one another at a given time, but represent a long-lived association of galaxies.

4.3 Clusters of galaxies

Although the number of galaxies in a cluster can vary by a large factor, clusters do not vary so much in physical size: the typical cluster size of a few megaparsecs is not much different from the diameter of the Local Group. Thus *richer* clusters (those with more members) also tend to be more densely packed. Since we can observe galaxies out to distances of hundreds of megaparsecs, clusters are still very small structures on the overall scale of the observable Universe.

The most obvious way to study the distribution of galaxies is simply to photograph large areas of the sky and then to analyse the pattern of galaxies seen in the images. Historically, obtaining suitable images for such studies was a challenge: galaxies tend to be faint objects, so large aperture telescopes and long exposure times were required.

Technical difficulties aside, however, this approach is straightforward enough, and it has been applied since the 1930s. The first major survey was carried out by Harlow Shapley (Figure 1.26) who, together with Adelaide Ames in 1932, catalogued the positions of 1250 galaxies. The images taken by Shapley and Ames showed the first strong indication that galaxies are not distributed randomly in space: they found a number of compact regions containing significantly higher than average densities of galaxies. This survey thus provided early evidence for the clustering of galaxies.

Further surveys followed, adding more clusters to the total. Below we will discuss one of the most important – the Abell catalogue – in some detail. This survey is important because it was the first to introduce a classification scheme for clusters. More recent surveys have extended both the number of clusters discovered and the volume of space surveyed, but the Abell catalogue is still the starting point for astronomers embarking on a study of these objects.

Figure 4.3 shows images of two well-known nearby clusters – the Coma cluster, and the Virgo cluster. These two examples illustrate some of the diversity and variety of clusters. The Coma cluster, which is at a distance of about 100 Mpc from our Galaxy, is a spherically symmetrical cluster consisting mainly of elliptical galaxies. By contrast, the Virgo cluster, which is about 20 Mpc distant, is much more irregular in shape, and contains a mixture of ellipticals and spirals. A feature that is common to both clusters however is the fact that each contains over a thousand galaxies.

Although clusters can contain many galaxies, it is important to appreciate that not all galaxies reside in clusters. In fact, the vast majority of galaxies exist outside of clusters. A galaxy that is not part of a cluster is called a **field galaxy**, and as we will see, care must be taken to identify and exclude these from cluster surveys.

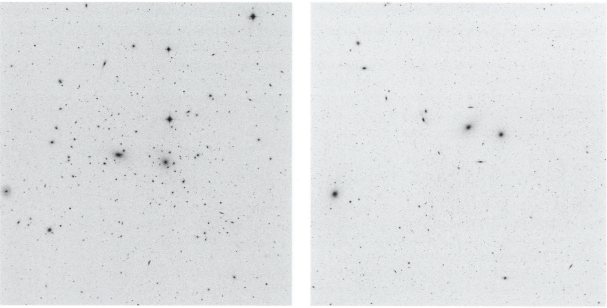

(a) Coma   (b) Virgo

Figure 4.3 Optical images (visual band) of clusters of galaxies. (a) The Coma cluster of galaxies which lies at a distance of about 100 Mpc from our Galaxy. (b) The Virgo cluster of galaxies, which is about 20 Mpc from the Milky Way. Both the Coma and the Virgo clusters of galaxies contain over a thousand galaxies. The fields of view of these images are relatively wide: $0.75° \times 0.75°$ in (a), and $2.5° \times 2.5°$ in (b). (Digitized Sky Survey/STScI)

4.3.1 The identification of clusters from imaging surveys

In 1958 George Abell (Figure 4.4) published a catalogue of 2712 clusters of galaxies that was the starting point for detailed study of these objects. In this section we will review the process by which Abell constructed this catalogue since it highlights some of the difficulties that have to be overcome by astronomers who endeavour to survey the Universe on large scales.

Abell's survey used plates taken using a special type of telescope called a Schmidt telescope (or Schmidt camera) that is well suited to taking images that cover large areas on the sky. During the mid-1950s the 48-inch Schmidt telescope at the Mount Palomar Observatory (Figure 4.4) had been used to create a detailed photographic atlas of the sky. The images on these plates – each with a field of view just over 6 degrees square – together covered approximately 75% of the celestial sphere, including most of the northern sky and part of the southern. Abell used 879 of the 935 plates of the full survey as the basis for his search for clusters of galaxies. He examined the survey plates by eye to look for regions containing larger than average concentrations of galaxies. Later, he and his co-workers extended the catalogue to include more of the southern sky.

Abell's catalogue is significant because, for the first time, it contained a sufficiently large sample of clusters to allow a meaningful comparison of their different characteristics. The scale and extent of the survey also allowed the spatial distribution of clusters to be analysed for the first time.

Based on their visual differences, Abell was able to classify clusters according to various criteria. The most important of these is one that describes how many galaxies there are within a cluster. Abell called this property the **richness** of a

GEORGE OGDEN ABELL (1927–1983)

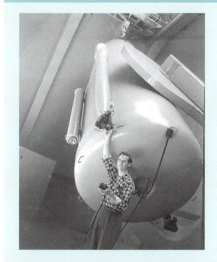

Figure 4.4 George Abell standing next to the Palomar 48-inch Schmidt telescope. This design of telescope has a very wide field of view, which facilitates the surveying of large areas of the sky. (California Institute of Technology)

George Abell (Figure 4.4) began his astronomical career as an observer on the Palomar Sky Survey (the survey on which his study of clusters of galaxies was based), and one of his early studies was to use the survey plates to examine low surface brightness planetary nebulae. Along with Peter Goldreich, Abell helped to establish the connection between these objects and the final stages of life of red giant stars.

For most of his career, he held a faculty position at the University of California, Los Angeles. He was an enthusiastic and popular teacher and subscribed to the view that in teaching science, it is more important to explain how we establish scientific knowledge than to simply present students with surprising or remarkable facts. He was committed to bringing science education to a broad audience: one example of this was his involvement in the production of a series of television programmes about relativity and cosmology in a collaboration between the Open University and the University of California.

cluster. Rich clusters are those that contain relatively high numbers of galaxies. However, as you saw in Chapter 2, it is difficult to detect faint galaxies, so a meaningful study of the number of galaxies in a cluster has to be based on the number of galaxies that exceed a certain threshold luminosity.

A vital piece of information that was needed in this process of detecting and classifying clusters was the distance of each. Abell had no direct means to measure the distances to all the galaxies in the survey, but he was able to use the apparent magnitudes of galaxies in a given cluster as the basis for estimating its distance. Specifically, he used the results of previous studies that indicated that the tenth brightest galaxy in each cluster should have about the same intrinsic luminosity. Thus the tenth brightest galaxy could be taken as a form of *standard candle*. This method did not allow Abell to calculate precise distances for each cluster – the values obtained were still very rough estimates. They were sufficient, however, to distinguish between clusters that were nearby and those that were more distant.

Abell's method of defining and selecting clusters was based on counting the galaxies within a circle of a certain radius on the photographic plate. Abell assumed that all clusters had roughly the same physical size: he estimated that clusters had a radius of about 2 Mpc. Subsequent studies have shown that this was a good assumption, and this 'standard' cluster radius is now known as the **Abell radius R_A**.

QUESTION 4.1

A cluster with an angular diameter of 1.9° is estimated to lie at a distance of 120 Mpc from the Earth.

(a) Calculate the diameter (in Mpc) of this cluster.

(b) What would the angular diameter of the same cluster be if it were at a distance of 420 Mpc?

Abell's work on defining clusters was very methodical – he was well aware that the presence of field galaxies and chance alignments between galaxies along a particular line of sight could give rise to spurious identifications of clusters. Since he wanted to minimize the number of false identifications in his catalogue he developed tests to identify clusters that are, to a high degree of statistical certainty, genuine associations of galaxies. The actual criterion used to define such a cluster was that the cluster must contain more than fifty members that exceeded a certain luminosity, and that these galaxies were located within a volume of space with radius R_A. Out of his original sample of 2712 suspected clusters, he identified 1682 cases that were statistically very likely to be genuine. Subsequent studies have shown that the vast majority of these objects are true clusters.

Abell also classified clusters of galaxies according to their symmetry, grading them on a scale running from *regular* to *irregular*. The regular clusters tend to be giant systems with spherical symmetry and a high degree of central concentration, and irregulars tending to be more open with low central concentrations and a significant amount of 'clumpiness' or subclustering.

Another type of study that can be based on imaging surveys relates to the morphological types of galaxies that exist within clusters. It appears that the proportion of galaxies of different morphological type depends on the symmetry of the cluster. Regular clusters such as the Coma cluster contain relatively few spiral galaxies, and are rich in lenticular and elliptical galaxies. This tendency is not exhibited by irregular clusters (such as the Virgo cluster), which seem to contain a higher proportion of spiral galaxies.

Abell's survey generated a number of interesting results: by adopting a methodical approach and introducing a classification scheme, Abell could do far more than merely catalogue the positions and richness of the clusters. He was also able to consider how the *distribution* of clusters – both over the surface of the sky and as a function of distance – might give information about the existence of larger scale structures.

One difficulty in mapping the distribution of clusters of galaxies is that it is not possible to observe the entire sky: external galaxies can only be seen in the part of the sky not obscured by our own Galaxy. This can be appreciated from maps that display the entire sky, such as Figure 4.5, which shows the positions of Abell clusters. Note that this map shows the celestial sphere in *equatorial* coordinates: the upper and lower halves of the map represent the northern and southern celestial hemispheres respectively. When a map of the celestial sphere is shown in such a way, the plane of the Galaxy snakes around the sky in an 'S'-shaped curve, as shown in the inset to Figure 4.5. It can be seen that only clusters that lie well away from the Galactic plane are visible.

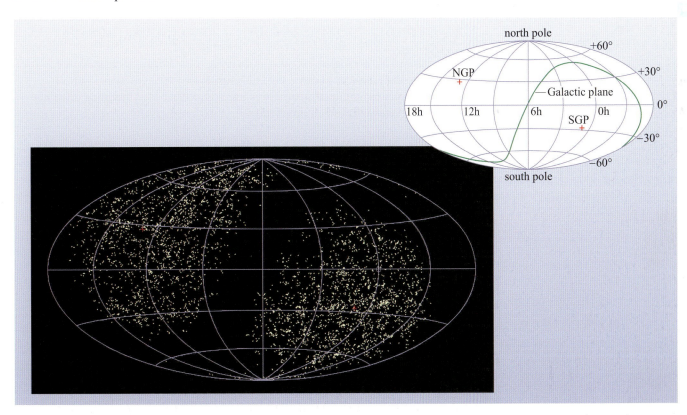

Figure 4.5 This map of the whole sky shows the distribution in equatorial coordinates of Abell clusters. Note that no clusters can be observed close to the Galactic plane because of obscuration due to the interstellar medium of the disc of our Galaxy. (The clusters are those identified in Abell's 1958 study of the northern hemisphere, and a similar study of the southern hemisphere that was published by Abell, Corwin and Olowin in 1989.) The red crosses show the North Galactic Pole (NGP) and the South Galactic Pole (SGP).

As mentioned above, another difficulty was that it was only possible for Abell to make rough estimates of distance. Modern redshift measurements together with improvements in the determination of the Hubble constant have allowed distances to clusters to be determined much more accurately. Within these constraints it was nevertheless possible to reach some overall conclusions on cluster distribution. Abell did not find much variation in the distribution of clusters with distance – there were just as many clusters at large distances as at smaller distances. Looking at the distribution *across* the sky, however, it was apparent that clusters are themselves *not* scattered randomly: although clusters were found in all parts of the observable sky their distribution, as can be seen from Figure 4.5, is far from uniform. Abell's data therefore suggested the existence of structure on larger scales than individual clusters.

QUESTION 4.2

The most distant Abell cluster has a redshift of 0.25. Approximately how far away is this cluster? (Assume that $H_0 = 72$ km s^{-1} Mpc^{-1}.)

Comment: as you will see in Section 4.4.1, the simple relationship between redshift and distance that we have used here becomes increasingly inaccurate for redshifts greater than about 0.2. However, this result can still be considered as a rough estimate of the distance probed by the Abell catalogue.

More recent surveys have probed out to much greater distances than Abell's catalogue. As we look out to greater and greater distances from the Earth, galaxies become fainter, but they also become more numerous. For both these reasons it has become impractical to carry out such surveys by manual inspection of photographic plates, and modern surveys use computer-based methods in which digital images are analysed automatically.

Photographic and digital imaging surveys have catalogued many thousands of clusters and clearly show that galaxies are not distributed randomly. But this type of survey has some limitations. In particular, imaging surveys are essentially *two-dimensional*: they show the positions of the galaxies as projected onto the celestial sphere but do not directly provide information on the *distance* from the Earth.

Space, however, is of course three-dimensional. In order to build up a clear picture of the large-scale structure of the Universe it is necessary to add *accurate* distance information to the two-dimensional position information. This allows a fully three-dimensional map to be built up showing the volume distribution of galaxies throughout space, as we shall see in Section 4.4.

In order to work towards the ultimate aim of understanding how clusters form and evolve, it is necessary to learn something about the physical properties of clusters such as mass and composition, and it is to these two aspects of clusters of galaxies that we now turn.

4.3.2 The masses of clusters of galaxies

In common with most astronomical objects, it is the total mass of a cluster of galaxies that is the single most important physical property that astronomers are interested in determining. Once the mass of a cluster is known, it becomes possible to understand the gravitational influence that the cluster has on its environment. Furthermore, knowledge of the total mass of the cluster will allow estimates to be made of the relative proportions of luminous and dark matter.

Mass, however, is not a property that can be directly measured: instead it has to be inferred from measurements of observable quantities – such as the wavelengths of spectral lines. The following three sections describe different methods of determining the masses of clusters of galaxies. Although these techniques are based on different physical effects, it is found that all three give very similar answers for cluster masses.

Estimation of cluster mass using velocity dispersion

In Chapter 2 you learned about the *virial theorem*, which is used to determine the masses of galaxies. The virial theorem states that the magnitude of the total gravitational potential energy of a bound system is equal to twice the total kinetic energy. In this way, the distribution of *velocities* (which is related to the kinetic energy) can be related to the overall *mass* of the system (which is related to the gravitational potential energy).

Clusters are collections of galaxies rather than stars, but the same principle applies – individual galaxies within a cluster will move under the influence of the gravitational field of the total mass in the cluster. The significance of this is that the velocities of galaxies within a cluster are observable quantities – they can be measured using Doppler shifts and this gives us a method for estimating the total cluster mass.

This is exactly the same principle as the method described in Section 2.3.2 for determining the masses of elliptical galaxies using the velocity dispersion of individual stars within the galaxy. In the case of clusters, the only modification is that it is the velocity dispersion of galaxies within the cluster that is used rather than that of stars within a galaxy.

- ■ What assumptions need to be made for the virial theorem to hold?

- □ The system must be virialized – the cluster must be in a steady state – neither expanding nor contracting and the distribution of velocities of the galaxies must be unchanging. (See Box 2.1.)

When it first begins to form, a cluster may be far from being virialized. Over time, collisions and other interactions between the individual galaxies, gas and dark matter within the cluster will cause their energy to be redistributed. Eventually the motions will settle down into a steady state where further interactions do not change the distribution of kinetic and potential energies. This state is sometimes referred to by describing a cluster as being *relaxed* or in a state of *dynamic equilibrium*. Some clusters – especially those that show a high degree of symmetry – are thought to be virialized. Clusters that appear irregular are far less likely to have reached this state, and so it may not be appropriate to apply this method of mass determination in such cases.

The redshifts of galaxies within clusters can be used to determine their velocities in the radial direction (i.e. along the line of sight). As for the case of stars within elliptical galaxies (Chapter 2) the kinetic energy can be characterized by the velocity dispersion. Then the mass of the cluster is given by:

$$M = \frac{R_A (\Delta v)^2}{G} \tag{4.1}$$

where R_A is the Abell radius and Δv the dispersion in the line of sight velocities of the cluster members.

Clearly, it is important to ensure that only galaxies belonging to the cluster are included. Care must be taken to ensure that foreground or background galaxies along the line of sight are identified and excluded from the velocity dispersion measurements.

QUESTION 4.3

In the Virgo cluster the (elliptical) galaxies show a velocity dispersion Δv of 550 km s^{-1} (this value is given to a precision of 2 significant figures). Calculate the mass of this cluster. Express your answer in solar masses.

The Swiss-American astronomer Fritz Zwicky was the first to apply the virial theorem, using it in the 1930s to estimate the mass of the Coma cluster. Surprisingly, he found that the mass was much larger than the sum of the masses of the individual member galaxies. Historically, this was one of the first indications of the presence of dark matter in the Universe.

We saw in Chapters 1 and 2 that there is evidence for the existence of dark matter within our Galaxy and other galaxies. The results of studies of the masses of clusters of galaxies indicate that a cluster as a whole must include a large amount of dark matter surrounding the galaxies. The conclusion that has been reached is that the luminous matter in clusters (including hot, X-ray emitting gas that we will discuss shortly) accounts for only a small proportion of the mass, with the remaining 70% to 90% of the total cluster mass provided by dark matter.

FRITZ ZWICKY (1898–1974)

Fritz Zwicky (Figure 4.6) was born in Varna, Bulgaria of Swiss parents. He studied in Switzerland and retained his Swiss nationality even when he moved to California Institute of Technology (Caltech), USA in 1925.

Zwicky originally trained as a crystallographer but became interested in the advances being made in astronomy at Caltech and nearby observatories. He remained professor of astronomy at Caltech until he retired in 1968. Zwicky was full of self-belief and came up with a lot of revolutionary ideas, largely intuitively.

His work on clusters of galaxies led to the idea of the existence of large amounts of dark matter in the Universe. Not only was Zwicky the first to infer the presence of dark matter but he also went on to suggest that the *gravitational lensing* of background galaxies would be the most direct way to probe the dark matter in the Universe.

Figure 4.6 Fritz Zwicky. (California Insitute of Technology)

Cluster mass from X-ray emission

Clusters of galaxies are not only visible in the optical part of the spectrum: they also produce strong X-ray emission. In 1971, results from the pioneering X-ray satellite Uhuru revealed that clusters of galaxies are among the brightest X-ray sources in the sky. This X-ray emission arises from a vast quantity of very hot gas (typically at temperatures of between 10^7 and 10^8 K) that pervades the intergalactic space within the cluster.

In recent years, advances in X-ray astronomy have allowed clusters to be identified from X-ray surveys, often more efficiently than using optical imaging methods. Optical observations of clusters suffer from the problem of distinguishing true members of a cluster from other galaxies that are not associated with the cluster, but happen to lie along the same line of sight. This is much less of a problem for X-ray observations because there are far fewer X-ray sources that could be incorrectly attributed to emission from a cluster. A cluster appears as an extended region of diffuse X-ray emission (Figure 4.7) that shows no variability with time. Other extragalactic sources of X-ray emission are normal and active galaxies and these are quite distinct from clusters of galaxies. As you saw in Section 3.2.2, normal galaxies typically have very low luminosities at X-ray wavelengths. Active galaxies, which can be luminous X-ray sources, are unresolved sources (i.e. point-like) that are often variable.

So how are the X-rays produced? Closer examination reveals that the X-rays from clusters form a broad continuous spectrum, with some emission lines. The broad continuous spectrum is characteristic of a mechanism known as **thermal bremsstrahlung**, which is normally associated with very hot ionized gas (Box 4.1, overleaf).

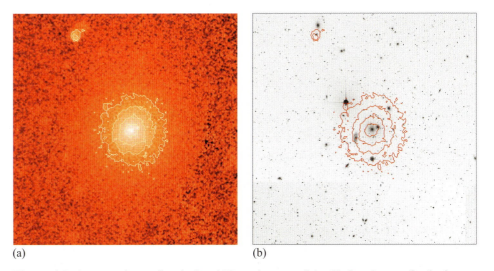

(a)   (b)

Figure 4.7 A comparison of optical and X-ray images of the Hydra cluster of galaxies, which is about 50 Mpc from Earth. (a) An X-ray image of the Hydra cluster, as observed using the ROSAT X-ray Observatory. The image shows the emission from a large cloud of gas, several Mpc across, that is at a temperature of about 3×10^7 K. (b) An optical image in the visual waveband of the Hydra cluster (shown as a photographic negative), with contours of X-ray emission overlaid. It can be seen that the hot X-ray emitting gas fills the space between the galaxies in the cluster. (Both (a) and (b) show the same field of view, and have an extent of $1° \times 1°$.) ((a) NASA; (b) Optical data from the Digitized Sky Survey/STScI)

The presence of bremsstrahlung X-ray emission is evidence for the presence of large quantities of hot ionized gas within the cluster. This gas is predominantly composed of hydrogen and helium, although it does contain a small fraction of heavier elements. As can be seen from Figure 4.7b this hot **intracluster medium**, or **ICM**, is present between the galaxies, permeating the cluster out to a radius of a few megaparsecs. This makes clusters appear as extended, diffuse areas of X-ray emission.

BOX 4.1 THERMAL BREMSSTRAHLUNG

Thermal bremsstrahlung is an X-ray emission mechanism that typically takes place in a high temperature, low-density plasma. As a free electron passes close to an ion in the gas it is deflected (Figure 4.8) without being captured. As a result of this acceleration the electron emits a photon while at the same time losing a corresponding amount of kinetic energy and slowing down a little.

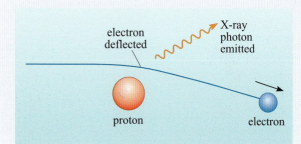

Figure 4.8 The mechanism for Bremsstrahlung emission.

X-rays generated in this way are known as *bremsstrahlung* (a German word that means *braking* or *deceleration* radiation). This type of X-ray emission

is also called *free-free* emission because the electron moves freely both before and after the encounter with the ion.

The energy ε_{ph} of a photon generated by this process depends on the average thermal energy of the electrons and is given approximately by:

$$\varepsilon_{ph} \sim kT$$

where k is the Boltzmann constant (1.38×10^{-23} J K^{-1}) and T is the temperature of the gas. It is common in X-ray astronomy to express photon energies in terms of kilo-electronvolts (1 keV = 1.60×10^{-16} J). Photon energies in the range 1 to 10 keV are typical for X-ray emission from clusters, corresponding to temperatures of about 10^7 to 10^8 K.

Because the X-rays are produced by the acceleration of *free* electrons, the spectrum of bremsstrahlung emission is a smooth continuum (Figure 4.9). This is characteristic of a fully ionized gas where the electrons are not bound to individual atoms. Bremsstrahlung spectra are thus distinct from the *line spectra* that are produced when electrons make transitions between the energy levels of atoms.

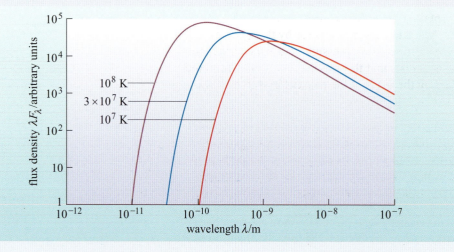

Figure 4.9 The spectral energy distribution due to hot gas (at three temperatures as indicated) emitting by the mechanism of thermal bremsstrahlung. Note the broad continuous spectrum that peaks at shorter wavelengths for higher gas temperatures.

With the benefit of X-ray images, our view of a typical cluster now becomes a large cloud of hot (10^7 to 10^8 K) ionized gas with the galaxies embedded in it.

■ The temperature of the intracluster medium at the centre of a cluster is typically around 10^7 to 10^8 K – similar to that found at the centre of the Sun. Why, then, do we not see strong X-ray emission from the hot plasma within the core of the Sun and other stars?

☐ X-rays generated in the core of a star do not escape directly: in the dense environment of the solar interior, photons are repeatedly scattered and reach thermal equilibrium with matter. Eventually, the energy that was originally in the form of X-ray photons escapes from the outer layers of Sun – the relatively cool photosphere – at ultraviolet, visible and infrared wavelengths.

Although the total mass of gas in a cluster is much larger than that of the Sun, it is spread out over a vast volume of space. The overall density of the ICM is therefore many orders of magnitude lower than the density of material in a star, and this allows the X-rays to escape. In this respect the intracluster gas is similar to the tenuous gas in the Sun's corona – although it should be noted that intracluster gas is many orders of magnitude less dense than gas in the solar corona.

To appreciate how this X-ray emission can be used to estimate the mass of a cluster it is necessary to consider how the intracluster gas supports itself. An important idea here is the concept of *hydrostatic equilibrium* similar to that used to model the stability of a star. A spherical region of gas will tend to collapse under the influence of its own gravity. In the case of a cluster, the gravitational field is produced not only by the mass of the gas itself, but also by the mass of the galaxies in the cluster, together with the mass of any dark matter.

The ICM will be supported against this gravitational collapse by the pressure of the gas. In equilibrium, the pressure changes with distance from the centre of the cluster in a way that exactly balances the effect of gravity. The temperature and density of intracluster gas can be measured from X-ray observations. The pressure of the gas can be calculated once the temperature and density are known, and by using the relationship that balances gravity against the pressure gradient, the total mass of the cluster can be inferred.

The cluster masses that have been obtained from X-ray observations are typically much higher than the mass that can be accounted for by the galaxies and intracluster gas alone. For a given temperature, pressure gradients within the ICM are *higher* than expected. This suggests that the gravitational field within the cluster is *stronger* than that provided by the mass of the galaxies and intracluster gas alone. Total cluster masses of 10^{14} to $10^{15} M_\odot$ are typical.

Again, this is much greater than the mass that can be accounted for by the galaxies in the cluster – which only constitute about 10% of the total mass. Furthermore, X-ray observations also allow an estimate to be made of the mass of gas in the intracluster medium and this is typically found to account for between 10% and 30% of the total mass of a cluster. The remaining mass is believed to be made up mainly of dark matter.

Cluster mass from gravitational lensing

A completely different approach to measuring cluster masses is based on the effect that gravity has on light. Einstein's general theory of relativity makes a number of predictions. One of these predictions is that, in addition to affecting the paths of objects such as planets, gravity can also affect light: the path of light will be bent if it passes close enough to a sufficiently massive object. This prediction was confirmed in 1919 by an expedition to the island of Principe, off the coast of West Africa, led by Sir Arthur Eddington. By measuring the positions of stars during a total solar eclipse (Figure 4.10), the bending of starlight passing close to the surface of the Sun was measured and found to be in agreement with Einstein's theory.

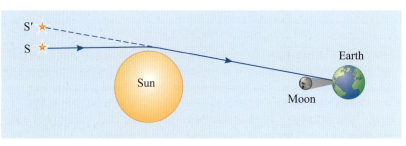

Figure 4.10 A schematic illustration of an eclipse observation of the gravitational bending of starlight. In order to be deflected by a significant amount, a ray of light from a distant star must pass very close to the surface of the Sun. Normally it would not be possible to see this effect, but during an eclipse the Moon obscures the extremely bright solar disc, allowing stars to be observed close to the limb of the Sun. Light from star S is deflected by a small amount, making its position appear to shift to S'.

This bending of light is a weak effect: Eddington's measurement was only possible when light passed very close to the surface of the Sun, and even then the angular deviation of just 1.74 arcsec was so small as to be barely measurable.

The deflection of rays of light becomes larger as the mass of the deflecting object increases. Since a cluster of galaxies typically has a high mass, we might expect that it could act as a **gravitational lens** and bend the paths of light rays from an object lying behind it, as shown in the simple arrangement in Figure 4.11. Rays of light from the galaxy at position S are deflected as they pass close to the cluster, and the observer at O sees an image at position I. As well as this change in apparent position, the gravitational lens also causes the image of the background galaxy to be distorted, and typically produces multiple images. This effect is termed **strong gravitational lensing**.

Many cases of strong gravitational lensing have been observed. In addition to causing multiple or distorted images, gravitational lenses can also increase the apparent brightness of distant objects, in much the same way as a conventional lens focuses light. By concentrating light in this way, one effect of gravitational

Figure 4.11 Schematic diagram of a gravitational lens. Light from the distant galaxy S is deflected as it passes a cluster of galaxies, causing a distorted image of the original galaxy to appear at I. Depending on the distribution of mass in the cluster, several images may be formed in different positions. It is important to note that unlike a glass lens, a gravitational lens does not focus light, it merely re-directs rays of light.

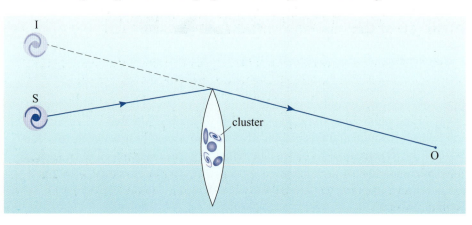

lensing is that it allows observation of distant objects that would normally have been too faint to be seen, and this is of particular interest to astronomers interested in studying galaxies at very high redshifts.

An example of strong lensing is shown in Figure 4.12. Lensing by the cluster CL 0024+1654 has produced five images of a more distant galaxy. The images of this galaxy are seen as blue elongated rings, one near the centre of the image and four others further out.

How can gravitational lensing be used to measure the masses of clusters of galaxies? Clearly the amount of distortion seen will depend in some way on the mass of the cluster that is acting as a lens. Calculating the exact distribution of mass in the lensing cluster can be quite difficult: just as with a glass lens, the exact shape of the lens will determine the nature of the image distortion and magnification. In cases like CL 0024+1654 the pattern of distortions is very complicated – rather like looking at the distant galaxy through the bottom of a glass bottle.

In order to illustrate how strong lensing is used to measure cluster mass, it is useful to consider simpler situations than CL 0024+1654. Occasionally the multiple images resulting from gravitational lensing form almost symmetrical arcs surrounding the centre of the lensing cluster. A particularly spectacular example of this can be seen in the cluster Abell 2218 (Figure 4.13).

Figure 4.12 Lensing by CL 0024+1654. Several distorted images of a distant blue galaxy can be seen encircling the yellower galaxies within the cluster. (W. N. Colley and E. Turner (Princeton University), J. A. Tyson (AT & T Bell Labs, Lucent Technologies) and NASA)

Figure 4.13 Gravitational lensing in Abell 2218. Note the arcs of concentric circles formed by the lens. (NASA, A. Fruchter and the ERO team)

Figure 4.14 shows a schematic view of this type of lensing situation. Here the distribution of mass in the lensing cluster is symmetrical and concentrated at the centre. The distant galaxy S is located exactly along the centreline. The paths of light rays passing either side of the cluster are distorted equally, resulting in symmetrical rings or arcs similar to those seen in Abell 2218.

If the alignment of source and lens is perfect, then the resulting image takes the form of a complete ring surrounding the lens. Such a ring is known as an **Einstein ring** and several examples like B 1938+666 shown in Figure 4.15 (overleaf) have been discovered.

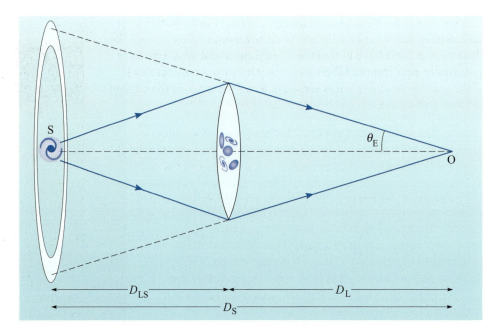

Figure 4.14 The geometry of a source imaged as a ring with angular radius θ_E by a foreground mass (assumed to be symmetrical).

■ For a given cluster-to-source distance, how would you expect the angular radius of the Einstein ring to depend on the mass of the lensing cluster?

☐ The greater the mass of the cluster, the more the path of the light rays would be bent. So a more massive cluster would produce a ring of greater angular radius.

Approximating the mass distribution in Figure 4.14 as a point mass, the image of the distant object would appear as a complete ring with an angular radius θ_E that is given by:

Note that distance measurements in an expanding Universe need to be carefully defined. The distances used in Equation 4.2 refer to distances defined by the apparent angular sizes of distant objects.

$$\theta_E = \sqrt{\frac{4GM}{c^2} \frac{D_{LS}}{D_L D_S}} \qquad (4.2)$$

where D_S is the distance between the observer and the distant source galaxy, D_L is the distance from the observer to the gravitational lens, and D_{LS} is the distance from the gravitational lens to the source.

Using Equation 4.2, the mass M of the lensing cluster can be estimated. (Note that similar calculations can be carried out for the less symmetrical situations where the lensing has not produced a complete ring.)

QUESTION 4.4

The largest lensed arcs in the image of Abell 2218 shown in Figure 4.13 have an angular radius θ_E of approximately 1.0 arcmin. This cluster is one of the most distant in the Abell catalogue: with a redshift of $z = 0.17$, it lies at a distance of approximately 700 Mpc from Earth. Using these values, and assuming that Equation 4.2 can be applied to this gravitational lens, estimate the mass of Abell 2218 in solar masses. Assume that the cluster is mid-way between the distant background galaxies and the Earth, and use a point mass approximation.

In practice, the point mass approximation is not valid, so astronomers use more sophisticated descriptions of the foreground mass distribution.

Gravitational lensing has an advantage over the other methods discussed in that it relies only on the distribution of mass within the cluster – we don't have to make assumptions about virialization or hydrostatic equilibrium. Of course, since lensing is a direct result of the gravitational field, it is sensitive to *all* the mass in the cluster, whether from galaxies, gas or dark matter.

Clusters also affect the paths of light rays from a wider field of background galaxies rather than from just the one or two that might typically lie almost directly behind the lensing mass. This causes a weaker gravitational lensing effect, where fluctuations in the density of matter cause the images of background galaxies to be very slightly stretched and distorted. This effect is known as **cosmic shear** and is similar to looking at the tiled bottom of a swimming pool through the ripples on the surface (Figure 4.16).

■ What effect would dark matter have on the passage of light if it were distributed uniformly throughout the Universe?

☐ None: if the distribution of dark matter were completely smooth, the light would not be deflected one way or the other. By analogy, the bottom of swimming pool would not appear distorted if the surface were smooth – it is the ripples that cause the distortion. Only if there is a non-uniform distribution of mass would deflections be seen.

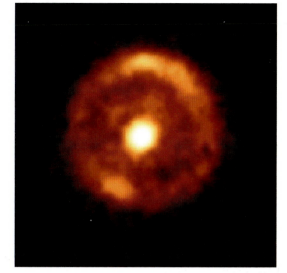

Figure 4.15 The Einstein ring B 1938+666 as observed with the Hubble Space Telescope. In this case the foreground lens is a galaxy seen as the bright spot in the centre. (L. J. King (University of Manchester) and NASA)

In the example of the swimming pool, the grid pattern of the tiles on the bottom of the pool makes even small distortions easy to see. The square tiles become distorted by the ripples on the surface: this type of distortion is known as a *shear*, hence the term *cosmic shear*. The sky, of course, does not have a convenient grid painted on it! Instead, use can be made of very distant galaxies that are revealed by sensitive surveys: if you look to a great enough distance there are a vast number of background galaxies in every tiny patch of sky.

This **weak gravitational lensing** technique is very powerful when applied to clusters of galaxies as it potentially allows not only the total cluster mass to be determined, but also a mapping of the distribution of mass in the cluster. An example of this technique in practice is shown in Figure 4.17, which shows a density map (projected onto the plane of the sky) of the cluster Abell 1758, and compares it to the image of X-ray emission from this cluster. This is quite a

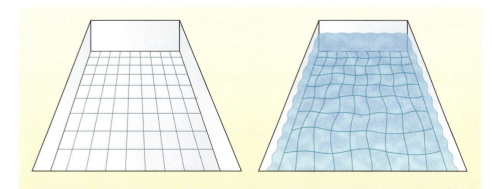

Figure 4.16 The bottom of a swimming pool can appear distorted when viewed through ripples on the surface. In a similar way, light from distant galaxies can be distorted as it passes through a non-uniform distribution of matter.

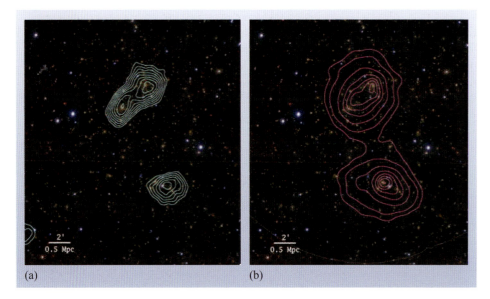

Figure 4.17 The double cluster Abell 1758 (redshift $z = 0.279$). (a) The projected surface density as determined from measurements of the cosmic shear of images of background galaxies. (b) The X-ray emission from the cluster as measured with the Chandra Observatory. In both (a) and (b), the respective maps are superimposed on a 3-colour optical (visual and near-IR) image. (von der Linden *et al.*, 2014)

complex cluster – it is clear from the images that it is a double cluster, and detailed analysis shows that in each of these clusters a merger of two smaller components is taking place. However, the key point is that the density map as determined by weak lensing, closely matches the distribution of hot X-ray emitting gas.

Masses of clusters estimated from strong and weak gravitational lensing lie within the range of 10^{14} to $10^{15}M_{\odot}$, in line with the other methods of mass determination. In those cases where detailed comparisons can be made against masses determined from velocity dispersion or X-ray emission techniques, the masses from gravitational lensing studies appear to agree well.

4.3.3 The composition of clusters

As we have seen, cluster masses can be estimated by three independent methods: velocity dispersion, X-ray emission, and gravitational lensing. The results from these methods are all fairly consistent, typically agreeing within a factor of two. Typical masses range from less than $10^{14}M_{\odot}$ for the smallest clusters and groups to $10^{15}M_{\odot}$ for the richest clusters. Furthermore, the fact that three different techniques based on completely different physical principles are in such good agreement is compelling evidence for the existence of dark matter.

In addition to measuring mass, these three methods also enabled us to learn something about the different constituents of a cluster – galaxies, gas and dark matter. The results suggest that only a small percentage of the total mass of a cluster is contained in the galaxies themselves, with about 10% to 25% in the intracluster medium, and the remaining 70% to 90% as dark matter (see Table 4.1). The picture that has emerged is of a diffuse cloud of gas and dark matter that surrounds the galaxies and permeates the space between them.

Table 4.1 Relative contributions to the total mass of a cluster of galaxies due to its three major constituents.

Constituent	Contribution to total mass
galaxies	<10%
intracluster gas	10–25%
dark matter	70–90%

As described in Chapter 2, each individual galaxy has its own complement of gas (in the form of *interstellar* medium) and dark matter. We could therefore think of a cloud of gas and dark matter filling the whole cluster, with individual denser halos around each galaxy. The density of material is highest around and within the member galaxies, but the large volume of the cluster compared to that of the galaxies means that the intracluster gas and dark matter – spread between individual galaxies – makes up most of the total mass of the cluster.

The fact that galaxies, intracluster gas and dark matter seem to occupy the same locations in space is not surprising, but allows for an alternative interpretation – that there is no dark matter and the effects that we attribute to it are actually due to a modification to the gravitational force law, or the way matter moves in response to gravity, on very large scales (a theory called Modification of Newtonian Dynamics (MOND) is one such alternative explanation). In situations where normal matter and dark matter are closely coincident, it turns out to be rather difficult to rule out explanations based on modified gravity or modified dynamics. A strong test of this alternative hypothesis would therefore be provided if there were to be a situation in which the intracluster gas (which, of course, is the dominant form of luminous matter in the cluster) and the dark matter could become separated.

It seems that some (rare) clusters of galaxies may provide such an environment, and collisions are the key to understanding these clusters.

■ Imagine that two large clusters of galaxies undergo a head-on collision. What would you expect might happen to the intracluster gas from each cluster during such a collision?

☐ The two clouds of intracluster gas would collide, and this collision would slow down their progress.

Dark matter, on the other hand, would only interact through gravitational effects, and hence would not 'feel' the changes in intracluster gas pressure involved in such a collision. So it would be expected that the dark matter associated with the two clusters will pass through each other without collision. Remarkably, there are a small number of clusters that have undergone recent collisions, which can be investigated to see whether the X-ray emitting gas has become separated from the dark matter.

A very clear example is provided by the cluster 1E 0657-558, also known as the Bullet cluster (Figure 4.18). The two subclusters underwent a head-on collision about 100 million years ago. The projected density of these subclusters has been mapped using weak lensing, (Figure 4.18a). However, X-ray imaging (Figure 4.18b) shows that the collision of the intracluster gas from each subcluster has slowed this component down – essentially it has been left behind in the collision. The fact that gravitational lensing reveals that the dominant masses in the system have become separated from the intracluster gas rules out a model of modified gravity or dynamics, and provides very strong evidence for the existence of dark matter.

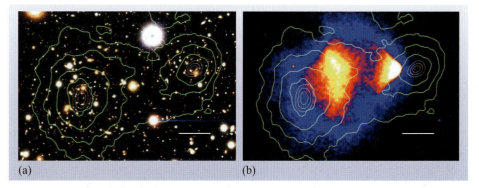

(a)   (b)

Figure 4.18 The Bullet cluster, 1E 0567-588 (redshift $z = 0.296$). The white scale bar represents a distance of 200 kpc. (a) An optical image (in B, V and R bands) with the projected density map as determined from weak gravitational lensing overlaid in green. (b) The X-ray emission as measured using the Chandra X-ray Observatory – brightest regions are shown in yellow/red, with faint emission in blue. Green contours show the weak lensing map. The brightest X-ray emission is in locations that are offset from the peaks in the density map, indicating the presence of dark matter that has become spatially separated from the intracluster gas. (Clowe *et al.*, 2006)

4.4 The large-scale distribution of matter

Clusters of galaxies, with typical radii of 2 Mpc, are very much smaller than the overall scale of the Universe. As shown by Abell's survey (Figure 4.5), the distribution of clusters across the sky is not uniform – other surveys since the 1950s have confirmed this finding: clusters are not the largest-scale objects in the Universe but are themselves organized into larger structures.

Much of this section will be concerned with the techniques that are currently being employed to map out the Universe on scales of hundreds of megaparsecs. However, before starting this discussion it is useful to take a quick tour of our 'local' part of the Universe beyond the Local Group.

Our Galaxy and the Local Group are part of a much larger structure: a **supercluster** – in this case, called rather unimaginatively, the **Local Supercluster** (Figure 4.19). This supercluster is centred on the Virgo cluster (Figure 4.3b), and is approximately 30 Mpc across. As suggested by Figure 4.19, this supercluster is a much more loosely organized structure than the individual clusters. Based on velocity measurements of the galaxies making it up, the Local Supercluster appears not to be gravitationally bound in the same way that clusters of galaxies are bound, and is certainly not a virialized system.

The Local Supercluster is not unique: as we shall shortly see, other superclusters in our vicinity have also been mapped with typical extents of a few tens of megaparsecs. Any structure that is on the scale of superclusters or above is often referred to by the generic term of **large-scale structure**.

Since there appears to be structure on the scale of superclusters, the obvious question is whether this organization continues hierarchically to larger and larger structures. It seems that this isn't the case. Moving upwards in distance scale, surveys have found that superclusters are not themselves organized into ever larger clusters of superclusters, but instead are distributed in a vast network consisting of high-density regions connected by filaments and sheets wrapped around (relatively) empty **voids.** This interconnected structure is often referred to as the **cosmic web**.

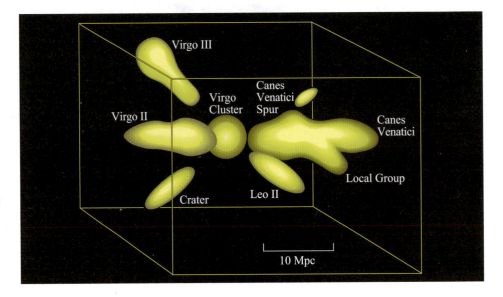

Figure 4.19 The Local Supercluster represented by a surface that separates high-density regions from lower-density regions in the Universe, hence the 'empty' regions in this diagram correspond to locations where the density of galaxies is low but not zero. (R. Brent Tully)

An impression of the large-scale distribution of matter in the Universe can be obtained from a remarkable map (Figure 4.20) of galaxy positions that was generated by automated scanning of plates from a Schmidt telescope survey. This map was developed by a group of astronomers using the Automatic Plate Measuring (APM) facility at the University of Cambridge from plates taken at the Anglo-Australian Observatory. Although it is a two-dimensional map, the filamentary structure of the distribution of galaxies is evident.

In the remainder of this section we shall discuss how the distribution of matter can be surveyed on large scales, using a variety of different tracers such as galaxies, cool intergalactic gas and dark matter. The aim of such studies is to better understand the structures in the Universe around us and inform the debate about how they were formed.

4.4.1 Redshifts of galaxies

The APM map (Figure 4.20) shows the distribution of galaxies or clusters of galaxies as a two-dimensional projection onto the celestial sphere. A comparison with constellations is appropriate here: as you may recall, the constellations do not generally represent physical groupings of stars: individual stars within a constellation may be at greatly differing distances from the Earth. Similarly, it is possible that the increased concentrations of galaxies seen in photographic surveys could be the result of coincidental alignments of galaxies at greatly different distances that just happened to be aligned along the same line of sight.

- How would you confirm that the galaxies that appear on a photographic plate to be within a cluster are indeed physically associated with each other, and not merely grouped as the result of chance alignment?

- □ You would need to show that the galaxies are all at a similar distance from the Earth.

This could be done by measuring the *redshift* of each individual galaxy within the cluster. If the galaxies are genuinely associated they should all have similar

Figure 4.20 The APM map of galaxy positions. The map was generated by using automated routines to find galaxies on photographic plates. The survey contains about two million galaxies and covers an area of the southern sky that is about 4000 square degrees in extent. Note that the gaps in the map are areas that could not be scanned because of the presence of bright stars. Structure in the large-scale distribution of matter in the Universe is clearly evident in this map. (S. Maddox, W. Sutherland, G. Efstathiou and J. Loveday)

values of redshift. Foreground and background galaxies could be identified and eliminated by having either smaller or greater redshifts than those in the cluster. Historically, this was a difficult task due to the vast number of galaxies that had to be observed to build up a three-dimensional picture of the Universe around us. At the time that Abell carried out his initial work on clusters in the 1950s the spectra of galaxies had to be measured one at a time using a large telescope and recorded using photographic film.

Thanks to improvements in detector technology, modern telescope and detector systems can measure redshifts much more quickly. A dedicated programme of such observations can, in a matter of years, measure tens or hundreds of thousands of redshifts. Such surveys provide a basis for mapping the Universe in three dimensions. Before discussing some of the major programmes for mapping the Universe we first need to take a closer look at how redshifts are related to distance.

The simple relationship $z = (H_0/c)d$ between redshift and distance introduced in Chapter 2 (Equation 2.7) is really only valid for small redshifts of up to $z \approx 0.2$: at larger redshifts this relationship no longer holds. As noted in Chapter 2, the redshift is a measure of the amount that the Universe has expanded since the light was emitted. At large distances the redshift increases more rapidly with distance than implied by Equation 2.7. For example an object with a redshift of 2.0 is not believed to be twice as far away as an object with redshift of 1.0.

There is little doubt that distance increases with redshift, but the exact relationship depends on a number of factors (or cosmological parameters) that characterize the behaviour of the expansion of the Universe. Different models of the cosmological expansion, and their consequences, will be discussed in more detail in Chapter 5. For now, it is sufficient to note that the precise relationship between redshift and distance depends on the model of the expansion used. The graph shown in Figure 4.21 represents a relationship between redshift and distance that is widely accepted by astronomers and will be used for the remainder of this chapter. It is also worth noting here that distances need to be defined carefully in an expanding Universe to avoid ambiguity. The distance plotted in Figure 4.21 is the distance that a galaxy, seen at a redshift of z, is at the present time. This is further away than the distance the galaxy was at when the light was emitted, because the Universe has been expanding in the meantime.

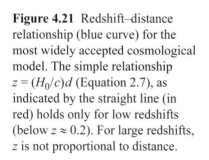

Figure 4.21 Redshift–distance relationship (blue curve) for the most widely accepted cosmological model. The simple relationship $z = (H_0/c)d$ (Equation 2.7), as indicated by the straight line (in red) holds only for low redshifts (below $z \approx 0.2$). For large redshifts, z is not proportional to distance.

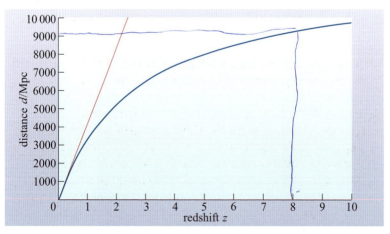

4.4.2 Mapping the Universe in three dimensions

The surveys discussed in this section aim to provide accurate distance information by measuring redshifts of large numbers of distant objects over large areas of the sky. While instruments such as the Hubble Space Telescope are useful for observing faint objects, they are ill-suited to surveying large-scale structure as the following question shows.

■ The celestial sphere covers about 40 000 square degrees. The Hubble Space Telescope ACS (Advanced Camera for Surveys) instrument has a field of view that is about 3 arcmin square. If one quarter of the sky were to be mapped using the HST, how many observations would be required?

☐ Each square degree would require 400 (20×20) observations, so to map one-quarter of the celestial sphere (10^4 square degrees) would require 4×10^6 observations.

Surveying a significant fraction of the entire sky using the Hubble Space Telescope is clearly totally impractical – it would require taking millions of observations. For survey work it is necessary to use telescopes with a much larger field of view – up to several square degrees. These are the natural successors to the Palomar Schmidt telescope that George Abell used for his pioneering work on clusters, except that the photographic survey plates have now been replaced by cameras equipped with very large arrays of sensitive CCD detectors.

The practical difficulties of imaging large numbers of galaxies are a challenge, but measuring their distances represents an even greater problem. The key issue is one of collecting spectra from a sufficient number of galaxies to enable a three-dimensional map to be created. Historically, redshift surveys had to compromise between sky coverage and distance probed, and it was necessary to choose between imaging large areas of the sky to small redshifts, or small areas of sky to great distances.

Surveys carried out in the 1980s and 1990s required painstaking observing campaigns, but eventually yielded maps of the local Universe. Notable examples of these early surveys were the Harvard–Smithsonian Center for Astrophysics (CfA) survey (Figure 4.22a) and the IRAS (Infrared Astronomical Telescope) Point Source Catalogue z-survey (PSCz – Figure 4.22b) – which concentrated on measuring distances to galaxies within large volumes of the local Universe out to about 200 Mpc.

More recent redshift surveys, such as the Two-degree Field (2dF) Galaxy Redshift Survey and the Sloan Digital Sky Survey (SDSS) have used optical fibre systems to collect the spectra of hundreds of galaxies simultaneously. This means that the redshift measurements from many galaxies (typically hundreds) in a single field of view can be collected at the same time. Compared to older techniques in which only one spectrum could be measured at a time, modern fibre optic systems give a great advantage in the rate at which data can be collected. In addition, telescopes such as that used for the SDSS are specifically designed to be operated in an automated survey mode, which again increases the rate at which they can gather data.

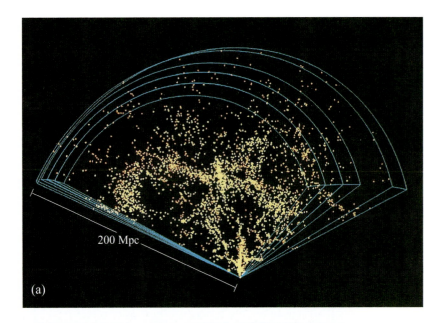

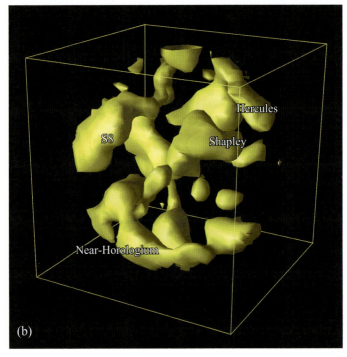

Figure 4.22 Examples of the structures revealed on scales of 100 Mpc by (a) the CfA survey, and (b) the PSCz survey showing the density distribution in the Universe around us to a distance of about 250 Mpc. The locations of four superclusters are indicated: Shapley, Hercules, S8 and Horologium. Note that as in Figure 4.19, the distribution of matter is represented by a surface that separates high- from low-density regions. Also as in Figure 4.19, the 'empty' regions in this diagram correspond to locations where the density of galaxies is low but not zero. ((a) M. Geller and J. Huchra, SAO; (b) figure by L. Teodoro, based on data described in Saunders *et al.*, 2000)

The initial survey to map large-scale structure using the SDSS (now called the SDSS Legacy Survey) simultaneously collected data using different tracers of the mass distribution in the Universe and the spectral survey included samples of normal galaxies, high luminosity galaxies and quasars. Consequently the Legacy Survey data can be used to map different scales in the Universe. The size of the surveys enabled by such automated programmes is far larger than the pioneering work of the CfA and PSCz surveys, with the SDSS Legacy Survey measuring the spectra from a million galaxies and 10^5 quasars.

The 2dF and SDSS results (Figure 4.23) clearly show that the web or sponge-like structure suggested by earlier studies extends out to great distances, forming a

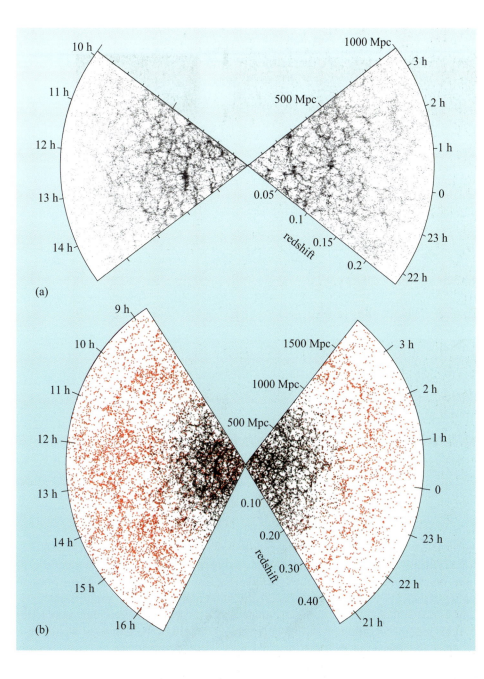

(a)

(b)

Figure 4.23 (a) Results from the 2dF Galaxy Redshift Survey. This data release (May 2002) shows the positions of over 220 000 galaxies plotted in plan view on the two wedges covered by the survey. (b) Results from the Legacy Survey of the SDSS, showing the distribution of normal galaxies (black) and luminous red galaxies (red). Note the difference in redshift scale between (a) and (b). ((a) M. Colless, (Mount Stomlo Observatory); (b) M.R. Blanton, (New York University)/SDSS)

cosmic network. The densest points within this network are clusters of galaxies containing hundreds or thousands of galaxies. Loose collections of clusters form superclusters of perhaps 30 to 50 Mpc in size. On larger scales, there are low-density voids, of up to about 60 Mpc in diameter, separating the higher density collections of clusters. These voids are separated by filaments of galaxies strung out in long chains, and by two-dimensional sheets that enclose the voids rather like the pores in a sponge. Although difficult to discern from representations such as Figure 4.23, the density of galaxies in the filaments is very much less than that in the clusters, and only a factor of two or three greater than the density in the voids themselves. The largest structures appear to be on a scale of approximately 200 Mpc; above this, the Universe becomes uniform in the sense that one 200 Mpc region looks much like another.

QUESTION 4.5

Summarize the scales of different cosmic structures by completing Table 4.2. Note that only approximate values, or ranges of values, are required.

Table 4.2 The scales of different types of cosmic structures – for use with Question 4.5.

Feature	Distance or length/Mpc
Milky Way (diameter of the stellar disc)	
Distance to Large Magellanic Cloud	
Distance to the Andromeda Galaxy	
Extent of the Local Group	
Typical diameter of a cluster	
Distance to nearest rich cluster (Virgo)	
Extent of a typical supercluster	
Extent of voids	
Scale on which the Universe appears uniform	

4.4.3 Cool intergalactic gas: quasars and the Lyman α forest

While redshift surveys of galaxies provide a direct way of mapping the visible three-dimensional structure in the Universe, we also know that much of the normal matter in the Universe is in the form of gas in intergalactic space, and this gas need not be associated with luminous matter such as galaxies. You might conclude that this gaseous component of the large-scale structure of the Universe would be invisible and hence inaccessible to astronomers. However, this gas may still absorb light, which makes it possible to measure the large-scale structure of the Universe in an independent way: by measuring the absorption by intervening gas clouds along the line of sight to a distant background galaxy or quasar.

As you saw in Chapter 3, quasars are very bright, point-like objects with very high redshifts: quasars with redshifts up to about $z = 7$ have been discovered. Since they lie at such large distances from us, the electromagnetic radiation that is emitted by most quasars will cross vast tracts of the intergalactic medium as it travels towards the Earth. As this electromagnetic radiation passes through this medium we might expect that a certain amount of absorption may occur and give rise to spectral absorption lines at specific wavelengths (Figure 4.24).

The most common element in the Universe is hydrogen, and the distribution of intergalactic gas can be mapped by making use of the spectrum of this element.

The spectrum of hydrogen consists of several series of spectral lines. Of these, the **Lyman series** has the highest energy (and hence shortest wavelengths). Within this series, the most prominent spectral line is the Lyman α line corresponding to the transition between $n = 2$ and $n = 1$ levels of the hydrogen atom. This line, which is often abbreviated to Lyα, lies in the ultraviolet part of the spectrum with a wavelength of 121 nm and is easily identified, even when red-shifted.

In the spectra of the most distant quasars the Lyman α spectral line gets shifted all the way from the ultraviolet, through the visible part of the spectrum and into the infrared (Figure 4.25). In the original spectrum from the quasar, this line is present as a bright *emission line*, which appears above the continuous spectrum produced by the AGN (see Figure 3.16).

As the electromagnetic radiation from the quasar passes through intergalactic space it encounters clouds of cool gas. Because the gas is cool, absorption will occur, and this will be prominent at the wavelength of the Lyman α line. To an

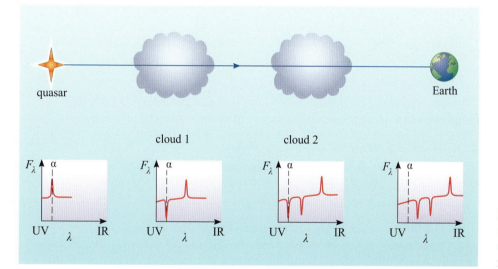

Figure 4.24 Intervening clouds in the line of sight from a quasar. Cloud 2 will have the smallest redshift when viewed from Earth.

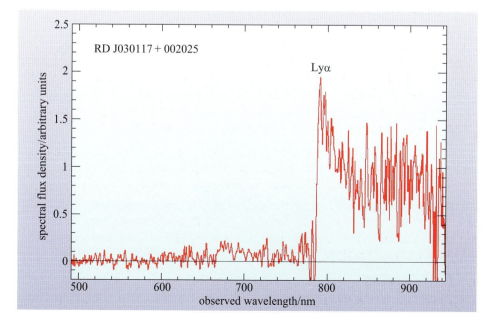

Figure 4.25 The spectrum of quasar RDJ 030117+002025 with redshift $z = 5.5$. The redshift of this object is so large that the Lyman α emission line has been shifted from the ultraviolet (121 nm) to the infrared (786 nm). (Stern *et al.*, 2000)

observer within one of these clouds the spectrum from the quasar will appear red-shifted, so the original Lyman α emission line will have a longer wavelength than 121 nm. Ultraviolet light that arrives at the cloud with a wavelength of 121 nm would have been emitted by the quasar at wavelengths shorter than the Lyman α line. The cloud will absorb radiation at a wavelength of 121 nm and so an absorption line will be formed. This process of absorption will occur for every cloud that the light from the quasar passes through. However, these clouds are at differing distances from the quasar, so according to observers situated on these clouds, the observed emission from the quasar will be red-shifted by differing amounts. Consequently, as the electromagnetic radiation from the quasar passes through a series of clouds on its way to Earth, it produces a set of spectral lines at progressively shorter wavelengths. The distances to the clouds can be found from the redshifts of these absorption lines. An example of a spectrum from the quasar Q 0149+336, which displays such absorption features is shown in Figure 4.26.

Because of the many closely packed absorption lines, this structure in a spectrum is frequently referred to as the **Lyman α forest**.

■ What does the presence of discrete absorption lines in the Lyman α forest suggest about the distribution of the material that the light has passed through? What would be seen if the absorbing material were distributed *uniformly* along the line of sight?

☐ In order to produce a distinct absorption line at a given wavelength, the light must have passed through a concentration of material at the distance corresponding to that redshift. If the material were distributed uniformly, light would be absorbed gradually all along the line of sight with continuously varying redshift, giving constant absorption at all wavelengths rather than a 'forest' of individual lines.

So the fact that there are many Lyman α absorption lines in the spectrum from a distant quasar tells us that the intergalactic medium is not smoothly distributed – but instead is in the form of 'clumps' or clouds. For example, an analysis of the absorption lines in the spectrum of the quasar Q 0149+336 (Figure 4.26) reveals seven clouds along the line of sight whose presence is confirmed by absorption lines due to elements heavier than hydrogen or helium ('metals'). Each of these clouds has a distinct redshift (ranging from about 0.5 to about 2.2 in this case). Note that there are many other lines, which are also due to absorption by intervening clouds, but that it is not possible to confirm the presence of any single cloud unless a *pattern* of spectral lines (Lyα and metal absorption lines) can be discerned in the spectrum.

You will see later on that it is possible to use the regions *between* the Lyman α forest lines to make an important and fundamental inference about the ionization state of almost the entire Universe.

QUESTION 4.6

In Figure 4.26, the spectrum shows a strong absorption line at a wavelength of 372 nm. Assuming this to be a red-shifted Lyman α line, calculate the redshift and hence estimate the distance of the cloud responsible for this absorption line.

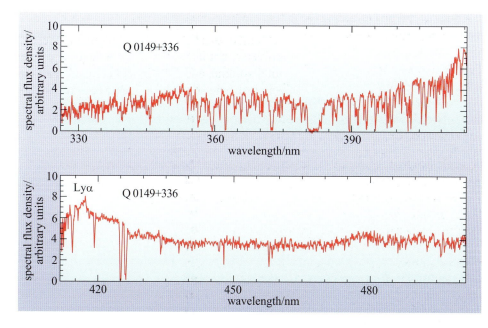

Figure 4.26 The spectrum of quasar Q 0149+336 ($z = 2.431$). The upper panel shows the spectrum from a wavelength of about 325 nm to 415 nm, the lower panel shows wavelengths from about 410 nm to 500 nm. The red-shifted Lyα line has an observed wavelength of 417 nm. At wavelengths shorter than the (red-shifted) Lyα line, there are a large number of absorption lines – due mainly to Lyα absorption by clouds of intergalactic gas at lower redshift than that of the quasar. At wavelengths longer than the red-shifted Lyα line there are relatively few absorption lines (the lines that are present are not due to Lyα absorption, but to absorption by elements heavier than hydrogen or helium in the intergalactic medium). (Wolfe *et al.*, 1993)

So how does this distribution of intergalactic gas correspond to the distribution of matter as traced by galaxies? This is an ongoing area of research, but it does seem as though the Lyman α absorbing clouds are more uniformly distributed in space than the luminous matter. For instance, it has been mentioned that there are 'empty spaces' or voids in the galaxy distribution, but clouds of absorbing gas do seem to be present in these voids. There is no inconsistency here: most Lyman α absorbing clouds have very low densities – too low to be sites of star, and hence galaxy formation, so it need not be the case that the distribution of intergalactic gas should exactly follow the distribution of galaxies.

Although the majority of Lyman α absorbing clouds are of low density, there are some clouds where the density enhancement is about a factor of 10^6 above the mean density of the intergalactic medium. The absorption lines from these clouds are very deep and broad, and are called **damped Lyman α systems** (the terminology 'damped' refers to the physical effect that gives these lines their characteristic shape). The trough at roughly 380 nm in Figure 4.26 is an example of this effect. Damped Lyman α systems are of particular interest because, unlike the low-density Lyman α clouds, they *are* a plausible source of material for star formation.

The fact that radiation from a distant quasar can reach us at all appears at first sight surprising. In Chapter 2, we saw that the big bang model suggests a Universe that is initially filled with a smooth distribution of hydrogen and helium that cools and forms neutral atoms. This is then followed by the growth of density perturbations by gravitational collapse. The matter that does not take part in this process might be expected to remain as un-ionized gas. The quantity of neutral gas that might naïvely be expected to remain in the intergalactic medium is such that it would cause very strong absorption of the light from distant quasars. This expected absorption by a neutral intergalactic medium is called the **Gunn–Peterson effect**. The fact that the absorption, as seen through the Lyman α forest, is much lower than expected suggests one of two possibilities: either that the expected hydrogen is not present, or (more likely) that any hydrogen is present but not as neutral hydrogen, rather it is in an *ionized* state, which prevents it from absorbing the radiation.

For light to reach us from distant quasars, intergalactic hydrogen must have been ionized by the time the light was emitted. The event that caused most of the neutral hydrogen in the Universe to become ionized is referred to as **reionization**. The time at which this occurred is called the **epoch of reionization**, and to be consistent with observations of high-redshift quasars this must have been when the Universe was less than 10% of its current age. So what could have caused this Universe-wide change? The most plausible explanation is that sources of ultraviolet radiation suddenly 'turned on' at this time.

■ Can you suggest two possible sources for this ultraviolet radiation?

☐ Star-forming regions and active galactic nuclei are both strong sources of ultraviolet radiation.

As we saw in Section 2.5.5, galaxies seen at high redshifts appear to be sites of energetic bursts of star formation. This was happening when the Universe was only about 10% of its present age, and could have provided the energy required to ionize the intergalactic hydrogen. Alternatively, you saw in Section 3.6.1 that in the past the density of quasars was much higher than it is now. So it is also possible that active galaxies provided a source of ultraviolet radiation that ionized the bulk of the neutral hydrogen in the Universe. At present there is no consensus about which of these two processes was the more important as a cause of ionization of the neutral intergalactic medium.

If it were possible to look back to times earlier than the epoch of reionization, then we should expect quasar spectra to show strong absorption from neutral hydrogen. The search for the Gunn–Peterson effect at these early epochs is currently an active area of research. Observations of the most distant quasars are beginning to show signs that the expected absorption has indeed been detected at redshifts greater than $z \sim 6$, and that we are now able to directly probe this epoch of reionization. However, while we can see the evidence for this process, it is still matter of intense debate as to whether the Universe was lit up by active galaxies or intense bursts of star formation.

4.4.4 The large-scale distribution of dark matter

So far in our discussion of large-scale structure we have concentrated on the distribution of galaxies. This is understandable, since galaxies are readily observed and can have their redshifts measured. However, as we know from the study of galaxies (including our own) and clusters of galaxies, the dominant contribution to the mass on large scales seems to come from dark matter. It is of fundamental interest then to know whether the large-scale structure as revealed by visible galaxies represents an underlying distribution of dark matter.

■ Which technique can be used to map the density of dark matter in clusters of galaxies?

☐ Weak gravitational lensing and the resulting cosmic shear can be used as a basis to map dark matter in clusters of galaxies.

Gravity is the one known way in which dark matter makes its presence felt. Because the distortions of background galaxies are caused by gravitational fields, cosmic shear provides one of the most direct means of mapping the distribution of dark matter.

To illustrate the basis of the weak lensing technique, Figure 4.27 shows an imaginary situation in which a distant regular array of circular galaxies is viewed through a web of dark matter. Note how distortions indicate the presence of a high density of intervening dark matter. In other parts of the image, the 'galaxies' are relatively undistorted – showing that light rays have passed through a region with little variation in density. In reality, the distortions are much smaller – only one or two per cent, and the outlines of real galaxies are not necessarily circular to start with, so the distortion of an individual galaxy can be difficult to measure. But the images of galaxies that are near to one another will tend to be stretched in the same direction, and each small area of sky contains many galaxies. So even if the distortions are small they can still be measured by averaging over many neighbouring galaxies. This average distortion of galaxies can be used as a measure of the intervening dark matter.

The practical application of the weak lensing technique to mapping the dark-matter distribution is a challenge because of the vast number of background galaxies that need to be observed and because observational uncertainties can easily distort the measured cosmic shear signal. Despite this, the technique has been successfully applied, as illustrated in Figure 4.28, which shows a projected mass map over a region that is about 72 square degrees in extent, and is based on the weak lensing

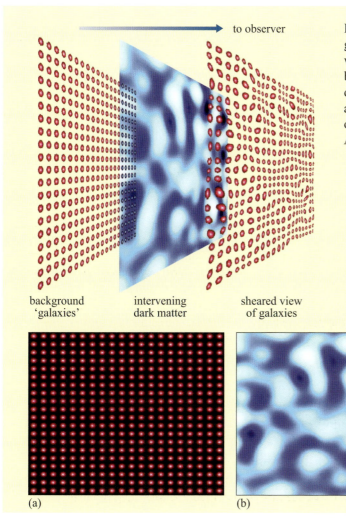

background 'galaxies'   intervening dark matter   sheared view of galaxies

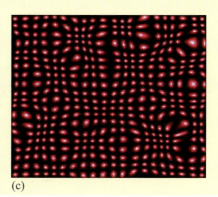

Figure 4.27 Cosmic shear: as light from a field of distant galaxies (a) passes through the web of dark matter (b) on its way to Earth, the image of each galaxy is slightly distorted by the non-uniform gravitational field (c). Although the distortion of each galaxy is small, the *average* distortion in a region of the image can be used to map the distribution of dark matter. (Adapted from figures produced by A. Refrieger, University of Cambridge)

(a)   (b)   (c)

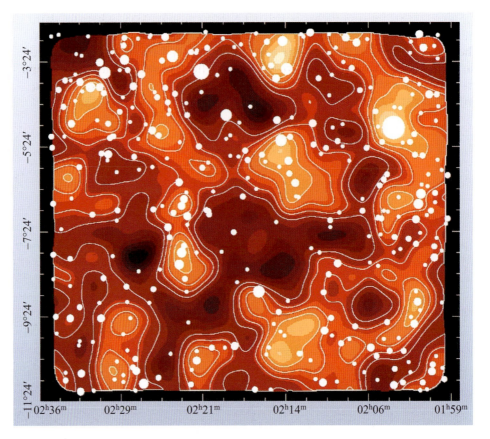

Figure 4.28 The smooth contours show the projected mass map as determined from analysis of weak lensing measurements from an optical survey carried out with the Canada–France–Hawaii Telescope. The galaxies imaged in the survey have a typical redshift of $z \sim 0.7$. The brighter regions show locations where the mass density is higher. The white circles indicate peaks in the mass distribution as would be predicted from the distribution of galaxies in this field. The correspondence between the bright- and dark-matter distributions is good although there are some discrepancies. Most notable, however, is the presence of voids with an angular scale of up to 2–3 degrees with a low density of matter. (Van Waerbeke *et al*., 2013)

of galaxies of a typical redshift of about 0.7. The map as determined from weak lensing can be compared with the mass distribution that is expected from the distribution of galaxies, and there is fairly good correspondence. It is particularly notable that the mass map reveals the presence of large voids. This provides further evidence that large-scale structure is dominated by a cosmic web of filaments of dark matter, and that galaxies more or less follow this underlying distribution.

4.5 The formation of clusters and large-scale structure

In this chapter we have seen that dark matter provides the dominant component to the mass of clusters and large-scale structure. This builds on the idea developed in Chapters 1 and 2 that most of the mass in our Galaxy, and other galaxies, is in the form of dark matter. We also saw in Chapter 2 that galaxy formation is believed to have progressed by a bottom-up scenario that is a consequence of the behaviour of cold dark matter. Clearly, such a model should also provide a natural explanation for the formation of clusters and the present-day appearance of the large-scale structure. We have already alluded to the similarity between the observed cosmic web and the outcomes of numerical simulations (such as discussed in Section 2.5.2), but it is important to note that the statistical properties of simulations can be compared in great detail to observations of the distribution of matter in the Universe. Simulations of the growth of structure typically follow the evolution under gravity of a large number of mass points representing the behaviour of dark matter, and it not uncommon to consider volumes enclosed by length scales of about 1000 Mpc. In many of these models, dark matter condenses first, creating a network within which galaxies form at locations where the dark matter is most densely concentrated. As noted in Chapter 2, a full simulation of the relatively small-scale processes involved in galaxy formation is not possible, so these simulations use approximate models to link the development of structure in the dark matter distribution to its likely effect on the formation and evolution of galaxies.

Figure 4.29 shows several views of the results of one such simulation that illustrates the expected structure on a range of scales. On scales greater than a few hundred megaparsecs, the structure is fairly uniform, but the network of filaments of dark matter is a good match to the observed structure in the Universe. Moving to smaller

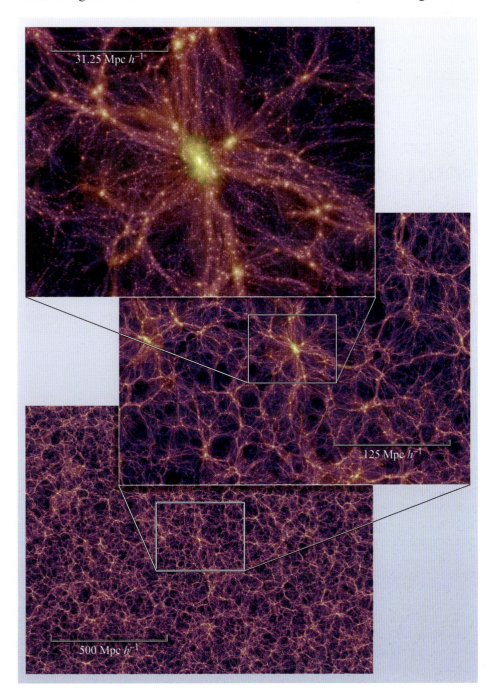

Figure 4.29 The results of a computer simulation of the growth of structure on large scales. The series of images shows the distribution of dark matter at the present time following a process of gravitational collapse over the age of the Universe. Each panel shows a scale bar – note that the factor h ($=H_0/(100 \text{ km s}^{-1} \text{ Mpc}^{-1})$) in these plots is believed to have a value of about 0.7, so the scale bars, from the top panel downwards, indicate lengths of about 45,180 and 700 Mpc. Note that the scale of the voids in the cosmic web is a good match to that seen in three-dimensional surveys of galaxies, and that rich clusters (bright yellow) form at the intersections of many dark-matter filaments. (Springel *et al.*, 2005)

scales, and zooming in on the point where several dark-matter filaments intersect, the results of this simulation show that the high density of dark matter expected at this location has resulted in the formation of a rich cluster of galaxies.

Having now surveyed the Universe to very large scales, we are now in a position to consider how the Universe as a whole can be subject to scientific investigation – through the field of cosmology.

4.6 Summary of Chapter 4

The large-scale distribution of galaxies

- Clusters are gravitationally bound systems of galaxies. Clusters containing small numbers of galaxies are called groups.

- Our Galaxy belongs to the Local Group of galaxies consisting of about 100 members but dominated by the Milky Way and Andromeda galaxy (M31).

- Medium-scale three-dimensional surveys confirm the existence of superclusters, loose collections of clusters that are about 30–50 Mpc in extent. Superclusters are not virialized systems.

- Our Local Group is at the outer edge of the Local Supercluster, which is centred on and dominated by the Virgo cluster.

- Redshifts can be combined with positions on the sky to obtain surveys of galaxies in three dimensions.

- The introduction of efficient electronic detectors and computer automation has resulted in a huge increase in the efficiency of redshift measurements. Recent surveys have mapped large portions of the sky to great depth to look at the distribution of structure on scales of hundreds of megaparsecs.

Clusters of galaxies

- Most clusters of galaxies have a radius of about 2 Mpc (the Abell radius). The mass of a typical cluster is of the order of 10^{14} to $10^{15} M_\odot$.

- Cluster masses can be estimated by three main methods: velocity dispersion, X-ray emission and gravitational lensing. The masses obtained by these methods typically agree within a factor of two or three.

- A cluster is said to be virialized if it is a gravitationally bound system and is in dynamical equilibrium. The mass of such a cluster can be obtained from the dispersion of the line of sight velocities (Δv) using $M \approx R_A (\Delta v)^2 / G$.

- Rich clusters are strong X-ray emitters due to the presence of hot intracluster gas. X-ray observations can be used to estimate the total mass of the cluster and the mass of X-ray emitting gas.

- A cluster can act as a gravitational lens of a distant galaxy, producing distorted multiple images. This is a means of detecting distant objects and can also be used to estimate the mass of the intervening cluster.

- Estimates of the masses of clusters from all three methods (virial theorem, X-ray emission and gravitational lensing) indicate that there is far more matter in clusters than the sum of the individual galaxy masses. This suggests the presence of dark matter in clusters. Typically galaxies will contribute less than about 10% of the mass of a cluster. The intracluster gas may constitute up to about 25% of the total mass. Dark matter makes up between 70% and 90% of the total mass of the cluster.

Cool intergalactic gas

- The spectra of some quasars and distant galaxies contain multiple absorption lines indicating the presence of gas (mainly neutral hydrogen) at different distances along the line of sight. Such absorption features are called the Lyman α forest, and are indicative of the large-scale distribution of neutral gas in intergalactic space.

- The absence of the Gunn–Peterson effect in the spectra of quasars at redshifts below $z \approx 6$ indicates that much of the gas in the intergalactic medium is in an ionized state. The increased opacity at higher redshifts is interpreted as showing that this gas was more neutral on average at early times. The transition between neutral and ionized states, which occurred at a time called the epoch of reionization, may be due to active galaxies or intense bursts of star formation.

The distribution of dark matter

- Dark matter forms the dominant contribution to the mass of clusters of galaxies (70–90%). Within clusters, the dark matter is distributed more smoothly than the matter that is present in the form of galaxies.

- The large-scale distribution of dark matter can be mapped by looking at small distortions in background galaxies – this effect is known as cosmic shear.

The formation of clusters and large-scale structure

- The formation of structures on the largest scales can be investigated by computer simulations. Models based on the behaviour of dark matter over cosmic history reproduce the cosmic web and clusters of galaxies in good accordance with observations.

Questions

QUESTION 4.7

If the Universe is 13.8 billion years old, calculate the fraction of this age at which we see the Virgo cluster at the present time.

QUESTION 4.8

The masses of clusters of galaxies can be measured using methods based on three different physical processes. Name these methods and state what assumptions must be made about the physical state of the cluster in order for the individual methods to be applied.

CHAPTER 5
INTRODUCING COSMOLOGY –
THE SCIENCE OF THE UNIVERSE

5.1 Introduction

Cosmology is the branch of science concerned with the study of the Universe as a whole. It involves questions such as: 'What is the composition of the Universe?'; 'What is its structure?'; 'How did it originate?'; 'How is it evolving?' and 'What is its ultimate fate?' These are obviously very challenging questions. They have been subjects of religious and philosophical speculation for thousands of years, but the development of scientific cosmology has brought them into the mainstream of astronomical debate over the past hundred years or so, and there is now real hope that their answers are coming into view.

Cosmology is a huge subject, and much of it concerns vast scales of time and distance. However, as you will see, cosmology also has to concern itself with the physics of the very small through the physics of subatomic particles. It is this combination of the very small and the very large – the microscopic and the macroscopic – that gives modern cosmology its distinctive flavour. It is also this combination that enables cosmology to cast new light on the fundamental nature of matter, and this gives it a vital role to play in answering one of the hardest questions in contemporary astronomy – 'Where did the galaxies come from?'

This chapter provides a broad introduction to scientific cosmology. Section 5.2 sets the scene by drawing together a number of observational facts about the Universe, most of which you have met in earlier chapters. Section 5.3 is concerned with 'modelling' the Universe: the process of formulating simplified descriptions of the Universe, usually expressed in mathematical form, that are consistent with modern physics, particularly with Einstein's general theory of relativity – his theory relating gravity to space and time. The process of modelling the Universe involves a number of important cosmological parameters, such as the age of the Universe, the total density of matter in the Universe, and the Hubble constant, H_0. Section 5.4 concludes this introduction to cosmology by highlighting these cosmological parameters and considering the relationships between them that are predicted by the most popular models of the Universe. Later chapters build on this introduction by considering, in turn, the nature of the early Universe and the big bang (Chapter 6), the challenges and results of measuring the key cosmological parameters (Chapter 7), and the many important questions that are still unanswered at the current stage in the development of cosmology (Chapter 8).

5.2 The nature of the Universe

This section briefly introduces some of the main facts about the Universe, as revealed by astronomical observations. Many of these facts should already be familiar to you from earlier chapters, but some will be new. In neither case will much be said here about how the information was obtained. The main aim is simply to catalogue the basic facts that must be explained by any theoretical account of the origin and evolution of the Universe. Where necessary, greater detail is given later.

Before giving the observational 'facts' there is one important point that deserves special emphasis. All of our observations of the Universe are carried out from points on or near the Earth. In interpreting astronomical observations we have learned from experience *not* to assume that the Earth is in any particularly privileged position. We are not at the centre of the Solar System, nor are we at the centre of the Milky Way. It seems reasonable, therefore, to suppose that we are not at the centre of the Universe either. (This is absolutely opposed to the ancient pre-scientific view that placed us at, or close to, the centre of the Universe.) The assumption that we do not occupy a privileged position in the Universe is often referred to as the **Copernican principle**, and is usually invoked in interpreting observational data. If, for example, we find that distant galaxies are heading away from us in all directions (as we do), the Copernican principle tells us that the observations do not mean that we have the privilege of being at the fixed centre of an expanding Universe, but rather that the nature of cosmic expansion is such that the recession of distant galaxies is what would be observed from *any* typical point in the Universe.

The term 'Copernican principle' recalls the work of Nicolaus Copernicus (1473–1543), who proposed a Sun-centred model of the Solar System at a time when the prevailing view favoured an Earth-centred model.

5.2.1 The matter in the Universe

One of the most obvious facts about the Universe is that it contains matter. We humans are made of matter, as are the planets, stars, nebulae and galaxies that we observe. All of these visible objects are basically composed of oppositely charged *electrons* and *nuclei*. In some cases, such as ourselves and most of the Earth, the electrons and nuclei are combined together to form electrically neutral *atoms*, but the major part of the visible matter – most of that in stars for example – takes the form of a *plasma* in which the electrons and nuclei are separate and distinct, although the plasma as a whole remains electrically neutral.

Wherever large bodies of visible matter are found, whether they are composed of atoms or plasma, it is always the case that their mass is mainly accounted for by the *protons* and *neutrons* that make up nuclei, since these particles are far more massive than the electrons that accompany them.

■ The masses of the electron, proton and neutron are: $m_e = 9.109 \times 10^{-31}$ kg, $m_p = 1.673 \times 10^{-27}$ kg and $m_n = 1.675 \times 10^{-27}$ kg. Use these values to evaluate the following ratios: m_p/m_n, m_p/m_e and m_n/m_e. Roughly what fraction of the mass of a helium atom is attributable to the protons and neutrons contained in its nucleus?

☐ The required ratios are $m_p/m_n = 0.9988$, $m_p/m_e = 1837$ and $m_n/m_e = 1839$. Taking the view that the helium nucleus has the combined mass of two protons and two neutrons, while the helium atom has the combined mass of the nucleus and two electrons, we should expect the fraction of atomic mass attributable to protons and neutrons to be

$$(2m_p + 2m_n)/(2m_p + 2m_n + 2m_e) = 6696/6698 = 0.9997$$

(This is only an approximate value since we have ignored the effects of *binding energy*, which causes the mass of the nucleus to be slightly less than that of two protons and two neutrons.)

As you saw in Chapter 1, both the proton and the neutron (but not the electron) belong to a family of elementary particles called baryons, and the term *baryonic matter* is used to refer to all forms of matter in which the mass is mainly attributable to baryons. All the familiar atomic and molecular gases, liquids and solids, whatever their chemical composition, and all the plasmas found in stars, nebulae and galaxies, are therefore examples of baryonic matter.

Analyses of the chemical compositions of stars, nebulae and galaxies indicate that the most common form of baryonic matter is actually hydrogen plasma, and the second most common form is helium plasma. Although there are other significant forms of baryonic matter, we can, very crudely, say that the Universe has about 75% of its baryonic mass in the form of hydrogen nuclei and about 25% in the form of helium nuclei. We refine this crude recipe in later chapters, but it's worth remembering it as a first approximation to the true chemical composition of the Universe. Explaining the relative abundance of hydrogen and helium is a major challenge for any theory of the Universe, and is addressed in Chapter 6 .

■ Accepting that 75% of the baryonic mass of the Universe is due to hydrogen and 25% is due to helium, what are the relative numbers of hydrogen and helium nuclei in the Universe? In particular, how many hydrogen nuclei would you expect to find for each helium nucleus?

☐ Representing the masses of the hydrogen and helium nuclei by m_H and m_{He}, respectively, and using the symbols n_H and n_{He} to represent the number densities of hydrogen and helium nuclei, the question implies that

$$\frac{m_H \, n_H}{m_{He} \, n_{He}} = \frac{75}{25} = 3$$

Making the approximation that $m_{He} = 4m_H$ it follows that

$$\frac{n_H}{4n_{He}} = 3, \text{ or equivalently, } \frac{n_H}{n_{He}} = 12$$

So, there should be 12 hydrogen nuclei for each helium nucleus.

It's clear that the Universe contains a great deal of baryonic matter, since it contains a lot of luminous stars, nebulae and galaxies. But earlier chapters have emphasized that many independent observations also indicate the presence of a great deal of non-luminous dark matter that has so far been detected only through its gravitational influence. As noted in Chapter 4, some scientists have suggested that dark matter may not actually exist at all, and that it may be our understanding of gravity that is at fault. However, the majority view at the time of writing is that dark matter does exist, and that it is far more common than baryonic matter. As stated in Chapter 1, it is expected that some of the dark matter is nonluminous *baryonic dark matter*, but there are good reasons to believe that most of it is non-baryonic, and that this *non-baryonic dark matter* is, at least in terms of density, the dominant form of matter in the Universe.

Recent cosmological observations indicate that the average density of non-baryonic dark matter is between five and six times greater than that of baryonic matter (see Figure 5.1). These measurements are discussed in Chapter 7, while the reasons for expecting only a limited amount of baryonic matter are considered in Chapter 6.

Figure 5.1 Galaxies represent large concentrations of matter. The visible parts of galaxies are certainly composed of baryonic matter, but most of the matter in galaxies is thought to be dark matter and the majority of that is expected to be non-baryonic. (Hubble Heritage Team (AURA/STScI/NASA))

QUESTION 5.1

For every 10 neutrons in the Universe, how many protons would you expect to find? Explain the assumptions you have made in arriving at your answer.

5.2.2 The radiation in the Universe

Another important fact about the Universe is that it is essentially filled with *electromagnetic radiation*. All that we know of distant galaxies and clusters has been learnt by observing electromagnetic radiation (mainly radio waves, microwaves, infrared radiation, visible light, ultraviolet radiation, X-rays and γ-rays) that has originated in those galaxies or clusters and then travelled through space until detected here on Earth. Of course, this direct observation only shows that there is a lot of radiation travelling towards the Earth, but the Copernican principle tells us that the Earth does not occupy any special place in the Universe, hence the belief that all parts of the Universe receive radiation from their cosmic surroundings, and the statement that the Universe is essentially filled with such radiation.

One of the characteristic properties of electromagnetic radiation is its wavelength, λ, and an important feature of any observed radiation is its *spectral energy distribution* (defined in Chapter 3), which determines the amount of energy delivered per second and per unit area by any narrow range of wavelengths belonging to the radiation concerned. The radiation that reaches the Earth from space comes mainly from the Sun, and is dominated by the visible light that our eyes have evolved to observe. However, this dominant role of visible light in our neighbourhood is a result of our close proximity to the Sun. Observations indicate that, in the Universe as a whole, the spectral energy distribution of radiation is actually dominated by microwave radiation, which occupies a range of wavelengths around 1mm, between radio waves and infrared radiation.

The predominance of microwave radiation is indicated in Figure 5.2, which shows the spectral energy distribution of 'extragalactic background radiation' at various wavelengths. This background radiation does not come from any identified source (such as the Sun or even the Milky Way), and is thought to be 'universal' in the sense that a similar background would be observed anywhere in the Universe at the present time. The highest peak that represents a universally dominant form of background radiation is in the microwave region. This contribution to the background radiation is usually referred to as the **cosmic microwave background (CMB)**.

The discovery of the CMB, in 1965, was one of the most important events in the development of scientific cosmology. As you will see in Chapter 6, early measurements of the CMB's general properties did a great deal to establish the 'big bang' as the best supported theory of cosmic evolution. More recently, increasingly detailed CMB studies, mainly discussed in Chapter 7, have become the source of some of the most precise and reliable information we possess about the nature, composition and evolution of the Universe.

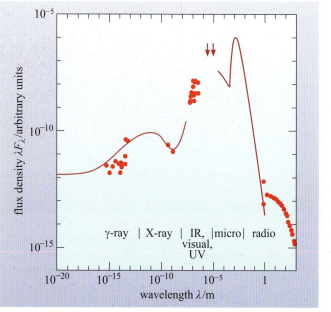

Figure 5.2 The spectral energy distribution of background radiation. The downward pointing arrows represent upper limits to the background radiation at the wavelengths indicated. (Adapted from Sparke and Gallagher, 2000; based on work by T. Ressell and D. Scott)

5.2.3 The uniformity of the Universe

It is an obvious fact that what is normally above your head (probably some air, a ceiling, a roof, a lot more air and then outer space) is different from what is below your feet (probably a floor, some building foundations and the whole of the Earth). This provides clear evidence that, locally at least, the Universe is *not* uniform. However, rather than concentrate on your immediate surroundings, think instead about a sufficiently large region of the Universe, such that an astronomer or cosmologist might regard it as a 'typical' or 'representative' sample. In practice this means considering a region that is large enough to contain a few superclusters of galaxies, along with the voids found between them. Such a region might well have a diameter of several hundred megaparsecs, and might represent, say, a millionth of the total volume of the observable part of the Universe. Provided the temptation to think about regions smaller than this is resisted, the view of modern cosmologists is that any part of the Universe is pretty much like any other at the present time. That is to say, if we consider any two regions of the present Universe, each sufficiently large to be representative, then those regions will have the same average density, pressure, temperature, etc. In this sense, cosmologists say, provided we consider sufficiently large size scales, the Universe is uniform.

Belief in the large-scale uniformity of the Universe has always played an important part in scientific cosmology. In the early days of the subject this belief was based on assumption and the absence of any contradictory evidence, but in recent years it has come to rest more and more on positive observational support. One thing that observation certainly makes clear is that, on the large scale, the Universe is pretty much the same in all directions. This is fairly clear just from the large-scale distribution of galaxies, which can be seen to be reasonably even in all directions that are not obscured by parts of the Milky Way (see Figure 4.20). However, even better evidence is provided by the cosmic microwave background (see Figure 5.3), which is observed to come with equal intensity (to about a few parts in 10^5) from all regions of the sky, once allowance has been made for the 'local' effects of the Earth's motion.

By combining the evidence that the large-scale Universe appears the same in all *directions* when observed from our location, with the Copernican principle that we are not in any 'privileged' location, it follows that the large-scale Universe should appear to be the same in all directions from *every* location. This in turn implies that it should be the same everywhere, that is to say it should be uniform at any given time. This combination of observation and assumption is quite convincing in itself, but in recent years even more support has been provided by the increasingly ambitious surveys of galaxy redshifts that were described in Chapter 4. These really do provide evidence that galaxies are distributed uniformly on the large scale. There are signs of clustering and superclustering on scales of tens or hundreds of megaparsecs, but there is no sign of any larger scale structure. It is *not* the case, for example, that the galaxies on one side of the Earth are significantly more clustered than those on the other side. The evidence for large-scale uniformity in the distribution of galaxies was discussed in detail in Chapter 4 and some results for quasars from the 2dF survey are shown in Figure 5.4.

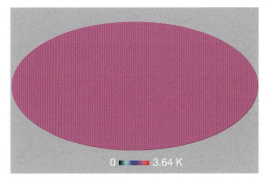

0 ▬▬▬▬ 3.64 K

Figure 5.3 The observed intensity of the cosmic microwave background radiation across the whole sky. These data come from space-based measurements by the COBE satellite (Cosmic Background Explorer) and have been corrected to compensate for the motion of the detector. They are based on measurements made with an angular resolution of a few degrees and are sensitive to intensity variations of about one part in a thousand. (Douglas Scott)

Figure 5.4 The distribution of quasars in two thin, wedge-shaped slices of the Universe. The quasars are observed out to such large distances that evolutionary effects allow changes in the number of quasars per unit volume to be observed as distance from the Earth increases. This is why the numbers of quasars drop off towards the centre, because quasars have become rarer in recent cosmic time. Also, only the brightest quasars can be seen at the greatest distances, which is partly why the numbers taper off at redshifts around three. However, at any given distance the data give strong support to the claim that the large-scale distribution of quasars is the same in all directions. (Robert Smith, 2dF Quasar Redshift Survey)

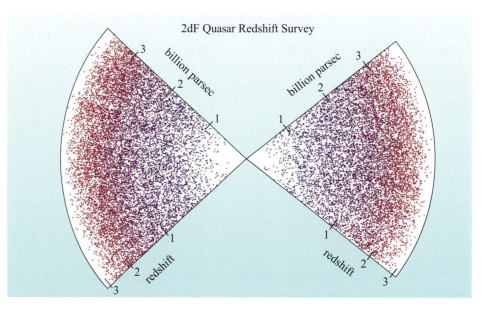

QUESTION 5.2

In assuming that we can use the Copernican principle to interpret our observations of the CMB we are assuming that the CMB is a truly cosmic phenomenon, rather than a purely local one such as, say, sunlight. Describe a piece of observational evidence that makes it plausible to suppose that the CMB is a cosmic phenomenon, whereas sunlight is only a local one.

5.2.4 The expansion of the Universe

The nearest galaxies to the Milky Way are mainly dwarf galaxies that appear to be 'satellites' orbiting our own Galaxy. The closest large spiral galaxy – the Andromeda Galaxy, M31 – is actually heading towards us. However, looking deeper into space it is found that all distant galaxies have a component of velocity that is directed away from the Earth, as revealed by the redshifts in their spectra. This finding, originally by Vesto Slipher (1875–1969), paved the way for the discovery of Hubble's law which was described in Chapter 2. As explained in Chapter 2, *Hubble's law*, applies to galaxies with redshifts up to about 0.2, and implies an approximate proportionality between redshift and distance that can be written as

$$z = \frac{H_0}{c}d \quad \text{(for } z \lesssim 0.2\text{)} \tag{5.1}$$

where H_0 is Hubble's constant and c is the speed of light in a vacuum. The redshift z is related to the observed and emitted wavelengths, λ_{obs} and λ_{em}, of some identified spectral line by the equation (this is a repeat of Equation 2.6)

$$z = \frac{\lambda_{obs} - \lambda_{em}}{\lambda_{em}} \tag{5.2}$$

The two parts of Figure 5.5 provide a visual reminder of the meaning of redshift, as well as an indication of some of the observational data that support Hubble's law.

For reasons that will become clear later, it would be a mistake to interpret the redshift seen in the spectra of distant galaxies as a simple Doppler effect. Nonetheless, it is true that the observed redshifts do indicate that all distant galaxies are receding even though the Doppler formula can't generally be used to determine the speed of that recession. As mentioned earlier, such an observed recession is not thought to prove that the Earth is at the centre of an expanding cloud of galaxies, but rather that the whole Universe is in a state of expansion, with every galaxy, on average, moving away from every other *distant* galaxy. This overall expansion, described by Hubble's law, is sometimes called the **Hubble flow**. Galaxies provide imperfect tracers of the Hubble flow, since they interact gravitationally and may therefore be disturbed by the presence of nearby galaxies or other local effects. These local disturbances manifest themselves as movements relative to the Hubble flow, and account for the *peculiar motions* of galaxies that were introduced in Chapter 2. They are one reason why a plot of redshift against distance, even for distant galaxies, is not a perfect straight line.

The current rate of cosmic expansion is measured by the Hubble constant, H_0. The uniformity of the Universe therefore implies that the Hubble constant should be a

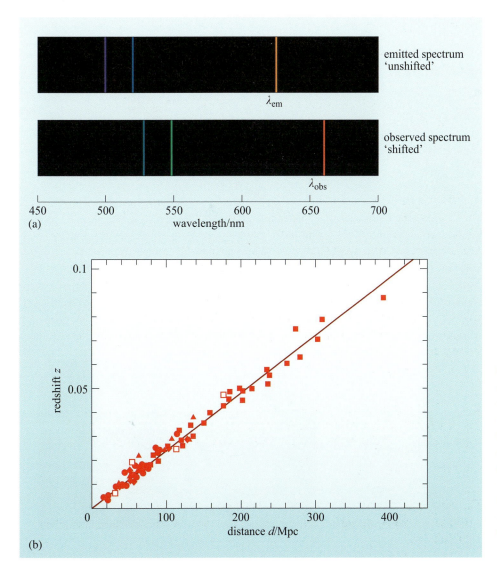

Figure 5.5 (a) The redshift z of a spectral line observed at wavelength λ_{obs} is a fractional measure of the line's displacement from the wavelength λ_{em} at which it was emitted by its source. (b) Observational evidence in support of Hubble's law. The best fitting straight line corresponds to a value of H_0 of 72 km s^{-1} Mpc^{-1}. ((b) Adapted from Freedman *et al.*, 2001)

'universal constant' with a value that is independent of where it is measured. Note, however, that the uniformity of cosmic expansion does not mean that the rate of expansion is constant over time. Since the late 1990s, when observations of Type Ia supernovae in distant galaxies were used to systematically extend independent measurements of distance and redshift well beyond $z = 0.2$, it has become clear that changes in the rate of cosmic expansion over billions of years cause detectable deviations from Hubble's law (see Figure 5.6a). Some such deviations at large redshift were expected since it was widely thought that cosmic expansion must be slowing over time. However, when the measured deviations from Hubble's law were first announced in 1998 the results created widespread astonishment. Two independent research teams both found that the rate of cosmic expansion had actually been *increasing* for at least several billon years. The cosmic expansion was accelerating!

The discovery that cosmic expansion is accelerating has had a great effect on cosmology. At the very least it implies that some other influence, apart from matter and radiation, is mainly determining the rate of cosmic expansion. The leading candidates at the present time are a uniformly distributed 'dark energy' or

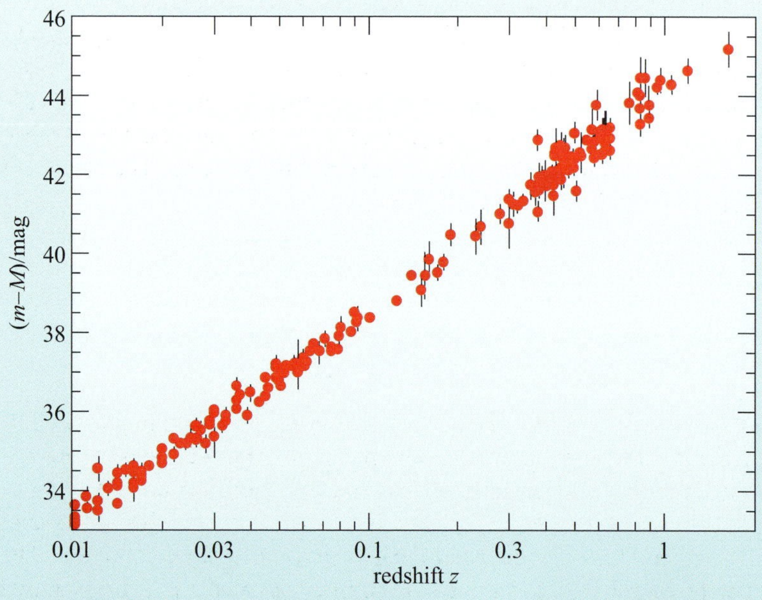

Figure 5.6 (a) Independent measurements of redshift and distance reveal deviations from Hubble's law at large redshifts, $z > 0.2$ say. Since the distance modulus $(m-M)$ is a logarithmic measure of distance, the redshift is also plotted on a logarithmic axis. The very distant galaxies are dimmer (i.e. $(m-M)$ is larger) and therefore further away than expected on the basis of Hubble's law. This provides evidence that cosmic expansion is accelerating. (b) The leaders of the teams responsible for the discovery that cosmic expansion is accelerating: Saul Perlmutter (1959–), Brian Schmidt (1967–) and Adam Reiss (1969–). ((a) Adapted from Kirshner, 2004; (b) (Left) University of California, Lawrence Berkeley National Laboratory. (Middle) SIPA USA / REX. (Right) Will Kirk.)

a universally effective 'cosmological constant', both of which will be introduced in the next section although neither can be said to be well understood. The full implications of accelerated cosmic expansion will undoubtedly take some time to work out. However, the significance of the discovery has already been recognized by the award of the 2011 Nobel Prize in Physics to the leaders of the teams responsible for the breakthrough (Figure 5.6b).

■ Summarize the four main facts about the present state of the Universe that have been discussed in this section, giving detailed clarification where appropriate.

☐ The Universe contains matter. The matter is mainly non-baryonic dark matter, but about 1/5th or 1/6th is baryonic matter, mainly hydrogen (~75%) and helium (~25%).

The Universe contains radiation. The radiation is mainly cosmic microwave background radiation.

The Universe is uniform. All regions that are sufficiently large to be representative currently have the same average density, wherever they are located. This is consistent with the observed distributions of matter and radiation.

The Universe is expanding. For redshifts of ≈ 0.2 or less, the expansion is well described by Hubble's law, implying that the current rate of expansion is given by the Hubble constant. Galaxies provide somewhat imperfect tracers of this expansion since they have 'peculiar' local motions relative to the large-scale Hubble flow. Observations at larger redshifts show that the rate of cosmic expansion has been accelerating for several billion years. The cause of this acceleration is still under investigation.

5.3 Modelling the Universe

Physicists generally take the view that a scientific understanding of a phenomenon has been achieved when that phenomenon can be accurately described in terms of a few concise statements, or better still a well-formulated equation. A typical example is provided by the flow of electric current I through a sample of electrically conducting material in response to an applied voltage V. This is described by a relationship known as *Ohm's law*, which is expressed by the equation $V = IR$, where R is a parameter, called the *resistance*, that characterizes the electrical properties of the sample. Ohm's law provides a **mathematical model** of the process of current flow, implying that, for any given sample, V is proportional to I, but requiring the value of R to be determined before it can supply quantitative predictions.

Cosmologists adopt a similar view regarding the understanding of the Universe. One of the central concerns of modern cosmology is the formulation and investigation of mathematical models of the Universe. These are called **cosmological models**. They usually take the form of a few *equations* that imply general relations between observable quantities, but they also involve *parameters*, such as the Hubble constant, that must be determined by observation before the model can be used to provide detailed quantitative predictions.

ALBERT EINSTEIN (1879–1955)

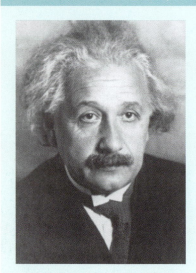

Figure 5.7 Albert Einstein. (Science Photo Library)

Albert Einstein (Figure 5.7) was born in Ulm, Germany in 1879. From 1896 to 1901 he lived in Zurich, Switzerland, where he was a student at the Federal Institute of Technology (ETH). In 1905 he was working in the patent office in Bern when he completed some of the most important papers in the history of physics. These included a paper on the photoelectric effect, a paper on Brownian motion, and two papers on the special theory of relativity, the second of which introduced the equation $E = mc^2$. About ten years later, as Professor of Physics in Berlin, Einstein completed his general theory of relativity, a generalization of the 1905 theory that related gravity to the geometry of space and time. General relativity received its first systematic exposition in 1916, and was first applied to cosmology in 1917.

Observations made in 1919, during a total eclipse of the Sun, confirmed one of the key predictions of general relativity, that starlight passing close to the edge of the Sun should undergo a gravitational deflection of 1.74 arcsec (see Section 4.3.2). The success of this prediction was front-page news, and made Einstein an international celebrity. He was awarded the Nobel Prize for physics in 1921 (mainly for his work on the photoelectric effect), and received many other honours and awards. In 1932, shortly before the Nazis came to power, he left Germany for the United States where he took up a post at the Institute for Advanced Study in Princeton. With war approaching, in 1939, Einstein was persuaded to sign a letter to President Roosevelt, pointing out the military implications of atomic power. In his later years Einstein became a prominent commentator on world affairs, but had little direct impact on the development of science. He pursued a bold but fruitless search for a unified field theory that would unite gravitation and electromagnetism, and in 1952 he was offered the Presidency of Israel, which he declined. He died in Princeton in 1955.

The aim of this section is to introduce you to some of the simplest but most important cosmological models that are currently in use, and to explore some of their implications. All of these models are based on **general relativity**, the theory of gravity published by Albert Einstein in 1916. For this reason, our discussion begins with a consideration of relativity and relativistic cosmology.

5.3.1 The relativistic Universe

The modern era of cosmology can be dated from the day in 1917 when Einstein's paper 'Cosmological Considerations of the General Theory of Relativity' was published in the Proceedings of the Prussian Academy of Science. An insight that is fundamental to this paper, and indeed to the whole of Einstein's theory of general relativity, is the crucial role that **space** and **time** must play in any attempt to model the Universe.

Before the advent of Einstein's theory of relativity, the view of most physicists was that space and time simply provided a sort of container for matter and radiation. Every particle of matter or radiation occupied some point in space at each instant of time. Moreover, space and time were supposed to be *passive*. They provided a setting for the drama of physics, but they were not themselves players in that drama. The properties of space and time were not thought to be in any way affected by the properties of the matter and radiation they contained.

Einstein changed this view radically and forever. Already, in 1905, his special theory of relativity (essentially a restricted form of the general theory that ignored gravity) had shown that the three dimensions of space and the single dimension of time should be melded together to form a unified four-dimensional entity usually referred to as **space–time**. But the general theory of 1916 went much further by showing that the geometric properties of this four-dimensional space–time were affected by the matter and radiation it contained, and that this could account for the cosmically important phenomenon of gravitation.

According to the Newtonian theory of gravity, introduced in 1668, gravitational phenomena, such as the attraction between the Sun and the Earth, were due to a 'force' that acted instantly, between one body and another, across any intervening space (see Figure 5.8). Newton was able to describe the strength and direction of this force in terms of the masses of the bodies and the displacement (i.e. distance and direction) of one body from the other. However, he was not able to explain the origin of the gravitational force. He did not know the 'mechanism' that actually caused one body to influence another body at a remote location in space. He tried to explain his mysterious gravitational force in terms of something called 'vortex theory' that was popular with European scientists at the time, but his efforts only convinced him that this would not work, so he contented himself with *describing* the gravitational force and saying that as far as its origin was concerned 'I frame no hypothesis.' (In the Latin of his great work *Principia Mathematica*: 'hypotheses non fingo'.)

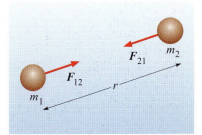

Figure 5.8 Newton's view of gravity: one body attracts another by means of a force that acts instantly across the intervening space.

Two hundred and fifty years later Einstein was able to go much further in accounting for gravitational phenomena. According to Einstein there is no gravitational force. In Einstein's view, a body such as the Sun acts on the space–time in which it is located, giving rise to a geometrical distortion usually referred to as a curvature of space–time. Bodies moving in the vicinity of the Sun, such as the Earth, respond to this **curvature** by moving in a way that is different from the way they would have moved if the Sun had been absent and the space–time undistorted. In this way, a body such as the Sun is able to gravitationally influence the behaviour of a body such as the Earth, even though there is no 'gravitational force' acting between them. Gravitation, in Einstein's view, is a result of space–time curvature – a geometric phenomenon – and general relativity is Einstein's 'geometric' theory of gravity. This is illustrated schematically in Figure 5.9, where 'space' is represented by a two-dimensional sheet and gravity is indicated by the 'curvature' of that sheet.

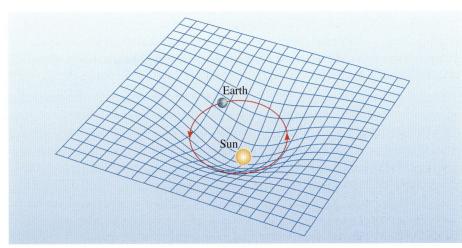

Figure 5.9 Einstein's view of gravity: a massive body (such as the Sun) significantly distorts the space–time in its vicinity. This space–time distortion (curvature) influences the motion of other bodies (such as the Earth) moving through that region of space–time, giving the impression that a 'force' is acting, although in reality there is only distorted space–time and motion in response to that space–time distortion.

Einstein was able to show that his theory of gravity agreed with all the correct predictions of Newton's theory. But general relativity went further, explaining anomalies for which Newton's theory had no account, and predicting entirely new phenomena that were outside the scope of Newtonian theory. Observations have consistently supported the novel predictions of general relativity, such as the gravitational deflection of starlight, which is why we now regard Einstein's theory as having superseded Newton's. Of course, we still use Newton's theory and even continue to speak of gravitational 'force', but we do so only because it is convenient, simple and sufficiently accurate for most purposes.

What has all this to do with cosmology? Well, according to general relativity, what determines the curvature of space–time in any region is not simply the presence of massive bodies in that region, but rather the associated distribution of **energy** and **momentum** throughout the region. The notion that particles of matter and radiation have energy will already be familiar, so the idea that a distribution of matter and radiation can be associated with a distribution of energy should not seem strange. The theory of special relativity, however, adds new depth to this idea since the relation $E = mc^2$ implies that even a stationary particle can be associated with a certain amount of energy. The idea of momentum may be less familiar, but in essence it is simply another physical quantity that, like energy, can be associated with any particle of matter or radiation once the mass and velocity of that particle are known. (Particles of radiation, such as photons, have zero mass but, according to special relativity, a photon of energy E carries momentum of magnitude $p = E/c$.) As far as cosmology is concerned, you have already seen that matter and radiation are spread throughout the Universe, so you should expect there to be some corresponding large-scale distribution of energy and momentum associated with all that matter and radiation. This large-scale distribution of energy and momentum, together with the equations of general relativity, allow us to obtain a mathematical description of space–time curvature on the large scale. This is the basis of relativistic cosmology.

QUESTION 5.3

On the basis of what you learned in Section 5.2 about the large-scale distribution of matter and radiation, what word would you expect to characterize the large-scale distribution of the associated energy and momentum?

Although the discussion in this section has been very qualitative and general up to this point, some important ideas have been introduced, so it's worth summarizing them here.

- The important ingredients of the Universe include space and time as well as matter and radiation.

- Einstein's special theory of relativity taught us to regard the three-dimensional space and one-dimensional time with which we are familiar as a four-dimensional space–time, in which all matter and radiation is contained.

- Einstein's general theory of relativity taught us that space–time has geometrical properties (e.g. curvature) that are determined by the distribution of energy and momentum associated with matter and radiation.

- By combining our beliefs about the large-scale distribution of energy and momentum with the general theory of relativity it should be possible to obtain a mathematical model of space–time on a large scale.

The next section explores the meaning of the phrase the 'geometric properties' of space–time, and the way in which those properties can be summarized mathematically. The section after that considers the large-scale distribution of energy and momentum. Having dealt with those two topics, the central equations of general relativity − Einstein's field equations − are introduced. We can then discuss the mathematical models of space–time that are consistent with our observations of matter and radiation, and the associated distribution of energy and momentum.

5.3.2 The space and time of the Universe

Imagine you were asked to describe space: not just the outer space beyond the atmosphere, but space in general, including the space you are occupying right now, and the space in which you might wave your arms without leaving your seat. You might say that space is big, or that it had three dimensions (i.e. three independent directions in which things can move), but what else could you do to describe space?

The chances that you will be asked to describe space may be slim, but for a cosmologist the question is crucial and the conventional answer is well known. For cosmologists the description of space is essentially a matter of **geometry**. According to dictionaries, geometry is 'the study of the properties and relations of lines, surfaces and volumes in space'. It is by studying the properties and relations of objects *in* space that we learn about space itself. Now, geometry is a big subject, but 19th century mathematicians such as Carl Friedrich Gauss (1777–1855) and Bernhard Riemann (1826–1866) found powerful ways of summarizing the whole of geometry in just a line or two of mathematics. Gauss in particular, certainly one of the greatest mathematicians who ever lived, initiated this development by recognizing the exceptional importance of **Pythagoras's theorem** about right-angled triangles.

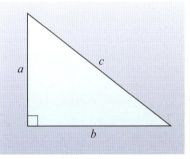

Figure 5.10 Pythagoras's theorem concerns the lengths, *a*, *b* and *c*, of the sides of a right-angled triangle (i.e. a closed, three-sided figure with one of its interior angles equal to 90°). In the notation of the figure, $c^2 = a^2 + b^2$.

According to Pythagoras's theorem, the square of the length of the longest side of a right-angled triangle is equal to the sum of the squares of the lengths of the other two sides. In symbols (see Figure 5.10), this can be expressed as

$$c^2 = a^2 + b^2 \tag{5.3}$$

Gauss realized that if this result was applied to the smallest conceivable right-angled triangles, then it could be used as the starting point for mathematical proofs of all the other known truths concerning the geometry of a two-dimensional plane. So, if we imagine an infinitesimally small version of Figure 5.10, and if we indicate its smallness by representing the lengths of its sides by d*s*, d*x* and d*y* rather than *c*, *a* and *b*, then we can say that the whole of two-dimensional plane geometry is implicitly contained in the single expression

$$(ds)^2 = (dx)^2 + (dy)^2 \tag{5.4}$$

where d*s* is the distance between two points whose *x*- and *y*-coordinates differ by the infinitesimal amounts d*x* and d*y* (see Figure 5.11). As far as a mathematician is concerned, Equation 5.4 is a complete answer to the question: 'Tell me all about the geometry of a two-dimensional plane.'

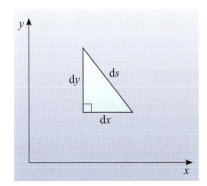

Figure 5.11 An infinitesimal version of Pythagoras's theorem.

The form of Equation 5.4 immediately suggests an answer to the question 'Tell me all about the geometry of three-dimensional space.' If we describe any point in space by using a three-dimensional coordinate system with mutually perpendicular axes x, y and z (see Figure 5.12), then you can easily imagine that the distance ds between two points separated by infinitesimal coordinate differences dx, dy and dz is given by a three-dimensional generalization of Pythagoras's theorem:

$$(\mathrm{d}s)^2 = (\mathrm{d}x)^2 + (\mathrm{d}y)^2 + (\mathrm{d}z)^2 \tag{5.5}$$

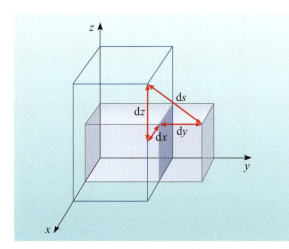

Figure 5.12 Two points in three-dimensional space with position coordinates that differ by the infinitesimal amounts dx, dy and dz. The points are separated by a distance ds.

This equation provides a basis for three-dimensional geometry, just as Equation 5.4 provides a basis for two-dimensional geometry. However, it's important to note that since Equation 5.4 only applies to a plane, it describes geometry on a *flat* surface. It does not, for example, describe the geometry of shapes drawn on the *curved* two-dimensional surfaces shown in Figure 5.13. Concise mathematical statements that summarize the geometry of these curved two-dimensional surfaces can be written down, but those equations are inevitably somewhat more complicated than Equation 5.4. Similarly, for all its power, Equation 5.5 only describes the geometry of what is confusingly called a 'flat' three-dimensional space. Mathematicians are familiar with similar but more complicated expressions that describe the geometry of 'curved' three-dimensional spaces, but we shall not write them down here. It is very difficult to try to picture what a 'curved' three-dimensional space would be like, but it is worth emphasizing that in a flat three-dimensional space familiar geometric results hold true, while in a curved three-dimensional space those same results may cease to be true. For example in a flat three-dimensional space the interior angles of any triangle add up to 180°, but in a curved three-dimensional space this may not be true. Similarly, 'straight' lines that are initially parallel remain parallel in a flat space but not in a curved space.

Figure 5.13 Some curved two-dimensional surfaces, viewed in three-dimensional space. The angles of triangles drawn on the curved surfaces do not add up to 180°, due to the curvature of the surfaces. Nor do lines that are initially parallel necessarily remain parallel.

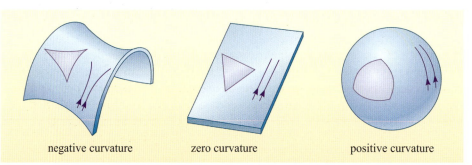

negative curvature   zero curvature   positive curvature

While Gauss was considering the possibility that three-dimensional space might be curved he was also involved in a land survey of the Kingdom of Hanover. Part of the survey involved measuring a large triangle between three mountain peaks (Figure 5.14). This particular measurement did not give any evidence that the space we inhabit is curved; the interior angles of the triangle did sum to 180° to the accuracy attainable with the surveying equipment. Nonetheless, Gauss realised that the true geometry of space was something that must be determined experimentally. Mathematics could determine the range of possible self-consistent geometries (flat or curved) but only observation would reveal which of those possibilities was actually realised in the real world.

It has already been stressed that Einstein's theory of general relativity explains gravitational phenomena in terms of the curvature of space–time. So you shouldn't be surprised that having discussed flat and curved three-dimensional spaces we now need to discuss the geometry of flat and curved four-dimensional space–times. Don't panic! As far as flat space–time is concerned all we need do is generalize Equation 5.5 so that instead of considering two narrowly separated *points* in space, we consider two neighbouring *events* in space–time. The two events can still differ in position by the infinitesimal amounts dx, dy and dz, but, being events, we can also choose them so that the times at which they occur differ by the infinitesimal amount dt. Investigations based on Einstein's theory of special relativity, which does not include the effects of gravity and therefore concerns flat space–time, show that the appropriate generalization of Equation 5.5 is

$$(ds)^2 = (dx)^2 + (dy)^2 + (dz)^2 - c^2(dt)^2 \qquad (5.6)$$

where c is the speed of light in a vacuum, a fundamental physical constant that would have been 'discovered' in the theory of relativity even if it had not already been known from studies of light.

It would be inappropriate to spend time justifying the precise form of Equation 5.6, but, as before, you should realize that Equation 5.6 describes the geometry of a 'flat' (i.e. zero curvature) four-dimensional space–time. The corresponding equation for a curved space–time will inevitably be more complicated.

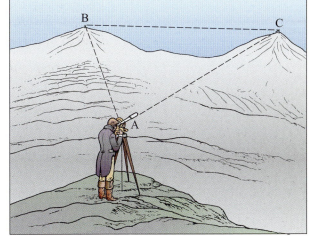

Figure 5.14 Survey measurements of the interior angles of a large triangle, with vertices on three mountain tops, showed no significant deviation from 180°. Nonetheless, Gauss realized that mathematics provided many possible 'geometries' so the true geometry of our space, flat or curved, could only be determined by experiment. (Adapted from Kittel *et al.*, 1965)

Events play the same role in space–time that *points* play in space. Whereas a point in space can be specified by three position coordinates (x, y, z), an event in space–time requires three position coordinates and a time coordinate (x, y, z, t). A point is an idealized location; an event is an idealized occurrence.

5.3.3 The distribution of energy and momentum in the Universe

As explained in Section 5.2.3, the observation that the Universe appears to be the same in all directions, combined with the Copernican principle – that we are not observing from a privileged position – implies that the Universe should be the same everywhere, on the large scale. As was also stated, the implied uniformity in the large-scale distribution of matter and radiation is now being increasingly well confirmed by observations, particularly the deep redshift surveys that are being carried out. Even so, because the observational evidence is necessarily limited, it is still appropriate to treat uniformity on the largest scales as an assumption supported by increasingly good evidence rather than a proven fact.

Cosmologists usually call this assumption of large-scale uniformity the **cosmological principle**, and sometimes state it in the following way.

> On sufficiently large size-scales (i.e. averaged over regions that are several hundred megaparsecs across) the Universe is **homogeneous** (i.e. the same everywhere) and **isotropic** (i.e. the same in all directions).

The technical terms 'homogeneous' and 'isotropic' make precise the rather loose notion of 'uniformity' that we have been using up to this point. Both terms are needed because it is possible for a distribution to be homogeneous without being isotropic. For instance, a universe in which there was a homogeneous magnetic field that everywhere pointed in the same direction would not be isotropic, although it would be homogeneous.

■ What is the precise feature of the cosmological principle that rules out the uniformly magnetized universe that has just been described? Explain your answer.

☐ Although the magnetized universe would be homogeneous it would not be isotropic. No point in the universe would be distinguished from any other point, as homogeneity demands, but at any point it would be possible to identify a 'preferred' direction by using a compass to determine the direction of the magnetic field. The fact that all directions are not equivalent in that universe shows that it is not consistent with the requirement for isotropy in our Universe.

In the simplest cosmological models that are consistent with the cosmological principle it is usually imagined that the Universe is completely filled with a uniform gas or fluid. (You can think of superclusters of galaxies as being the equivalent of 'atoms' in this cosmic fluid.) One advantage of taking such a simplified view of the contents of the Universe is that describing the state of the gas at any time t only involves specifying the *density* and *pressure* at the relevant time. These two properties of the gas determine all the other properties, such as the temperature. Density and pressure are usually denoted by the symbols ρ (the Greek letter 'rho') and p, but in an expanding Universe the density and pressure should be expected to change with time and we can indicate this dependence on time by writing the density and pressure at any time t as $\rho(t)$ and $p(t)$. Cosmological discussions are often further simplified by assuming that the pressure is negligible. This seems to be a reasonable assumption throughout much of cosmic history, since there is no evidence of superclusters colliding and rebounding in the way that atoms in a gas are supposed to do. We shall assume that pressure is negligible throughout most of this chapter but not in Chapter 6, which deals with the hot, dense, early Universe.

Thanks to the simplifying assumptions outlined above, it is quite easy for a cosmologist to write down a mathematical description of the large-scale distribution of energy and momentum in the Universe. The uniformly distributed cosmic gas gives rise to a cosmic distribution of energy and momentum that is also homogeneous and isotropic. Armed with the mathematical description of this energy–momentum distribution, cosmologists are able to use the equations of general relativity to determine the large-scale geometry of space–time and hence formulate a cosmological model. The first models of this kind are discussed in the next section.

5.3.4 The first relativistic models of the Universe

According to general relativity, the distribution of energy and momentum determines the geometric properties of space–time, and, in particular, its curvature. The precise nature of this relationship is specified by a set of equations called the *field equations* of general relativity. In this book you are not expected to solve or even manipulate these equations, but you do need to know something about them, particularly how they led to the introduction of a quantity known as the *cosmological constant*. For this purpose, Einstein's field equations are discussed in Box 5.1.

BOX 5.1 EINSTEIN'S FIELD EQUATIONS OF GENERAL RELATIVITY

When spelled out in detail the **Einstein field equations** are vast and complicated, but in the compact and powerful notation used by general relativists they can be written with deceptive simplicity. Using this notation, the field equations that Einstein introduced in 1916 can be written as

$$[G_{\mu v}] = \frac{-8\pi G}{c^4}[T_{\mu v}] \qquad (5.7)$$

Different authors may use different conventions for these equations. The symbols $[G_{\mu v}]$ and $[T_{\mu v}]$ represent complicated mathematical entities called *tensors* that it would be inappropriate to explain here, except to say that $[G_{\mu v}]$ describes the curvature of space–time while $[T_{\mu v}]$ describes the distribution of energy and momentum, and that both these quantities may vary with time and position. The other symbols just represent numerical constants and have their usual meanings, $G = 6.673 \times 10^{-11}$ N m^2 kg^{-2} is Newton's gravitational constant and $c = 2.998 \times 10^8$ m s^{-1} is the speed of light in a vacuum.

Given the distribution of momentum and energy at all points in space and time (i.e. given $[T_{\mu v}]$), the field equations determine the geometrical quantity $[G_{\mu v}]$, from which it may be possible to derive a detailed description of space–time geometry along the lines of the infinitesimal generalizations of Pythagoras's theorem that were discussed in Section 5.3.2. In his 1916 paper, Einstein used these equations to investigate planetary motion in the Solar System and to predict the non-Newtonian deflection of light by the Sun.

Einstein published his first paper on relativistic cosmology in the following year, 1917. In that paper he tried to use general relativity to describe the space–time geometry of the whole Universe, not just the Solar System. While working towards that paper he realized that one of the assumptions he had made in his 1916 paper was inappropriate in the broader context of cosmology. This led him to introduce another term into the field equations, a term he had deliberately chosen to ignore in 1916. With this extra term included the field equations used in the cosmology paper of 1917 can be written

$$[G_{\mu v}] + \Lambda[g_{\mu v}] = \frac{-8\pi G}{c^4}[T_{\mu v}] \qquad (5.8)$$

As you can see, the extra term takes the form $\Lambda[g_{\mu v}]$, where Λ is the upper case Greek letter 'lambda'. The $[g_{\mu v}]$ here is another of these tensor quantities that was actually already implicitly involved in $[G_{\mu v}]$, while Λ represents a new physical constant called the **cosmological constant**. Provided Λ is sufficiently small (which it is) the presence of the $\Lambda[g_{\mu v}]$ term does not invalidate any of the results that Einstein obtained in the 1916 paper but, in the context of cosmological calculations, a positive value of Λ implies the action of a long-range repulsion that might, under appropriate circumstances, balance or even overwhelm the otherwise attractive influence of gravity.

Note that the left-hand side of Equation 5.8 still describes the curvature of space and time, while the right-hand side describes the content of space and time. Some cosmologists opt to subtract $\Lambda[g_{\mu v}]$ from both sides, to yield this alternative formulation of Equation 5.8:

$$[G_{\mu v}] = \frac{-8\pi G}{c^4}[T_{\mu v}] - \Lambda[g_{\mu v}]$$

The effect is to regard the cosmological constant as one of the contents of space–time, rather than an intrinsic property of space–time itself. What sort of fluid or substance could have the effect of Λ? A *very* strange one, it turns out. For example, we will see that it has a negative pressure! This hasn't stopped cosmologists speculating about it, and to go beyond Einstein in supposing that this strange fluid could possibly change from place to place, or might vary over cosmic time (unlike Einstein's *constant* cosmological constant). This strange fluid is also known as *dark energy*, of which more later in this chapter.

In 1917 there was no evidence to indicate that the Universe was either expanding or contracting. So, in developing the first relativistic cosmological model, Einstein sought a value for the cosmological constant Λ that would ensure everything was constant and unchanging with time. He was also guided by the cosmological principle, so he required that the average density of matter in the Universe, ρ, should be homogeneous (i.e. independent of position) as well as constant (i.e. independent of time). He ignored the possibility of pressure, effectively assuming $p = 0$. Using these assumptions together with the modified field equations (Equation 5.8), Einstein constructed the first relativistic cosmological model, which is now known as the **Einstein model**. The need to balance the gravitational attraction of matter and the repulsive effect of the cosmological constant led Einstein to the relation

$$\Lambda = \frac{4\pi G\rho}{c^2} \tag{5.9}$$

Because it is not expanding, the Einstein model is no longer thought to represent the Universe we actually inhabit. In fact, after the expansion of the Universe had been discovered, Einstein himself described his first use of the cosmological constant as his 'greatest blunder'. Nonetheless, it is worth exploring the geometrical properties of the Einstein model since it can provide insight into some of the extraordinary possibilities of relativistic cosmology.

The universe described by the Einstein model is **static**, neither expanding nor contracting. In this model universe, space is **finite**, having a total volume that is proportional to $\Lambda^{-3/2}$. Despite being finite, space in the Einstein model is also **unbounded**, that is to say, you can travel as far as you like in any direction without ever hitting a wall or encountering anything like an 'edge' of space. However, if you were to travel far enough in a straight line, you would eventually find yourself back at your starting point.

How is this possible? How can a straight line close back upon itself? Very simply, because what we are discussing here is a straight line in a *curved* space. In the Einstein model, which is homogeneous and isotropic, the curvature of space must be the same everywhere and in all directions. In addition, this uniform curvature has a positive value at every point, which means that a 'straight' line will be uniformly bent back upon itself all along its length. The actual value of the positive curvature depends on the value of the cosmological constant Λ, with the consequence that a line that is as straight as it can be in any region (the kind of line defined by a light ray, say) will close on itself after a distance that is proportional to $\Lambda^{-1/2}$. Figure 5.15 attempts to provide some idea of the geometry of space–time in the Einstein model.

If we did inhabit the kind of universe described by the Einstein model, we might expect astronomical observations to reveal distant images of the Earth, Sun or Milky Way, due to light that had travelled along the space–time paths that the model implies.

■ If we did live in the universe described by the Einstein model, how might we determine the value of the cosmological constant?

☐ By measuring the average density of matter on the large scale, ρ, and then using Equation 5.9 to determine Λ.

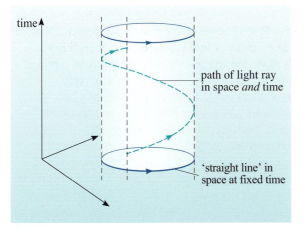

time

path of light ray
in space *and* time

'straight line' in
space at fixed time

Figure 5.15 In this attempt to represent the four-dimensional space–time of the Einstein model, time is measured along the vertical axis. The whole of space at any time must therefore be represented in the horizontal plane, but because of the need to indicate that space is intrinsically curved, the circle that you see in the horizontal plane actually represents a 'straight' line. The helical path drawn above the circle might represent the path of a pulse of light that follows a 'straight' line through space while time passes. Because of the nature of this space–time diagram, the Einstein model is sometimes called 'Einstein's cylindrical world', but it's important to realize that the four-dimensional universe described by the Einstein model is no more akin to a real 'cylinder' than the 'flat' space of special relativity is akin to a pancake. (Adapted from Raine, 1981)

As you have seen, an important feature of the Einstein model is that the curvature of space is everywhere positive. This is conventionally indicated by introducing a **curvature parameter**, k, that may take the value +1, 0 or −1, and by saying that in the Einstein model $k = +1$. You will shortly be meeting other cosmological models with $k = 0$ and $k = -1$, as well as more models with $k = +1$. The curvature parameter is one of the most important characteristics of these models, since it strongly influences the large-scale geometric properties of the model. The value of k immediately determines whether space is finite ($k = +1$) or infinite ($k = 0$ or -1). As Figure 5.16 indicates, it also determines whether the interior angles of cosmically large triangles will add up to be less than, equal to or greater than 180°, how the circumference of a circle is related to its radius, and whether parallel lines in space remain parallel. The significance of this should soon become apparent, because the next model we are going to discuss has zero curvature everywhere and is characterized by $k = 0$. In this model, space is infinite, and 'straight' lines do not close back upon themselves.

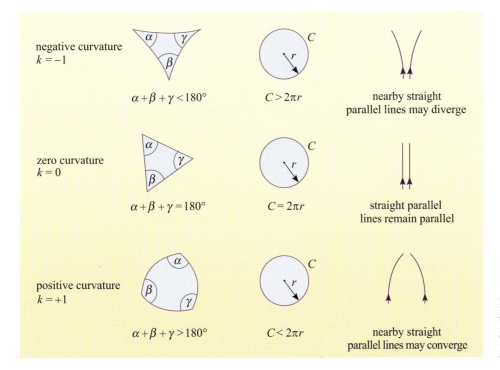

negative curvature
$k = -1$

α γ β

$\alpha + \beta + \gamma < 180°$

$C > 2\pi r$

nearby straight
parallel lines may diverge

zero curvature
$k = 0$

α γ β

$\alpha + \beta + \gamma = 180°$

$C = 2\pi r$

straight parallel
lines remain parallel

positive curvature
$k = +1$

α β γ

$\alpha + \beta + \gamma > 180°$

$C < 2\pi r$

nearby straight
parallel lines may converge

Figure 5.16 The effect of the curvature parameter, k, in determining the large-scale geometry of a cosmological model.

Figure 5.17 Willem de Sitter was a Dutch astronomer who devised the second relativistic cosmology in association with his colleague Paul Ehrenfest (1880–1933). He realized that, in his model, observers would find that the light from distant sources was red-shifted, but, due to a misinterpretation, he described the outward radial velocity this indicated as 'spurious'. (Science Photo Library)

Within a year of Einstein's publication of the first relativistic cosmological model, a radically different model was proposed by the Dutch astronomer Willem de Sitter (1872–1934; Figure 5.17). As in the case of the Einstein model, the mathematical details of the **de Sitter model** can be found by solving the field equations (Equation 5.8). Like Einstein, de Sitter assumed that the Universe was homogeneous and isotropic, in accordance with the cosmological principle, but he did not require it to be static. Instead he took the view that the effect of matter was negligible, implying that both the average pressure and the average density could be taken to be zero. The geometric properties of space would therefore be determined by the cosmological constant alone. From a modern perspective we can now see that in de Sitter's model $k = 0$, and a positive value of Λ results in the never-ending expansion of space at an accelerating rate. What little matter was actually present in the Universe would simply be carried along by the expansion of the space in which it was located (see Box 5.2) and this would result in observable redshifts. Unfortunately, this interpretation was not clear at the time of the de Sitter's discovery, and for many years it was mistakenly believed that what had been found was another static model of the Universe with the strange property that stationary sources of radiation, provided they were sufficiently distant, would have red-shifted spectral lines.

One way of describing the expansion of space mathematically uses a coordinate grid that can expand or contract along with space. Such coordinates are said to be **co-moving** or **comoving** (we will use the latter spelling). Such coordinates are said to be comoving and are widely used in cosmology. This sort of coordinate system is indicated schematically in Figure 5.18, although for the sake of clarity only a two-dimensional grid is shown, rather than the three-dimensional grid that would really be required to label all the points in space. Due to the use of comoving coordinates, typical points in an expanding space have unchanging coordinates, even though those points are moving apart. Since the coordinates themselves do not describe the expansion of space, it is necessary to introduce another parameter that does. This is called the **scale factor** and is represented by $R(t)$, where the parenthesized t indicates that the scale factor can change with time, increasing or decreasing as the Universe expands or contracts. If $R(t)$ increases with time, so that its value at time t_2 is greater than its value at some earlier time t_1, then the physical distance between two points with fixed coordinates will also increase, as 'expansion' implies that it should (this is the case shown in Figure 5.18). If $R(t)$ decreases with time then the distance between typical points is reduced, and space may be said to be contracting.

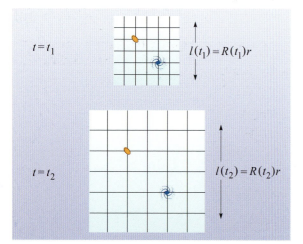

Figure 5.18 When comoving coordinates are used to identify points, the expansion (or contraction) of space can be indicated by the behaviour of a scale factor, $R(t)$. If two points are separated by a comoving coordinate distance r, the physical distance (in metres, say) between those points at time t, will be $l(t) = R(t)r$.

It's important to realize that the use of comoving coordinates removes any direct relationship between the coordinate differences dx, dy and dz of two narrowly separated points and the actual physical distance, ds, between those points. In order to find the real physical distance between the two points (measured in metres, say), at time t, the scale factor $R(t)$ must be taken into account. In the simplest case of an expanding flat space with zero curvature, this can be indicated by writing

$$(ds)^2 = [R(t)]^2[(dx)^2 + (dy)^2 + (dz)^2] \qquad (5.10)$$

BOX 5.2 GALAXY RECESSION AND THE EXPANSION OF SPACE

At first sight, the observed recession of distant galaxies suggests that those galaxies are moving through space. This, however, neglects the general relativistic view that space (or more properly space–time) has intrinsic properties that are influenced by the contents of the Universe. A useful way of thinking about space in general relativity is to imagine it as the three-dimensional analogue of a rubber sheet that may expand or contract as it is stretched or released.

If you picture galaxies as something like buttons placed on the rubber sheet, then it is clear that *one* way of causing them to separate is to move them across the sheet. This is the analogue of galaxies moving *through* space. But a second way of increasing the separation between buttons is to stretch the rubber itself. It is this latter view that is most helpful when trying to interpret a phrase such as 'matter would be carried along by the expansion of space'.

The galaxies we actually observe can be thought of as moving for two reasons. On the one hand they are being carried along *by* space as a result of the uniform cosmic expansion described by the Hubble flow. On the other hand those same galaxies are also moving *through* space due to local effects, such as the gravitational influence of nearby concentrations of matter. The local effects can be dominant on the small scale, explaining why some nearby galaxies are moving towards us. On larger scales, however, local effects are always overwhelmed by cosmic expansion, so *all* sufficiently distant galaxies are moving away from us, carried away by the expansion of space.

Equation 5.10 implies that if the 'coordinate distance' between the two points is dr, then the physical distance between them at time t is $ds = R(t)\,dr$, where

$$dr = \sqrt{(dx)^2 + (dy)^2 + (dz)^2}\ .$$

In the case of the de Sitter model, the field equations show that the scale factor grows exponentially with time, as indicated in Figure 5.19. The steepness of the curve increases at a rate that is determined by the value of the cosmological constant, since it is Λ that drives the expansion. In fact, it can be shown that in the de Sitter model

$$R \propto e^{Ht} \qquad \text{where } H = \sqrt{\frac{\Lambda c^2}{3}} \qquad\qquad (5.11)$$

Moreover, it turns out that in this model the constant H is also equal to the Hubble constant that observers would discover if they had any distant galaxies to observe.

So, if the density of matter in our Universe had been negligible, and de Sitter's model (when correctly interpreted) had provided a good description of cosmic expansion, then observations of distant galaxies would have revealed the value of H in Equation 5.11, and hence the value of the cosmological constant Λ.

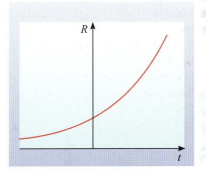

Figure 5.19 The behaviour of the scale factor in the de Sitter universe.

QUESTION 5.4

The scale factor was not discussed in the context of the Einstein model but, if it had been, what could have been said about its behaviour?

5.3.5 The Friedmann–Robertson–Walker models of the Universe

Alexander Friedmann (Figure 5.20) was a Russian mathematician who worked at the University of St Petersburg. In 1922 and 1924 he published two important

ALEXANDER FRIEDMANN (1888–1925), HOWARD PERCY ROBERTSON (1903–1961) AND ARTHUR GEOFFREY WALKER (1909–2001)

Alexander Friedmann (Figure 5.20) was born and educated in St Petersburg and returned there in 1920 to work at the Academy of Sciences. Known for his work on theoretical meteorology, Friedmann became interested in general relativity and used his mathematical talents to undertake a highly original exploration of the cosmological consequences of the theory. His researches led him to the equation that determines the evolution of the scale factor $R(t)$ in a homogeneous, isotropic universe that is uniformly filled with matter. This is now known as the Friedmann equation.

Howard Percy Robertson (Figure 5.21) was an American mathematical physicist who specialized in the application of general relativity to practical situations. In 1929, using general arguments that did not depend on specific assumptions about the properties of matter, Robertson deduced the general expression for the

Figure 5.20
Alexander Friedmann.
(Science Photo Library)

Figure 5.21
Howard Percy Robertson.

Figure 5.22
Arthur Geoffrey Walker.

separation of events in the space–time of any universe that is spatially homogeneous and isotropic at all times (see Equation 5.12).

Arthur Geoffrey Walker (Figure 5.22) spent most of his academic career at the University of Liverpool, initially as a lecturer and later as Professor of Mathematics. His expression for the separation of events in a homogeneous and isotropic universe was published in 1936, and was based on a somewhat different approach from the earlier work of Robertson .

papers that had the effect of showing that the cosmological models of Einstein and de Sitter were special cases of a much wider class of expanding or contracting models, all of which were consistent with the field equations of general relativity and with the cosmological principle. However, Friedman was unable to exploit this breakthrough due to his early death in 1925. Later, Howard P. Robertson (Figure 5.21) and Arthur G. Walker (Figure 5.22) independently found improved ways of describing these models and ensuring their generality. It was the work of these three, Friedmann, Robertson and Walker, that resulted in the general mathematical framework still used today when discussing relativistic cosmological models of a homogeneous and isotropic Universe.

The geometric properties of space–time in any of the **Friedmann–Robertson–Walker** models (usually abbreviated to **FRW models**) can be deduced from a single expression for the space–time separation $\mathrm{d}s$ of two events whose coordinates differ by the infinitesimal amounts $\mathrm{d}x$, $\mathrm{d}y$, $\mathrm{d}z$ and $\mathrm{d}t$, and which are located at a coordinate distance $r = \sqrt{x^2 + y^2 + z^2}$ from the origin.

There are several different but equivalent ways of expressing this relationship. The form given here is not the most conventional but has been chosen for simplicity.

$$(\mathrm{d}s)^2 = \frac{[R(t)]^2}{\left(1 + \dfrac{kr^2}{4}\right)^2} \left[(\mathrm{d}x)^2 + (\mathrm{d}y)^2 + (\mathrm{d}z)^2\right] - c^2(\mathrm{d}t)^2 \tag{5.12}$$

BOX 5.3 THE FRIEDMANN EQUATION: EPISODE I

It should be clear from what has already been said that the Friedmann equation is of the utmost importance in the process of cosmological modelling. The Friedmann equation determines the form of $R(t)$, and thereby fixes the evolutionary history of a model universe. In this sense, the fate of the Universe is determined by the Friedmann equation, and many cosmologists would say that it is certainly the most important equation in cosmology.

All of this might make you wonder why we have not actually written down the Friedmann equation at this point. The reason is simple. The Friedmann equation involves mathematical notation and concepts with which you may not be familiar at this stage, but which arise naturally in the next section. We are therefore delaying the explicit introduction of the Friedmann equation until then.

If you really can't wait to see it, take a look at Episode 2 in Section 5.4.3 (in Box 5.4).

Equation 5.12 is one form of what is known as the **Robertson–Walker metric**. We shall not be using this expression as the basis of any detailed arguments, but you should notice three things about it. First, it is a generalization (to the case of uniformly curved space–time) of Equation 5.6, which provided a complete description of the geometric properties of a flat space–time. Second, it contains the curvature parameter k, which helps to characterize the curvature and can take the values −1, 0 or +1. Third, it contains the scale factor $R(t)$ that describes the expansion or contraction of space as a function of time.

Equation 5.12 applies to all the FRW models, but in order to work out the details of any particular model it is necessary to specify the value of k and to determine the precise form of $R(t)$. In the case of a universe uniformly filled with pressure-free (i.e. $p = 0$) matter of density ρ, the form of $R(t)$ can be determined by solving a complicated equation known as the **Friedmann equation** (Box 5.3). This important equation relates the value of R, and the rate of change of R, to the curvature parameter k, the cosmic density ρ and the cosmological constant Λ. We shall not go into the details of its solution, but different values of k and Λ can lead to quite different forms for $R(t)$, and these are indicated schematically in Figure 5.23. The implications of these solutions are discussed below.

Figure 5.23 (overleaf) contains a great deal of information and deserves careful study. The first thing to notice is that each of the small graphs of R against t corresponds to different values (or ranges of values) of the curvature parameter k and the cosmological constant Λ. For each set of values, the small graph shows the history of spatial expansion or contraction in a homogeneous and isotropic universe uniformly filled with pressure-free matter. For example, the left-hand column (the column headed $\Lambda < 0$) contains all the cases where the cosmological constant is less than zero. The uppermost of the three graphs corresponds to $k = +1$, the middle graph corresponds to $k = 0$, and the lowest graph corresponds to $k = -1$. In all three of these cases the graphs are similar: R is 0 at $t = 0$, increases up to some maximum value, and then decreases to zero again after a finite time. In other words, all these models describe universes with a finite lifetime that expand, reach a state of maximum expansion, and then contract again. It is important to remember that the quantity R plotted in these graphs represents the 'scale' of the universe, not its radius. Only in the case where $k = +1$ does space have a finite total volume, and even in that case, as in all the others, the model is homogeneous and isotropic so there can be no 'centre' nor any 'boundary' or 'edge' of space. Recalling that R is a *scale factor*, and that it characterizes the changing separation of typical (comoving) points in a uniform universe, will help you to avoid making misleading statements about the 'radius' or 'diameter' of the universe.

The second column in Figure 5.23 is especially interesting; it contains the models that have a vanishing cosmological constant ($\Lambda = 0$). Until recently, these were believed to be the most realistic models of the Universe we actually inhabit. The first of the models in this class has $k = +1$, implying that space is finite, and the corresponding graph of R against t once again indicates a cycle of expansion and contraction with a finite lifetime. In this and other models that begin with $R = 0$ at time $t = 0$, the early part of the expansion is now known as the **big bang**; the collapse that takes place at the other end of the cycle is known as the **big crunch**. Among the $\Lambda = 0$ models, all start with a big bang but only the $k = +1$ model ends with a crunch: it is known as the **closed model**.

In the other two models of the $\Lambda = 0$ class, space is infinite and expands forever. The $k = -1$ model is called the **open model**. In this model, as t approaches infinity the relationship between R and t approaches the simple form $R \propto t$. The $k = 0$ model represents a special case between the open and closed models and is known as the **critical model**. In this case, the relationship between R and t takes the form $R \propto t^{2/3}$ for all values of t. Somewhat confusingly, the critical model is also known as the **Einstein–de Sitter model**, even though it has no direct relation to either the Einstein model or the de Sitter model. (The alternative name became popular because Einstein and de Sitter agreed to support this model in 1932.)

In all of the remaining models of Figure 5.23 the cosmological constant is greater than zero ($\Lambda > 0$). It's best to regard all these models as occupying a single column, even though when $k = +1$ there are actually several quite distinct cases to discuss. But, before considering any of these models in detail, answer the following question.

■ (a) In Figure 5.23, locate the graph of R against t that characterizes the Einstein model, and write down the corresponding values of k and Λ.

(b) Can you see any graph in Figure 5.23 that corresponds to the de Sitter model?

☐ (a) The R against t graph for the Einstein model is the flat line shown in the middle of the top row of $\Lambda > 0$ models. According to Figure 5.23 it corresponds to $k = +1$ and $\Lambda = \Lambda_E$. (Note that Λ_E represents a particular value of the cosmological constant Λ. It follows from Equation 5.9 that, for a static universe of density ρ, that value is $\Lambda_E = 4\pi G\rho/c^2$.)

(b) There is no graph in Figure 5.23 that is obviously identical to the R against t graph of the de Sitter model shown in Figure 5.19. However, as you will see below, the de Sitter model is present in Figure 5.23 as a 'limiting case' of the $\Lambda > 0$, $k = 0$ model.

Let's examine the $\Lambda > 0$ models in turn, starting with the case where $k = +1$ and Λ is greater than zero, but less than Λ_E (i.e. $0 < \Lambda < \Lambda_E$). In this case the graph

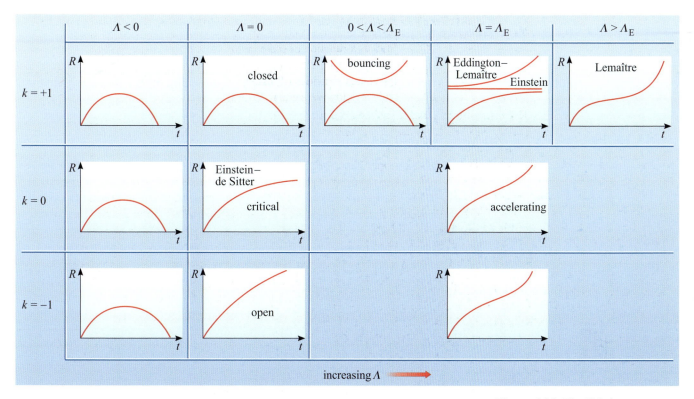

Figure 5.23 The Friedmann–Robertson–Walker models, classified according to the values of k and Λ. In each case the model is represented by a small graph of R against t, which encapsulates the history of spatial expansion and/or contraction implied by the model. Note that Λ_E represents the value of the cosmological constant in the Einstein model, $4\pi G\rho/c^2$. (Adapted from Landsberg and Evans, 1977)

indicates two possible kinds of behaviour. One is the now familiar situation in which R starts from zero at time $t = 0$, increases up to some maximum value and then decreases to zero again in a finite time. The alternative behaviour, indicated by the higher of the two curves, is one in which there is no obvious time to choose as $t = 0$, since there is no equivalent of the big bang. Rather, this is an infinitely old model in which an infinitely long period of contraction leads to a 'bounce' (when the scale factor reaches its minimum value) followed by an infinitely long period of expansion. If the Universe we actually inhabit is represented by a Friedmann–Robertson–Walker model with $k = +1$ and Λ in the range $0 < \Lambda < \Lambda_E$, then its actual behaviour – whether it follows the upper curve or the lower one – will be determined by its behaviour in the distant past. If the Universe really started with a big bang, the upper curve would be ruled out.

The next case to consider is that in which the cosmological constant has the particular value Λ_E that allows the model to be static. We have already noted that the static behaviour of the Einstein model is one of the allowed modes of behaviour in this case, and that is indicated by the presence of the flat line in the R against t graph. But other kinds of behaviour are also possible, as indicated by the other two curves in this part of the figure. One possibility, shown by the lower curve, is that R starts from zero and increases, gradually approaching the value specified by the Einstein model. The other possibility, represented by the upper curve, describes a universe that starts out in something very close to the static state, but expanding just a little. Even the tiniest initial expansion of this kind will eventually lead to a perpetual expansion, making it possible that an expanding universe might have had an indefinitely long history before the expansion really took hold, a possibility that many cosmologists have found attractive. This last kind of behaviour characterizes the **Eddington–Lemaître model**. The model was introduced in 1927 by Georges Lemaître (1894–1966; Figure 5.24), a Belgian cleric who made several contributions to relativistic cosmology. Initially neglected, the model was elaborated and strongly advocated in a 1930 paper by the Cambridge astrophysicist Sir Arthur Eddington (1882–1944; Figure 5.25), who felt that it might well describe the real Universe.

Despite his pioneering work on the Eddington–Lemaître model, Georges Lemaître is particularly associated with the $k = +1$ model in which Λ is greater than Λ_E, which is known as the **Lemaître model**. Lemaître advocated this model in the 1930s, when a number of scientists became interested in the origin of the chemical elements, or, more specifically, the origin of the nuclei of the various elements. The Lemaître model describes a universe that is homogeneous and isotropic, in which space has a finite total volume at any time, and where R starts from zero at time $t = 0$ but increases without limit. In this model the expansion passes through a 'pseudo-static' phase in which the R against t graph becomes almost flat, so that, for a while at least, it resembles a static universe even though it is never truly static. Lemaître argued that the late stages of this model could represent the expanding Universe we actually observe, that the intermediate 'coasting' phase provided the necessary time for the formation of stars and galaxies, and that the early, highly compressed state would have been so hot and dense that it could account for the 'cooking up' of some of the nuclei that certainly exist in the Universe. Although the details of Lemaître's argument are no longer accepted, his notion that the first nuclei were formed in a hot, dense phase of the early Universe is widely accepted, and Lemaître is therefore generally credited with recognizing the importance of the 'big bang' even though he did not use that specific term.

In Figure 5.23, only one graph corresponds to $\Lambda > 0$ and $k = 0$. As you will see later, a range of astronomical evidence from several independent sources now favours models in which k is compatible with zero and $\Lambda > 0$. So, this is now regarded as the model that most closely represents the real Universe. As you can see from the R against t graph, this model describes a uniform universe that starts with a big bang and goes on expanding forever. The expansion undergoes a 'slowdown' at some stage, but does not exhibit the sort of 'pseudo-static' behaviour seen in the Lemaître model. After the slowdown, the rate of expansion increases continuously, and for this reason the model is sometimes referred to as the **accelerating model**.

The final graph in Figure 5.23 corresponds to $\Lambda > 0$ and $k = -1$. In a uniform universe of this type, space is infinite and has the kind of negatively curved geometry that causes cosmically large triangles to have interior angles that sum to less than 180°. This is another model that starts with a big bang. As in the accelerating model, the scale factor grows from zero, slows its growth temporarily and then accelerates again.

We have now discussed each of the FRW models, but we have still not found the de Sitter model among them. This is rather surprising, since Figure 5.23 should contain *all* the homogeneous and isotropic models that are filled by a pressure-free fluid. Where is the de Sitter model in this family? Well, the de Sitter model has $\Lambda > 0$ and $k = 0$, so we might expect to find it in the box that contains the accelerating model, and in fact that is where it is, but it is only present as a 'limiting case' of the behaviour that is illustrated in that part of the figure. The de Sitter model has a negligible amount of matter, so it corresponds to $\rho = 0$ as well as $p = 0$, whereas the graph that represents the $\Lambda > 0$ and $k = 0$ model in Figure 5.23 shows the general case in which ρ may have a non-zero value. In this general case the density of matter decreases with time. As t increases the matter is eventually

GEORGES LEMAÎTRE (1894–1966) AND ARTHUR EDDINGTON (1882–1944)

Georges Lemaître (Figure 5.24) was a Belgian cosmologist who was also a Catholic priest. He initially trained as a civil engineer, but after serving in World War I he entered a seminary, became a priest, and subsequently studied solar physics in Cambridge. While there he met Arthur Eddington, and then visited America where he became familiar with the work of Hubble and Shapley. After returning to Belgium, Lemaître became a part-time lecturer at the University of Louvain in 1925. In 1931 he formulated his notion of an ultra-dense 'primeval atom', the explosion of which might start the observed expansion of the Universe.

Figure 5.24 Georges Lemaître. (Science Photo Library)

Figure 5.25 Sir Arthur Stanley Eddington. (Royal Astronomical Society Library)

Sir Arthur Stanley Eddington (Figure 5.25) was a Cambridge-based astrophysicist who made several important contributions to the study of stellar structure, particularly through his recognition of the significance of radiation pressure in maintaining (or destroying) equilibrium. Eddington was an early supporter of Einstein's general theory of relativity, and was the author of the first important book about the theory to appear in English. In 1919 he led the celebrated expedition that provided experimental support for the theory by observing the predicted deflection of starlight passing close to the edge of the Sun during a total eclipse.

so thinly spread that such a universe increasingly resembles an essentially empty de Sitter model in which the cosmological constant is solely responsible for the expansion. As a result, the graph of R against t increasingly approaches the de Sitter form ($R \propto e^{Ht}$) as t approaches infinity. So, the de Sitter model is implicitly present in Figure 5.23, as a limiting case of what is shown. Though we shall not bother to discuss them, there are similar limiting cases elsewhere in Figure 5.23.

QUESTION 5.5

In the context of the Friedmann–Robertson–Walker models, which values or ranges of the parameters k and Λ correspond to universes with the following characteristics?

(a) The universe is neither homogeneous nor isotropic.

(b) There is no possibility of a big bang.

(c) A big bang is possible, but there is at least one other possibility (assume $\rho > 0$).

(d) The particular point in space where the big bang happened can still be determined long after the event.

(e) At any time, the large-scale geometrical properties of space are identical to those of a three-dimensional space with a flat geometry.

(f) Space has a finite volume, and 'straight' lines that are initially parallel may eventually meet.

(g) There is a big bang, but the volume of space is infinite from the earliest times.

5.4 The key parameters of the Universe

At the beginning of the last section it was stated that a cosmological model typically consists of:

- *equations* that imply general relations between observable quantities, together with

- *parameters* that must be determined by observation before the model can be used to provide detailed quantitative predictions.

You should fully appreciate the significance of this assertion now that you have examined the class of Friedmann–Robertson–Walker (FRW) models. You have just seen that in those models there is a general expression (Equation 5.12) that describes the geometry of space–time in terms of the separation of events. The form of this equation is enough to show a cosmologist that the universe being described is homogeneous and isotropic. However, a detailed appreciation of the properties of space–time in such a universe involves determining the parameters that arise in the model, specifically the curvature parameter k and a scale factor $R(t)$. Only when these are known does it become possible to evaluate quantities such as the curvature of space, which is determined at time t by the quantity $k/[R(t)]^2$. The importance of observable parameters is further emphasized by recalling that the behaviour of the scale factor is determined by the Friedmann equation, which involves the value of the curvature parameter k, the cosmological constant Λ and the average density of matter ρ, all of which are, in principle, observable parameters at any given time.

This section is concerned with the parameters that arise in the FRW models (basically k and $R(t)$), and their relationship to the observational parameters (such as the Hubble constant) that characterize our Universe. By determining and exploiting these relationships, it should be possible to use astronomical observations to determine which of the many cosmological models most closely resembles the Universe in which we live. This is one of the central challenges of the branch of cosmology known as *observational cosmology*.

Although this section concerns measurable parameters, its emphasis is on relationships rather than values. The values of the observable parameters, and the best ways of determining those values, are discussed much more fully in Chapter 7, which is entirely devoted to the subject of observational cosmology.

5.4.1 Hubble's law, the Hubble constant and the Hubble parameter

One observational result that finds a very natural explanation in the context of FRW cosmology is Hubble's law. You will recall Hubble's law describes the general tendency for the redshift z of a galaxy to increase in proportion to its distance d from the observer, as described by the equation

$$z = \frac{H_0}{c} d \tag{5.1}$$

where the constant of proportionality, H_0/c, is made up of *Hubble's constant*, H_0, and the speed of light in a vacuum, c. You will also recall that for any particular galaxy the redshift z in Equation 5.1 can be related to the observed and emitted wavelengths, λ_{obs} and λ_{em}, of some identified spectral line by the equation

$$z = \frac{\lambda_{obs} - \lambda_{em}}{\lambda_{em}} \tag{5.2}$$

By measuring the redshifts of distant galaxies, and independently measuring the distances of those galaxies, it is possible to use Equation 5.1 to determine the value of the Hubble constant H_0, since the speed of light is known. Hubble himself did this, although his result was wildly inaccurate due to limited and incorrectly interpreted data. More modern determinations have greatly improved the situation. A major advance came in 2001 with the results of a HST Key Project that yielded

$$H_0 = (72 \pm 8) \text{ km s}^{-1} \text{ Mpc}^{-1}$$

This is the value that is used throughout this book. A more recent (2012), and widely quoted result is

$$H_0 = (74.3 \pm 2.1) \text{ km s}^{-1} \text{ Mpc}^{-1}$$

This latter result indicates an uncertainty in H_0 of less than 3%. However, it should be noted that another determination made in 2013 using the Planck satellite found the value $H_0 = (67.3 \pm 1.2) \text{ km s}^{-1} \text{ Mpc}^{-1}$. The two results, though close, are not consistent given their associated uncertainties and indicate that some care is still needed when interpreting measurements of H_0. This will be discussed further in Chapter 7.

H_0 is one of the most important observational parameters in cosmology, but how does it relate to the FRW models, where the obvious parameters are k and $R(t)$, and there is no H_0 to be seen? This is what we must now investigate.

Figure 5.26 indicates the basis of the relationship. The figure shows two snapshots of an expanding FRW universe, with a growing scale factor $R(t)$. Two galaxies, A and B, happen to be located at the grid points of a set of comoving coordinates that expands with the universe. The first snapshot represents a time t_{em} at which some light is emitted from galaxy A, and the second snapshot represents a later time t_{obs} at which that same light is observed in galaxy B.

While the light is travelling from A to B, the (comoving) coordinates of the galaxies do not change, but the physical distance between the galaxies does increase because it is proportional to $R(t)$, and $R(t_{obs})$ is greater than $R(t_{em})$. Whatever the distance from A to B at time t_{em}, it will have increased by a factor of $R(t_{obs})/R(t_{em})$ by the later time t_{obs}. Now, this expansion factor $R(t_{obs})/R(t_{em})$ represents the growth of space itself, so it will also influence the wavelength of

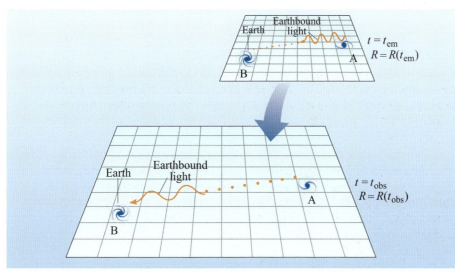

Figure 5.26 Cosmic expansion, measured by the increasing scale factor $R(t)$, as a cause of the cosmological redshift of distant galaxies. The 'stretching' of space also stretches light waves travelling from one galaxy to another, causing the observed light to be red-shifted relative to the emitted light. (Adapted from Finkbeiner, 1998)

the light that is moving freely between the two galaxies. As a result, light that is emitted from A at time t_{em} with wavelength λ_{em} will be observed at B at time t_{obs} with the longer wavelength $\lambda_{obs} = \lambda_{em} \times R(t_{obs})/R(t_{em})$. This increase in wavelength will, of course, be interpreted as a redshift by the observer in galaxy B.

■ According to an observer in galaxy B at time t_{obs}, what is the redshift of galaxy A? Express your answer in terms of the expansion factor $R(t_{obs})/R(t_{em})$.

☐ Rearranging Equation 5.2

$$z = \frac{\lambda_{obs} - \lambda_{em}}{\lambda_{em}} = \frac{\lambda_{obs}}{\lambda_{em}} - \frac{\lambda_{em}}{\lambda_{em}}$$

$$z = \frac{\lambda_{obs}}{\lambda_{em}} - 1$$

Replacing $\lambda_{obs}/\lambda_{em}$ by the equivalent expansion factor $R(t_{obs})/R(t_{em})$, we find that

$$z = \frac{R(t_{obs})}{R(t_{em})} - 1 \tag{5.13}$$

Note that according to the FRW model, the redshift of a distant galaxy is primarily caused by the expansion of space, it is *not* a Doppler shift due to movement through space. These expansion-based redshifts are usually referred to as **cosmological redshifts** in order to distinguish them from Doppler shifts. Of course, real galaxies do not necessarily behave like the idealized galaxies of Figure 5.26. Real galaxies may have some 'peculiar' motion of their own relative to the grid of comoving coordinates, and this peculiar motion can give rise to Doppler shifts that cause the observed redshifts of galaxies to differ somewhat from the cosmological redshifts implied by a smoothly expanding FRW model.

We have now seen how redshifts can arise from expansion in a Friedmann–Robertson–Walker model, but the real point of Hubble's law is that the redshift z of distant galaxies *increases in proportion to their distance*. How do FRW models account for this? Very simply, as it turns out. The greater the distance of a galaxy, the longer the light takes to reach us from that galaxy. The greater the time the light spends 'in flight' between the moments of emission and observation, the greater is the expansion factor $R(t_{obs})/R(t_{em})$, and the greater the redshift of the light, $z = [R(t_{obs})/R(t_{em})] - 1$.

The time-of-flight argument provides a qualitative explanation of Hubble's law in an expanding FRW model, but the explicit nature of the FRW models allows us to be even more precise about the exact nature of the relationship. In particular, it is possible to derive an equation that relates Hubble's constant to the scale factor. To see this, consider two galaxies separated by a relatively *small* distance d at time t when the scale parameter is $R(t)$. Because these galaxies are close together, the flight-time for light passing from one to the other, d/c, is also small and is represented by the quantity Δt. It follows from Equation 5.13 that the observed redshift of one of these galaxies, when observed from the other, is

$$z = \frac{R(t + \Delta t)}{R(t)} - 1 \tag{5.14}$$

Due to the expansion of the Universe, $R(t + \Delta t)$ is greater than $R(t)$, and we can indicate this by writing $R(t + \Delta t) = R(t) + \Delta R(t)$, where the new quantity $\Delta R(t)$ represents the small increase in scale factor that occurs during the short time Δt. Note that $\Delta R(t)$ represents a single quantity, it is not the result of multiplying together quantities such as Δ and $R(t)$.

Replacing $R(t + \Delta t)$ in Equation 5.14 by the alternative expression $R(t) + \Delta R(t)$, shows that

$$z = \frac{R(t) + \Delta R(t)}{R(t)} - 1 \qquad (5.15)$$

and this can be rewritten as

$$z = 1 + \frac{\Delta R(t)}{R(t)} - 1 \qquad (5.16)$$

that is,

$$z = \frac{\Delta R(t)}{R(t)} \qquad (5.17)$$

Now for the crucial step: $\Delta R(t)$ – the change in scale factor that occurs during the short time interval Δt – will be equal to the result of multiplying the interval Δt by the rate of change of R at the time t. (This is like saying that during a time Δt a car travelling with velocity v will change its position by an amount $\Delta t \times v$, since v is the rate of change of position.) It is conventional to represent the rate of change of the scale factor at time t by the symbol $\dot{R}(t)$ (read as 'R dot at time t'), so $\Delta R(t) = \Delta t \times \dot{R}(t)$, and to rewrite Equation 5.17 as:

If you are familiar with differential calculus, you may recognize $\dot{R}(t)$ as a common shorthand for the derivative $dR(t)/dt$.

$$z = \frac{\Delta t \times \dot{R}(t)}{R(t)} \qquad (5.18)$$

Since $c/c = 1$, we can rewrite this as

$$z = \frac{c\Delta t}{c} \times \frac{\dot{R}(t)}{R(t)} \qquad (5.19)$$

But, $c\Delta t = d$, the distance between the two galaxies. Using this, we can write

$$z = \frac{1}{c} \times \frac{\dot{R}(t)}{R(t)} \times d \qquad (5.20)$$

Now, you should be able to see that this prediction of any FRW model is similar to Hubble's law, according to which, at the present time t_0

$$z = \frac{H_0}{c} d \qquad (5.1)$$

This similarity suggests that we should identify the time-dependent quantity $\dot{R}(t)/R(t)$ in Equation 5.20 as a time-dependent **Hubble parameter** that we can denote $H(t)$. Thus,

$$H(t) = \frac{\dot{R}(t)}{R(t)} \qquad (5.21)$$

The value of this Hubble parameter varies with time, but the precise way that it varies depends on the precise way in which $R(t)$ varies with time, and will therefore differ from one FRW model to another. However, in any model that provides a good description of the real Universe, we expect to find that if we evaluate the model's Hubble parameter at the present time t_0, then the value obtained should equal the observed Hubble constant, that is

$$H(t_0) = \frac{\dot{R}(t_0)}{R(t_0)} = H_0 \qquad (5.22)$$

As you can see, we have now managed to relate an observational parameter, H_0, to the scale factor $R(t)$ – a parameter in the FRW model.

The function $\dot{R}(t)$ represents the rate of change of R at time t, so $\dot{R}(t)/R(t)$ represents the 'fractional' rate of change of the scale factor. Using this terminology we can say that in Friedmann–Robertson–Walker cosmology, the Hubble constant represents the fractional rate of change of the scale factor evaluated at the present time, t_0. More succinctly, we can say that the observed Hubble constant represents the current value of the model's Hubble parameter.

QUESTION 5.6

Figure 5.27 is a plot of redshift against distance for a (fictitious) sample of galaxies. The plot includes a 'best fit' line drawn through the data. Assuming that the Universe can be well represented by an expanding Friedmann–Robertson–Walker model, state the significance of the gradient of the line, evaluate that gradient from the graph, and hence determine the value of the Hubble constant.

QUESTION 5.7

What is the rate of change of R in the Einstein model? What does your answer imply about the Hubble parameter in the Einstein model?

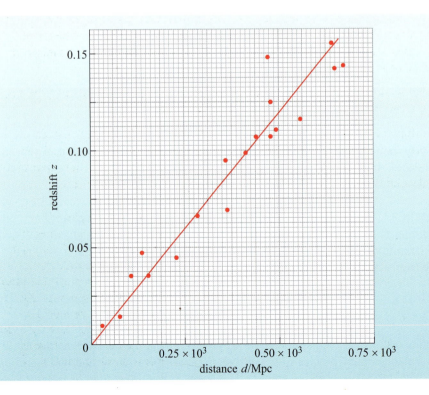

Figure 5.27 A plot of redshift against distance for a number of galaxies, together with a best-fit line through the data. (These data are artificial; real data can be seen in Figure 5.5.)

5.4.2 Systematic deviations from Hubble's law, and the deceleration parameter

According to the FRW models, at any time t, the Hubble parameter measures the fractional rate of expansion of the Universe and depends on the rate of change of R, which is indicated by the slope of the R against t graph. However, in most FRW models the expansion either speeds up or slows down as time progresses – that is to say the expansion is accelerating or decelerating – and this is indicated by the curvature of the R against t graph (see Figures 5.23 and 5.28).

■ Referring back to Figure 5.23 (and the accompanying text), identify three named FRW models that might correspond to the curves A, B and C in Figure 5.28.

☐ A could represent the accelerating model ($k = 0$, $\Lambda > 0$)
   B could represent the open model ($k = -1$, $\Lambda = 0$)
   C could represent the critical model ($k = 0$, $\Lambda = 0$).

Just as the rate of change of R is indicated by $\dot{R}(t)$, so the rate of change of $\dot{R}(t)$ is indicated by $\ddot{R}(t)$ (read as 'R double dot at time t'). If the expansion of the Universe is speeding up at time t then $\ddot{R}(t)$ will be positive, if the expansion is slowing down, $\ddot{R}(t)$ will be negative, and if there is no acceleration or deceleration $\ddot{R}(t) = 0$.

In the context of the FRW models, it turns out that a useful way of characterizing an increasing or decreasing rate of expansion is in terms of a quantity called the **deceleration parameter**. This varies with time, and is conventionally denoted by the symbol $q(t)$. It is defined as follows

$$q(t) = \frac{-R(t)}{[\dot{R}(t)]^2}\ddot{R}(t) \tag{5.23}$$

Note the negative sign in this definition; this implies that if $\ddot{R}(t)$ is positive (i.e. if the expansion is accelerating) then the deceleration parameter will be negative. In the cases shown in Figure 5.28, at large values of t the deceleration parameter would be negative for curve A, zero for curve B, and positive for curve C.

■ What would you expect the corresponding results to be for small values of t in the three FRW models named in the last question?

☐ As shown in Figure 5.23, in all three models the R against t graph curves downwards at early times, similar to the behaviour of the Einstein–de Sitter model at very early times. This indicates that in its early phases the expansion is decelerating, and implies that $q(t)$ will be positive in all three cases.

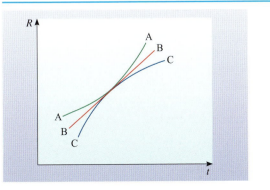

Figure 5.28 A plot of R against t, for large values of t, in three different FRW models denoted A, B and C. The upward curvature of A indicates acceleration, the downward curvature of C indicates deceleration and the relative lack of curvature in B indicates an almost steady expansion.

The detailed argument presented in the last section showed that the expanding FRW models predict a direct proportionality between z and d of just the kind described by Hubble's law. However, that argument was based on the behaviour of galaxies that were sufficiently close together for the time of flight of light passing from one to the other to be considered 'short'. When more distant galaxies are taken into account, the FRW models predict that the direct proportionality between z and d will break down, and systematic deviations from Hubble's law will be observed. Moreover, the models show that the systematic deviations from Hubble's law depend on the value of the deceleration parameter at the time of observation.

In fact, the FRW models predict that present-day observations of galaxy redshifts and distances should show, to a first approximation, that

$$d = \frac{cz}{H_0} \tag{5.24}$$

which is just a rearrangement of Hubble's law, and agrees with observations for redshifts of less than 0.2 or so. But the FRW models also predict that, to a better approximation

$$d = \frac{cz}{H_0}\left[1 + \frac{1}{2}(1 - q_0)z\right] \tag{5.25}$$

where q_0 is the current value of the deceleration parameter. Figure 5.29 illustrates this relationship by showing the kind of systematic deviations from Hubble's law that might be expected for various values of q_0 out to a redshift of about 1. (There are several different ways of deducing the distance of a galaxy from observations, so it should be noted that the distance d in these relationships is the so-called 'luminosity distance' deduced from observations of the apparent magnitude m of a source of known absolute magnitude M.)

In principle then, given sufficiently good observational data, the straight part of the redshift against distance graph can be used to determine the current value of the Hubble parameter, H_0, and the observed deviations from straightness can be used to determine the current value of the deceleration parameter, q_0. The determination of these two values, H_0 and q_0, was, for many years, the primary objective of observational cosmology. In fact, the American astronomer Allan Sandage (1926–2010), a former assistant of Hubble, once famously described observational cosmology as 'the search for two numbers', a characterization that was largely true until the 1970s.

Although there is now far more to observational cosmology than the determination of H_0 and q_0, the determination of those two numbers is still of very great importance. As indicated earlier, recent observations have reduced the uncertainty in the value of H_0 to about 3% or so. However, the determination of q_0 presents a greater challenge. In order for the deviations from Hubble's law to be seen it is necessary to observe galaxies at large redshifts (up to $z = 1$ say) and to independently measure their distances. Finding high redshift galaxies is relatively easy, but accurately determining their distances is very hard. A number of recent observations (see Chapter 7 for details) have established that q_0 is negative, implying that the expansion of the Universe is accelerating. If we do indeed live in a Universe broadly described by an accelerating FRW model, then it must also be the case that the cosmological constant is greater than zero.

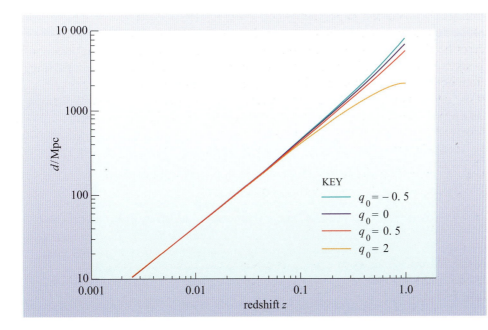

Figure 5.29 The graph of redshift against distance for galaxies, as predicted by expanding FRW models. At small redshifts the models predict a straight line determined by the Hubble constant (in this case $H_0 = 72$ km s^{-1} Mpc^{-1} has been assumed). However, at larger redshifts the models predict deviations from a straight line that depend on q_0, the current value of the deceleration parameter. These model predictions can be compared with the observational data shown in Figure 5.6a.

KEY

— $q_0 = -0.5$
— $q_0 = 0$
— $q_0 = 0.5$
— $q_0 = 2$

QUESTION 5.8

Justify the assertion made in the last sentence above.

5.4.3 The critical density and the density parameters

An important parameter in any FRW model is the average density of cosmic matter. In an expanding or contracting FRW model this quantity will change with time as the matter becomes increasingly dispersed or compressed, so it is represented by the symbol $\rho(t)$. Observationally we might hope to determine the current value of the density, $\rho(t_0)$, by adding together the masses of all the galaxies and clusters in some sufficiently large region of the Universe, and dividing that sum by the volume of the region. Of course, this has been attempted many times, but the answers are fraught with uncertainties, partly due to the problems of measuring large distances and observing faint galaxies, but also due to the very great problem of determining the total mass of dark matter in any region. For this reason, more indirect approaches to the determination of ρ are needed.

When discussing the cosmic density a useful reference value is the density of matter in the FRW model with $k = 0$ and $\Lambda = 0$. You will recall from our discussion of Figure 5.23 that this particular model is often referred to as the *critical model*, since it sits exactly on the borderline between the open and closed $\Lambda = 0$ models. It turns out that, in order to maintain this precarious position, the density of matter in the critical model must be precisely related to the value of the Hubble parameter at all times. In fact, if we denote the density of matter in the critical model at time t by $\rho_{crit}(t)$, the Friedmann equation (see Box 5.4) implies that

$$\rho_{crit}(t) = \frac{3H^2(t)}{8\pi G} \qquad (5.26)$$

Note that it is conventional to write $(H(t))^2$ as $H^2(t)$.

where G is Newton's gravitational constant. The quantity $\rho_{crit}(t)$ is known as the **critical density** at time t. Note that both $\rho_{crit}(t)$ and $H(t)$ vary with time, but, in

BOX 5.4 THE FRIEDMANN EQUATION: EPISODE 2

This is a good place to finally write down the explicit form of the crucially important Friedmann equation that was first discussed in Section 5.3.5. The equation makes use of the symbol $\dot{R}(t)$ to represent the rate of change of the scale factor R, and may be written as

$$\dot{R}^2 = \frac{8\pi GR^2}{3}\left(\rho + \frac{\Lambda c^2}{8\pi G}\right) - kc^2 \qquad (5.27)$$

We have omitted the parenthesized t that should follow R and ρ for the sake of simplicity, but it is still the case that R and ρ are time-dependent quantities. Given the values of k and Λ, Equation 5.27 may be used to determine the behaviour of R, although in order to work out the precise details it is also necessary to know the value of ρ at some particular time. In fact, it can be shown that in a pressure-free universe $\rho(t)[R(t)]^3$ has a constant value, D say, so the density information is often provided by specifying the value of this constant, and replacing ρ in Equation 5.27 by D/R^3.

Note that in the case of the critical model, where $k = 0$ and $\Lambda = 0$, the Friedmann equation implies that

$$\dot{R}^2 = \frac{8\pi GR^2}{3}\rho \text{ (critical model only)} \qquad (5.28)$$

Identifying ρ as ρ_{crit} in this case, and recalling that $[H(t)]^2 = \dot{R}^2/R^2$, the above equation can be rearranged to give Equation 5.26, $\rho_{crit}(t) = 3H^2(t)/8\pi G$.

It is worth pointing out that on the basis of Newtonian physics (rather than general relativity), the Friedmann equation (Equation 5.27) can be derived by considering the total energy of an expanding spherical distribution of galaxies. In such a derivation kc^2 is related to the total energy of the sphere, $\dot{R}^2$ is related to the kinetic energy of the sphere, and the term involving G is related to the gravitational potential energy of the sphere. On this basis the Friedmann equation is sometimes referred to as the 'energy equation' of the Universe.

the critical model, Equation 5.26 always relates their variations. So, it is always possible to work out the current value of the critical density from the current value of the Hubble parameter.

■ Until the 1990s it was widely believed that our Universe was well represented by the critical model. If this belief had been correct what would have been a reasonable estimate of the current value for the density of the Universe? (Assume $H_0 = 72$ km s^{-1} Mpc$^{-1} = 2.3 \times 10^{-18}$ s^{-1}.)

☐ Under these conditions the current value of the density, $\rho(t_0)$, would be the current value of the critical density, $\rho_{crit}(t_0) = 3H_0^2/8\pi G$. So

$$\rho_{crit}(t_0) = \frac{3 \times (2.3 \times 10^{-18})^2}{8 \times \pi \times 6.67 \times 10^{-11}} \text{ kg m}^{-3} \approx 1 \times 10^{-26} \text{ kg m}^{-3}.$$

Using the critical density as a reference value, we can express the actual density of cosmic matter at any time as a fraction of the critical density at that time. This fraction is called the **density parameter for matter**, and may be represented by the symbol $\Omega_m(t)$, so

Ω is the upper case Greek letter 'omega'.

$$\Omega_m(t) = \frac{\rho(t)}{\rho_{crit}(t)} \quad \text{where } \rho_{crit}(t) = \frac{3H^2(t)}{8\pi G} \qquad (5.29)$$

If, at some time t, the density of matter in the Universe was half the critical value, then $\Omega_m(t) = 0.5$; if the actual density was one-quarter of the critical density then $\Omega_m(t) = 0.25$, and so on. Note that this parameter includes *all* kinds of matter: dark matter as well as luminous matter, baryonic as well as non-baryonic.

Interestingly, it is possible to represent the value of the cosmological constant in a similar way. If you look at the Friedmann equation (Equation 5.27), you will see that the constant $\Lambda c^2/8\pi G$ enters the equation in a similar way to the matter density

ρ. This suggests that we can interpret the term $\Lambda c^2/8\pi G$ as a sort of 'density' associated with the cosmological constant. Of course, $\Lambda c^2/8\pi G$ is a very odd sort of density because it is expected to remain constant while the Universe expands, whereas the matter density ρ is expected to decrease in proportion to $1/R^3$ in an expanding Universe. Nonetheless, if we use the symbol ρ_Λ to represent $\Lambda c^2/8\pi G$, then we can define the **density parameter for the cosmological constant** as

$$\Omega_\Lambda(t) = \frac{\rho_\Lambda}{\rho_{\text{crit}}(t)} \quad \text{where } \rho_\Lambda = \Lambda c^2/8\pi G \qquad (5.30)$$

Note that although Λ and ρ_Λ are constants, the density parameter $\Omega_\Lambda(t)$ does depend on time, because it involves the critical density, and that is time dependent.

Even though ρ_Λ doesn't make much sense as a matter density, if we multiply it by c^2 we obtain a quantity that can be measured in units of J m^{-3} (i.e. joule per cubic metre). This quantity, $\rho_\Lambda c^2 = \Lambda c^4/8\pi G$, can be interpreted as an *energy density* and this way of representing the cosmological constant has led to some interesting physical speculations.

One line of thought is that the energy density representing the cosmological constant arises from the energy associated with quantum fluctuations that occur naturally in empty space. This is known as **vacuum energy** and would neatly explain why the expansion of space did not decrease the effect of the cosmological constant. Rather, expanding space simply produces more empty space and more vacuum energy with the consequence that the energy density associated with the cosmological constant remains unchanged while the energy densities associated with all kinds of matter and radiation are reduced. Unfortunately, simple estimates of the energy density expected to come from quantum fluctuations in the vacuum are much too high (by a factor of about 10^{120}) to be realistic. Consequently those who favour this approach have to provide a plausible reason why the simple estimates are so unrealistic.

A different approach is to consider a wider range of possibilities that may mean that $\rho_\Lambda c^2$ is not truly constant, nor even completely uniform in its distribution. In this approach the energy density $\rho_\Lambda c^2$ is described as the average density of **dark energy** and is attributed to some hitherto unidentified cosmic ingredient with a range of exotic properties. If the source of dark energy was an exotic fluid with a negative average pressure $p = -\rho_\Lambda c^2$, as well as an average energy density $\rho_\Lambda c^2$, then its cosmic effect would be indistinguishable from that of a cosmological constant. However some departures from this particularly simple kind of behaviour are still compatible with observations, leaving open the possibility of a range of exotic cosmic constituents. Note, however, that the dark energy cannot be simply associated with dark matter since the latter tends to slow cosmic expansion while the former is responsible for its acceleration.

Although the source of dark energy continues to be a mystery, it has become common practice to refer to $\rho_\Lambda c^2$ as the **dark energy density**, and to call $\Omega_\Lambda(t)$ the **density parameter for dark energy**. Meanwhile, observers continue to seek evidence that the source of dark energy, whatever it may be, behaves in some way that distinguishes it from the effect of a cosmological constant. We will return to this point in Chapter 8.

A lot of observational effort has been devoted to determining the current values of $\Omega_m(t)$ and $\Omega_\Lambda(t)$ (which may be denoted $\Omega_{m,0}$ and $\Omega_{\Lambda,0}$) as you will see in Chapter 7. These two parameters are hugely significant since, together with the observed value of the Hubble constant, they can be used to distinguish the various FRW models, as indicated in Figure 5.30. At all points on the red line, $\Omega_{m,0} + \Omega_{\Lambda,0} = 1$. If the observed values of $\Omega_{m,0}$ and $\Omega_{\Lambda,0}$ satisfy this condition, then the geometry of space will be flat, and it must be the case that $k = 0$ (Question 5.9 asks you to prove this from the Friedmann equation). On the other hand, if $\Omega_{m,0} + \Omega_{\Lambda,0} > 1$ then $k = +1$, or if $\Omega_{m,0} + \Omega_{\Lambda,0} < 1$ then $k = -1$. Thus, the geometric properties of space depend crucially upon the sum of $\Omega_{m,0}$ and $\Omega_{\Lambda,0}$. As Figure 5.30 also indicates, another condition involving $\Omega_{m,0}$ and $\Omega_{\Lambda,0}$ (represented by the blue line) determines whether the Universe will eventually collapse or continue expanding forever. And, if the fate of the Universe is perpetual expansion, yet another condition involving $\Omega_{m,0}$ and $\Omega_{\Lambda,0}$ (represented by the green line) will determine whether the expansion will accelerate or decelerate.

The current values of these density parameters are not easy to determine. However, several recent measurements have been compatible with $k = 0$, making $\Omega_{m,0} + \Omega_{\Lambda,0} = 1$ a popular choice. In fact, current estimates (see Chapter 7 for details) indicate that $\Omega_{m,0} \approx 0.3$ and $\Omega_{\Lambda,0} \approx 0.7$. These figures imply that most of the energy in the Universe is currently dark energy, space has a flat geometry, and cosmic expansion will continue forever at an accelerating rate.

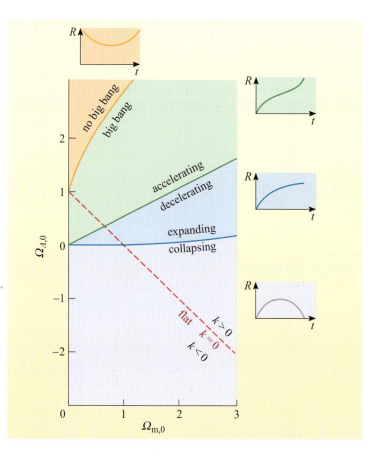

Figure 5.30 A plot of $\Omega_{\Lambda,0}$ against $\Omega_{m,0}$. The values of these two quantities determine important characteristics of a Friedmann–Robertson–Walker cosmological model, such as the sign of the curvature parameter k (red line), whether the Universe will expand forever and eventually collapse (blue line), whether that expansion will accelerate or decelerate (green line), and whether or not there was a big bang (orange line).

QUESTION 5.9

Show that the Friedmann equation can be rewritten as

$$H^2 = \frac{8\pi G}{3}\left(\rho + \frac{\Lambda c^2}{8\pi G}\right) - \frac{kc^2}{R^2}$$

and that this may itself be rewritten as

$$\Omega - 1 = \frac{kc^2}{R^2 H^2} \quad \text{(where } \Omega = \Omega_m + \Omega_\Lambda\text{)}$$

Hence justify the statement that if $\Omega_m + \Omega_\Lambda = 1$, then $k = 0$.

QUESTION 5.10

The Friedmann equation, together with the relation $\rho R^3 = $ constant, may be used to show (you are not expected to demonstrate this)

$$2\dot{R}\ddot{R} = -\frac{8\pi G}{3}(R\dot{R}\rho - 2R\dot{R}\rho_\Lambda)$$

Use this, and the definitions given above for $H(t)$, $q(t)$ and $\rho_{crit}(t)$ to show that at any time t

$$q(t) = \frac{\Omega_m(t)}{2} - \Omega_\Lambda(t)$$

Use this, together with the data given above, to estimate the value of q_0.

5.4.4 The Hubble time and the age of the Universe

The critical model not only provides a useful reference value for the cosmic density, it also provides useful insights into the age of the Universe. The usefulness of the critical model stems from the fact that its scale parameter varies with time in a very simple way

$$R(t) = At^{2/3} \quad \text{(where } A \text{ is a constant)}$$

(This relationship is found by solving the Friedmann equation with $k = 0$ and $\Lambda = 0$.)

Combining this with the definition of the Hubble parameter, $H(t)$, it can be shown that in the critical model

$$H(t) = \frac{2}{3t} \quad \text{(critical model only)}$$

It follows from this that observers living in a universe that was well described by the critical model would find that, after their universe had been expanding for a time t_0, their observations of distant galaxies would indicate that the Hubble constant had the value $H_0 = 2/3t_0$. This means the observers in such a hypothetical universe would be able to deduce the age of their universe by measuring H_0 and using the relation

$$t_0 = \frac{2}{3H_0} \quad \text{(critical model only)}$$

Since H_0 may be expressed in units of s^{-1}, the quantity $1/H_0$ may be expressed in time units, such as seconds or years. The quantity $1/H_0$ is known as the **Hubble time** and is often used as a reference value in discussions of cosmic age, just as ρ_{crit} is a useful reference value in discussions of cosmic density. The exact value of the Hubble time is somewhat uncertain, due to the uncertainties that still exist in measurements of H_0, but it is thought to be about 4.3×10^{17} s or, if you prefer, about 14 billion years (i.e. 1.4×10^{10} yr).

■ If our Universe was well represented by the critical model, how old would it be?

☐ According to the critical model the age of the Universe, t_0, is two-thirds of the Hubble time. Since the Hubble time for our Universe is about 14 billion years, it follows that the age of our Universe, if it were well represented by the critical model, would be about 9 billion years. (This is too short to be realistic, so our Universe cannot be described by the critical model.)

The critical model is unusual in providing such a simple relationship between the age of the Universe and the observed value of the Hubble constant. Similar relationships exist in other FRW models, but they are generally more complicated. Rather than trying to write down those relationships it is much easier to represent them graphically. First however, take a look at Figure 5.31, which should give you a general feel for what you can expect to see later.

Figure 5.31 shows the growth of the scale factor for four different FRW models; the models are numbered 1 to 4, and are, respectively,

1   a closed model with $\Lambda = 0$ and $k = +1$

2   a critical model with $\Lambda = 0$ and $k = 0$

3   an open model with $\Lambda = 0$ and $k = -1$

4   an accelerating model with $\Lambda > 0$ and $k = 0$

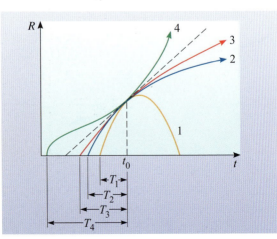

Figure 5.31 The evolution of the scale factor in closed, critical, open and accelerating FRW models that have the same Hubble constant at time t_0. The accelerating model, the only one of the four to have a non-zero cosmological constant, has the greatest age.

The general behaviour of these four models was shown in Figure 5.23, but as redrawn in Figure 5.31 the curves have been shifted horizontally so that they all have the same values for R and $\dot{R}$ at the time t_0 that corresponds to the present. This amounts to saying that the curves have been drawn in such a way that they all indicate the same value for the Hubble constant. Given that the four curves in Figure 5.31 all correspond to the same Hubble constant, what do they tell us about the ages of these four kinds of universe? Well, the age of each model universe is represented by the time that elapses between the moment when R was first equal to zero and the time t_0. These times are different in the four models and are indicated by the values T_1, T_2, T_3 and T_4. As you can see, each is larger than its predecessor, with the closed universe having the smallest age and the accelerating universe the greatest.

You saw earlier that in our Universe the observed value of the Hubble constant is so large that a critical model would imply an unrealistically small age for the Universe. A number of cosmologists have taken this as *prima facie* evidence that the cosmological constant is non-zero.

Now look at Figure 5.32; this puts the implications of Figure 5.31 into a more general context by plotting curves that show the relationship between the age of the Universe t_0 (i.e. the length of time it has been expanding) and the Hubble constant (i.e. the Hubble parameter at t_0) in three different FRW models. The curves represent critical and open models with $\Lambda = 0$, and an accelerating model with the same density as the open model, but with a positive cosmological constant. It can clearly be seen that, for a given value of the Hubble constant, the accelerating model is always the one that has been expanding longest.

Finally, Figure 5.33 provides an even broader context by showing the age of the Universe for any value of the Hubble constant and for a wide range of possible values for $\Omega_{m,0}$ and $\Omega_{\Lambda,0}$. The possible values of the Hubble constant do not appear explicitly anywhere in the diagram, but the curved lines that sweep across the diagram indicate the ages of various models measured in multiples of the Hubble time, and this latter quantity is just the reciprocal of the Hubble constant. Note that the axes of Figure 5.33 are $\Omega_{\Lambda,0}$ and $\Omega_{m,0}$, so the condition for flat space ($k = 0$) is represented by the red line where $\Omega_{m,0} + \Omega_{\Lambda,0} = 1$.

It was mentioned earlier that the currently favoured values of the density parameters for matter and dark energy are $\Omega_{m,0} \approx 0.3$ and $\Omega_{\Lambda,0} \approx 0.7$. These values mean that our Universe is roughly represented by the red dot in Figure 5.33. This indicates an expanding, accelerating Universe that began with a big bang and now has an age that is slightly less than the Hubble time. Based on current estimates of the Hubble time, this implies that the age of our Universe is about 13.8 billion years, which, pleasingly, is a little older than the oldest known astronomical bodies, the ancient globular cluster stars discussed in Chapter 1.

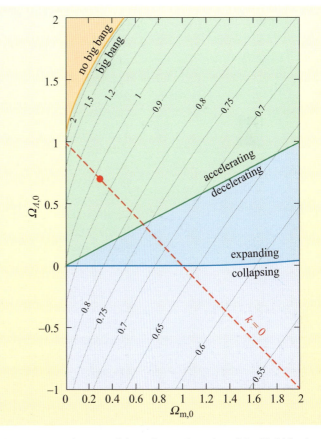

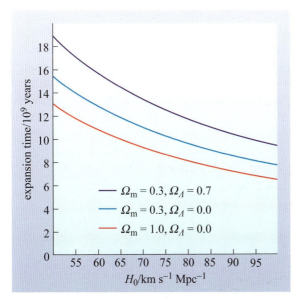

Figure 5.32 The age of the Universe plotted against the Hubble constant for critical, open and accelerating universes of various densities.

Figure 5.33 The age of the universe in units of the Hubble time (i.e. $1/H_0$) is indicated by the various curves that cross this plot of $\Omega_{\Lambda,0}$ against $\Omega_{m,0}$. (Adapted from a figure by Hannu Kurki-Suonio.)

5.5 Summary of Chapter 5

The nature of the Universe

- The Universe contains matter. About 5/6ths of it is believed to be dark matter that may be non-baryonic, and the remaining 1/6th is baryonic matter. The baryonic matter is mainly hydrogen (~75% by mass) and helium (~25% by mass).

- The Universe contains (electromagnetic) radiation. Much of it is visible light, but the major part of the energy is contained in the cosmic microwave background (CMB) radiation.

- The Universe is uniform. That is to say, all regions that are sufficiently large to be representative have the same average density and pressure, wherever they are located. This claim is consistent with the observed distributions of matter and radiation.

- The Universe is expanding. Distant galaxies provide somewhat imperfect tracers of the expansion since they have 'peculiar' local motions relative to the large-scale Hubble flow. For those with redshifts of ~0.2 or less, the expansion is well described by Hubble's law, implying that the current fractional rate of expansion is given by the Hubble constant H_0. Galaxies with larger redshifts show that the rate of cosmic expansion has been accelerating for several billion years. The cause of this acceleration is generally attributed to a cosmological constant or some widely distributed but enigmatic source of 'dark' energy.

Relativistic cosmology and models of the Universe

- According to Einstein's theory of general relativity the geometric properties of space–time are related to the distribution of energy and momentum within that space–time. The precise relationship is described by the field equations of general relativity, which provide the basis for Einstein's theory of gravity and for relativistic cosmology.

- The geometric properties of space–time include curvature. In a curved space, geometric results can take on unfamiliar forms. The interior angles of a triangle may have a sum that is different from 180°, straight lines may close upon themselves, and pairs of straight lines that are initially parallel may converge or diverge. The geometric properties of any particular space–time can be summarized by writing down an appropriate four-dimensional generalization of Pythagoras's theorem. In the case of a static (i.e. non-expanding), flat (i.e. zero curvature) space–time this takes the form

$$(ds)^2 = (dx)^2 + (dy)^2 + (dz)^2 - c^2(dt)^2$$

- The distribution of energy and momentum throughout space–time is believed to be uniform on the large scale. This assertion is given precise form by the cosmological principle according to which, on sufficiently large-size scales, the Universe is homogeneous and isotropic. Simple cosmological models that are consistent with this principle assume that a gas uniformly fills the Universe. Describing the state of this gas involves specifying its density and pressure, $\rho(t)$ and $p(t)$, both of which are expected to change with time due to the expansion or contraction of the Universe.

- In applying general relativity to cosmology, Einstein introduced a cosmological constant Λ. Thanks to this he was able to formulate a relativistic cosmological model that is neither expanding nor contracting, and in which space has a uniform positive curvature. Later, de Sitter presented a model devoid of matter, which was eventually recognized as describing a universe in which the curvature of space was zero, but there was perpetual expansion.

- The work of Friedmann, Robertson and Walker resulted in the specification of the class of cosmological models that are consistent with general relativity and with the cosmological principle. These models involve a curvature parameter k, that characterizes the geometry of space, and a scale factor $R(t)$ that describes the expansion or contraction of space. The full range of FRW models includes cases that are closed, critical, open and accelerating. The Einstein model arises as a special case, and the de Sitter model as a limiting case.

- The behaviour of the scale factor in a pressure-free universe is determined by the Friedmann equation, and depends on the values of k, Λ and the density ρ at some particular time.

Key parameters of the Universe

- The FRW models provide a natural interpretation of the redshifts of distant galaxies as cosmological redshifts caused by the stretching of light waves while they move through an expanding space.

- The Hubble parameter, $H(t)$, provides a measure of the rate of expansion of space in any FRW model. It is defined by $H(t) = \dot{R}/R$, where $\dot{R}$ denotes the rate of change of R. Observations of distant galaxies are predicted to show that, to a first approximation, $d = cz/H_0$ (i.e. a rearrangement of Hubble's law), where H_0 represents the value of the Hubble parameter at the time of observation.

- The deceleration parameter, $q(t)$, provides a measure of the rate of decrease of the rate of cosmic expansion in an FRW model. It is defined by $q(t) = -R\ddot{R}/\dot{R}^2$, where $\ddot{R}$ denotes the rate of change of $\dot{R}$. Observations of very distant galaxies are predicted to show systematic deviations from Hubble's law described by

$$d = (cz/H_0)[1 + (1 - q_0)z/2]$$

where q_0 represents the value of the deceleration parameter at the time of observation.

- The density parameters Ω_m and Ω_Λ provide a useful way of representing the cosmic matter density and the density associated with the cosmological constant at any time. The parameters are defined by $\Omega_m = \rho/\rho_{crit}$ and $\Omega_\Lambda = \rho_\Lambda/\rho_{crit}$ respectively, where ρ is the cosmic matter density at the time of observation, $\rho_\Lambda = \Lambda c^2/8\pi G$ is a 'density' associated with the cosmological constant, and $\rho_{crit} = 3H^2(t)/8\pi G$ is the density that the critical universe would have at the time of observation. The quantity $\rho_\Lambda c^2$ is an energy density and might be physically associated with a uniform distribution of 'vacuum energy' (arising from the quantum physics of empty space) or perhaps with some more general 'dark energy' arising from an unknown cosmic constituent (though not dark matter). In a Universe with a flat space (i.e. $k = 0$), the Friedmann equation implies that $\Omega_m + \Omega_\Lambda = 1$ at all times.

- The various cosmological parameters are not all independent. The Friedmann equation implies that $\Omega_m + \Omega_\Lambda - 1 = kc^2/(R^2H^2)$, and it may also be shown that $q = (\Omega_m/2) - \Omega_\Lambda$.

- The age of the Universe, t_0, may be conveniently expressed in terms of the Hubble time, $1/H_0$ in any FRW model. In the case of the critical model $t_0 = 2/3H_0$. In other models t_0 may be a different fraction of the Hubble time, depending on the values of Ω_m and Ω_Λ. Increasing the value of Ω_Λ increases the age of the Universe for a given value of the Hubble constant.

Questions

QUESTION 5.11

A number of important events in the history of cosmology have been mentioned in this chapter. Compile a chronological listing of these events, starting with the publication of Einstein's theory of general relativity in 1916.

QUESTION 5.12

List the assumptions that underpin the Friedmann–Robertson–Walker models and the Friedmann equation.

QUESTION 5.13

Describe some of the possible consequences of positive curvature in a three-dimensional space, in the context of the FRW models.

QUESTION 5.14

The detailed argument given in Section 5.4.1 showed that the behaviour described by Hubble's law is an expected consequence of expansion in a FRW model. However, in Section 5.4.2 it was stated that this argument was only approximately true because it ignored the acceleration or deceleration of the expansion. Carefully reread the argument in Section 5.4.1 and identify the key step at which acceleration is ignored.

QUESTION 5.15

List the values that have been assigned to all the observational parameters mentioned in Section 5.4. Where a quantity is expressed in more than one unit system, confirm the equivalence of all the given values.

CHAPTER 6
BIG BANG COSMOLOGY –
THE EVOLVING UNIVERSE

6.1 Introduction

In Chapter 5 we saw how general relativity can be used to construct models of the Universe. These models describe how the scale factor varies in a Universe that is filled with a smooth distribution of matter and radiation, but say very little about the properties and behaviour of these components. So, for instance, they do not account for the consequences of microscopic processes such as the interactions between the particles that make up the matter within the Universe.

However, such interactions play an important role in cosmology. One example that we shall see later in this chapter is that cosmological theories can offer an explanation for the observation that most of the stars in the Universe have a composition that is approximately 75% hydrogen and 25% helium (by mass). A fundamental aspect of the process that forms helium is that it involves reactions between nuclei and particles at an early stage in the history of the Universe. Thus physical processes on small scales *must* be taken into consideration in order to develop an understanding of the evolution of the Universe. This chapter starts by examining the conditions under which particles interacted in the early Universe (Section 6.2). The main focus, however, is to follow a chronological sequence from very early times in the history of the Universe (Section 6.3) through the formation of the first nuclei (Section 6.4) and the first neutral atoms (Section 6.5) to the stage at which gravitational clustering gives rise to the large-scale structure that we observe in the present-day Universe (Section 6.6).

At first sight, it might seem impossible to use the models in Chapter 5 to make any predictions about the small-scale behaviour of matter. Cosmological models describe how the scale factor varies with time, but at the present time any change in the scale factor certainly does not have any effect on, for instance, the atoms that make up your body. However one of the major assumptions made in Chapter 5 was that the matter in the Universe is smoothly distributed. This assumption of a uniform distribution of matter is a key to linking the large-scale dynamical behaviour of the Universe to small-scale effects.

To see why this is so, consider a volume of the Universe that, at some particular time, is bounded by an imaginary cube, as shown in Figure 6.1. Let us further suppose that we want to follow the evolution of the matter within this cube at all times using some particular Friedmann–Robertson–Walker model with scale factor $R(t)$. To do this, the edges of the cube must follow the expansion (or contraction) of the model universe.

■ Each edge of the cube has an associated length l. How must the length of each edge change with time if the cube is to follow the expansion or contraction of the model universe?

☐ The length of each edge of the cube must be proportional to the scale factor, i.e. $l \propto R(t)$.

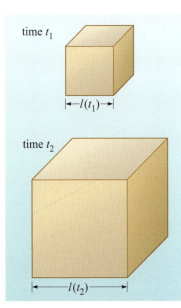

time t_1

$|{\leftarrow}l(t_1){\rightarrow}|$

time t_2

$|{\leftarrow}\qquad l(t_2)\qquad{\rightarrow}|$

Figure 6.1 An imaginary cubic volume (with sides of length l) that evolves with the expansion of a model universe such that the mass within the volume is constant. Any volume that behaves in this way, whatever its shape, is called a comoving volume.

Thus the volume ($= l^3$) of the cube at any time t is proportional to $R(t) \times R(t) \times R(t) = (R(t))^3$. This is illustrated by the change in volume shown in Figure 6.1. Although we have chosen to discuss a cubic volume here, a volume of *any* shape that follows the expansion (or contraction) of such a model universe will have a volume $V \propto (R(t))^3$. Such a volume is called a **comoving volume**.

Because the volume V of a comoving region, such as the cube, changes as the scale factor changes, but the mass M within it is constant, the density of matter within the comoving volume, which we denote by ρ_m, also varies with scale factor. In fact,

$$\rho_m = \frac{M}{V} \propto \frac{1}{(R(t))^3} \tag{6.1}$$

Now, the density of matter is an important physical parameter in determining how interactions between particles progress on a microscopic scale. For example, the rate at which molecules in a sample of gas collide with one another increases as the density of the gas is increased. Thus, the large-scale behaviour *is* related to small-scale effects.

This still may not appear to be a great help in understanding the real Universe as opposed to a cosmological model, since we know that the matter in the Universe at the present time does not have a uniform density. Specifically, we saw in Chapter 4 that the distribution of matter is homogeneous only when we consider scales greater than about 200 Mpc: if we look at the Universe on smaller scales, we see large density variations. Thus the average density of the matter in the Universe would seem to be a quantity of limited practical use. However, we shall see later that there is good evidence that at times in the distant past the matter in the Universe was much more smoothly distributed than it is at present – even on relatively small scales. At such times, the average density *does* relate to the small-scale behaviour of matter.

The relationship between density and scale factor that is described by Equation 6.1 holds true for any of the cosmological models described in Chapter 5. The majority of these models are characterized by a scale factor $R = 0$ at time $t = 0$. As was noted in Section 5.3.5 the early expansion phase of any such model is referred to as the *big bang*. Many lines of cosmological evidence strongly favour a model of the Universe that experienced a big bang phase – and for the remainder of this chapter we shall only consider big bang models. An immediate consequence that can be noted from Equation 6.1, is that in a big bang model, early stages in the history of the Universe (when $R(t)$ was very small) are characterized by high densities. You may also have noticed that, strictly speaking, the mathematical relationship $\rho_m \propto 1/R^3$ implies an infinite density when $R = 0$. We shall return to consider the significance of such infinite quantities later.

Finally, we should make a brief note about terminology: much of the discussion of this chapter refers to specific times in the history of the Universe. A convention that we adopt throughout the chapter is that the age of the Universe is denoted by t (so, for instance, $t = 1$ s denotes the time at which the Universe was 1 second old). The time now, at which we observe the Universe, is referred to as t_0. So, t_0 represents the current value of the age of the Universe.

6.2 The thermal history of the Universe

A key physical parameter in particle interactions is temperature. It may seem to make no sense to talk of the 'temperature' of the Universe. The temperature of matter seems to range from a few degrees above absolute zero within giant molecular clouds, up to temperatures of over 10^7 K that are found in extreme astrophysical environments. However, we will see shortly that there *is* a cosmic temperature; it does not refer to the temperature of *matter* within the Universe, but to the *radiation* that pervades the Universe. This radiation, which is observable today as the cosmic microwave background (CMB), plays an extremely important role in modern cosmology. We shall discuss several aspects of the CMB in this, and in the following chapter, but in this section we shall consider how the existence of this background radiation in an expanding Universe allows us to determine how temperature has changed over cosmic history. The starting point for this discussion is to consider the cosmic microwave background in a little more detail.

6.2.1 The temperature of the background radiation

In Section 5.2.2 you saw that the most significant contribution to the radiation content of the Universe is the cosmic microwave background radiation. The spectral flux density of the CMB peaks at a wavelength of about 1 mm – this is illustrated in the spectrum of the background radiation that is shown in Figure 6.2. The form of this spectrum is highly significant: it is, to a very good approximation, a **black-body spectrum** – implying it can be associated with a particular temperature.

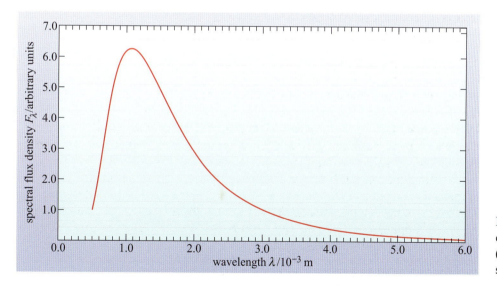

Figure 6.2 The spectrum of the cosmic microwave background. (Note that this spectrum shows the spectral flux density F_λ.)

The characteristic temperature T indicated by any given black-body spectrum is related to the wavelength λ_{peak} at which the spectral flux density (F_λ) is a maximum. According to Wien's displacement law,

$$(\lambda_{peak}/m) = \frac{2.90 \times 10^{-3}}{(T/K)} \tag{6.2}$$

■ Calculate the characteristic temperature of the cosmic microwave background radiation.

☐ Rearranging Equation 6.2

$$(T/K) = \frac{2.90 \times 10^{-3}}{(\lambda_{peak}/m)}$$

and using $\lambda_{peak} \approx 1\ mm = 1 \times 10^{-3}\ m$

$$(T/K) = \frac{2.90 \times 10^{-3}}{(1 \times 10^{-3})} = 2.90$$

So, to one significant figure, the temperature of the cosmic microwave background radiation is 3K.

Detailed spectral measurements have been used to determine the temperature of the cosmic microwave background to a high degree of accuracy, with a value of $T = 2.725 \pm 0.002\ K$ being widely accepted.

The fact that the CMB follows a black-body spectrum is, at first sight, puzzling. Black-body spectra are formed when photons are continually absorbed and re-emitted by matter. However, matter in the nearby Universe is transparent to cosmic microwave background photons. Thus there is essentially no interaction between matter and photons, and so nearby matter could not give rise to the observed black-body spectrum. So, if the CMB did form by the interaction of radiation and matter – how could this have occurred? To answer this question we have to consider the effect of the expansion of the Universe on the photons of the cosmic microwave background.

6.2.2 The evolution of the temperature of background radiation

A clue to the origin of the microwave background lies in an effect that was introduced in Chapter 5 – the cosmological redshift of photons. In Section 5.4.1 you saw that the effect of the expansion of the Universe on a single photon was to increase its wavelength. The relationship between the wavelength λ_0 of a photon that is observed now (when the scale factor has the value $R(t_0)$) and the wavelength λ that the photon had when the scale factor was $R(t)$ is

$$\frac{\lambda}{\lambda_0} = \frac{R(t)}{R(t_0)} \tag{6.3}$$

Thus, when the scale factor was smaller than it is at present, the wavelengths of photons that are now seen in the cosmic microwave background were all correspondingly smaller. In fact, the background radiation that is now observed as the cosmic *microwave* background, would, when the scale factor was much smaller, have had a peak in another part of the electromagnetic spectrum. For this reason we shall use the term **cosmic background radiation** to denote this radiation at any time in cosmic history. The cosmic microwave background is just the observable form of the cosmic background radiation at the present time.

- If a microwave background photon currently has $\lambda = 1$ mm, what wavelength would it have had when the scale factor was 1000 times smaller than its present-day value? In which part of the electromagnetic spectrum does this wavelength lie?

□ Using Equation 6.3, with values of $\lambda_0 = 1$ mm and $R(t)/R(t_0) = 1/1000$ gives

$$\lambda = 10^{-3} \text{ m}/1000 = 10^{-6} \text{ m}$$

So when the scale factor was 1000 times smaller than at present, photons that are currently at the peak of the cosmic microwave background had a wavelength of 10^{-6} m, which lies in the near-infrared part of the spectrum.

Thus, at high redshift, the wavelengths of photons in the cosmic background radiation would have been much shorter than at present, and consequently interactions between photons and matter would have been much more likely. However, before discussing this interaction, we need to consider the form of the red-shifted spectrum in a little more detail.

An important feature of the black-body spectrum is that if the photons that make up such a spectrum are all red-shifted by the same amount, then it will remain a black-body spectrum. Photons that are currently at the wavelength at which the spectrum has a peak, will always be at the peak, but the wavelength of that peak will change. The way in which this wavelength, λ_{peak}, changes with scale factor is given by Equation 6.3. $R(t_0)$ and λ_0 are the current values of $R(t)$ and λ respectively, and so can be considered as constants in Equation 6.3. Thus Equation 6.3 can be written as

$$\lambda_{peak} \propto R(t) \tag{6.4}$$

However, the temperature of a black-body spectrum is related to the wavelength of the peak of emission by Wien's displacement law (Equation 6.2), which can be rearranged and expressed as

$$T \propto \frac{1}{\lambda_{peak}} \tag{6.5}$$

Using the relationship between λ_{peak} and the scale factor (Equation 6.4) gives

$$T \propto \frac{1}{R(t)} \tag{6.6}$$

The temperature of the cosmic background radiation at any time is inversely proportional to the scale factor at that time.

This relationship is important because, in principle, it allows us to calculate the temperature of the background radiation at any given epoch for any cosmological model. Remember from Chapter 5 that different cosmological models provide different relationships for the scale factor R as a function of time (see, for example, Figure 5.23).

Even if we do not know the exact way in which the scale factor varies with time, Equation 6.6 shows that if the scale factor was once much smaller than it is at present, then the temperature of the background radiation at that time would have been much higher than it is at present.

■ Use Equation 6.6 to express the ratio of the temperature at two times (t_1 and t_2) in terms of the scale factor at those two times.

☐ Equation 6.6 can be expressed as

$$T(t) = \frac{\text{constant}}{R(t)}$$

So for times t_1 and t_2 we can write

$$T(t_1) = \frac{\text{constant}}{R(t_1)} \quad \text{and} \quad T(t_2) = \frac{\text{constant}}{R(t_2)}$$

respectively. Dividing the first of these equations by the second gives

$$\frac{T(t_1)}{T(t_2)} = \frac{R(t_2)}{R(t_1)}$$

There is, however, a problem in applying Equation 6.6, which is highlighted by the following question:

■ What is the predicted temperature of the Universe if the scale factor has a value of zero?

☐ Since $T \propto 1/R(t)$, if $R = 0$, the predicted temperature would be infinite!

A prediction of an infinite value of any physical quantity is treated with great suspicion by most physicists. Rather than taking this infinite temperature at face value, it is assumed that our understanding of physical processes is incomplete. The limits at which our knowledge of physical laws break down will be discussed briefly in Section 6.3 and taken up again in Chapter 8. However, for the present discussion, the important point is that at times when the scale factor was very small, the temperature would have been very high.

EXAMPLE 6.1

For the Lemaître cosmological model (in which $\Lambda > \Lambda_E$) and $k = +1$, use Figure 5.23 and Equation 6.6, to sketch a corresponding curve $T(t)$ that shows approximately how the temperature of the cosmic background radiation varies with time.

SOLUTION

The curve that shows how the scale factor R varies with time in the Lemaître cosmological model is shown in Figure 5.23 and is reproduced here as Figure 6.3a.

In order to draw a sketch of how the temperature T varies with time, we need to make use of the relationship between the temperature and the scale factor. This relationship is given by Equation 6.6, $T \propto 1/R(t)$.

The question asks for a *sketch* of how T varies with time. The implication of this is that the curve that shows $T(t)$ does not have to be exact, but that it should show the most important features of how the temperature varies with time. A way of doing this is to consider a few times (labelled A, B, C and D) on the corresponding curve of $R(t)$ as shown in Figure 6.3a. At each time, we shall use Equation 6.6 to deduce how T is behaving and use this information to help us draw a sketch of $T(t)$. The deductions that can be made about T at these times are shown in Table 6.1.

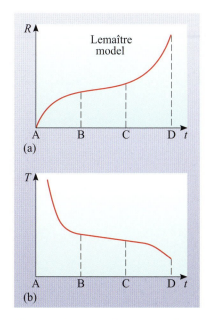

(a)

(b)

Figure 6.3 (a) Scale factor, and (b) temperature as functions of time for the Lemaître cosmological model. A, B, C and D are times that are referred to in Table 6.1.

Table 6.1 The behaviour of the scale factor R at various times indicated on Figure 6.3a and the inferred behaviour of the temperature T at those times.

Time	Behaviour of R at this time	Behaviour of T at this time
A	$R = 0$	$T = 1/R = \infty$
B	R has increased to some value and now does not vary much with time	T must decrease to some value and also only change slowly with time
C	R has a value that is slightly higher than that at B	T must have a value that is slightly lower than that at B
D	R is increasing to very high values	T must decrease to very low values

We can now use the deductions about the way in which T varies at these times to draw a sketch. Starting with time A, we clearly cannot plot an infinite temperature at A, so we simply show T as having a very high value as we approach $t = 0$. At time B, we simply choose a finite value of temperature, and note that the temperature at time C is slightly lower than at B. Finally, at time D, the temperature decreases to very low values. These points are shown on Figure 6.3b, and the final stage is to draw a smooth curve through these points to complete the sketch.

QUESTION 6.1

For all the Friedmann–Robertson–Walker models with $k = 0$, shown in Figure 5.23, use Equation 6.6 to draw a corresponding curve $T(t)$ that shows approximately how the temperature of the cosmic background radiation would vary with time.

The answer to Question 6.1 illustrates the point that in any big bang model (i.e. one that has $R = 0$ at $t = 0$) the temperature of the cosmic background radiation would have been very high in its early stages. Such a scenario is often referred to as the **hot big bang**.

The change in wavelength of the cosmic microwave background has a profound effect on the way in which photons interact with matter. At times when the temperature of the cosmic background radiation was very high, the typical photon energy would have been greater than the ionization energy of the hydrogen atom. Under these conditions, the baryonic matter in the Universe would have been in the form of a plasma.

The *opacity* of a medium is a measure of the extent to which the medium is opaque to radiation.

■ What is the qualitative difference between the opacity of a plasma and that of an un-ionized gas?

□ A plasma tends to have a much higher opacity than an un-ionized gas, i.e. un-ionized gases tend to be much more transparent than plasmas.

The reason for the dramatic difference in opacity between a plasma and an un-ionized gas is the presence of free electrons in the plasma. Photons interact with a plasma primarily by scattering from the free electrons in the plasma (Figure 6.4 – this process is called *Thomson scattering* after the discoverer of the electron J. J. Thomson). The degree of interaction between photons and electrons in a plasma can be very high, and this offers a clue as to the origin of the near perfect black-body spectrum of the background radiation. The conditions for forming a black-body spectrum are that there must be many collisions between the material that makes up a thermal source and the photons that are radiated by it. So an interpretation of the black-body spectrum of the cosmic microwave background is that it was formed at a time when the Universe consisted of a hot plasma, and so there were many collisions between the photons and the free electrons. As the Universe has expanded, the wavelengths of the photons have increased, and the black-body spectrum has shifted to longer wavelengths. Consequently, the temperature associated with this black-body spectrum has dropped with the expansion of the Universe.

Figure 6.4 The interaction between photons and free electrons in a plasma.

6.2.3 The evolution of energy densities in the Universe

So far, we have only considered one physical property of the cosmic background radiation – its temperature. However, to establish whether this background radiation plays an important role in the evolution of the Universe, it is necessary to consider its *energy density* and how this quantity varies with scale factor.

Recall that in Chapter 5, the behaviour of cosmological models was shown to depend on the density of matter ρ_m, and on the cosmological constant Λ. In the models that we considered there, we simply assumed that electromagnetic radiation was a minor constituent of the Universe. However, we now want to question this assumption – so we must compare the importance of these three components: matter, cosmological constant, and electromagnetic radiation. The physical parameter that determines the importance of any one of these components within a cosmological model is the energy density, i.e. the energy per unit volume due to that component. If the energy density of any one component far exceeds that of the other two, then it is this component that will have the dominant effect on the dynamical behaviour of the Universe.

So, let us now consider the current energy densities due to radiation u_r, matter u_m and the cosmological constant u_Λ:

The energy density of the cosmic microwave background is the total energy of all the microwave background photons per cubic metre of space. Note that the energy density (like the mass density) is defined per cubic metre, i.e. for a physical volume of space that is *not* comoving. In an expanding universe, the energy density can therefore be expected to decrease as the universe expands.

■ What are the SI units in which energy density should be expressed?

☐ Since the SI unit of energy is the joule, and the unit of volume is m^3, the SI unit of energy density is J m^{-3}.

The current energy density of the cosmic background radiation can be found from measurements of the CMB and has a value of $u_{r,0} \approx 5 \times 10^{-14}$ J m^{-3}. ($u_{r,0}$ is a shorthand way of writing $u_r(t_0)$ – the value of u_r at the present time.)

The energy density of matter $u_{m,0}$ can be found from density of matter in a straightforward way as the following question illustrates.

QUESTION 6.2

The current *average mass density* of all matter, both luminous and dark, is estimated to be about $\rho_{m,0} \approx 3 \times 10^{-27}$ kg m^{-3}. By using the equivalence between energy E and mass m given by $E = mc^2$, calculate the current average *energy density* due to matter.

The answer to Question 6.2 shows that at the present time, the energy density of matter is $u_{m,0} \approx 3 \times 10^{-10}$ J m^{-3}. Thus at the present time, the energy density due to matter exceeds the energy density in the cosmic microwave background by a factor of several thousand.

Finally, we consider the energy density due to the cosmological constant. In Chapter 5, it was noted that the cosmological constant Λ has an associated density

$$\rho_\Lambda = \Lambda c^2 / 8\pi G \qquad (6.7)$$

■ Give an expression for the *energy* density (u_Λ) of the vacuum in terms of Λ.

☐ The energy density of the vacuum is obtained by multiplying Equation 6.7 by c^2,

$$u_\Lambda = \rho_\Lambda c^2$$

and so

$$u_\Lambda = \Lambda c^4 / 8\pi G \qquad (6.8)$$

As observed in Chapter 5, the energy that may be associated with the cosmological constant is often referred to as *dark energy*. Consequently the quantity u_Λ can be interpreted as the energy density of dark energy.

As you saw in Chapter 5, the underlying reason for the cosmological constant being regarded as an energy density is due to the way Einstein's theory of gravity, general relativity, is constructed. Einstein's equations equate something that measures the curvature of space–time, with something that measures the mass and momentum and energy within space–time. This was shown in Equation 5.7:

$$[G_{\mu v}] = \frac{-8\pi G}{c^4} [T_{\mu v}]$$

However, Einstein found he had the freedom to introduce an additional term into the space–time curvature side of the equation, which was shown in Equation 5.8:

$$[G_{\mu v}] + \Lambda[g_{\mu v}] = \frac{-8\pi G}{c^4} [T_{\mu v}]$$

The effect of this term is to give space an in-built tendency to expand or contract, depending on the sign of Λ. An alternative approach is to subtract $\Lambda[g_{\mu v}]$ from both sides, which effectively regards the Λ term as a consequence of some as-yet-unknown content of space–time. Equation 5.8 then looks like this:

$$[G_{\mu v}] = \frac{-8\pi G}{c^4} [T_{\mu v}] - \Lambda[g_{\mu v}]$$

This substance would have to have many strange properties. Einstein's cosmological constant is just that – constant – but going beyond Einsten's theory, some astronomers have supposed that if it is a substance of some sort, then perhaps it could vary from place to place or vary with time. It is this strange substance or fluid that is referred to as dark energy, and the simplest type of dark energy is Einstein's originally envisaged cosmological constant. The nature of this dark energy is still a mystery, but recent observations imply that u_Λ has a value of about 9×10^{-10} J m^{-3}. So, rather surprisingly, dark energy makes the dominant contribution to the total energy density of the Universe at the present time.

It might seem then that the cosmic background radiation is an insignificant component of the total energy density of the Universe. However, this was not always the case. To see why, it is necessary to compare the way in which the three energy densities u_r, u_m and u_Λ change with scale factor.

We start with the simplest case of the three, which is the energy density of the dark energy. By inspecting the terms on the right-hand side of Equation 6.8 we can see that this energy density depends only on values of physical constants (c, G and Λ). Thus, this energy density does not change with scale factor. As noted above, u_Λ currently has a value of about 9×10^{-10} J m^{-3}, and this value has been constant throughout the history of the Universe (with perhaps one brief, but important, exception that we shall discuss later). However, the fact that u_Λ is constant and relatively large, does not mean that it has always been the most important factor in determining how the Universe evolves.

Next, let's consider how the energy density due to matter changes with scale factor. This is found from the (normal) density of matter.

■ For a cosmological model in which matter is uniformly distributed, write down an equation that describes how the density of matter changes with scale factor.

□ We have already seen that the density of matter $\rho_m(t)$ varies according to Equation 6.1

$$\rho_m(t) \propto \frac{1}{R(t)^3} \tag{6.1}$$

The energy density of matter is related to the density of matter by $u_m = \rho_m c^2$, but c is a constant, so we can write

$$u_m(t) \propto \frac{1}{R(t)^3} \tag{6.9}$$

The behaviour of the energy density of *radiation* can be analysed by taking a similar approach to that taken when we examined the way in which the density of matter changes with scale factor. The first step is to consider the number of photons per cubic metre, i.e. the *number density* of photons $n(t)$. Assuming that cosmic background photons are neither created nor destroyed during the relevant part of the expansion we can expect that

$$n(t) \propto \frac{1}{R(t)^3} \tag{6.10}$$

So the number density of cosmic background photons behaves in a similar way to the density of matter. But what about the energy density? Here there is a difference. The energy of a photon of frequency f is given by

$$\varepsilon_{ph} = hf \tag{6.11}$$

where h is the Planck constant. The frequency f and wavelength λ of electromagnetic radiation are always related by $f = c/\lambda$, so we can also say

$$\varepsilon_{ph} = \frac{hc}{\lambda} \tag{6.12}$$

But as the Universe expands (i.e. as $R(t)$ increases) the wavelength of a photon will also increase (Figure 6.5). The photon wavelength is proportional to the scale factor

$$\lambda \propto R(t) \tag{6.13}$$

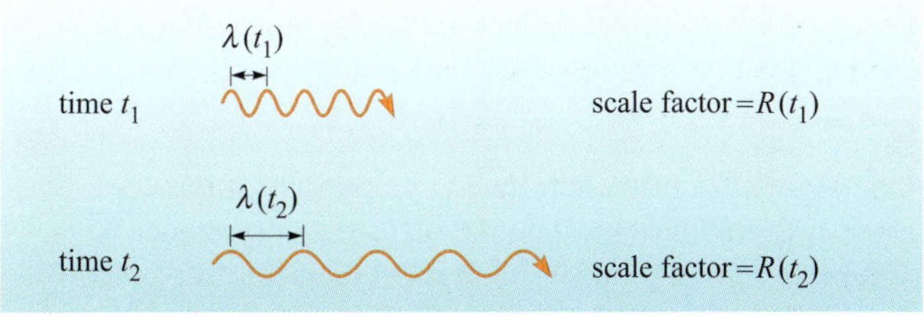

Figure 6.5 As the Universe expands (from scale factor $R(t_1)$ to $R(t_2)$), the wavelength of any photon will increase in proportion to the scale factor.

Hence for each photon in the cosmic background radiation

$$\varepsilon_{ph} \propto \frac{1}{R(t)} \tag{6.14}$$

Now, the energy density of radiation $u_r(t)$ is given at any time t by

$$u_r(t) = n(t) \times \varepsilon_{ph}(t) \tag{6.15}$$

It follows that

$$u_r(t) \propto \frac{1}{R(t)^4} \tag{6.16}$$

Comparing this with Equation 6.9 shows that the energy density of radiation behaves differently from the energy density of matter. Specifically, the energy density of radiation is inversely proportional to the *fourth* power of the scale factor, whereas the energy density of matter is inversely proportional to the *third* power of the scale factor.

Figure 6.6 The energy densities of matter (blue line) and radiation (red line) as a function of scale factor. At a time when $R(t)/R(t_0) \approx 10^{-4}$ the energy densities of matter and radiation were equal. Prior to this time, the energy density of radiation exceeded that of matter – during this era the dynamical evolution of the Universe was determined by its radiation content. After this time, the energy density of matter was greater, so it was the matter in the Universe that controlled its dynamical evolution. The behaviour of the energy density due to the cosmological constant is also shown (purple line) – this does not vary with redshift and is exceeded by the energy densities in matter and radiation at early times.

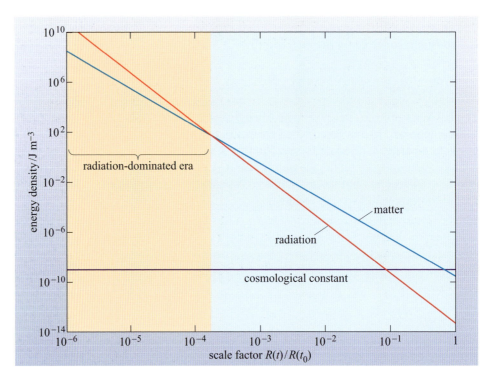

Figure 6.6 shows how all three energy densities (u_Λ, u_m, and u_r) vary as a function of scale factor. At the present, the energy densities of the dark energy (9×10^{-10} J m^{-3}) and of matter (3×10^{-10} J m^{-3}) are far in excess of the energy density of radiation (5×10^{-14} J m^{-3}). As we look back to earlier times however, when the value of the scale factor was smaller, we can see that both the energy densities of radiation and of matter were greater in the past than at present. When the value of $R(t)/R(t_0)$ was less than about 0.1, both the energy density of matter and of radiation exceeded the energy density of the dark energy.

Furthermore, as shown in Figure 6.6, the energy density of radiation has declined more rapidly than the energy density of matter. Indeed there was a time when the value of $R(t)/R(t_0)$ was such that these two energy densities were equal. This appears to have occurred when $R(t)/R(t_0) \approx 10^{-4}$. For most plausible cosmological models this corresponds to a time when the age of the Universe was a few times 10^4 years. As we look back to even earlier times, when $R(t)/R(t_0)$ was even smaller, we see that the energy density of radiation exceeded that of matter – this period of the history of the Universe is called the **radiation-dominated era**.

The key points of this discussion so far can be summarized as follows:

1   At the present time, the energy density due to radiation is much lower than the energy density of matter or the energy density of dark energy.

2   As the Universe expanded, the energy density of radiation decreased more rapidly than the energy density of matter. The energy density of dark energy has remained constant with time.

3   At times when the scale factor was less than about 10^{-4} of its current value, the energy density of radiation would have exceeded the energy density due to matter. At this time, the energy density of dark energy would have been negligible in comparison to the energy densities of radiation or matter.

As you will shortly see, the existence of an early radiation-dominated era has a profound effect on the dynamical evolution of the Universe.

At this point you may be wondering how is it that the energy density of matter and radiation change in different ways? The numbers of photons and particles within a comoving volume remain constant, so this cannot be the origin of the difference. The answer lies in the fact that the energies of photons change with the expansion of the Universe, whereas the masses (and hence energies) of particles such as protons and electrons and of any cold dark-matter particles remain constant.

6.2.4 A radiation-dominated model of the Universe

We have just seen that in the early Universe, the dominant energy density is that due to the radiation within the Universe. The Friedmann equation that was described in Chapter 5 (Box 5.4) can be solved for such conditions and the way in which the scale factor varies with time for such a model is shown in Figure 6.7. One important feature of such a model is that the scale factor varies in the following way:

$$R(t) \propto t^{1/2} \tag{6.17}$$

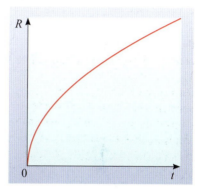

Figure 6.7 The evolution of the scale factor with time in a cosmological model in which the dominant contribution to the energy density arises from the radiation within the Universe (i.e. during the radiation-dominated era).

Because the energy density of radiation is dominant for times when $R(t)/R(t_0) < 10^{-4}$, all cosmological models that start with $R(t = 0) = 0$ will go through a phase that is well described by this radiation-dominated model. Thus we are in the rather remarkable position that regardless of which type of cosmological model best describes the Universe at the present, we can be reasonably confident that we know how the scale factor varied with time in the first few tens of thousands of years of the big bang.

However, the temperature of the background radiation varies with scale factor according to $T(t) \propto 1/R(t)$ (Equation 6.6). It follows that during the radiation-dominated era the temperature of the background radiation varies with time according to

$$T(t) \propto t^{-1/2} \tag{6.18}$$

This describes how temperature changes with time in an expanding universe where the energy density of radiation is the dominant component. Of course, to use Equation 6.18 to predict the temperature, it is necessary to know the constant of proportionality between T and $t^{-1/2}$. In fact, this can be derived from the Friedmann equation and Equation 6.18 becomes

$$(T/K) \approx 1.5 \times 10^{10} \times (t/s)^{-1/2} \tag{6.19}$$

(Note that T is measured in kelvin and t in seconds.) Equation 6.19 is an approximate relationship. As you will see later, other physical processes can change the temperature of the radiation in the real Universe during the radiation-dominated phase of its expansion.

■ What is the temperature when the age of the Universe is one second?

☐ By substituting a value of $t = 1$ s in Equation 6.19 the temperature is 1.5×10^{10} K.

Thus the temperature of the Universe in the first few seconds of the big bang was higher than the highest temperatures that are found in the cores of the most massive stars (where temperatures may reach about 10^9 K). This immediately suggests that nuclear reactions may have occurred in any matter that was present at this time. We will look into such processes in more detail in Section 6.4, but in the next section we will discuss even more extreme conditions: we will consider the processes that occurred when the Universe was less than 1 second old.

QUESTION 6.3

Rearrange Equation 6.19 to find an expression for the age of the Universe for a given temperature. What was the age of the Universe when the temperature was 10^6 K? Express your answer in terms of years.

6.3 The early Universe

We have seen that when considered together, the cosmic microwave background and the expansion of the Universe imply that there was an early phase of the history of the Universe that was characterized by high temperatures and high densities. In particular, the temperature at any time t can be estimated using Equation 6.19. A natural question to ask then is how far back towards $t = 0$ can we go in understanding processes in the Universe? This will be the first major question that we shall consider in this section, and we will find that there is a limit to our knowledge of the evolution of the Universe. The remainder of the section will be concerned with understanding the processes that occurred before the Universe was about 1 second old.

6.3.1 Cosmology and the limits of physical theory

We have seen that the Friedmann equation gives a model for a radiation-dominated Universe that is characterized by the scale factor having a value of zero at the instant of $t = 0$. As we said in Section 6.2.2, the naïve interpretation of this is that the Universe came into existence with an infinitely high temperature; the truth of the matter is that we don't really understand the physical processes in the very early Universe. So, how early in the history of the Universe can we be confident that our physical theories really do apply? There are essentially two answers to this question, which reflect two levels of certainty in physical theory. The first approach is to say that theories are only well tested for the ranges of physical conditions that can be explored by experiments. Thus, we may have a good deal of confidence in describing the Universe at times when the particle energies were similar to the highest values that can be imparted in large accelerator experiments. At present, this limit corresponds to being able to describe physical processes in the Universe that occurred after the temperature fell to below 10^{15} K, which corresponds to a time of $t \sim 10^{-9}$ s.

An alternative approach is to apply physical theories to conditions that never have been, and probably never will be, tested in the Earth-bound laboratory and to look for observable consequences in nature. Clearly, this is a somewhat more speculative approach than having to rely on 'tried-and-tested' physical theory. However, it is one way in which physical theories can be explored and developed, and is a very exciting field for cosmologists.

While it might be expected that physical theories could be extrapolated to describe processes at ever increasing temperatures, it turns out that there is a well-recognized limit to our theoretical understanding of the processes of nature. This limit arises because of a surprising incompatibility between the physical theory that is used to describe the interactions of subatomic particles and the theory that describes gravity. The interactions of subatomic particles are described by a branch of quantum physics called the **standard model** of elementary particles. The gravitational interaction is described, as you saw in Chapter 5, by Einstein's general theory of relativity.

The general theory of relativity describes effects that were not explained by Newton's theory of gravity, and as far as the theory can be tested, there have been no observations or measurements to suggest that the theory is incorrect. The standard model of elementary particles is much more amenable to being tested by experiment than is general relativity, and its predictions have been well tested by laboratory measurements. Despite the fact that both theories appear to be sound, it has proven impossible to join them together to form a single consistent theory.

Thus physicists expect that neither general relativity nor the standard model offers a full description of the fundamental interactions of nature, and propose that there must be a unifying 'theory of everything' that is yet to be discovered. In particular, such a theory is needed to describe processes in the very extreme conditions that occurred when the Universe was less than about 10^{-43} s old. This limiting time is called the **Planck time** and represents the limit of how far back in time towards $t = 0$ can be investigated using current physical theory. (The Planck time is $t_{Planck} = (Gh/2\pi c^5)^{1/2} = 5.38 \times 10^{-44}$ s.)

6.3.2 Conditions and processes in the early Universe

To set the scene for our account of the evolution of the Universe from a time of about 10^{-43} s, it is necessary to review some important physical concepts and processes. A key feature of the early Universe is that the radiation and matter were interacting so much that they were in a state of **thermal equilibrium**. This means the temperature of matter (as defined by the distribution of particle energies) and the temperature of radiation (as defined by the black-body spectrum) were equal.

At the high temperatures that existed in the early Universe, the composition of the Universe, in terms of the particles that were present, was determined by the typical energy that was available in particle interactions. This energy is termed the **interaction energy**, and is related to the temperature by

$$E \sim kT \tag{6.20}$$

where k is the Boltzmann constant ($k = 1.38 \times 10^{-23}$ J K^{-1}).

(The '~' sign is used to indicate a very approximate relationship. Note also that some books use $E \sim 3kT$, rather than Equation 6.20 given here.)

Note that it is common practice to express the interaction energy in terms of electronvolts (eV) where 1 eV = 1.60×10^{-19} J. The energies involved are usually expressed in MeV (1 MeV = 10^6 eV) or GeV (1 GeV = 10^9 eV).

■ Calculate the interaction energy when the temperature is 10^{14} K. Express your answer in joules and GeV.

☐ The interaction energy is found using Equation 6.20

$$E \sim kT = 1.38 \times 10^{-23} \text{ J K}^{-1} \times 10^{14} \text{ K} = 1.38 \times 10^{-9} \text{ J}$$

In terms of GeV

$$E = 1.38 \times 10^{-9} \text{ J}/(1.60 \times 10^{-19} \text{ J eV}^{-1}) = 8.63 \times 10^{9} \text{ eV} = 8.63 \text{ GeV}$$

So the interaction energy is 9 GeV (to one significant figure).

The fundamental interactions and their evolution

A key idea in the cosmology of the very early Universe relates to the four *fundamental interactions* of nature: these are gravitational, electromagnetic, weak and strong interactions (see Box 6.1). In the context of the present-day Universe, these interactions seem quite distinct. They operate over quite different ranges – the weak and strong interactions act only over distances that are comparable to the size of an atomic nucleus. Furthermore, the 'strength' of these interactions can be defined in a way that allows sensible comparison. The weakest interaction

BOX 6.1 FOUR FUNDAMENTAL INTERACTIONS

At a fundamental level nature has just four types of interaction. This means that any physical process, for example, the scattering of a photon off an electron, or the generation of electricity in a nuclear power plant, can be analysed in terms of one or more of these four interactions.

The fundamental interactions are:

1 The *gravitational interaction*, which for instance, keeps the Earth in orbit around the Sun. This acts over large distances, so it is part of our everyday experience.

2 The *electromagnetic interaction*, which for instance, keeps electrons bound to atoms. Like the gravitational interaction, this interaction also acts over long distances.

3 The *strong interaction*. This interaction only acts over distances comparable to the diameter of a nucleus. An example of the effect of the strong interaction is the binding of the protons and neutrons together in the nucleus of an atom. The strong interaction overcomes the mutual repulsion that acts between the positively charged protons in a nucleus.

4 The *weak interaction*. This is also a short-range interaction, which acts only on scales comparable to that of the nucleus. An example of the effect of the weak interaction occurs in the transformation of a neutron to a proton in β^--decay.

The standard model of elementary particles explains the operation of the strong, weak and electromagnetic interactions in terms of so-called 'exchange particles' that carry energy and momentum between interacting particles and thereby account for the action of a 'force' in a fundamental way. The exchange particles for the various interactions are as follows:

• The **photon**: the exchange particle of the electromagnetic interaction.

• The W^+, W^- and Z^0 *bosons*: the exchange particles of the weak interaction. The masses of these particles are responsible for the short range of the weak interaction.

• The *gluons*: a family of eight similar particles that are responsible for the strong interaction. These particles are confined within the protons or neutrons that comprise a nucleus.

The gravitational interaction is described by the general theory of relativity. Within this theory gravity arises from the curvature of space–time rather than from a particle interaction.

is gravity, and then in ascending order of 'strength', there follows the weak interaction, the electromagnetic interaction, and appropriately enough, the strong interaction.

It is suspected that all four interactions may be different manifestations of a single fundamental type of interaction. The reason for such a belief is partly philosophical and partly experimental. The 'philosophical' justification for the unification of interactions is that this type of approach – reducing the physical world to what appears to be the minimum number of particles and processes – has been outstandingly successful, and physicists see this as the next logical advance. If this sounds wildly idealistic, then the 'experimental' justification should offer some reassurance. A key idea in demonstrating that two interactions are linked is that under certain physical conditions they should behave in the same way. So, for instance, the strength of two interactions may become the same.

Experiments using particle accelerators have revealed that the strength of interactions depends on the interaction energy. In particular, the strengths of the electromagnetic and weak interactions are observed to become closer to one another at high interaction energies. At interaction energies of about 1000 GeV, the strengths of these two interactions are predicted to be the same, and the electromagnetic and weak interactions should appear as different manifestations of a single underlying *electroweak* interaction.

It is believed that the unification of the other interactions occurs at very much higher interaction energies. The unification of the strong interaction with the electroweak interaction – which is termed 'grand unification' is predicted to occur at an interaction energy of about 10^{15} GeV. The theoretical framework that is used to describe this unified interaction is called a **grand unified theory** or **GUT**.

QUESTION 6.4

Calculate the temperature corresponding to the minimum interaction energy required for grand unification. Hence calculate the age of the Universe when the strong and electroweak interactions became distinct.

(It is appropriate to quote the results as order-of-magnitude estimates, i.e. to the nearest whole number power of ten.)

The energy at which the gravitational interaction might become unified with the other interactions, if such a thing happens at all, is expected to be higher still – about 10^{19} GeV. At such extreme interaction energies the gravitational interaction might become important for interactions between particles (at lower energies, the gravitational interaction has a negligible effect on particle interactions). In terms of the evolution of the Universe, an interaction energy of 10^{19} GeV corresponds to the Planck time ($\sim 10^{-43}$ s). As has already been mentioned, there is no accepted 'theory of everything' that allows the processes that occurred in this **Planck era** to be understood.

The interaction energies associated with GUT interactions and the Planck era are extreme – there is probably no accessible environment in the present-day Universe in which particles interact with such energy. Thus, it is unlikely that direct experimental verification will ever be made of theories that describe interactions at such high energies.

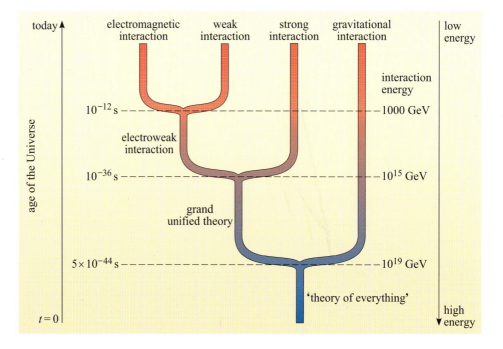

Figure 6.8 The evolution of the fundamental interactions with cosmic time.

The expected behaviour of the fundamental interactions over the first few moments of the history of the Universe can be summarized as follows. Prior to $t \sim 10^{-43}$ s all four fundamental interactions may have been unified. After this time, the gravitational interaction became distinct from the GUT interaction. Some time later, at $t \sim 10^{-36}$ s (see the answer to Question 6.4) the strong interaction and the electroweak interaction became distinct from one another. Finally, at $t \sim 10^{-12}$ s, when the typical interaction was about 1000 GeV, the weak and electromagnetic interactions took on the form in which they act in the present-day Universe. This evolution of the fundamental interactions is illustrated schematically in Figure 6.8.

Note that the earliest 'branch' shown in this diagram represents the end of the Planck era. The second branch corresponds to the time that grand unification ends – as we shall see later, it is speculated that this change is associated with a dramatic cosmological event, known as *inflation*. The third branch is associated with the separation of the electroweak interaction into the weak and electromagnetic interactions. This is the only one of the three branches that can currently be regarded as experimentally supported; the others are still very speculative.

Particle–antiparticle pair creation

An important process that occurs when interaction energies become very high is that particles and antiparticles can spontaneously form in pairs. This process is known as **pair creation**, and it has an important effect on the composition of the Universe at early times. Interactions obey a set of conservation rules: the conserved quantities include energy, electric charge, and baryon and lepton numbers (See Box 6.2). The energy available in an interaction plays a vital role in determining which particles may form. Provided that all other conservation rules are obeyed, a particle that has a mass m can be formed if the available energy is equal to or exceeds its *mass energy*, which is given by $E = mc^2$.

For example, when the age of the Universe was 10^{-12} s (i.e. about the time at which electroweak unification ended) the temperature was 10^{16} K and the typical energy of a photon or particle was about 10^3 GeV. Thus any interactions that occurred could easily supply 10^3 GeV to create a new particle. This energy exceeds the mass energy of all of the quarks and the leptons (see Box 6.2). As a result, the material content of the Universe at this time includes all types of lepton and quark and their respective antiparticles.

BOX 6.2 QUARKS AND LEPTONS

The ultimate building blocks of matter are two families of fundamental particles called quarks and leptons. There appear to be six members of each family, as shown in Tables 6.2 and 6.3. Note that for each fundamental particle shown here there exists a corresponding antiparticle with the opposite charge, but the same mass.

Table 6.2 The six quarks

	Name	Symbol	mass $\times c^2$/GeV
quarks with electric charge of $+2e/3$	up	u	5×10^{-3}
	charm	c	1.5
	top	t	1.8×10^2
quarks with electric charge of $-1e/3$	down	d	8×10^{-3}
	strange	s	0.16
	bottom	b	4.25

Table 6.3 The six leptons

	Name	Symbol	mass $\times c^2$/GeV
leptons with electric charge $-e$	electron	e^-	5.11×10^{-4}
	muon	μ^-	0.106
	tauon	τ^-	1.78
leptons with zero electric charge	electron neutrino	ν_e	$< 1.5 \times 10^{-9}$
	muon neutrino	ν_μ	$< 1.7 \times 10^{-4}$
	tauon neutrino	ν_τ	$< 2.4 \times 10^{-2}$

Quarks are the fundamental particles that make up baryons, such as the proton and the neutron. A proton comprises two up quarks and a down quark, while a neutron comprises two down quarks and an up quark. Quarks are never found in isolation in laboratory experiments – they are always confined in clusters consisting of three quarks or three antiquarks, or in a pair comprising a quark and antiquark. The particles formed by such combinations of quarks are generally termed **hadrons**. A hadron that consists of three quarks is a baryon, and a hadron that comprises three antiquarks is an **antibaryon**. A quantity that is conserved in all known particle interactions is the **baryon number** – the baryon number of each quark is 1/3 while that of each antiquark is −1/3. So the baryon number of a baryon is +1, and that of an antibaryon is −1. The baryon number of all other particles is 0.

The family of **leptons** includes the electron (e^-) and two other charged particles: the muon (μ^-) and the tauon or tau particle (τ^-). The other members of the lepton family are the three types of neutrino – there is one type of neutrino for each of the charged leptons (ν_e, ν_μ, ν_τ). As in the case of quarks, for each type of lepton there exists a corresponding antilepton. Unlike quarks, leptons are not confined and *can* be found in isolation in laboratory experiments.

In interactions that involve leptons, there is a conserved quantity called the **lepton number**. The lepton number of each of the leptons shown in Table 6.3 is +1 and that of each of the antileptons is −1. The lepton number of all other particles is 0.

QUESTION 6.5

Consider the following reaction in which a free neutron decays into a proton, an electron and an electron antineutrino ($\bar{\nu}_e$) (such a reaction is an example of β^--decay).

$$n \rightarrow p + e^- + \bar{\nu}_e$$

(a) What is the baryon number (i) before, and (ii) after this process? Hence show that baryon number is conserved in this reaction.

(b) What is the lepton number (i) before, and (ii) after this process? Hence show that lepton number is conserved in this reaction.

(c) What combination of quarks constitutes (i) a neutron, and (ii) a proton? Hence express β^--decay as a reaction involving quarks and leptons only.

The effect of the other conservation rules in determining which particles may be formed in an interaction is profound. Of particular importance are the conserved quantities known as total baryon number and total lepton number (described in Box 6.2). Consider a simple interaction in which two energetic photons interact to form particles

$$\gamma + \gamma \rightarrow \text{'particles'}$$

■ What is (a) the total lepton number of the two photons; (b) the total baryon number of the two photons?

☐ Photons have a lepton number of zero and a baryon number of zero. Thus (a) the total lepton number of the two photons is zero, and (b) the total baryon number of the two photons is zero.

Since lepton and baryon number are conserved, the two-photon reaction can only form products whose total lepton and baryon number is zero. This does not mean that the lepton and baryon number of each particle that is formed must be zero, but that the *sum* of the lepton and baryon numbers for all the particles that are formed must be zero. For instance, the reaction may result in the production of an electron (lepton number +1) and a positron (an antielectron, lepton number −1)

$$\gamma + \gamma \rightarrow e^+ + e^- \tag{6.21a}$$

Since electrons and positrons have a baryon number of zero, this reaction clearly conserves baryon number.

The pair-creation reaction described by Equation 6.21a is reversible – a positron and an electron can combine to produce two photons according to

$$e^+ + e^- \rightarrow \gamma + \gamma \qquad (6.21b)$$

This process, in which a particle and its corresponding antiparticle interact and disappear is called **annihilation**. This process can occur for any particle–antiparticle pair, and the total energy of the photons can be found using the mass energy equivalence relation ($E = mc^2$).

QUESTION 6.6

Calculate the minimum interaction energy required for electron–positron pair production (Equation 6.21a). Express your answer in electronvolts. At what temperature is electron–positron pair production likely to occur?

Note that given sufficiently energetic photons, a two-photon reaction could generate any lepton–antilepton pair. Similarly, a two-photon reaction could generate a quark–antiquark pair

$$\gamma + \gamma \rightarrow q + \bar{q} \qquad (6.22)$$

(where the symbol $\bar{q}$ represents an antiquark.) The photon–photon interaction is just one of many types of interaction that could occur, but without going into detail about these, we can see that the conservation rules will dictate that, provided a sufficiently high interaction energy is available, the Universe will be populated by a mixture of quarks and antiquarks and of leptons and antileptons.

We have concentrated here on quarks and leptons, but there are other particles too that were present in the early Universe. There are two categories of particles that deserve mention. The first are the exchange particles that act to transmit the fundamental interactions of nature (in the parlance of particle physics, these particles 'mediate' the interactions). The most familiar of these is the photon – a massless particle that mediates the electromagnetic interaction. In addition, there are other particles, as described in Box 6.1, that mediate the strong and weak interactions.

From an astronomical point of view, the other important category of particle comprises the massive, stable particle (or particles) that make up dark matter. As has been mentioned, the nature of dark matter is not known, but it is believed to be in the form of particles that are neither baryons or leptons. Presumably, such particles must have been present in the early Universe, but until we have a better idea of what they are, their origin remains a mystery. As far as this chapter is concerned, we shall assume that there are dark-matter particles present in the early Universe, but we shall also assume they are essentially non-interacting, so we shall not need to mention them. We will however, consider the role of dark matter at later times – when structure begins to form as a result of gravitational collapse.

Having now reviewed some of the important processes in the early Universe, we can begin a chronological account of the evolution of the Universe. Throughout the following discussion we shall indicate the time t, and the temperature T and interaction energy E at these times (note that in most cases, these are order of magnitude estimates only).

So, let's start at (almost) the very beginning.

6.3.3 The Planck era

$$t < 5 \times 10^{-44} \text{ s}, \, T > 10^{32} \text{ K}, \, E > 10^{19} \text{ GeV}$$

We have already noted that there is no physical theory to describe processes of the Planck era ($t < 5 \times 10^{-44}$ s). When the age of the Universe was less than the Planck time, it is believed that the fundamental interactions would have had similar strengths, but without a consistent physical theory very little can be predicted about what would happen at this time. We shall simply assume that the Universe was in an extremely hot and dense state. We shall return to consider these very early times again in Chapter 8 when we look at the way in which theoretical physicists are attempting to develop theories that describe this era.

6.3.4 Inflation and the end of grand unification

$$t \sim 10^{-36} \text{ s}, \, T \sim 10^{28} \text{ K}, \, E \sim 10^{15} \text{ GeV}$$

When the age of the Universe was about 10^{-36} s, the high-energy conditions under which the strong and electroweak interactions were unified came to an end. After this time, these two types of interaction would become distinct from one another.

Although the typical interaction energies at this time ($\sim 10^{15}$ GeV) are far in excess of laboratory experiments, physicists do have a theoretical framework within which some predictions about this era of cosmic history can be formulated. In fact, it was this approach that, in 1980, resulted in a significant advance in cosmological theory. A theoretical physicist named Alan Guth was tackling a problem: as the grand unified era came to an end, it seemed as though a vast number of particles called *magnetic monopoles* should be formed. Such a particle should not decay, and so, if the theory was correct, these magnetic monopoles should be easily detectable in the present-day Universe. The problem was that no such particle had ever been found.

The solution that Guth proposed was quite remarkable. It was based on an analysis of the behaviour of the vacuum. We saw in Chapter 5 that the dark energy may arise from the energy of the vacuum. Although we have no well-developed theory to explain the energy of the vacuum, Guth found that at the end of the grand unified era the vacuum could have had a different, and substantially higher energy density than it does in the present-day Universe. This peculiar state is referred to as the *false vacuum* to distinguish it from the *true vacuum*. This situation would not last for long – in a very short time the energy of the vacuum would drop to the value that it is observed at today – but during that time, something very dramatic would happen.

A high value of the energy density of the vacuum has the same physical effect as a high value of the cosmological constant. (Strictly speaking, the energy density of the false vacuum should not be referred to as being due to a 'cosmological constant' since it is hardly constant!) You saw in Chapter 5 that the cosmological model in which the cosmological constant plays a dominant role is the de Sitter model. The evolution of this model is described by Equation 5.11

$$R \propto e^{Ht} \quad \text{where } H = \sqrt{\frac{\Lambda c^2}{3}} \tag{5.11}$$

In this model the scale factor undergoes exponential expansion. What Guth proposed was that in a very short interval of time – maybe only lasting from $t \sim 10^{-36}$ to 10^{-34} s – the Universe underwent a period of exponential expansion. During this time the scale factor increased by an enormous amount. The theory was not sufficiently well developed to say by exactly how much the scale factor would have increased, but such a process could have caused the scale factor to increase by a factor of at least 10^{27}, and possibly much more (Figure 6.9). This dramatic episode of expansion is termed **inflation**.

So how would inflation solve the monopole problem? Well, the rapid expansion of space during inflation would result in any particles being swept apart from one another. Even if monopoles were abundant prior to inflation, the rapid expansion of space would spread them out so much that there would be a negligible chance of our detecting one in the present-day Universe.

The inflationary scenario has an important consequence for the material content of the Universe. As the period of inflation came to an end, the energy of the vacuum had to drop to the level we see today. So, as the vacuum made a transition to its 'true vacuum' state, energy was released and formed particle–antiparticle pairs. According to the inflationary model, the vast majority of particles in the Universe were created from the energy released as inflation came to an end. Thus, the matter now in the Universe may be a product of inflation.

It should be stressed that Guth's original formulation of the inflationary model should not be considered to be a complete and consistent theory. Rather, it should be viewed as a starting point for exploring a new paradigm in cosmology. In fact, the exact mechanism behind inflation is essentially unknown. In view of this it might seem odd that such prominence is given to the inflationary model until it is appreciated that the *effects* of inflation – regardless of the underlying mechanism – provide solutions to a series of cosmological problems. We shall discuss these problems, and how inflation resolves them, in Chapter 8, but for now we shall simply assume that the process of inflation did occur.

After the process of inflation, the energy released by the false vacuum would have eventually formed all types of quarks and leptons (including their antiparticles).

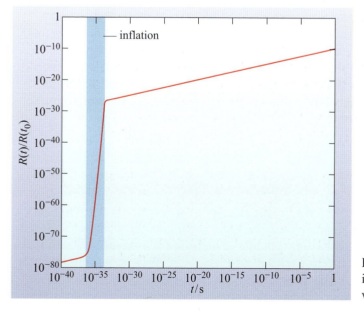

Figure 6.9 The evolution of the scale factor with time including the process of inflation. Note that the numerical values shown here are highly speculative.

Furthermore, the numbers of quarks and antiquarks would have been almost, but not quite, equal, and a similar condition would have held true for leptons and antileptons. This slight inequality between matter and antimatter in the Universe is thought to have originated in physical processes that occurred at this time. (More will be said about this in Chapter 8.) As we shall shortly see, if this imbalance had not existed, then the present day cosmos would have contained no baryonic matter, and we would not be here to speculate on the origin of the Universe!

Thus, the time at which grand unification came to an end is suspected of playing a major role in the evolution of the Universe. However, from this time at around $t \sim 10^{-34}$ s to the time that the electromagnetic and weak interactions became distinct at $t \sim 10^{-12}$ s, no new physical processes occurred. This interval is sometimes referred to as *the desert*.

6.3.5 The end of electroweak unification

$$t \sim 10^{-12} \text{ s}, \ T \sim 10^{16} \text{ K}, \ E \sim 10^3 \text{ GeV}$$

The desert came to an end as electromagnetic and weak interactions became distinct. In contrast to the end of the grand unification, it is not thought that this caused any effects akin to inflation. The constituents of the Universe immediately after this time would have continued to be all types of quark and lepton and their antiparticles. There would also have been photons, and particles that mediate the strong interaction between quarks. However, the temperature was now too low for the creation of W^+, W^- and Z^0 bosons, so the particles that mediate the weak interaction would essentially disappear, thus separating the weak and electromagnetic interactions.

6.3.6 The quark–hadron transition

$$t \sim 10^{-5} \text{ s}, \ T \sim 10^{12} \text{ K}, \ E \sim 1 \text{ GeV}$$

In the present-day Universe, quarks are never seen in isolation – they are always confined within particles called hadrons (see Box 6.2). In the high energy conditions of the early Universe however, quarks were not bound into hadrons; they existed as free individual particles. The existence of free quarks and antiquarks came to an end when the Universe cooled to such an extent that the typical interaction energy was about 200 MeV. At this stage the Universe underwent a phase transition (a process akin to the freezing of water to form ice) in which the quarks became bound into hadrons. This particular phase transition is called the **quark–hadron phase transition**. Although many different types of hadron were formed in this process, there are only two types of hadron that are stable enough to have any long-lasting effect on the composition of the Universe: the proton and the neutron.

The proton (and its antiparticle – the antiproton $\bar{p}$) is, as far as is known, a stable particle. Both the proton and antiproton can participate in reactions with other particles but, left to themselves, no proton or antiproton has ever been observed to spontaneously decay. The fact that hydrogen exists in copious amounts in the present-day Universe is testament to the stability of the proton. In fact, there is some belief that the proton may decay on very long timescales – but experimental searches for proton decay have shown that its half-life must exceed 10^{33} years. The current age of the Universe is about 1.4×10^{10} years, so as far as this discussion is concerned we can assume the proton to be a stable particle.

Unlike the proton, the neutron (and its antiparticle – the antineutron ñ) is unstable: an isolated neutron will undergo the β^--decay reaction

$$n \rightarrow p + e^- + \bar{v}_e \qquad (6.23)$$

However, the half-life of the neutron is 615 s, and this is a very long time in comparison to the timescale on which the Universe is changing (remember that we are discussing processes that occur within about 10^{-4} s). So to a good approximation, the effect of this decay process can be ignored at this time – and we can say that the neutron is relatively stable.

The mass energies of the proton and the neutron are 938 and 940 MeV respectively. At the time that free quarks became bound into hadrons, the typical interaction energy was too low for proton–antiproton pairs to be produced. Thus, protons and antiprotons would have disappeared from the Universe as they annihilated one another according to the reaction

$$p + \bar{p} \rightarrow \gamma + \gamma \qquad (6.24)$$

while there would have been no significant counter-conversion of photons into proton–antiproton pairs.

A reaction similar to that shown in Equation 6.24 would also have occurred for neutrons and antineutrons. Prior to this time, baryonic matter was in the form of particles and their antiparticles (either quarks or baryons), but at around this time the majority of such particles annihilated one another. Now we can appreciate the significance of the slight imbalance between matter and antimatter that had been present since the grand unified era. If there had been no imbalance, then at this time, all of the protons would have annihilated with an equal number of antiprotons, and a similar annihilation process would have resulted in the disappearance of all neutrons and antineutrons. The result would have been a Universe that contained no baryonic matter. Within such a Universe there would, of course, be no galaxies, stars, planets or life, so the fact that there was such an imbalance between matter and antimatter was vital to the Universe ending up as we observe it today.

The magnitude of this imbalance between the number of particles and antiparticles is small, but it can be measured from present-day observations. Neither the number of baryons nor the number of photons in a comoving volume has changed significantly since this time. The present-day ratio of the number of CMB photons to the number of baryons thus provides an estimate of the imbalance between baryons and antibaryons at this time.

At present, there are approximately 10^9 photons in the cosmic microwave background for every stable baryon (proton or neutron) in the Universe.

Thus for every 10^9 baryon–antibaryon annihilation reactions that occurred there would have been one proton or neutron left over.

We have seen that quarks became confined into hadrons and that these hadrons decayed or annihilated one another, leaving a residual number of relatively stable protons and neutrons. However, these protons and neutrons were not inert – they could undergo the following reactions that transformed one into the other.

$$\bar{v}_e + p \rightleftharpoons n + e^+ \qquad (6.25a)$$

$$v_e + n \rightleftharpoons e^- + p \qquad (6.25b)$$

When the age of the Universe was 10^{-2} s, there were large numbers of neutrinos, antineutrinos, electrons and positrons available for such reactions, and the rate of these reactions was high. Consequently the temperature of the neutrinos (as defined by the distribution of their energies) would have been the same as the temperature of the baryonic matter and the temperature of the radiation (as defined by the black-body spectrum).

In the following section (Section 6.4) we shall consider situations in which protons and neutrons participate in fusion reactions. The outcome of these reactions depends on the ratio of the number density of neutrons to the number density of protons (n_n/n_p), and so it is of interest to follow how this ratio varies with time. While the reactions described by Equation 6.25 were occurring, the ratio n_n/n_p depended on the difference in mass energy between these two types of baryon. The proton has a rest mass energy of 938.27 MeV whereas the neutron has a rest mass that is 1.29 MeV greater than this. When the interaction energy was much greater than this difference, i.e. much more than about 1 MeV, then the number densities of protons and neutrons would have been equal. However, once the interaction energy became similar to this energy difference, the number density of neutrons fell below that of protons. At $t = 10^{-2}$ s, when the interaction energy was 10 MeV, the value of $(n_n/n_p) \approx 0.9$, but by $t = 0.1$ s, the typical interaction energy had fallen to 3 MeV and the neutron to proton ratio was $(n_n/n_p) \approx 0.65$.

6.3.7 Neutrino decoupling and electron–positron annihilation

$t \sim 1$ s, $T \sim 1.5 \times 10^{10}$ K, $E \sim 1$ MeV

By the time that the Universe reached an age of 0.7 s, conditions had changed to such an extent that some of the reactions described in Equation 6.25 no longer occurred. In particular, the probability of a neutrino (or antineutrino) interacting with another particle dropped as the density of the Universe decreased. Consequently the reactions shown in Equations 6.25a and 6.25b would only operate from right to left. This was the last occasion on which the bulk of the neutrinos in the Universe underwent any interaction apart from being influenced by gravitational fields. The effective end of the interaction between neutrinos and other particles is termed **neutrino decoupling**. As a result of this, huge numbers of neutrinos, usually referred to as *cosmic neutrinos*, started to travel unimpeded through the Universe. They are thought to have been doing so ever since.

Just after neutrino decoupling, when the age of the Universe was about 1 second, the falling temperature of the Universe corresponded to a mean interaction energy of about 1 MeV, which is the energy required for the formation of an electron–positron pair (see Question 6.6). As the temperature fell further, electrons and positrons began to disappear because no new $e^+ e^-$ pairs were being created, whereas the annihilation reaction

$$e^+ + e^- \rightarrow \gamma + \gamma \qquad (6.21b)$$

was continuing. The number of electrons and positrons decreased in a dramatic fashion. As was the case when baryons annihilated, there was a slight excess of matter over antimatter – a surplus of one electron for every 10^9 or so annihilation events. The number of negatively charged electrons that were left over is believed to be exactly the number to balance the charge of all the positively charged protons that were left over earlier, thus making the matter in the Universe electrically neutral overall.

An important effect of electron–positron annihilation is that energy was released and this would have been rapidly shared-out amongst the photons, baryons and remaining electrons. Because of this release of energy there was a short interval in which the temperature did not decrease as rapidly as Equation 6.19 would predict, and this is one reason why that equation was described as being approximate.

The process of electron–positron annihilation also leads to a prediction about cosmic neutrinos.

■ Would the energy that is released by electron–positron annihilation be transferred to the neutrinos?

☐ No. We have just seen that neutrinos effectively stop interacting with the other constituents of the Universe just before electron–positron annihilation occurs.

So cosmic neutrinos do not gain any energy from the process of electron–positron annihilation. Consequently the temperature of cosmic neutrinos should be slightly lower than that of the background radiation. It is predicted that at the present time the cosmological background of neutrinos should have a temperature of about 1.95 K. Experimental confirmation of this would provide strong evidence that the big bang scenario that is described here is correct, but unfortunately, the detection of such low-energy neutrinos is unfeasible at present.

The disappearance of all of the positrons and most of the electrons further restricted the reactions shown in Equation 6.25 that converted protons into neutrons and vice versa. At the time that these reactions stopped completely, the ratio of the number density of neutrons to the number density of protons had a value of $(n_n/n_p) \approx 0.22$, i.e. for every 100 protons in the Universe there were 22 neutrons.

■ There was however one reaction that causes neutrons to transform into protons which did not stop. Which reaction was this, and why didn't it stop?

☐ The reaction that continues is the β^--decay of the free neutron (Equation 6.23). It did not stop because, unlike the reactions in Equation 6.25, it does not require any other reactant apart from the neutron itself.

Thus, starting from a value of $(n_n/n_p) \approx 0.22$, the number of neutrons started to drop. Unless some new process intervened, the neutrons would have all decayed and we would have a Universe in which the only element that could form would be hydrogen. The way in which this fate was avoided is the next part of our story.

6.4 Nucleosynthesis and the abundance of light elements

$t <$ a few hundred seconds, $T >$ a few $\times 10^8$ K, $E >$ a few $\times 10^4$ eV

We have already seen that conditions in the early Universe led to a situation, such that at $t \sim 1$ s, the temperature was about 10^{10} K and the baryonic matter in the Universe was in the form of protons and neutrons. At this time the physical conditions became suitable for the onset of nuclear fusion reactions that lead to the formation of nuclides with a higher atomic mass than hydrogen. Such a process is believed to have occurred and is called **primordial nucleosynthesis** – a term that distinguishes it from the processes of stellar nucleosynthesis that create elements within stars.

There are some distinct differences between the nucleosynthetic processes that could have occurred in the early Universe and those that occur within stars.

One difference is that the conditions in the Universe were changing rapidly, as the following question illustrates.

QUESTION 6.7

Find the time t at which the temperature was (a) 10^9 K, and (b) 5×10^8 K.

As the answer to Question 6.7 shows, the temperature of the Universe dropped markedly in the first few hundred seconds after $t = 0$. In order for nuclear fusion reactions to have had a significant effect they must have progressed at a rapid rate, and this would have required temperatures in excess of 5×10^8 K. This is in marked contrast to the conditions in the cores of stars where fusion reactions progress at a relatively leisurely rate in lower temperature conditions.

As time progressed in the early Universe, one nuclear reaction that did not require high temperatures, the β^--decay of free neutrons, was proceeding. However, the presence of a large number of free neutrons highlights another difference between the early Universe and stellar cores – that of composition. As we shall now see, it is the declining number of free neutrons that plays an important role in determining how many nuclei can be formed before fusion reactions become ineffective at a temperature of about 5×10^8 K.

6.4.1 The formation and survival of deuterium

The first fusion reaction that could occur was that between a proton and a neutron to form a nucleus of deuterium (which is referred to as a **deuteron**). This is the neutron capture reaction:

$$p + n \rightleftharpoons {}_1^2\text{H} + \gamma \tag{6.26}$$

Note that this is a reversible reaction: the deuteron can be broken apart by γ-rays in a process called **photodisintegration**. In order to cause the photodisintegration of a deuteron, an incident photon must have an energy that exceeds 2.23 MeV. Although at $t = 1$ s the average interaction energy is less than this, there were so many photons in comparison to the number of baryons, that there was a sufficient number of photons with energies greater than 2.23 MeV (i.e. well above the

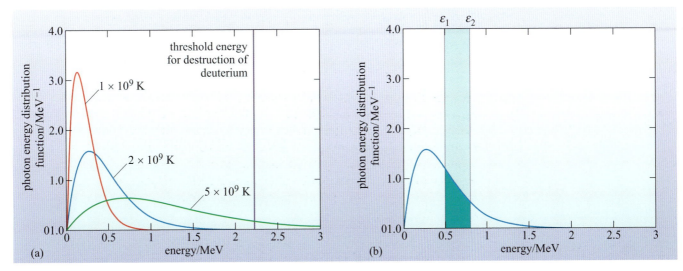

(a)

(b)

Figure 6.10 (a) Photon energy distributions at various temperatures. (b) The area under the curve of the photon energy distribution and between two energies, indicates the fraction of photons with energies in that range.

average value) to rapidly destroy any deuterium that formed. However, as the Universe continued to expand and cool, the average photon energy decreased. This decrease allowed deuterium to survive from about $t = 3$ minutes onwards.

To investigate this in more detail we need to know, for a given volume and at a given time (or temperature), what fraction of the photons have sufficiently high energy to cause the photodisintegration of deuterium. A quantity called the **photon energy distribution function** tells us this; it is defined as the fraction of the total number of photons that lie within a narrow energy range, divided by the width of that range. This definition may sound similar to the definition of the spectral flux density (F_λ) – and indeed there is a straightforward mathematical relationship between the two. As in the case of the spectral flux density, the photon energy distribution of a black-body source is a smooth function that has a peak value that depends on temperature.

Figure 6.10a shows the photon energy distribution expected over a range of temperatures in the early Universe. In all cases, of course, the photon energies follow a black-body distribution.

■ By inspection of Figure 6.10a state qualitatively how the energy at the peak of the photon energy distribution of a black-body source varies with temperature.

□ The peak energy of the photon energy distribution of a black-body source decreases as the temperature decreases.

The key feature to note about the energy distribution function is that it indicates the fraction of photons that have energies in a certain range. The fraction of photons with an energy between ε_1 and ε_2 is given by the area under the curve between these two limits as shown by the shaded area in Figure 6.10b. Note that the units of the photon energy distribution function are 'per unit energy interval', i.e. $(MeV)^{-1}$. Thus the area under the curve has units that are given by $(MeV) \times (MeV)^{-1}$: it has no units, as would be expected for a quantity that represents the fraction of photons.

Returning now to the question of the destruction of deuterium, we know that the ratio of photons to baryons is about 10^9. Thus, if the photon energy distribution were such that more than about 1 in 10^9 photons had an energy greater than 2.23 MeV, then there would have been a sufficient number of energetic photons to destroy any deuterons that may have formed.

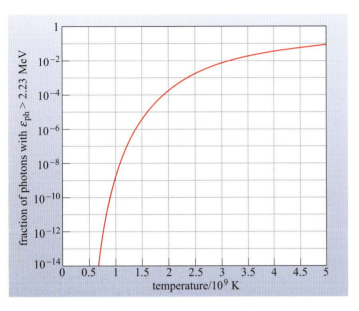

Figure 6.11 The variation with temperature of the fraction of photons with energies greater than 2.23 MeV in a black-body distribution of photon energies.

Conversely, we can say that deuterons only began to survive once the temperature was such that the fraction of photons with energies greater than 2.23 MeV became less than 10^{-9}. Figure 6.11 shows how this fraction varies with temperature.

■ From Figure 6.11 estimate the lowest temperature at which deuterium can survive, i.e. at which it does not undergo photodisintegration.

☐ From Figure 6.11 the temperature at which this fraction is 10^{-9} is 1.0×10^9 K.

Thus significant quantities of deuterium started to build up only after the temperature dropped below about 1.0×10^9 K. This temperature was reached when the age of the Universe was about 200 s (see answer to Question 6.7a).

It is interesting to compare the mean photon energy of a black-body distribution with the energy required for the photodisintegration of deuterons. For a black-body distribution of photons, the mean photon energy ε_{mean} is related to the absolute temperature T by the relation

$$\varepsilon_{mean} = 2.7kT \tag{6.27}$$

where k is the Boltzmann constant. So at the time that deuterium starts to build up, (when $T = 1.0 \times 10^9$ K) the mean photon energy is $\varepsilon_{mean} = 0.233$ MeV. The ratio between the energy that can cause the photodisintegration of a deuteron and the mean photon energy is therefore 2.23 MeV/0.233 MeV = 9.6. Thus, the process of photodisintegration did not stop until the mean photon energy was about a factor of ten lower than the photodisintegration energy. This highlights the fact that because there are vastly more photons than baryons, the very small fraction of photons that have energies much higher than the mean photon energy can have a significant effect on physical processes.

The survival of deuterium has been considered in some detail, since similar considerations about the relative numbers of energetic photons play an important role at a much later time in the history of the Universe – the epoch at which electrons and ions combined to form neutral atoms.

6.4.2 Primordial nuclear reactions

As soon as there was a significant build up in the abundance of deuterium, other nuclear reactions could then proceed. In particular, there were several series of reactions that form the very stable nuclide helium-4 (i.e. $^4_2$He).

For instance, an isotope of hydrogen called tritium ($_1^3$H) was formed by

The significance of the colour coding is explained below.

$$_1^2\text{H} + \text{n} \rightarrow {}_1^3\text{H} + \gamma \qquad\qquad (6.28\text{a})$$

$$_1^2\text{H} + {}_1^2\text{H} \rightarrow {}_1^3\text{H} + \text{p}$$

and the tritium thus formed could then undergo reactions to produce helium-4 as follows

$$_1^3\text{H} + {}_1^2\text{H} \rightarrow {}_2^4\text{He} + \text{n} \qquad\qquad (6.28\text{b})$$

$$_1^3\text{H} + \text{p} \rightarrow {}_2^4\text{He} + \gamma$$

However deuterium also reacted to produce helium-3

$$_1^2\text{H} + {}_1^2\text{H} \rightarrow {}_2^3\text{He} + \text{n} \qquad\qquad (6.28\text{c})$$

$$_1^2\text{H} + \text{p} \rightarrow {}_2^3\text{He} + \gamma$$

and this isotope of helium could undergo reactions to form helium-4 as follows

$$_2^3\text{He} + \text{n} \rightarrow {}_2^4\text{He} + \gamma \qquad\qquad (6.28\text{d})$$

$$_2^3\text{He} + {}_1^2\text{H} \rightarrow {}_2^4\text{He} + \text{p}$$

These were the dominant reactions that led to the production of helium-4. (Although the direct fusion of two deuterons to form helium-4 was also a possible reaction, this did not occur to any great extent.)

If the plethora of reactions bothers you, you may be relieved to note that there are only four types of reaction at work here. The reaction equations have been colour coded to illustrate this. The reactions are of the following types:

- A neutron is captured and a photon is emitted (colour-coded green).
- A proton is captured and a photon is emitted (colour-coded red).
- A deuteron is captured and neutron is emitted (colour-coded blue).
- A deuteron is captured and proton is emitted (colour-coded purple).

The major product of primordial nucleosynthesis was helium-4. The fact that nucleosynthesis did not progress to produce large quantities of nuclides with higher mass numbers is due to two factors. Firstly, the rate at which two nuclei will fuse together depends very strongly on the temperature, and higher temperatures are required to fuse nuclei of higher atomic number. Because the deuteron is easily photodisintegrated, the process of nucleosynthesis could only start once the temperature was relatively low. As a consequence, the rate of fusion reactions that involved nuclides other than hydrogen and helium would have been very low.

A second factor is the lack of any stable nuclide with mass number 5 or 8. The lack of a stable nuclide with mass number 5 means that helium-4 could not react with the two most abundant species – protons and neutrons. This hurdle could, however, be overcome by reactions that involve tritium or helium-3,

$$_2^4\text{He} + {}_1^3\text{H} \rightarrow {}_3^7\text{Li} + \gamma \qquad\qquad (6.29)$$

$$_2^4\text{He} + {}_2^3\text{He} \rightarrow {}_4^7\text{Be (unstable)}$$

$$_4^7\text{Be} + \text{e}^- \rightarrow {}_3^7\text{Li} + \nu_\text{e}$$

One further reaction that is worth noting is that lithium-7 can react with a proton, but the result is the destruction of the newly formed lithium and the formation of two nuclei of helium-4

$$\tfrac{7}{3}\text{Li} + \text{p} \rightarrow \tfrac{4}{2}\text{He} + \tfrac{4}{2}\text{He} \tag{6.30}$$

The yield of lithium was small. The relative amount of lithium formed can be quantified by the mass fraction, i.e. the proportion of the baryonic mass that was in the form of this element. Since lithium is the only element heavier than hydrogen or helium that is formed at this time, the mass fraction of lithium corresponds to the metallicity Z. Primordial nucleosynthesis created lithium such that the metallicity was less than 10^{-9}, but as we will see, the abundance of lithium provides a useful way of probing the conditions during the first few minutes of the big bang.

It should be noted that the fusion reactions that are outlined above only operated for a brief period of time. By the time the Universe reached an age of about 1000 s (i.e. about 17 minutes) and had a temperature of about 5×10^8 K, all such reactions effectively ceased. However, this brief spell of history left a signature, in terms of the abundances of light elements, that can be read today.

6.4.3 The primordial abundance of helium

We can now investigate how much helium would have been produced in the first few minutes of the big bang. The approach that we take here is to estimate the mass fraction of helium (Y) that would have been produced given the processes outlined above. The starting point for the calculation is the ratio of the number density of neutrons to the number density of protons.

■ What was the value of this ratio at the time that the reactions described by Equation 6.25 came to a halt. At what time did this happen?

☐ It was stated above that $n_\text{n}/n_\text{p} \approx 0.22$ at the time that these reactions stopped. This occurred a time $t \approx 0.7$ s.

The temperature at this time was much higher than the maximum temperature at which deuterium can survive.

■ Why is the temperature at which deuterons can survive relevant to the production of helium?

☐ The formation of deuterons is the first step in the sequence of nucleosynthesis reactions that leads to helium.

So there was a delay before helium synthesis could start. During this time the ratio n_n/n_p does not remain constant because the isolated neutron is not a stable particle. Free neutrons undergo β^--decay with a half-life of 615 s. The next step is to calculate how much time elapses from the instant at which $n_\text{n}/n_\text{p} \approx 0.22$ and the time at which deuterium can form.

■ At what time does the temperature drop to the point at which deuterium can survive? (*Hint*: See Question 6.7.)

☐ Deuterium survives once the temperature drops below 1×10^9 K. This temperature is reached at $t = 225$ s. (This was the answer to Question 6.7a.)

Thus there was an interval of a few hundred seconds before helium production could start in earnest. The way in which the neutron number dropped with time is illustrated in Figure 6.12, which shows the fraction of the original sample of neutrons that would have remained at a given time.

QUESTION 6.8

Calculate the value of n_n/n_p at the time when deuterium started to be formed in significant amounts. (Figure 6.12 provides a way of estimating the number of neutrons that decay in a given time. Take care in calculating n_n/n_p that you account for particles that are created as well as those that disappear!)

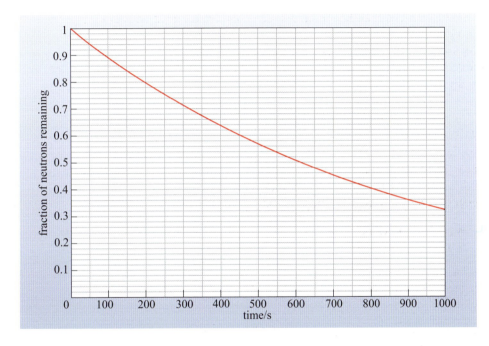

Figure 6.12 The fraction of the number of neutrons that would remain in a sample after a given interval of time. (For use with Question 6.8.)

The result of Question 6.8 shows that when deuterium began to be formed, the ratio of the neutron and proton number densities, n_n/n_p, had a value of about 0.16. The final stage of the calculation is to work out the mass fraction in helium that arises from this value. You have already seen that the major product of primordial nucleosynthesis was helium-4, and as far as calculating the helium abundance is concerned, it is a reasonable approximation to assume that *all* of the neutrons that were present ended up in nuclei of helium-4.

The quantity that we wish to calculate is the mass fraction contained in helium-4

$$Y = \frac{\text{mass of helium in a sample}}{\text{total mass of baryonic matter in the sample}} \qquad (6.31)$$

For the purposes of this calculation the sample can be taken to be the baryonic matter in any comoving volume. There are two contributions to the mass of this sample: that due to hydrogen and that due to helium (which, for simplicity, we assume here to be purely helium-4). If the number of hydrogen and helium nuclei present in our sample, after all nucleosynthesis reactions have stopped, are N_H and N_{He} respectively, then the mass fraction in helium can be written as

$$Y = \frac{N_{He}m_{He}}{N_H m_H + N_{He}m_{He}}$$

where m_H and m_{He} are the masses of the hydrogen and helium nucleus respectively. This equation can be simplified by making the approximation that $m_{He} \approx 4m_H$.

$$Y = \frac{4N_{He}}{N_H + 4N_{He}} \tag{6.32}$$

Since there are two neutrons in each helium nucleus, the number of helium nuclei is simply half the number of neutrons. The number of hydrogen nuclei is the number of protons minus the number of protons that are locked up in helium nuclei,

$$N_{He} = N_n/2$$

$$N_H = N_p - 2N_{He} = N_p - N_n$$

These expressions for N_H and N_{He} can be substituted into Equation 6.32 to give

$$Y = \frac{2N_n}{N_n + N_p} = 2\left(\frac{1}{1 + (N_p/N_n)}\right) \tag{6.33a}$$

The ratio of the number of protons to neutrons (N_p/N_n) in our comoving sample is the same as the ratio of the *number density* of protons to that of neutrons (n_p/n_n). Hence

$$Y = 2\left(\frac{1}{1 + (n_p/n_n)}\right) \tag{6.33b}$$

QUESTION 6.9

Using the value of n_n/n_p that you obtained in Question 6.8, calculate the value of the mass fraction in helium that you would expect from primordial nucleosynthesis.

The answer to Question 6.9 shows a remarkable result: the hot big bang model predicts that the mass fraction of helium-4 should have a value of about 28%. More refined calculations obtain a value that is lower than this – about 24%. This agrees very well with the observation that the mass fraction of helium-4 in low-metallicity interstellar gas seems to be about 24–25%. Until the development of the hot big bang model, the only other mechanism that was a plausible explanation for the production of helium was stellar nucleosynthesis. While it was known that this process does produce helium, it was a mystery how helium could have an almost identical abundance in every location that astronomers measured. The standard hot big bang model provides a more natural explanation for the abundance of helium-4. Such is the success of this outcome of the model, that it is generally interpreted as a key piece of evidence to support the big bang model.

6.4.4 Abundances of light elements as a cosmological probe

In the previous section we saw how the big bang model predicts the formation of helium-4 with an abundance close to that observed in the Universe. We have also seen that primordial nucleosynthesis forms other nuclides apart from helium-4.

■ Apart from helium-4, what other stable nuclides are formed by primordial nucleosynthesis?

☐ Deuterium, helium-3 and lithium-7 are the stable nuclides formed by primordial nucleosynthesis. (Note that tritium is unstable with a half-life of about 12 years.)

The abundances are sensitive to the density of matter that is in the form of protons and neutrons, i.e. the density of the baryonic matter in the early Universe. The density of baryonic matter at any time is usually expressed in terms of a baryonic density parameter $\Omega_b(t)$ which is defined as follows

$$\Omega_b(t) = \frac{\text{density of baryonic matter at } t}{\text{critical density at } t} \qquad (6.34)$$

where the critical density at any time $\rho_{crit}(t)$ is given by Equation 5.26

$$\rho_{crit}(t) = \frac{3H^2(t)}{8\pi G} \qquad (5.26)$$

The density of baryons in the early Universe can be related to the present-day density of baryonic material, which is expressed using the current value of the baryonic density parameter $\Omega_b(t_0)$. We shall refer to this value as $\Omega_{b,0}$.

The dependence of the abundances of the light elements on the value of $\Omega_{b,0}$ can be determined by detailed calculations: results of such calculations are shown in Figure 6.13. (The abundances also depend on the value of H_0: the calculation shown assumes a value of $H_0 = 72$ km s^{-1} Mpc^{-1}.) The first notable aspect of Figure 6.13 is that the abundance of helium-4 does not vary dramatically with baryon density. Thus, the helium abundance is not very sensitive to the baryon density.

In contrast to helium-4, the abundances of deuterium and helium-3 *are* sensitive to the baryon density. The mass fraction in both of these nuclides decreases substantially as the baryon density increases. The behaviour of the abundance of lithium is somewhat more complex – the curve shows a dip to a minimum value that occurs at a value of $\Omega_{b,0} \approx 0.02$.

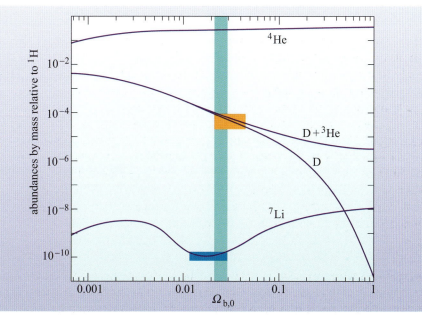

Figure 6.13 The abundances of the light elements as a function of the current value of the baryonic density parameter $\Omega_{b,0}$. Note that D stands for deuterium ($^2_1$H). The smooth curves show the prediction of nucleosynthesis calculations. The orange and blue boxes indicate the ranges of observed abundances. The vertical strip shows the values of $\Omega_{b,0}$ that are consistent with all observations. (Adapted from Walker *et al.*, 1991)

If the abundances of these light elements could be measured in material that has not undergone any change in composition since the time of primordial nucleosynthesis, then we could, in principle, measure the current baryonic density parameter of the Universe. A fundamental problem in this approach is highlighted by the following question.

QUESTION 6.10

Use Figure 6.13 to make an order of magnitude estimate of the maximum metallicity of material that has only undergone primordial nucleosynthesis, and has not been chemically enriched by stars. Is this value of metallicity consistent with the idea that the oldest observed stars were formed from primordial material?

The answer to Question 6.10 shows that even the oldest stars have undergone some chemical enrichment in a previous generation of stars; no stellar material appears to be left over from the big bang that has not undergone some subsequent nuclear processing by stars. Despite this complication, it is possible to make progress in this field by trying to work out how the abundances of deuterium, helium-4 and lithium may have been changed by nuclear processes that might have occurred since the end of primordial nucleosynthesis. When corrections of this sort are made, the result is a range of plausible values for the primordial abundances of deuterium, helium-4 and lithium. These ranges are shown as rectangular regions on Figure 6.13, and the overlap of these regions gives an allowed range of $\Omega_{b,0}$. The result is that $\Omega_{b,0}$ lies in the range 0.02 to 0.03. This immediately indicates that if the Universe has a flat geometry (implying $\Omega = 1$), then at most, 3% of the contribution to the density of the Universe can arise from baryonic matter. Given the difficulties in analysing primordial abundances, these figures should be treated with some caution, but it seems certain that baryonic matter can contribute no more than 5% to the total density of the Universe. As you will see in the next chapter, evidence is growing that $\Omega = 1$, which has the immediate implication that at least 95% of the density of the Universe is in a form that is different to the matter from which stars are made.

QUESTION 6.11

Suppose that the primordial abundances of light elements were measured in three 'hypothetical universes' as shown in Table 6.4. What would be the current value of the density parameter $\Omega_{b,0}$ of baryonic matter in each case? Could you determine the value of $\Omega_{b,0}$ from the lithium abundances alone?

Table 6.4 Some abundances in hypothetical universes – for use with Question 6.11.

Hypothetical Universe	Lithium abundance by mass relative to hydrogen (^1_1H)	Deuterium abundance by mass relative to hydrogen (^1_1H)
A	1×10^{-9}	7×10^{-4}
B	1×10^{-9}	1×10^{-5}
C	$< 1 \times 10^{-9}$	1×10^{-4}

6.5 Recombination and the last scattering of photons

$t \approx 3$ to 4×10^5 years, $T \approx 4500$ to 3000 K, $E \sim$ a few eV

After the nucleosynthesis of light nuclei that occurred in the first few hundred seconds after $t = 0$, the Universe expanded and cooled for several hundred thousand years before it underwent another dramatic change. This next big event is called *recombination* and occurred when nuclei and electrons combined to form neutral atoms. In this section we will examine this epoch of the Universe and see how processes that took place at that time account for the cosmic microwave background that is detectable today.

6.5.1 The Universe in the post-nucleosynthesis era

As the temperature of the Universe dropped below 10^8 K, the nuclear reactions that resulted in the formation of light elements came to a halt. The composition of the Universe at this time was: protons; nuclei of deuterium, helium and lithium; electrons; neutrinos; photons; and dark-matter particles (whatever they are!). The important interactions were those that shared energy out between the different constituents. In particular, photons and electrons interacted with one another to exchange energy. The electrons also collided with protons and nuclei and this also led to a sharing out of energy. These two types of interaction ensured that the radiation, electrons and nuclei remained in a state of thermal equilibrium.

■ Why were the neutrinos not in thermal equilibrium with the other particles at this stage in the evolution of the Universe?

□ Because the rate of neutrino interactions was very small. Most neutrinos did not interact with any particle after the time of neutrino decoupling.

As the Universe expanded, the temperature of the background radiation dropped with time. Because the photons greatly outnumbered the electrons and nuclei, the temperature of the electrons and nuclei was kept in step with that of the cooling radiation.

It has already been noted that at early times the expansion of the Universe was dominated by the energy density of radiation. However as you saw in Figure 6.6, the energy density of matter became equal to that of radiation when the scale factor attained about 10^{-4} of its current value, and this would have occurred when the age of the Universe was a few times 10^4 years. Following that event, the rate of expansion would have been dominated by the energy density of matter, and the effect of radiation would have progressively declined. Despite this change in the expansion rate, the temperature would still have been that of the background radiation due to the incessant interactions between photons, electrons and nuclei. This state of affairs would have persisted until the interactions between cosmic background photons and free electrons became negligible.

6.5.2 The era of recombination

An important type of interaction between electrons and nuclei is that which results in the formation of a neutral atom. As an electron becomes bound into an atom, energy is released in the form of a photon. This process is called **recombination**.

As the Universe expanded and cooled, conditions became favourable for recombination to occur. (In this context, the term 'recombination' may seem somewhat of a misnomer as electrons and nuclei had never been 'combined' as neutral atoms before this time!)

The process of recombination is the opposite of ionization. The reaction can be written as

$$p + e^- \xrightleftharpoons[\text{ionization}]{\text{recombination}} (^1_1 H)_{\text{neutral}} + \gamma \tag{6.35}$$

In the case of neutral hydrogen, the energy required to ionize the atom from its lowest energy (ground) state is 13.6 eV. Thus, if conditions were such that for every atom, there were many photons with an energy of at least 13.6 eV, then the neutral atom would not have survived for long, but would have soon been re-ionized. Alternatively, the atom may have undergone a collision with an electron or proton that could also have supplied the 13.6 eV of energy required to ionize the atom. Either way, at high temperatures, the reaction described in Equation 6.35 favours the production of protons and electrons, and the number density of neutral atoms is exceedingly low.

As the temperature of the Universe fell, the equilibrium of Equation 6.35 shifted to favour the production of neutral atoms and photons. The temperature at which this change occurred was subject to very similar constraints as those you met earlier in connection with the photodisintegration of deuterium. The number of photons exceeds the number of hydrogen atoms by a factor of about 10^9. Thus if only one in 10^9 photons had an energy of 13.6 eV, then this will be sufficient to ionize any neutral hydrogen.

QUESTION 6.12

By analogy with the case of deuterium photodisintegration, estimate the temperature at which recombination of hydrogen occurs.

In practice, Equation 6.35 is not perfectly symmetrical, because recombination sometimes occurs in multiple stages. (A reminder of hydrogen energy levels is shown in Figure 6.14.). Sometimes the recombination process would involve the electron first dropping down to an excited state, and then dropping down to the ground state. This would involve the emission of two photons, rather than one. Neither of these two photons would have sufficient energy on their own to ionize hydrogen. While a one-step recombination creates a photon capable of re-ionizing a hydrogen atom, each two-step (or more) recombination reduces the number of photons capable of ionizing, accelerating the recombination process.

Recombination started to occur when the temperature was about 4500 K. This temperature was reached when the Universe was about 3×10^5 years old. As neutral atoms formed, the density of free electrons (i.e. electrons that are not bound up in atoms) decreased. This had an important effect on the interrelationship between photons in the background radiation and matter in the Universe. Scattering interactions between free electrons and photons became infrequent.

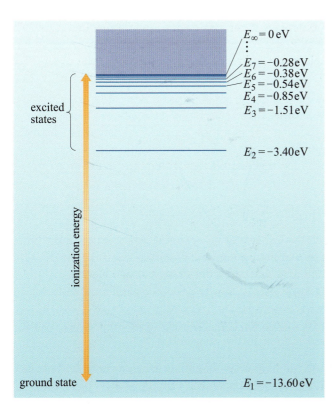

$E_\infty = 0\,\text{eV}$
$\vdots$
$E_7 = -0.28\,\text{eV}$
$E_6 = -0.38\,\text{eV}$
$E_5 = -0.54\,\text{eV}$
$E_4 = -0.85\,\text{eV}$
$E_3 = -1.51\,\text{eV}$

$E_2 = -3.40\,\text{eV}$

excited states

ionization energy

ground state $E_1 = -13.60\,\text{eV}$

Figure 6.14 The energy level diagram of the hydrogen atom. Electrons can only orbit the central nucleus with these energy levels, as a result of the laws of quantum mechanics. Atoms can absorb photons, and the photons' energy can then be put to use to move electrons to higher energy levels. Alternatively, electrons can sometimes drop down levels, and release energy as photons. There are infinitely many energy levels but those of highest energy are too closely crowded together to be shown separately on a diagram of this kind. Note that the energies are given in terms of electronvolts (1 eV = 1.602×10^{-19} J). A total of 13.60 eV of energy is needed to completely remove a ground-state electron from the atom. The process of removing electrons from atom is called ionization, while the reverse process of electron capture is called recombination.

By the time the temperature had dropped to 3000 K, the number density of free electrons was so low that the Universe essentially became transparent and photons could travel unhindered from this time on. As we shall shortly see, the radiation that we observe now as the cosmic microwave background was last scattered at the time of recombination.

The lack of scattering interactions had important thermal and dynamical effects. As we saw above, prior to recombination, the energy of photons was 'shared' with the thermal energy of matter in the Universe, and this ensured that matter had the same temperature as the background radiation. After recombination, the temperature of the matter and the background radiation evolved independently of one another. The temperature of the background radiation changed with scale factor according to Equation 6.6 – indeed it was from this relationship we were able to infer that the cosmic microwave background that is observed at present implies a hot early Universe.

The important dynamical role of the coupling between photons and electrons relates to the stability of over-dense regions against gravitational collapse. Prior to recombination, the radiation pressure played an important role in opposing the gravitational collapse of over-dense regions. After recombination, the rate of interactions between photons and electrons dropped dramatically, and this suddenly allowed over-dense regions that had been expanding with the Hubble flow to start to begin to contract under gravity. We will explore this aspect of the post-recombination Universe in more detail in Section 6.6, but first we will consider the evolution of the background radiation to form the cosmic microwave background.

6.5.3 The formation of the cosmic microwave background

The major change that occurred to the photons as they travelled after their last scattering was that they were red-shifted by the expansion of the Universe. As has already been mentioned, cosmological expansion causes the wavelengths of photons to increase, and if those photons are distributed according to a black-body spectrum, then this form of the spectrum is retained even though its characteristic temperature is reduced. We have now reached a detailed explanation for the formation of the cosmic microwave background: it is the cooled 'gas' of red-shifted photons that were in thermal equilibrium with matter at the time of last scattering.

> **QUESTION 6.13**
>
> Calculate the redshift at which the last scattering occurred. Assume that at the time of last scattering the temperature of the background radiation was 3.0×10^3 K. (*Hint*: Start by using Equation 6.6 to determine the change in scale factor.)

As the answer to Question 6.13 shows, the last scattering of photons in the cosmic background radiation occurred at a redshift of about 1100.

One way of thinking about how such photons should appear to us now is illustrated schematically in Figure 6.15: the diagram shows our vantage point – 'here and now' – at the centre, and the radial direction from this point represents a direction on the sky. The radial distance from our viewpoint on this diagram represents time, such that the present is at the centre whilst the outermost circle of the diagram is at $t = 0$, i.e. at the first instant of cosmic expansion. Whatever direction we look in we see the radiation from the time that the Universe became transparent. This transition, from being opaque to being transparent, occurred when the Universe was about 3×10^5 years old. At times before this, the Universe was opaque due to the scattering of photons by electrons.

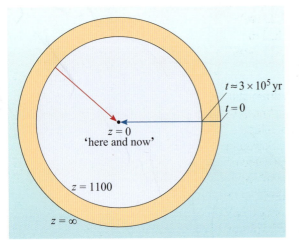

Figure 6.15 A schematic figure showing the origin of photons in the cosmic microwave background as observed from the Earth. In this diagram the radial direction from the Earth represents a direction on the sky and the radial distance from the Earth represents time.

Thus the cosmic background radiation appears to come from a spherical surface called the **last-scattering surface**, that is centred on us. Of course, this does *not* mean that we are at the centre of the Universe! An analogy may help here: if you go for a walk on a foggy day when the visibility is 30 m, then you will be able to see up to 30 m away in any direction.

Your view of the 'foggy world' is that you are at its centre, and that it has a radius of 30 m. Of course, this is just your view of the world; any other observer in the fog will have a similar view even though they might be somewhere else, see Figure 6.16. Thus the boundary defined by how far any single observer can see has no global significance. Likewise the last-scattering surface simply represents as far as we can see from our location in the Universe. The situation is somewhat different from viewing the world through a uniform fog – since in viewing the CMB we are essentially looking back in time to an era when the Universe was, in a sense, 'foggy'. However, the principle is the same: observers at another location in the Universe would also see a last-scattering surface, but it will not correspond physically to the last-scattering surface that we observe.

Figure 6.16 Observers walking on a foggy day can see a fixed distance in any direction. Despite the naïve claims of the observers, the boundary defined by this distance has no global significance!

6.5.4 Observing the cosmic microwave background

Now that we have developed an explanation for the cosmic microwave background in terms of red-shifted radiation that was last scattered by free electrons at a redshift of $z \approx 1100$, it is appropriate to consider its observed properties in more detail and how such observations are made.

The properties of the cosmic microwave background radiation that have already been described can be summarized as follows:

- The CMB has a black-body spectrum, with a characteristic temperature of 2.725 ± 0.002 K.

- The CMB is highly uniform in every direction that we observe (Section 5.2.3).

We shall shortly see that the CMB is not perfectly uniform, and that this is an important source of information about the Universe. However we shall begin by considering how measurements of the spectrum of the CMB are made.

The cosmic microwave background was discovered in 1965 by Arno Penzias and Robert Wilson (Box 6.3). Penzias and Wilson could not measure the spectrum of this signal since their receiver was tuned to work at a single wavelength (of 7.35 cm), so initially there were no data to support the idea that the background had the spectrum of a black body. However, the surface brightness of the radiation was characteristic of emission from a black body at a temperature of 3 K.

As we have already seen, the peak of emission from a black-body spectrum at a temperature of 3 K occurs at wavelengths of about 1 mm. The Earth's atmosphere partially absorbs radiation in this part of the electromagnetic spectrum, which makes it difficult to measure the spectrum of the cosmic microwave background using a ground-based antenna. The obvious solution to this is to place a detector above the Earth's atmosphere. During the 1970s and 1980s measurements were made using balloon-borne instruments that went some way to confirming that the cosmic microwave background has a black-body spectrum. But a major advance came from a satellite-based experiment called the Cosmic Background Explorer or COBE. This mission was launched in 1989, and was designed to measure various characteristics of the cosmic microwave background. One striking result from COBE was the measurement of the spectrum of the microwave background radiation as shown earlier in Figure 6.2. This spectrum follows the theoretical black-body curve to a very high degree – in fact, it is the best example of a black-body spectrum that has ever been observed in nature.

BOX 6.3 THE DISCOVERY OF THE COSMIC MICROWAVE BACKGROUND

In 1964, two researchers at Bell Laboratories in Holmdel, New Jersey, Arno Penzias and Robert Wilson (Figure 6.17), were charged with the task of calibrating a radio antenna to be used for telecommunication and galactic radio astronomy. The antenna, which is shown in the background of Figure 6.17, was in the form of a horn: radio waves entered through the aperture and were directed onto a detector at the narrow end of the horn. A feature of this type of design of radio antenna is that the signal that is received at the detector should be relatively free from any contamination that arises from sources that are outside the field of view (this is a problem that most other designs of antenna, such as a dish, suffer from). Soon after starting, Penzias and Wilson measured the signal from an airborne radio emitter, and found that there was an unexplained source of noise within the system. For the next year or so, they tried to track down the source of this noise; eliminating the possibility that it may be due to faulty electronics, rivets in the antenna – or even pigeon droppings in the horn. One characteristic of this noise was that it seemed to be the same wherever on the sky the antenna was pointed. Another was that the intensity of the signal was as would be expected if the whole sky were a black-body source at a temperature of 3 K.

In seeking an explanation for this signal, Penzias and Wilson were put in touch with Robert Dicke, a professor at Princeton University (which is only a few miles away from Holmdel). Dicke had realized several years earlier that the early stages of the big bang would have been characterized by very high temperatures, and that the radiation from this phase of the history of the Universe should be detectable today with a black-body spectrum and a temperature of a few kelvin. He had also realized that with the then current state of radio technology it should have been possible to detect such a signal. His research group were in the process of building a dedicated detector when Penzias and Wilson contacted him. Although Dicke and his team were beaten to the discovery of the signal, they were in a prime position

Figure 6.17 Arno Penzias (right) and Robert Wilson (left) in front of the horn antenna with which they discovered the cosmic microwave background. Penzias and Wilson were awarded the 1978 Nobel Prize in Physics for their discovery of the cosmic microwave background. (Bell Laboratories, AT&T)

to offer an interpretation of the result. The discovery was published as two papers in 1965; one by Penzias and Wilson that described only the observational result, and another by Dicke and his co-workers that offered an interpretation of the result as the signal being a relic of the hot big bang.

A further twist in the tale arose after the results were first published. Dicke received a terse letter from George Gamow, a Russian-born American physicist. In 1948, Gamow, with his collaborators, Ralph Alpher and Robert Herman had predicted that the temperature during the early stages of the big bang were sufficiently high that nuclear reactions would have taken place (Section 6.4). Furthermore, he suggested that a remnant of this high temperature phase should exist as background radiation with a temperature of about 5 K. Having made a prediction about the background radiation seventeen years earlier, Gamow was justifiably annoyed that his work had been overlooked. However, in the years since the discovery of the cosmic background radiation, Gamow's contribution to the implications of the hot big bang model have become widely recognized.

The second major aspect of the cosmic microwave background that has been subject to intense observational scrutiny is its uniformity. The largest variations in the microwave background as we look from one position to another correspond to a variation in temperature of less than one part in 10^3. It is difficult to make absolute measurements of the intensity of the CMB, and so experiments that measure variations in intensity do so by comparing signals from two different regions of the sky. Typical observing strategies involve making a large number of these comparison measurements across the area of sky that is of interest (which may be the entire celestial sphere). In this way, instrumental uncertainties are reduced, and levels of *relative* variation in the CMB can be measured to better than one part in 10^5. We will return to this topic in Chapter 7, but now we will take a first look at the implications of the measured uniformity of the cosmic microwave background.

6.5.5 Interpreting the cosmic microwave background

Much of this chapter has been based on interpreting one aspect of the cosmic microwave background – its spectrum. Thus we have studied the significance of the black-body form of the cosmic background radiation, and explored the consequences of the presence of such radiation in an expanding Universe. However there is a lot more to be learnt from the cosmic microwave background than can be deduced from its spectrum alone. In particular, the uniformity of the cosmic microwave background, and departures from this uniformity, also need to be explained in terms of cosmological processes. It is to this aspect of the cosmic microwave background that we now turn.

The uniformity of the cosmic microwave background

The view of the formation of the cosmic microwave background as illustrated by Figure 6.15 seems reasonable but gives rise to a puzzle. The temperature of the cosmic microwave background is, to a very high degree, uniform in whichever direction we look. The uniformity in the microwave background implies that at the time of recombination, the temperature at any point on the last-scattering surface was also highly uniform. The puzzling aspect of this is that the last-scattering surface is too large to have settled down to a uniform temperature in the 3 to 4×10^5 years that the Universe had been expanding.

A limit to the size of a region that can come to thermal equilibrium can be found by a similar argument to that used in Chapter 3 to constrain the size of AGN. The underlying principle is that a physical signal cannot propagate at a speed that is greater than the speed of light. In this case the 'physical signal' is the heat flow caused by a difference in temperature between two locations. If we consider one location in the Universe at the instant when the expansion of the Universe began, then at subsequent times, the most distant point that could possibly be in thermal equilibrium with our starting point would be at the distance that light could travel in the age of the Universe. This distance, which is the limiting size of a region that can be expected to be in thermal equilibrium at a given time, is called the **horizon distance**. This is illustrated schematically in Figure 6.18 – the point X represents an initial point and X' is the same point at recombination (i.e. X and X' have the same comoving coordinates). The maximum distance that could be covered by any signal at a given time prior to recombination is shown by the line from X to A, and the distance from X' to A is the horizon distance at the time of recombination. If the horizon distance is calculated for the last-scattering surface, it turns out that it

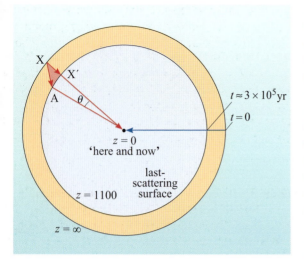

Figure 6.18 The horizon distance at the time of recombination as it appears on the last-scattering surface. Point X represents a location in the Universe just after $t = 0$. X′ is the location with the same comoving coordinates as X at the time of recombination. The distance from X′ to A is the maximum distance that a signal could travel prior to recombination. The angular separation of points A and X′ is about 2°. (Note that this diagram is not to scale.)

corresponds to an angle θ on the sky of about 2°. Regions of the last-scattering surface that are separated by more than this angle could not have affected each other. The fact that the microwave background has almost the same temperature over scales that are much greater than the horizon scale of 2° is a problem that the standard big bang model fails to address. The theory of inflation offers a potential solution: regions further apart than 2° on the CMB were originally *much* closer together prior to inflation, so they could have been in communication with each other before inflation. The uniformity of the CMB is then a consequence of the idea that our entire observable Universe was originally a much smaller region that suddenly expanded enormously during inflation (Figure 6.9).

Although the cosmic microwave background is remarkably smooth there are departures from perfect uniformity. These variations – or **anisotropies** – are at a very small level. As discussed in the following two subsections, there are two distinctly different mechanisms that give rise to anisotropies in the microwave background.

The dipole anisotropy

Mapping of the cosmic microwave background reveals that in one direction in the sky the temperature is 3.36 mK higher than the mean temperature, whereas in the opposite direction it is 3.36 mK lower than the mean. Between these two directions there is a smooth variation between the maximum and minimum temperatures. This is illustrated by the all-sky map shown in Figure 6.19 – the blue and red areas are below and above the average temperature respectively. Because the variation is symmetric about two opposite directions, or poles, on the celestial sphere, it is referred to as the **dipole anisotropy**.

The interpretation of the dipole anisotropy is that it arises from our motion relative to the comoving reference frame. Figure 6.20 indicates the effect of observer motion on the measured wavelength of cosmic microwave background photons.

Figure 6.19 A map from the NASA Wilkinson Microwave Anisotropy Probe (WMAP) showing the dipole anisotropy. The elliptical map area covers the entire sky. The blue and red coloured regions represent temperatures that are lower and higher, respectively, than the mean. The features across the centre are microwave emission from the Milky Way. (NASA/WMAP Science Team)

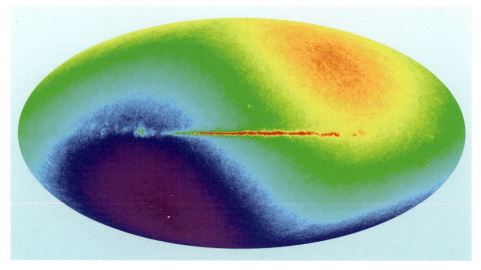

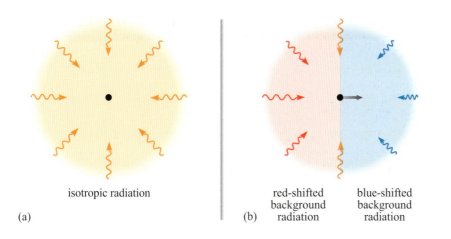

(a) isotropic radiation

(b) red-shifted background radiation   blue-shifted background radiation

Figure 6.20 The view of the cosmic microwave background according to (a) an observer who is stationary with respect to the frame of reference of comoving coordinates, and (b) an observer who is moving with respect to this frame of reference. According to the observer in case (a) the background is isotropic, whereas the observer in case (b) measures a blueshift in the direction of motion and redshift in the opposite direction.

Figure 6.20 concerns two observers at the same location in space, but who are moving with respect to one another. Observer (a) sees the cosmic microwave background as being perfectly uniform in all directions, whereas observer (b) sees the same last-scattering surface, but the radiation is Doppler shifted. The perfectly uniform case arises for an observer who is stationary in comoving coordinates, i.e. one who is perfectly following the Hubble flow of the expanding Universe. Any motion relative to this frame of reference, such as that of observer (b), would cause the observer to see the cosmic microwave background blue-shifted in the direction of motion and red-shifted in the opposite direction. When radiation with a black-body spectrum is subject to a Doppler shift, the spectrum retains its black-body form, but the peak wavelength is shifted and the characteristic temperature changes accordingly. Redshifts and blueshifts result in lower and higher temperatures respectively. Thus the direction on the sky in which the temperature increases the most corresponds to the direction in which we are moving with respect to the Hubble flow.

■ Why is it unlikely that the Earth would be stationary with respect to the frame of reference of comoving coordinates? List possible contributions to the Earth's velocity with respect to this frame?

☐ The Earth is unlikely to be stationary in comoving coordinates because (i) the Earth is in motion around the Sun, (ii) the Sun is in motion around the centre of our Galaxy, and (iii) the Galaxy has a peculiar motion with respect to the Local Group, and (iv) the Local Group is likely to have a random motion with respect to the Hubble flow.

Analysis of the dipole anisotropy shows that the Sun has a speed of about 365 km s^{-1} with respect to the local comoving frame of reference. When the Sun's motion around the Galaxy, and the Galaxy's motion with respect to the Local Group are accounted for, it is found that the Local Group has a speed of about 630 km s^{-1} with respect to the local comoving frame of reference.

Intrinsic anisotropies in the cosmic microwave background

Even when the effect of the dipole anisotropy is removed the cosmic microwave background is still not perfectly smooth – it exhibits smaller scale variations in temperature from one point to another. Figure 6.21 shows a map of these fluctuations over the whole sky. The magnitude of such variations is small: at

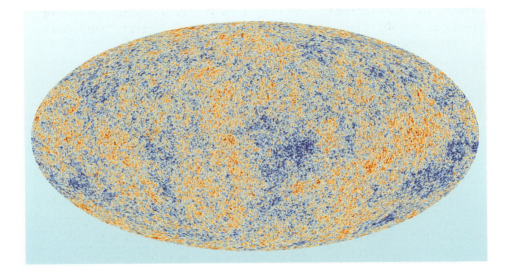

Figure 6.21 An all-sky map of the fluctuations in the cosmic microwave background as measured by the European Space Agency's Planck satellite. (ESA and the Planck Collaboration)

most they correspond to a fractional change in temperature of only a few parts in 10^5. The significance of these intrinsic anisotropies is that, although they mainly represent temperature variations in the Universe at the time of recombination, they are intimately linked to the small variations in *density* that are believed to have arisen at very early times in the history of the Universe. In particular, it is hypothesized that these density fluctuations may have their origin in the process of inflation that occurred at $t \sim 10^{-36}$ s. We shall consider the origin of these so-called *primordial fluctuations* in Chapters 7 and 8, but we note here that the analysis of intrinsic anisotropies in the cosmic microwave background can potentially provide information about processes in the very early history of the Universe.

Temperature variations in the cosmic microwave background are also useful because the photons that were last scattered at the time of recombination have been moving freely, apart from the effects of gravity, ever since. As a result, the anisotropy pattern can also provide information about the geometry of space–time. For this reason, the analysis of anisotropies in the cosmic microwave background is a key area of observational cosmology, and will be discussed more fully in Chapter 7. However, here we continue our discussion of the evolution of the Universe in the post-recombination era by considering how the small variations in density that were present at recombination helped to form structure in the Universe.

6.6 Gravitational clustering and the development of structure

The small amplitude of temperature anisotropies in the cosmic microwave background indicates that the distribution of baryonic matter in the Universe was very smooth at the time of recombination. This is in marked contrast to the present-day Universe in which the baryonic matter is concentrated into stars, galaxies, clusters and large-scale structure. An outline of how the Universe evolved from an almost, but not quite, uniform state into such structures was given in Chapters 2 and 4, and here we consider the formation of structure in the light of our account of the big bang model. Our starting point is to consider a simple, but rather unrealistic scenario in which all the matter in the Universe is assumed to be baryonic. We shall then see how the study of the growth of structure can provide some clues as to the nature of the non-baryonic matter that is actually believed to have been present.

6.6.1 Gravitational collapse in a baryonic matter Universe

You have already seen (Chapter 2) that structure is believed to form as a result of the gravitational collapse of a region that is initially denser than average. The early stages of this process are very gradual. Remember that the Universe is expanding and, initially, over-dense regions will simply expand at a slower rate than average. Furthermore, the density variations in the early Universe are at a low level. Since the density of the Universe is changing with time, it is useful to express the magnitude of density variation using the **relative density fluctuation**, which is defined as follows

$$\frac{\Delta\rho}{\rho} = \frac{\text{density within a fluctuation} - \text{mean density of the Universe}}{\text{mean density of the Universe}}$$

For clarity, we have used ρ rather ρ_m to denote the density of matter in this equation. We shall continue to use this simplified notation in the next chapter.

The level of the primordial fluctuations suggests that in the early Universe $\Delta\rho/\rho \sim 10^{-5}$.

Whether a particular density enhancement will grow by gravitational collapse depends on the balance between two effects. One effect is the self-gravity of the matter within the over-dense region that, of course, tends to cause collapse. The opposing effect is due to pressure that acts to stabilize over-dense regions against collapse. Which of these two effects is dominant under specified conditions depends on the mass of the region. The British scientist Sir James Jeans, who first analysed gravitational collapse in relation to the formation of stars, discovered that a key parameter is a quantity that is now known as the **Jeans mass**. The Jeans mass represents the boundary between two different types of behaviour. If the mass of an over-dense region exceeds the Jeans mass then it will collapse. However, a region that contains less than the Jeans mass would be supported by its internal pressure and hence be stable.

The horizon distance plays an important role in the discussion of stability against collapse. An over-dense region that is larger than the horizon distance cannot be supported by its internal pressure. This is because any changes in pressure are propagated at a speed that is lower than the speed of light. Nevertheless, the relative density fluctuation within this over-dense region *does* grow with time, albeit slowly, since the over-dense region is expanding at a lower rate than the Universe around it. However, the horizon distance increases with time, so eventually, the over-dense region will lie within the horizon distance and can respond to internal pressure changes. After this time, an over-dense region can be stable against collapse provided that the mass within this region is less than the Jeans mass.

The Jeans mass depends on the density and pressure of the region under consideration. In an expanding Universe, the density and pressure change as the scale factor changes. This causes the Jeans mass also to vary with time – this behaviour is shown in Figure 6.22. Note that the Jeans mass increases steadily with scale factor during the radiation-dominated phase, but it then flattens out and then drops at the time of recombination. To help in interpreting this diagram, it is useful also to indicate the criterion that the diameter of an over-dense region must be less than the horizon distance before it can become stable against collapse. To do this,

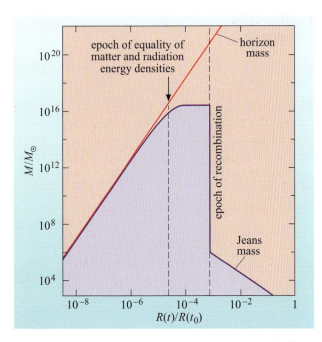

Figure 6.22 The variation of the Jeans mass (purple line) and the horizon mass (red line) as a function of the scale factor ($R(t)/R(t_0)$). Gravitational collapse only occurs for regions of the diagram that are shaded in pink. Regions that are shaded in lilac are stable against collapse. (Adapted from Longair, 1998)

Figure 6.22 also shows a quantity called the **horizon mass** that is the mass contained within a sphere with radius equal to the horizon distance. As expected, the horizon mass increases with time because the horizon distance increases with time (because light has had the time to travel further). The region above the horizon mass line represents over-dense regions that have a diameter greater than the horizon and, as mentioned above, these grow in a rather sedate fashion.

Figure 6.22 implies that, prior to the time when matter and radiation energy densities became equal, any over-dense region with a mass of up to about $10^{16} M_\odot$ will at some time be overtaken by the horizon – that is, the mass within the region will be overtaken by the horizon mass. However the horizon mass and the Jeans mass are very close to one another, so almost immediately, the mass of such an over-dense region will also be overtaken by the Jeans mass. Thus, the over-dense region will become stable against collapse, and the amplification of fluctuations as a result of gravitational contraction comes to a halt – at least temporarily.

An important change occurs at the time of recombination: the Jeans mass drops dramatically. Prior to recombination the major contribution to the pressure arose from the interaction of photons with free electrons. After recombination electrons become bound into neutral atoms that do not interact with the background photons. The radiation pressure becomes negligible and the only source of support against collapse comes from the thermal pressure of the gas. This thermal pressure is much smaller than the radiation pressure prior to recombination, so much smaller masses suddenly become unstable to gravitational collapse. The Jeans mass drops from a value of about $10^{16} M_\odot$ just before recombination to about $10^5 M_\odot$ just after. Thus, any fluctuations in the mass range from $10^5 M_\odot$ to $10^{16} M_\odot$ that had previously been stabilised suddenly find themselves, in a manner of speaking, without any visible means of support. Collapse then proceeds in earnest, and it is the smaller fluctuations, with masses close to $10^5 M_\odot$, that will most rapidly collapse to form virialized systems.

So far, this account of the formation of structure looks promising. Gravitational collapse is essentially arrested until the time of recombination, but soon after this the Jeans mass drops to a value of $10^5 M_\odot$, a mass that is typical for the oldest known stellar systems – the globular clusters. There is however a fundamental problem with this scenario that relates to the level of relative density fluctuation that would be required at recombination in order for this model to produce the structure that we observe in the present-day Universe. Detailed calculations show that this purely baryonic model would require fluctuations in density at recombination that would be detectable now as temperature fluctuations in the cosmic microwave background at the level of about 1 part in 10^3.

■ Why is this a problem for this model?

□ The observed amplitude of fluctuation in temperature in the cosmic microwave background is about 1 part in 10^5 – about a hundred times too small to give rise to the structures observed in the present-day Universe.

An alternative way of viewing the problem is that there has been insufficient time since recombination for fluctuations to grow from their observed value at that time (1 part in 10^5) to give the structure that we observe today.

We have seen throughout this book that dark matter plays an important dynamical role in many astrophysical systems, so it seems sensible to consider whether dark matter may help in resolving the problem of structure formation.

■ If the dark matter was baryonic in nature, could that help in resolving the problem of formation of structure?

☐ No. We have just seen that any baryonic matter – whether it ends up being luminous or dark – could not give rise to the structure that we see in the present-day Universe.

In the next section we will see how the formation of structure might be modified by the presence of non-baryonic dark matter.

6.6.2 Gravitational collapse with dark matter

In Chapter 2 we saw that numerical simulations of the formation of structure usually incorporate a dark-matter component. It was also noted that hot dark-matter scenarios typically cannot reproduce structures across the range of scales that are observed in the present-day Universe. Cold dark-matter models seem somewhat more successful, and here we will consider how cold dark matter could play a vital role in forming structure in the Universe.

In the context of cold dark matter, the term 'cold' refers to the fact that the particles have been moving at speeds much slower than the speed of light since very early times in the history of the Universe. This in itself does not help in the formation of structure – since this is essentially the same as the behaviour of the normal baryonic matter in the Universe. As you saw in the previous section, the formation of structure requires a higher degree of density fluctuation at the time of recombination than is observed. This could happen if the cold dark-matter particles do not interact to any significant extent with photons or with the baryonic matter. This would mean that the cold dark-matter particles would not be supported by the radiation pressure, and so gravitational collapse of density fluctuations of this cold dark matter could start before the time of recombination.

This lack of interaction with photons or electrons is an important characteristic of cold dark-matter particles. The term **weakly interacting massive particle** (or WIMP) is often used to denote the hypothetical cold dark-matter particles. The term 'weakly' refers to the fact that such particles only interact by the weak interaction and do not take part in electromagnetic or strong interactions (they do however respond to gravitational fields). The issue of what WIMPs might actually be is considered in Chapter 8.

In cold dark-matter scenarios, primordial fluctuations can start to grow at much earlier times than the epoch of recombination. There is ample time for the density fluctuations in this matter to reach a value of one part in 10^3 at the time of recombination. This level of fluctuation is not evident on the last-scattering surface because there is no significant interaction between WIMPs and photons.

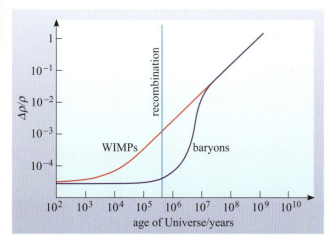
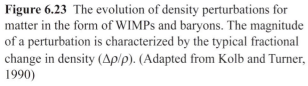

Furthermore the behaviour of the baryonic matter is dominated by radiation pressure up until recombination, and thus its distribution appears smooth to one part in 10^5, just as if the Universe contained only baryonic matter.

After recombination however, the dominant effect on baryonic matter is the gravitational attraction of regions in which over-densities of cold dark matter have accumulated. In a sense the baryonic matter 'falls into' the density enhancements of cold dark matter. This behaviour is summarized in Figure 6.23, which shows the evolution of density enhancements, characterized by the typical relative density fluctuation ($\Delta\rho/\rho$) of matter in the form of WIMPs and in baryons.

Figure 6.23 The evolution of density perturbations for matter in the form of WIMPs and baryons. The magnitude of a perturbation is characterized by the typical fractional change in density ($\Delta\rho/\rho$). (Adapted from Kolb and Turner, 1990)

Models of the formation of structure under the influence of cold dark matter have undergone scrutiny by conducting computer-based simulations and comparing the resulting structure to that which is measured in the real Universe. The comparison that is made between the outcome of a simulation and real observations usually involves some statistical measure of the structure. An added, and major, complication is that it is not clear how closely the visible mass in the Universe traces the distribution of dark matter.

Such studies are able to constrain some cosmological parameters relating to the distribution of cold dark matter. For instance, it is possible to rule out the cosmological model that has a mass density parameter $\Omega_m = 1$, no cosmological constant ($\Lambda = 0$), and in which the matter is predominantly in the form of cold dark matter – this is often referred to as the 'standard cold dark matter' (SCDM) model (a simulation of this case is shown in Figure 6.24a). However, other variants of cold dark-matter models do produce structure that is similar to that which occurs in the real Universe. For instance one cosmological model with cold dark matter, a non-zero cosmological constant and a flat geometry (the so-called ΛCDM model that was mentioned in Chapter 2) appears to be a good representation of reality (Figure 6.24b).

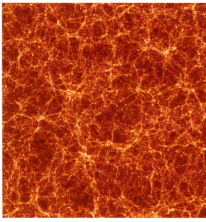
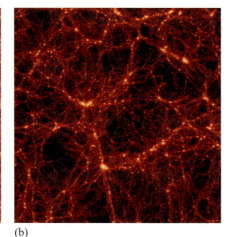

(a)

(b)

Figure 6.24 The results of numerical simulations of the formation of structure in two different dark-matter scenarios. Both of the models shown have cold dark matter as the dominant contribution to mass. The simulations show the distribution of matter at the present time.
(a) A model with no cosmological constant and $\Omega_m = 1$. (b) The ΛCDM model with a non-zero cosmological constant and with $\Omega_m + \Omega_\Lambda = 1$. The model in (a) is incompatible with the observed structure in the real Universe, whereas model (b) gives rise to structure that is similar to that which is observed. (Jenkins et al., 1998/ The Virgo Consortium)

So study of the formation of structure can provide useful information about the nature of dark matter. Measurements of the cosmic microwave background can also help constrain other cosmological parameters, and there are other observational techniques that also allow the cosmological parameters to be constrained. It is to this topic that we turn in the next chapter.

6.7 Summary of Chapter 6

The evolution of the cosmic background radiation

- The cosmic background radiation has a black-body spectrum and its current temperature is about 2.73 K.

- The temperature T of the cosmic background radiation varies with scale factor $R(t)$ according to

$$T \propto \frac{1}{R(t)} \tag{6.6}$$

- At times when the scale factor of the Universe was much smaller than at present, the temperature of the cosmic background radiation would have been much higher.

- At early times the dominant contribution to the energy density of the Universe was that due to radiation. At such times, the temperature is related to time by

$$(T/\mathrm{K}) \approx 1.5 \times 10^{10} \times (t/s)^{-1/2} \tag{6.19}$$

The very early Universe

- Current physical theory breaks down in describing events that took place at, or before, the Planck time ($t \sim 10^{-43}$ s).

- It is speculated that major physical effects could have arisen when grand unification ended ($t \sim 10^{-36}$ s). At this time, the strong and electroweak interactions became distinct. One such effect may be the process of inflation, which resulted in the scale factor increasing very rapidly for a short period of time.

- At early times, the content of the Universe would have been all types of quark and lepton and their antiparticles. There were also particles present that mediate the fundamental interactions (such as the photon), as well as dark-matter particles. There was a slight excess of matter over antimatter.

- At $t \sim 10^{-5}$ s free quarks became bound into hadrons. Most of these hadrons either decayed or annihilated with their antiparticles, leaving only protons and neutrons. For every 10^9 or so annihilation events that occurred, there would have been one proton or neutron left over.

- At $t \approx 0.7$ s neutrinos had their last significant interaction with other particles (apart from the effects of gravity). Shortly after this, electron–positron pairs annihilated, leaving only a residual number of electrons whose summed electric charges exactly balance the charge on the protons.

Primordial nucleosynthesis

- In the first few hundred seconds of the history of the Universe, the physical conditions were such that nuclear fusion reactions could occur. Such reactions led to the formation of deuterium, helium and lithium.

- The first step in the production of helium is the formation of deuterium. This nuclide is unstable to photodisintegration at temperatures above 10^9 K. The formation of helium did not start until $t \approx 225$ s. During this time, some neutrons decayed to protons, and this had an effect on the mass fraction of helium that was produced by primordial nucleosynthesis.

- The mass fraction of helium that is predicted by primordial nucleosynthesis is about 24%. This is in good agreement with measurements of the helium abundance in interstellar gas and stars, and provides very strong evidence to support the hot big bang model.

The scattering of photons and the cosmic microwave background

- The cosmic microwave background that is observed at the present time, is radiation that was last scattered at a redshift of about 1100 ($t \sim 3$ to 4×10^5 years). The radiation appears to originate from the last-scattering surface that is at this redshift.

- The scattering of background radiation photons stopped when the number density of free electrons became very low, and this occurred because of the recombination of electrons and nuclei to form neutral atoms.

- The observed high degree of uniformity of the cosmic microwave background leads to the horizon problem – which is that regions of the last-scattering surface that are more than about 2° apart could not have come into thermal equilibrium by the time that last scattering occurred.

- Our motion relative to comoving coordinates can be determined by analysis of the observed dipole anisotropy of the cosmic microwave background.

- The cosmic microwave background shows intrinsic anisotropies in temperature at a level of a few parts in 10^5. These anisotropies result from density variations in the early Universe.

The formation of structure

- The formation of structure in the Universe would have proceeded by gravitational collapse from density fluctuations in the early Universe.

- Prior to recombination, the high degree of scattering between photons and electrons prevented density fluctuations in baryonic matter from growing substantially.

- If all matter was baryonic in form, then the level of fluctuation that is observed on the last scattering surface is too small to explain the structure that we observe at the present time.

- The observed level of structure in the present-day Universe can be explained if density fluctuations in non-baryonic matter had begun to grow prior to recombination, and baryons were subsequently drawn into those collapsing clouds of dark matter.

Questions

QUESTION 6.14

Draw a 'time-line' for the history of the Universe that indicates the major events that occurred at different times from the Planck time to the present day. Include on this time-line an indication of the temperature at the times of these events.

QUESTION 6.15

Briefly summarize what is meant by a 'theory of everything'. Why is such a theory required to understand the processes that occurred in the very early Universe?

QUESTION 6.16

During the process of inflation, the scale factor would have increased by an enormous factor. What consequences would this have for the temperature at this time?

QUESTION 6.17

State what the qualitative effect would be on the mass fraction of helium-4 produced by primordial nucleosynthesis if the photodisintegration of deuterium required photons of much higher energy than 2.23 MeV.

QUESTION 6.18

Suppose we received a message from (hypothetical) astronomers in a galaxy that has a current redshift of $z = 2.5$. What would they say they found as the temperature of the cosmic microwave background at the time of their transmission?

CHAPTER 7
OBSERVATIONAL COSMOLOGY –
MEASURING THE UNIVERSE

7.1 Introduction

Observational cosmology is the branch of science concerned with measuring the parameters that characterize the Universe. Some of the most important of those parameters were introduced in Chapter 5, including the Hubble constant H_0, the current value of the deceleration parameter q_0, the age of the Universe t_0, and the current values of the density parameters for matter and for the cosmological constant (dark energy), $\Omega_{m,0}$ and $\Omega_{\Lambda,0}$, respectively. (The cosmological constant, Λ, can be given as an alternative to $\Omega_{\Lambda,0}$.) The study of the early Universe in Chapter 6, with its emphasis on nuclear and subnuclear processes, introduced several more parameters, including the current value of the density parameter for baryonic matter $\Omega_{b,0}$ (one of the contributions to $\Omega_{m,0}$). Other parameters may arise from any detailed proposal regarding the nature of dark matter. Yet more parameters can be associated with the cosmic microwave background radiation (CMB). Since the CMB has a black-body spectrum, one of the most important of these parameters is the present temperature of the CMB. This is currently one of the best determined of all the cosmological parameters: according to a widely quoted result, its value is $T_{cmb} = (2.725 \pm 0.002)$ K.

This chapter is mainly concerned with the determination of just a few of these parameter values, particularly H_0, q_0, $\Omega_{m,0}$, $\Omega_{\Lambda,0}$ and $\Omega_{b,0}$. The first four of these determine which of the various Friedmann–Robertson–Walker models (FRW models; introduced in Chapter 5) provides the best description of the Universe we actually inhabit. The fifth, the current value of the density parameter for baryonic matter, represents a particularly familiar contribution to the total matter density. The current 'best estimates' for these five parameters are discussed, along with the methods used to arrive at those values and the degree of uncertainty in each result. As you will see, there is now a confidence among observational cosmologists that they are finally closing in on the true values of these parameters. Ten years ago, many cosmological parameters were only known to within a factor of two, but now the term '**precision cosmology**' is used to express not only the accuracy and precision that has been achieved, but also the consistency between very different types of measurement.

One topic that arises at several different points in our discussion is the CMB. You have already seen that its absolute temperature is known to better than one part in 10^3, which is an extraordinary level of precision for any cosmological parameter. The precision of this result is an indication of what is now happening throughout observational cosmology. As is explained later, it is our ability to measure and understand the CMB that has made the largest single contribution to this new era. However, other methods are still important, and it is perhaps the sign of a mature scientific discipline that a diverse range of observational techniques all zoom in on similar values for the fundamental parameters. For this reason, the more traditional (non-CMB) methods are also discussed in this chapter.

Any account of observational cosmology is likely to become quickly outdated. Even as this chapter is being written, new observational data are being prepared from a space probe called Planck. Some of the Planck data are discussed in Section 7.4, and give just one indication of the need to be always alert for new findings and new developments in a subject as vibrant and active as observational cosmology. This chapter can provide nothing more than a snapshot of this fast-moving field.

7.2 Measuring the Hubble constant, H_0

The Hubble constant is in many ways the most fundamental of the cosmological parameters. Of them all, it is the one most easily related to the 'theoretical' parameters that characterize the FRW models, and the value of H_0 plays an important part in almost all determinations of the other constants. In this section we discuss two methods of determining H_0. A third method, based on the CMB, is discussed in Section 7.5. First, though, here are some questions to help you recall what you learned about the Hubble constant and the related Hubble parameter in Chapter 5.

■ Briefly describe the curvature parameter k, and the scale factor $R(t)$ that appear in the FRW models, and state which aspects of space–time they describe.

☐ The *curvature parameter k* may be -1, 0 or $+1$. It indicates the sign of the curvature of space, and helps to determine geometric properties such as the sum of the interior angles of a triangle, the relationship between the radius and the circumference of a circle, and whether neighbouring straight lines that are initially parallel will converge or diverge. The positive value, $k = +1$, indicates that space has a finite total volume; the values $k = 0$ and $k = -1$ indicate that space is infinite.

The scale factor $R(t)$ varies with time, t. It describes the uniform expansion or contraction of space. If the positions of galaxies are expressed in terms of comoving coordinates (i.e. ones that expand with the Universe) then the coordinate separation of two galaxies might have the fixed value r, but the physical distance between those galaxies (measured in metres, say) will be proportional to $R(t)$, and will increase as $R(t)$ grows and the Universe expands.

■ What is the precise relationship between the Hubble constant H_0 and the scale factor $R(t)$? You may find it useful to start by defining the Hubble parameter.

☐ The *Hubble parameter*, $H(t)$, is a fractional measure of the rate of change of the scale factor (see Section 5.4.1). In terms of symbols, this relationship may be written as

$$H(t) = \frac{\dot{R}(t)}{R(t)} \tag{7.1}$$

where $\dot{R}(t)$ represents the rate of change of R at time t. The *Hubble constant H_0* is simply the value of the time-dependent Hubble parameter $H(t)$ at the present time t_0; that is to say,

$$H_0 = H(t_0) = \frac{\dot{R}(t_0)}{R(t_0)} \tag{7.2}$$

■ Write down Hubble's law and suggest a way in which this can be used to determine the Hubble constant from observations.

☐ According to Hubble's law, the redshift (z) of galaxies increases in proportion to their distance (d) from us, provided the redshift is not too great (less than about 0.2, say). If the constant of proportionality is identified as H_0/c, then Hubble's law may be written

$$z = \frac{H_0}{c} d \qquad (7.3)$$

This suggests that if the redshifts of a sample of galaxies are measured, and if the distances of the same galaxies are determined independently of the redshifts, then those observational data can be used to determine H_0/c, and hence H_0, since c is well known. One way of doing this would be to plot a graph of z against d, determine the best straight line through the data points and then measure the gradient of that line, since this gradient should be H_0/c.

7.2.1 Determinations of H_0 using the Hubble diagram

The first determination of the Hubble constant was by Hubble himself, and was published in 1929. The determination appeared in the paper in which Hubble announced the law that is now named after him, and was based on measurements of the distances of 24 relatively nearby galaxies. The redshifts of these galaxies were already known thanks to the work of Vesto M. Slipher (1875–1969), who had systematically measured the redshifts of almost all the spiral galaxies that were known at the time. Hubble determined the distances of the 24 galaxies by using Cepheid variables and other distance indicators in the manner described in Chapter 2. For those galaxies where Cepheid variables could be seen, the distances were deduced as follows.

- The *period* of each Cepheid was measured by recording the rise and fall of its apparent magnitude and determining the time between successive maxima. Technically this amounts to measuring the *light curve* of the Cepheid – a plot of apparent magnitude against time over a complete period.

- A previously calibrated *period–luminosity relation* for Cepheids was then used to determine the luminosity or (equivalently) the *absolute magnitude* of each Cepheid.

- The absolute magnitude M determined in this way was then combined with a measurement of the apparent magnitude m to determine the *distance* of each of the Cepheids, and this was assumed to be the distance of the galaxy that contained the Cepheid.

Independently of these distance measurements, the formula $v = cz$ was used to assign a 'naïve' recession speed v to each of the 24 galaxies on the basis of its measured redshift z. We refer to v as the 'naïve' recession speed since the use of $v = cz$ was based on the assumption that the redshift z is the result of a simple Doppler shift, which is not true. The value of v differs from the real recession speed of the galaxy, and the discrepancy increases with the redshift. Nonetheless, the measurements made it possible for Hubble to represent each of the 24 galaxies by a single point on a plot of naïve recession speed against distance, as shown in Figure 7.1.

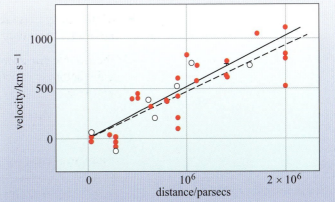

Figure 7.1 The original 'Hubble diagram', based on a figure that appeared in 1929. The 24 individual galaxies are represented by the red dots; the solid line represents the best straight-line fit to these data. The open circles and dashed line refer to an analysis in which these galaxies are grouped on the basis of proximity to one another. (Courtesy of the National Academy of Sciences)

The data points were somewhat scattered, but Hubble detected a linear trend, and on this basis claimed that the recession speed of galaxies increased in proportion to their distance from us. From the gradient of the straight line in Figure 7.1, Hubble deduced that the rate of increase of recession speed with distance was 500 km s^{-1} Mpc^{-1}. Using the notation and terminology of modern astronomers, the relation that Hubble proposed can be rewritten as $v = H_0 d$, where $v = cz$, and H_0 is Hubble's constant. From this we obtain the modern form of Hubble's law, $z = (H_0/c)d$, and we can see that 500 km s^{-1} Mpc^{-1} represents the first determination of the Hubble constant H_0. This value is far outside the range of values that are currently thought to be credible, but the measurement is historically important as the start of a long campaign to measure H_0.

There have now been hundreds of attempts to determine the Hubble constant. Like all observational results, these determinations are subject to various kinds of uncertainty. There are, for instance, the **random uncertainties** that beset any measurement, and which generally cause different determinations of a measured quantity to vary about some mean value. In addition there may be **systematic uncertainties**, which will always influence the measured value in the same way, no matter how many times the measurement is repeated. It is now known that there were substantial systematic uncertainties in the early determinations of the Hubble constant. Some of these were the result of studying only nearby galaxies where the motion was dominated by 'local' effects, such as the attraction of the local supercluster (see Chapter 4), rather than the large-scale Hubble flow. Others arose from the mistaken belief that two different kinds of variable star were actually the same kind of Cepheid. However, even after these sources of error were removed, different determinations of the Hubble constant still tended to disagree by substantial amounts. This is indicated in Figure 7.2, where the results of various determinations of H_0 are plotted against the year in which they were made. Note that uncertainty ranges, represented by vertical bars through the data points, have been used to show the range of values that are consistent with the measured results,

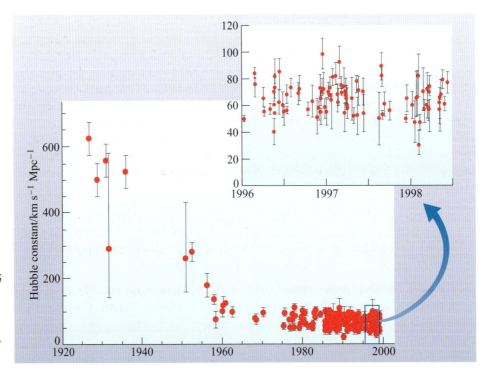

Figure 7.2 The evolution of the measured value of the Hubble constant H_0. Over 300 measurements performed since 1975 have yielded values between 50 and 100 km s^{-1} Mpc^{-1}. (Adapted from *Sky and Telescope*, based on work by J. Huchra and the HST Key Project on the Distance Scale)

rather than pretending that each measurement results in a unique value. Estimating the appropriate size for the uncertainty range in an astronomical measurement is often highly complicated and very time consuming, but it is vital if other scientists are to properly appreciate the significance of the results.

As the figure shows, by the 1980s a large number of determinations indicated that the Hubble constant was probably somewhere in the range between 50 and 100 km s^{-1} Mpc^{-1}, and for a while there were strong proponents of either end of that range. Perhaps the most notable were Gerard de Vaucouleurs who favoured a high value of the Hubble constant, and Allan Sandage who favoured a low value. The disagreement between these eminent astronomers was not caused by the inherent difficulty of making the observations, but rather by the different procedures they adopted when interpreting their observations and correcting them for various observational effects.

In an effort to reduce the uncertainty, one of the designated '*key projects*' of the Hubble Space Telescope (HST) in the 1990s was a programme to determine the Hubble constant to an accuracy of about 10%. The efforts of a team of astronomers, led by Canadian-born US astronomer Wendy Freedman, resulted in the value $H_0 = 72 \pm 8$ km s^{-1} Mpc^{-1}, which has now been broadly supported by more recent CMB-based measurements, of which more later.

The basis of the HST Key Project to measure H_0 was to use the Cepheid period–luminosity relationship to calibrate other methods of distance measurement. The period–luminosity relation that was used was based on studies of Cepheids in our neighbouring galaxy, the Large Magellanic Cloud (LMC), which is sufficiently close for large numbers of Cepheids to be observed. The calibration of the Cepheid period–luminosity method required the distance to the LMC to be known, and it was assumed, on the basis of measurements made by other researchers, that the LMC is at a distance of 50 kpc from the Sun.

The first stage of observations was to search for Cepheids in selected galaxies that lie between about 3 and 25 Mpc from the Milky Way. At these distances the 'local' motion of galaxies still has a significant effect on their measured redshifts, so the selected galaxies could not themselves be used to determine the Hubble constant. However, once the Cepheids had been found and measured, the distances of the selected galaxies could be precisely determined. Once this had been done, a variety of measurements made in those selected galaxies could be used to calibrate other methods of distance measurement that could be applied to more remote galaxies, which were more likely to represent the Hubble flow. The galaxies that were searched for Cepheids were carefully chosen so as to allow the calibration of not one, but five other methods of distance measurement, three of which were discussed in Chapter 2.

- On the basis of Chapter 2 and Figure 7.3, name three methods of galactic distance determination (excluding the use of Hubble's law) that are appropriate for measuring distances to about 100 Mpc. Briefly explain the nature of the 'calibration problem' that these methods must confront.

☐ The three methods appropriate for measuring distances to about 100 Mpc are: the Type Ia supernova method, and the methods based on the Tully–Fisher and fundamental plane relations.

The calibration problem consists of making measurements in relatively nearby galaxies at known distances that enable the 'relative' distances indicated by standard candles in more remote galaxies to be converted into 'absolute' distances that may be expressed in units such as megaparsecs.

In addition to the three methods mentioned above, the HST Key Project also used methods based on Type II supernovae and surface brightness fluctuations in galaxies. These five methods used to determine H_0, along with their results, are shown in Table 7.1. The numbers of measurements used to calibrate the five methods are shown in the second column. The fact that some techniques were calibrated using a small number of galaxies highlights the difficulty of the calibration problem: prior to the HST measurements, some of these methods had never before been calibrated against the Cepheid period–luminosity method.

Having calibrated the five methods, the next stage was to use those methods to determine the distances of galaxies that are sufficiently far away that their measured redshifts are dominated by the expansion of the Universe rather than by any 'local' effects. The typical uncertainty in redshift caused by local effects is roughly 10^{-3}, so in order to bring this uncertainty to below 10% of the measured redshift, that measured redshift must be at least $z \approx 10^{-2}$. The actual ranges in redshift that were sampled using the five different methods of distance determination are shown in Table 7.1, along with the number of measurements that were used to obtain a value of H_0.

The correlation between redshift and distance for a number of galaxies and clusters, based on the HST Key Project results, is shown in Figure 7.3. One of the remarkable outcomes, evident from the fifth column of Table 7.1, is that the five different methods, which are based on quite different physical principles, agree well with one another. The values obtained for H_0 vary from 70 to 82 km s^{-1} Mpc^{-1}. This level of agreement provides some reassurance that there are no gross systematic differences between the five methods that were adopted, and leads, after further analysis, to the result quoted earlier: $H_0 = 72 \pm 8$ km s^{-1} Mpc^{-1}.

Plots such as those in Figures 7.1 and 7.3 are examples of what are generally called **Hubble diagrams**. They still play an important part in the determination of the Hubble constant, although in modern versions, such as Figure 7.3, it is often the redshift z that is plotted along the vertical axis (rather than the 'naïve' recession speed) and the range of redshifts is generally much greater than in the original Hubble diagram of Figure 7.1. This last point is important since it is only by extending the Hubble diagram to sufficiently distant galaxies, as the HST Key Project did, that there is any hope of escaping the effects of local motion and determining the true rate of expansion of the Universe.

Table 7.1 A summary of the different methods used to measure H_0 from the HST Key Project.

Distance measurement method	Number of Cepheid calibration measurements	Number of measurements used to obtain H_0	Range of redshift in the measurements used to obtain H_0	Value of H_0 obtained/ km s^{-1} Mpc^{-1}
Type Ia supernovae	6	36	0.013–0.10	71
Type II supernovae	4	4	0.006–0.047	72
Tully–Fisher relation	21	21	0.003–0.030	71
fundamental plane	3	11	0.003–0.037	82
surface brightness fluctuations	6	6	0.013–0.019	70

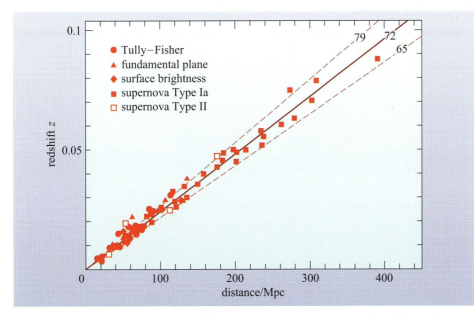

Figure 7.3 A modern version of the Hubble diagram from the HST Key Project, with recession speed replaced by redshift. Note that in this case the gradient of the graph will be equal to the quantity H_0/c. Various straight lines through the data are shown, together with the corresponding values of H_0 measured in km s^{-1} Mpc^{-1}. Surface brightness fluctuations and supernova Type II are other methods of distance determination not discussed in Chapter 2. (Adapted from Freedman *et al.*, 2001)

7.2.2 Other methods of determining H_0

Because it has been so difficult to obtain convergence amongst measurements of H_0 based on the use of the Hubble diagram, there has always been an interest in other methods that might be used to determine the Hubble constant. A method based on detailed observations of the cosmic microwave background radiation is discussed in Section 7.5. Yet another method involves the phenomenon of *gravitational lensing* that was introduced in Chapter 4.

You will recall from Chapters 4 and 5 that, according to general relativity, matter and radiation cause space–time to curve, and the path of a light ray responds to that space–time curvature. This means that concentrations of matter, such as galaxies or clusters of galaxies, can act as 'gravitational lenses', which can magnify and distort an observer's view of objects that lie beyond the lens. A possible outcome of gravitational lensing is the formation of multiple images of the kind shown in Figure 7.4. In this case, the gravitational lens consists of a large, elliptical galaxy and a surrounding cluster of galaxies, while the two-part image provides two views, A and B, of a single quasar, QSO 0957+561 (quasars and their variable luminosities were discussed in Chapter 3). The redshift of this quasar is 1.41, while that of the elliptical galaxy is 0.36. The way the lens forms this image is shown in Figure 7.5.

As Figure 7.5 indicates, the two images of the quasar reach the observer by different pathways that have different lengths. This means that any change that occurs in the quasar, such as a sudden brightening, will be observed in one image some time after it is observed in the other. In other words, as the two images fluctuate in response to changes in the quasar itself, there will be some time lag between the observed patterns of fluctuation of the two images due to the different path lengths

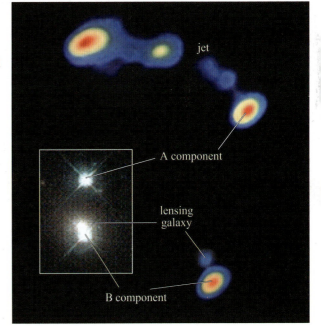

Figure 7.4 The two-component image of the gravitationally lensed quasar QSO 0957+561. The large, elliptical galaxy that is partly responsible for the lensing can be seen just above the B component of the quasar image. The main image was recorded by a radio telescope, the inset is an optical image. The jet is part of the quasar. (NRAO/AUI)

Figure 7.5 A schematic diagram, with an exaggerated angular scale, showing how the double image of QSO 0957+561 is formed. The distance to the lensing galaxy is a little over 1000 Mpc, about one-third of the distance to the quasar. Note that the two routes from the quasar to the observer are not of equal length. (Adapted from Henbest and Marten, 1983)

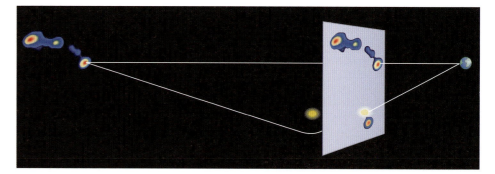

from the quasar. Unfortunately, the difference in path length is not the only cause of a time lag between fluctuations in the two images. The distribution of material within the gravitational lens determines the detailed geometry of space in the neighbourhood of the lens, and this influences the time of passage for light as it passes through the gravitational lens. Light passing through the denser part of the lens is generally delayed relative to light passing through the less dense parts. Since the light responsible for the two images passes through different parts of the lens, just such a density-dependent time-of-passage contribution to the observed time lag between fluctuations in the two images can be expected. The total time lag, including both the path-length and the time-of-passage effect, has been measured for QSO 0957+561 and for a few other gravitational lenses (see Figure 7.6). In the case of QSO 0957+561, the time lag is about 415 days.

In the 1960s, well before the discovery of the first gravitational lens, it was known from theoretical calculations that the path length contribution to the time delay would depend on the Hubble constant. So, in principle, this part of the observed total time lag can be used to determine the Hubble constant without the need for calibration. However, in order to carry out such a determination it is first necessary to calculate the time-of-passage delay and to subtract that from the observed total time lag. Performing such a calculation involves formulating a theoretical

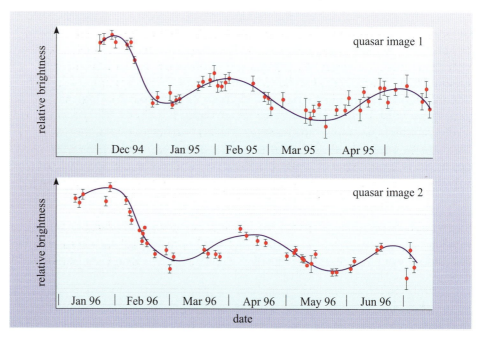

Figure 7.6 The observed patterns of brightness fluctuation in the two images of a double quasar differ by a fixed time lag. (Adapted from Schild and Thomson, 1997)

'model' of the lens and making a number of assumptions about the way that gravitating material is distributed within the lens. The need to model the lens in this way introduces a large degree of uncertainty into what might otherwise be a rather precise way of evaluating the Hubble constant. In the particular case of QSO 0957+561, the lens is dominated by a large, elliptical galaxy that is fairly simple to model. Even so, there are still several remaining uncertainties. At the time of writing, the best current estimate of H_0 from this particular source is 79 ± 8 km s^{-1} Mpc^{-1}, but this is about 30% different from the 'best' value provided by the same source just a few years ago – the difference being almost entirely due to changes in the modelling.

Because of this sensitivity to modelling assumptions it's probably best to regard the gravitational lensing technique as providing a useful cross-check on the broad value of the Hubble constant, but to still need considerable improvement (at least in the certainty of theoretical modelling) before it can match other methods.

QUESTION 7.1

Listed below are some possible sources of error and uncertainty in a determination of the Hubble constant based on gravitational lensing. Classify each as either random or systematic, justifying your decision in each case.

(a) Uncertainty about the distribution of dark matter in the lensing galaxy.

(b) Failure to detect a small companion galaxy located directly behind the lensing galaxy.

(c) Difficulty, due to incomplete data, in determining the total time lag between fluctuations in the two images of the lensed galaxy.

7.3 Measuring the current value of the deceleration parameter, q_0

In the context of the Friedmann–Robertson–Walker models, the Hubble constant, H_0, measures the rate of expansion of the Universe. However, the Hubble constant only represents the *current* rate of expansion: the *current* value of the more general Hubble parameter $H(t)$ that describes the rate of expansion at *any* time t. The rate of expansion might have been higher or lower in the past, and may, in principle at least, be greater or smaller in the future. The observable quantity that tells us whether the cosmic expansion rate is currently increasing or decreasing is the current value of the deceleration parameter q_0, the current value of the time-dependent deceleration parameter $q(t)$ that was introduced in Chapter 5. A positive value of q_0 would indicate that the expansion is slowing down, a negative value that it is speeding up.

The current value of the deceleration parameter is expected to influence the relationship between the distance and the redshift of distant galaxies. As you saw in Chapter 5, to a first approximation this relationship is described by Hubble's law:

$$d = \frac{cz}{H_0}$$

but a more accurate relationship, according to the Friedmann–Robertson–Walker models, is given by

$$d = \frac{cz}{H_0} \left[1 + \frac{1}{2}(1 - q_0)z \right] \tag{7.4}$$

As was shown in Figure 5.29, these two different expressions agree well at low redshift, but they differ significantly when z is greater than about 0.2.

Equation 7.4 provides a way of determining the current value of the deceleration parameter from observation. It implies that a plot of d against z for distant galaxies will show systematic deviations from the straight line implied by Hubble's law if q_0 has any value other than $q_0 = 1$. In principle, then, we can expect to determine q_0 by making accurate independent measurements of d and z and then determining which values of H_0 and q_0 in Equation 7.4 provide the best fit to the observed data. In practice, of course, this procedure is nothing like as simple as it sounds. You saw in the last section the enormous difficulty that astronomers have had in using this kind of approach to pin down the value of H_0, without the added complication of determining q_0 as well. With this in mind, you will not be surprised that many of the early attempts to measure q_0 were not particularly successful. Nonetheless, they deserve a brief discussion because of the light they shed on the observational challenges of measuring cosmic deceleration.

7.3.1 Early attempts to determine q_0

The direct proportionality between redshift and distance indicated by Hubble's law is in excellent agreement with the data obtained from galaxies out to redshifts of 0.1 or so. Hence any attempt to detect departures from Hubble's law must involve galaxies with redshifts greater than 0.1. Such measurements were not really possible until the late 1940s, when the 200-inch telescope on Mount Palomar (see Figure 7.7) was commissioned. Even then, observing individual bright stars in such distant galaxies was not possible, so distance determinations had to make use of clusters of stars or even whole galaxies.

Early attempts to measure deceleration using the Palomar telescope concentrated on whole galaxies. Hubble had pointed out in 1936 that, if galaxies had some natural upper limit to their brightness, then the brightest galaxies in clusters that contained hundreds or thousands of galaxies might be expected to be close to that limit. Thus the brightest galaxy in a cluster might represent a 'standard candle' (i.e. a source of fixed luminosity or absolute magnitude), and once that absolute magnitude had been determined it should be possible to work out the distance of any such galaxy from an observation of its apparent magnitude. (This was the basis of the technique used by George Abell to determine the distances of the rich clusters discussed in Chapter 4.) Accepting this idea, an extensive survey involving clusters with redshifts up to $z = 0.18$, published in 1956, found evidence of curvature in the Hubble diagram and led to the value $q_0 = 3.7 \pm 0.8$. However, another study soon gave $q_0 = 1.0 \pm 0.5$, and a reconsideration of earlier work led to a claim that $q_0 = 0.2 \pm 0.5$. Clearly, this approach was not yielding consistent results, although there did seem general agreement that q_0 was probably positive, as might be expected in a Universe where gravity gradually slowed the cosmic expansion. By the late 1970s extensive studies of galaxies, theoretical as well as observational, were showing that attempts to use the brightest galaxies in clusters as standard candles were being seriously undermined by evolutionary effects, since the galaxies themselves showed signs of changing over time. The larger the redshift of the galaxy, the greater the time its light had spent travelling to the Earth, and the 'younger' the galaxy in cosmic terms; these youthful, bright galaxies did not, it was becoming clear, have the same luminosity as their more evolved 'older' counterparts at lower redshifts. Since these evolutionary effects were overwhelming the effects due to cosmic deceleration, the brightest-galaxy approach became discredited and some other technique had to be used to determine q_0.

The alternative that came to the fore was based on the use of *supernovae*. As you know from earlier discussions, supernovae are highly luminous events that mark the explosive death of certain kinds of star. They can be classified into various types according to their spectra, with Type I supernovae being the most luminous. Astronomers are now keenly aware of the existence of various subclasses of Type I supernovae, such as the Type Ia supernovae that were discussed in Chapters 2 and 5. However, the early attempts to use supernovae to determine q_0 treated all Type I supernovae on the same basis.

Type I supernovae can easily be observed in galaxies with redshifts that are greater than 0.1, and even be detected at redshifts as great as $z = 1$ and beyond. This ensures that the galaxies containing the more distant supernovae are so far away that they are unlikely to be significantly influenced by local disruptions of the Hubble flow. Hence any value of q_0 based on observations of distant supernovae might be expected to be free of the systematic uncertainties that might affect more 'local' measurements.

Figure 7.7 The 200-inch telescope at the Mount Palomar observatory. The large, white structure that supports the open lattice-work of the telescope itself is aligned with the Earth's rotation axis to facilitate the tracking of stars. (Caltech Archives)

Making the assumption that all Type I supernovae attain the same maximum luminosity, attempts were made to use them to determine q_0. The basic approach was similar to that based on brightest galaxies in clusters: measurements of the supernova's apparent magnitude were combined with the assumed 'standard' value of the absolute magnitude to determine the distance, while a separate measurement yielded the supernova's redshift. The way that the redshift varied with distance, for an appropriate sample of Type I supernovae, was expected to reveal the value of q_0. Unfortunately this particular approach was recognized as unreliable when it was realized that the broad class of Type I supernovae actually included several quite distinct types of supernova that differed in intrinsic brightness. However, this failure paved the way for the more recent attempts to measure the current value of the deceleration parameter using the carefully studied subclass of supernovae known as Type Ia supernovae, as you saw briefly in Section 5.2.4. The results of this study have had such a significant impact on observational cosmology that they deserve to be discussed in a separate section.

QUESTION 7.2

Use Hubble's law to calculate the distance of a supernova in a galaxy with redshift $z = 1.0$. Explain why the distance you have estimated may not be a good estimate of the true distance to the galaxy. How does the distance you calculated compare with the size of the local supercluster, as discussed in Chapter 4?

7.3.2 Determinations of q_0 using Type Ia supernovae

Type Ia supernovae are believed to occur when a white dwarf star in a close binary system accretes so much matter from its binary companion that it exceeds a critical mass (about 1.4 times the mass of the Sun) and collapses under its own weight. In a rapid succession of events the collapse is transformed into an explosion that causes a rapid brightening of the supernova, followed by a more gradual decline that can typically be observed for a month or more.

Because the mass required to initiate the supernova is thought to be the same in all cases, and because the physical processes involved are believed to be always very similar, it is expected that all Type Ia supernovae attain approximately the same maximum brightness. This means that Type Ia supernovae might be used as 'standard candles' just like Cepheid variables with a known period.

In practice, simply expecting all Type Ia supernovae to have approximately the same luminosity is not a sufficiently good basis for a major observational research programme. Such a programme must also be supported by an extensive and painstaking investigation of Type Ia supernovae that enables any intrinsic differences between Type Ia supernovae to be studied, understood and, as far as possible, eliminated as a source of uncertainty. This constitutes the process of *calibration* that was described in general terms in Chapter 2. In this case it can be achieved by studying Type Ia supernovae that occur in galaxies whose distances can be determined independently of the supernovae. This work was effectively carried out in the early 1990s by a number of investigators, each pursuing their own goals. As a result, it was found that even within the tightly defined subclass of Type Ia supernovae there were still differences of around 35% in the maximum luminosity attained by different Type Ia supernovae. However, it was also found that the value of the maximum brightness correlates with the rate at which the brightness of the supernova declines from its maximum (see Figure 7.8). Thanks to this correlation it became possible to 'correct' the observed brightness of any particular Type Ia supernova in such a way that it behaved as though its maximum luminosity was actually within 15–20% of the maximum luminosity of any other Type Ia supernova. Unfortunately this correction lacks a sound theoretical basis, but it does appear to be what is needed to substantially reduce the intrinsic differences in luminosity between different Type Ia supernovae. Calculations indicated that 'corrected' observations of thirty or so Type Ia supernovae with redshifts roughly in the range 0.5–1 should provide a good indication of cosmic deceleration.

■ In general terms, what sort of observations would be necessary in order to 'correct' the brightness of any particular Type Ia supernova with the aim of minimizing the effect of intrinsic differences in luminosity?

☐ The apparent magnitude of the supernova would have to be measured repeatedly over a sufficiently long period (about a month) to determine the rate of decline of its light curve, and hence the supernova's peak luminosity relative to other Type Ia supernovae.

By the mid-1990s Type Ia supernovae were sufficiently well understood that they could reasonably be used to determine the current value of the deceleration parameter, and as you saw briefly in Section 5.2.4 , two independent teams of researchers set out to do this. The two groups – the High-z Supernova Search Team

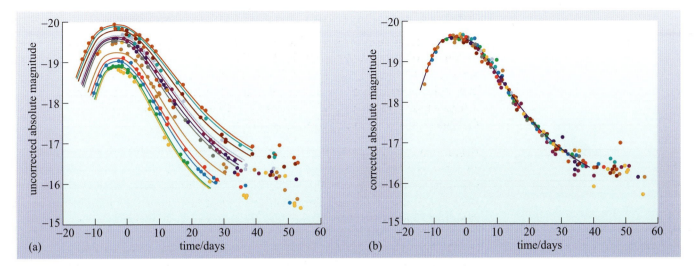

(a)   time/days

(b)   time/days

and the Supernova Cosmology Project – faced similar challenges. Each group had to detect a reasonable number of Type Ia supernovae, measure the redshift of each, and carry out sufficient follow-up observations over a period of several weeks to determine the rate of decline in brightness of each supernova. They would then be able to 'correct' their observations in order to minimize the effect of the intrinsic differences in maximum brightness that exist between different Type Ia supernovae.

This programme presented many observational challenges and required the involvement of several astronomers using a range of telescopes and detectors. The High-z Supernova Search Team, for example, included about two dozen astronomers from fifteen different institutions spread over four continents. Their search for distant Type Ia supernovae involved recording large-area images of the sky with sufficient sensitivity to detect the faint images of supernovae out to redshifts of about $z = 1$. This was initially done using large-area CCD detectors on the 4-metre telescope at the Cerro Tololo Inter-American Observatory (CTIO) in Chile, where the team was able to image about three square degrees of sky per night, down to a limiting magnitude of about 23. In order to identify candidate supernovae they imaged the same area of sky 21 days later and then compared the two images, looking for point-like objects that had changed in brightness over that time. (The period of 21 days was chosen to reflect the 'rise time' of a typical Type Ia supernova – the time it takes for the light curve to build up to a maximum, before starting its much slower decline.) The comparison was mainly an automatic process in which a computer program aligned the two images, compensated for various observational effects, and eliminated known stellar sources in our own Galaxy before presenting the astronomers with the data from which to make a final selection of candidate Type Ia supernovae. This resulted in about five to 20 candidates for each night of observation, most of which were later confirmed as Type Ia supernovae.

Once the candidates had been identified the detailed follow-up work could begin. This involved studying each candidate individually, both *spectroscopically* (to determine its redshift and to confirm that it really was a Type Ia supernova), and *photometrically* (to determine the light curve and hence the 'corrected' maximum brightness of the source). In the case of the High-z Supernova Search Team, the spectra were obtained using one of the 10-metre Keck telescopes in Hawaii (Figure 7.9).

Figure 7.8 The light curves in (a) show the absolute magnitude of relatively nearby Type Ia supernovae as a function of time. Note that the greater the maximum luminosity, the slower the decline in brightness from that maximum. (Crudely, the higher the curve, the wider it is.) Part (b) shows the effect of 'correcting' all the light curves according to a standard prescription derived from the observations. The 'corrected' light curves show much less intrinsic spread in maximum luminosity. (Adam Reiss)

Figure 7.9 The twin domes of the two 10-metre Keck telescopes in Hawaii. (WY'east Consulting, 2002)

A very large telescope was needed because many of the candidates were very faint, so recording the light in their spectra required a large-aperture telescope that could capture as much light as possible in a relatively short time. The photometric (brightness) studies were carried out using a range of telescopes in both the Northern and Southern Hemispheres. For some of the sources, even the Hubble Space Telescope was used. Accurate photometry generally involves comparing the object being studied with stars whose brightness has already been determined with great care, and often requires the observations to be made through a set of standard filters designed to pass light in specific ranges of wavelength. The use of filters allows the effects of obscuration (by dust in the host galaxy) to be determined and eliminated.

It's important to note that the procedure adopted by the High-z Team was not designed to accurately determine the absolute magnitude of the high redshift supernovae. Rather, the aim was to determine their brightness relative to one another (and relative to lower redshift Type Ia supernovae). This approach has the advantage of being simpler than methods requiring accurate measurements of absolute magnitudes, while still allowing deviations from Hubble's law to be detected.

The results from both the High-z Team and from the Supernova Cosmology Project began to appear in 1998. In both cases the crucial data took the form of a graph of 'corrected' apparent magnitude against redshift (see Figure 7.10), and indicated that distant Type Ia supernovae have smaller redshifts than would be expected on the basis of observations of nearer Type Ia supernova and Hubble's law. As you saw in Chapter 5, the results of supernova studies actually imply that the cosmic expansion is accelerating. So, in contrast to all the early determinations of q_0

Figure 7.10 A plot of the relative brightness against redshift for Type Ia supernovae that have been 'corrected' for various factors, including the intrinsic differences in brightness indicated by the differing rates of decline of their light curves. The vertical axis of the graph uses a magnitude-based relative brightness, similar to Figure 5.6. The lowest redshift data are nearer supernovae that were used to calibrate the Type Ia light curve. The curve is the prediction for the best-fit cosmological parameters discussed in this chapter, including a cosmological constant. The lower curve is the prediction for a flat universe with no cosmological constant. The data are clearly highly inconsistent with this latter model.

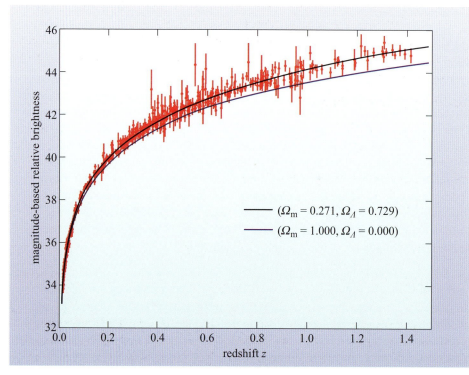

discussed in the last section, it appears that the current value of the deceleration parameter is negative, probably around $q_0 \approx -0.6$.

Further confirmation of cosmic acceleration has come from the observation of individual, very distant, supernovae such as one found in the Hubble Deep Field at a redshift of 1.7. These too indicate that the rate of cosmic expansion is currently greater than it has been in the past.

Despite some remaining concerns about systematic uncertainties, the Type Ia supernova results have led to a major reassessment of the viability of the various FRW models, and (as you saw in Chapter 5) a Nobel Prize. An accelerating cosmic expansion implies a non-zero cosmological constant, and in such a situation the cosmological implications of the Type Ia supernova measurements are not well described by simply quoting a value for q_0. The results are better described in terms of their implications for various contributions to the cosmic density, so it is to measurements of those quantities that we now turn.

QUESTION 7.3

Studies of Type Ia supernovae have been used in the attempt to measure the Hubble constant, but the high redshift measurements detailed in this section are not appropriate for this purpose. Why not? (*Hint*: look at the quantities that have been plotted in Figure 7.10.)

7.4 Cosmology from the cosmic microwave background

Chapters 5 and 6 introduced a number of density parameters that play an important role in characterizing the contents of the Universe. Each of these parameters was defined as the ratio of some kind of density to the critical density $\rho_{crit}(t)$, where $\rho_{crit}(t) = 3H^2(t)/8\pi G$ represents the density in a 'critical Universe' with Hubble parameter $H(t)$ where both the cosmological constant and the curvature parameter are zero (i.e. $\Lambda = 0$ and $k = 0$). The three density parameters that will be mainly of concern in this section are:

the density parameter for matter

$$\Omega_m(t) = \frac{\rho(t)}{\rho_{crit}(t)} \tag{7.5}$$

the density parameter for baryonic matter

$$\Omega_b(t) = \frac{\rho_b(t)}{\rho_{crit}(t)} \tag{7.6}$$

the density parameter for the cosmological constant

$$\Omega_\Lambda(t) = \frac{\rho_\Lambda}{\rho_{crit}(t)} \tag{7.7}$$

where $\rho(t)$ represents the average cosmic density of all kinds of matter (baryonic and non-baryonic), $\rho_b(t)$ represents the density of baryonic matter, and ρ_Λ has no simple physical interpretation but is probably best regarded as a convenient way of representing the cosmological constant, since $\rho_\Lambda = \Lambda c^2/8\pi G$. (You will recall that Ω_Λ is also referred to as the density parameter for dark energy.) Note that all three of these density parameters, $\Omega_\Lambda(t)$, $\Omega_b(t)$ and $\Omega_m(t)$, depend on time since $\rho_{crit}(t)$, $\rho_b(t)$ and $\rho(t)$ all vary with time (although ρ_Λ does not). As in earlier sections, our concern is with determinations of the current values of these time-dependent parameters, which we denote $\Omega_{\Lambda,0}$, $\Omega_{b,0}$ and $\Omega_{m,0}$.

7.4.1 Constraints on $\Omega_{\Lambda,0}$ and $\Omega_{m,0}$

In the context of the FRW models, the deceleration parameter is related to a combination of density parameters (see Section 5.4.3). If the very small contribution due to radiation is ignored, the Friedmann equation implies

$$q(t) = \frac{\Omega_m(t)}{2} - \Omega_\Lambda(t) \tag{7.8}$$

As this suggests, the results of the supernova measurements can be expressed in terms of constraints on the current values of $\Omega_\Lambda(t)$ and $\Omega_m(t)$. In fact, this is the best way of presenting those results, and both research groups chose to present their findings in this way. The results obtained by the two groups are very similar: their combined results are shown in Figure 7.11. On a plot of $\Omega_{\Lambda,0}$ against $\Omega_{m,0}$ the results appear as a set of ellipses, marked with figures that represent various **confidence levels**. This is a way of indicating the uncertainties that attend any experimental or observational result. The lightest-blue ellipse showing the 99.7% confidence level, for instance, indicates that, given the measured results, there is a 99.7% chance that the true values of $\Omega_{\Lambda,0}$ and $\Omega_{m,0}$ are located within the contour. The narrower contours represent increasingly tight constraints on the values of $\Omega_{\Lambda,0}$ and $\Omega_{m,0}$ and are presented with decreasing levels of confidence. All these confidence levels implicitly assume that all sources of uncertainty have been

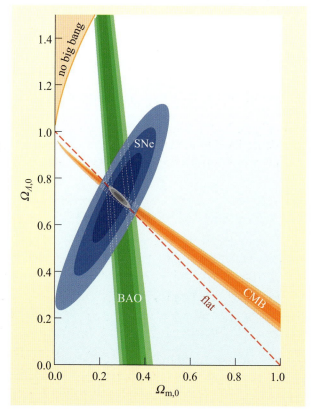

Figure 7.11 Results of the combination of the two supernovae surveys, the Supernova Cosmology Project and the High-z Supernova Team, plotted as a single joint constraint (indicated by confidence level) on the current values of Ω_Λ and Ω_m. There is a 68% confidence that our Universe lies in the darkest-blue region, 95% that it lies in the middle-blue region, and about 99.7% that it lies in the lightest-blue region. The results effectively rule out the kind of Universe in which $\Omega_{\Lambda,0} = 0$ and $\Omega_{m,0} = 1$ that was favoured by many cosmologists prior to the publication of the supernova data. Also shown are the independent constraints from the cosmic microwave background (orange regions); note that these are at almost 90 degrees to the supernovae constraints, so the combination of CMB and supernova data is much more useful than either data set in isolation. The green region shows another set of independent constraints, known as baryonic acoustic oscillations, which measures the clumping in the galaxy distribution remaining from the primordial structures in the CMB. These will be discussed later on in Section 7.4.4. (Adapted from Suzuki *et al.*, 2011)

properly taken into account. In the case of the supernova results there is, of course, some concern about the possibility of unrecognized systematic uncertainties, possibly due to evolution, but also possibly due to some other cause.

As you can see, the results of supernova cosmology are consistent with a whole range of values for $\Omega_{\Lambda,0}$ and $\Omega_{m,0}$. A significant feature of these results, however, is that they strongly exclude a FRW model in which $\Omega_{\Lambda,0} = 0$ and $\Omega_{m,0} = 1$, which was exactly the kind of model Universe favoured by many cosmologists until the supernova data were published.

Although the supernovae results alone do not determine $\Omega_{\Lambda,0}$ and $\Omega_{m,0}$ very precisely, the results have been used in conjunction with the outcomes of other studies to deduce tight constraints on these two values. Perhaps the most important of these studies is the Cosmic Microwave Background, because they constrain $\Omega_{\Lambda,0}$ and $\Omega_{m,0}$ to lie within a narrow track (shown in orange in Figure 7.11) that is roughly perpendicular to the major axis of the ellipse of allowed parameter values from the supernovae. Some of these additional studies are described later, but they broadly support the idea that the Universe has a flat spatial geometry, implying that $k = 0$ and, consequently, $\Omega_{\Lambda,0} + \Omega_{m,0} = 1$. In the context of Figure 7.11 this implies that the point representing the true values of $\Omega_{\Lambda,0}$ and $\Omega_{m,0}$ lies on the diagonal line marked 'flat geometry'. Accepting this and working to a reasonable level of confidence, the results of the Supernova Cosmology Project indicated that

$$\Omega_{m,0} = 0.28 \pm 0.09 \tag{7.9}$$

while the High-z Supernova Team separately found

$$\Omega_{m,0} = 0.32 \pm 0.10 \tag{7.10}$$

The agreement between their results (within the quoted uncertainties) is impressive, and while there must always be concerns about systematic uncertainties, there is now widespread agreement that $\Omega_{m,0}$ is within the range indicated by these results. Figure 7.11 shows their final, joint constraints on cosmological parameters. Further evidence of this is presented in Section 7.5.

QUESTION 7.4

From Figure 7.11, estimate the range of possible values of $\Omega_{m,0}$ at the 99.7% confidence level from the SNe data alone with no other assumptions. How does this improve if it is additionally assumed that $k = 0$?

7.4.2 The cosmic microwave background

The cosmic microwave background (CMB) radiation was introduced in Chapter 5 and then explored more fully in Chapter 6. Among the many features of the CMB that were described in those earlier discussions were the following.

1 The CMB radiation is intrinsically uniform to better than one part in 10 000. This means that, after correcting for effects due to the motion of the detector, the CMB comes with very nearly equal intensity from all directions.

2 The CMB is 'thermal radiation' with a characteristic temperature of 2.725 K. This means that the intrinsic spectrum of the radiation is well described by a black-body curve with a peak in intensity at a wavelength of about 1 mm.

3   The CMB radiation is believed to have originated at the time of recombination, when the mean temperature of cosmic matter became low enough to allow electrons and protons to form hydrogen atoms that were not immediately ionized again. This means that the photons that make up the CMB were last scattered about 3 to 4×10^5 years after the start of cosmic expansion, when the temperature was about 3000 K.

4   The radiation has been expanding and cooling since it was last scattered, during which time its temperature has decreased by a factor of about 1100. This means that the 'last-scattering surface' from which the radiation was released is now at a redshift of about 1100.

5   Despite its high degree of intrinsic uniformity, anisotropies in the intensity of the radiation have been observed, at a level of a few parts in 100 000, over a range of angular scales.

This section concentrates on the anisotropies in the CMB since it turns out that despite their low amplitude, they hold the key to determining many cosmological parameters. We will discuss the physical origin of these anisotropies, how they can be measured and described, and how comparison with detailed cosmological models allows astronomers to measure cosmological parameters. It also contains recent results that improve and make more precise some of the values quoted in earlier chapters and in the points listed above.

QUESTION 7.5

In the numbered list above, no explicit mention is made of the 'dipole anisotropy' that was introduced in Chapter 6. Briefly describe the dipole anisotropy, explain its origin and identify the words and phrases in points 1, 2 and 5 that imply the dipole anisotropy has been taken into account, despite the lack of any explicit reference to it.

7.4.3 Detecting anisotropies in the CMB

The discovery of the CMB by Penzias and Wilson in 1965 was a turning point in observational cosmology. Following the discovery, many observers reoriented their research in order to concentrate on the CMB and the other implications of big bang cosmology. Penzias and Wilson had made their serendipitous discovery using a ground-based radio detector working at a wavelength of about 70 mm, but much of the subsequent work tried to cover wavelengths closer to the expected peak in the CMB spectrum, at about 1 mm. At such relatively short wavelengths, radiation coming from space is absorbed by the Earth's atmosphere, so many of the observations were made using detectors carried by balloons or high-flying aircraft. The existence of the dipole anisotropy was first established in this way, in 1977, using a detector fitted to an American U-2 aircraft. (U-2s had been employed as spy planes during the Cold War, but were extensively used for scientific work from the mid-1970s.) Other high-altitude work gave a strong indication that the CMB spectrum was thermal (i.e. described by a black-body curve), but left open the possibility that there might be small departures from a pure black-body spectrum. However, none of the balloon or aircraft observations revealed any anisotropies in the CMB, apart from the large-scale dipole anisotropy. Small-scale anisotropies were predicted by big bang cosmology, but highly sensitive detectors with at least moderate angular resolution were needed to identify them. The observation of those small-scale anisotropies, and the confirmation of the purely thermal nature of the CMB spectrum, had to await the launch of the first space satellite to be dedicated to the study of the cosmic background radiation.

COBE, the Cosmic Background Explorer (Figure 7.12), was launched on 18 November 1989. COBE was about the size and mass of a large car, and carried three main experimental packages:

- DMR, a set of three differential microwave radiometers designed to search for anisotropies in the cosmic background radiation;

- FIRAS, the Far-Infrared Absolute Spectrophotometer, designed to measure the CMB spectrum; and

- DIRBE, the Diffuse Infrared Background Experiment, designed to search for cosmic infrared background radiation, and to assess local sources of diffuse infrared radiation, including the Milky Way.

Within a short time of starting its astronomical observations, the FIRAS detector gave powerful support to the big bang prediction of a very pure black-body spectrum. It is now known that, over the wavelength range from 0.1 mm to 5 mm, the CMB spectrum follows a 2.725 K black-body spectrum very precisely. Observations of small-scale anisotropies by the DMR took longer to emerge and longer still to be refined.

As mentioned in Chapter 6, absolute measurements of the intensity of the CMB are difficult to make, so the DMR was designed to look for differences in the intensity of the CMB from pairs of regions separated by 60° on the sky.

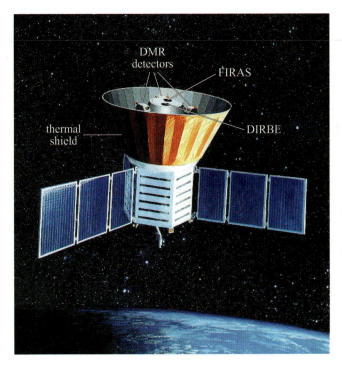

Figure 7.12 COBE, the Cosmic Background Explorer. COBE's three main scientific instruments, DMR, FIRAS and DIRBE, made their observations from behind a collapsible thermal shield that was deployed after the satellite was placed in orbit. COBE followed a relatively low Earth orbit that took it over the Earth's poles. The plane of the orbit precessed at the rate of about 1° per day; this allowed the shielded instruments to scan all parts of the sky over a period of six months. (NASA)

The same technique had been used to detect the dipole anisotropy in the 1970s, and the DMR on COBE was a direct descendant of the detector that had been carried to high altitude by a U-2 spy plane. The COBE DMR actually consisted of three pairs of detectors working at wavelengths of 3.3, 5.7 and 9.6 mm. By comparing observations at these three wavelengths, the effects of microwave emission from the Milky Way could be identified and eliminated, leaving just the cosmic signal, albeit contaminated by unavoidable detector 'noise'. Thanks to COBE's rotation (it turned on its axis once every 75 seconds), the satellite's 103-minute orbital period, and the gradual precession of the orbital plane, the DMR was eventually able to examine all parts of the sky and to produce the all-sky CMB anisotropy map shown in Figure 7.13.

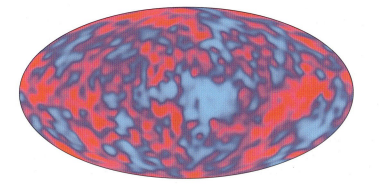

Figure 7.13 An all-sky CMB anisotropy map based on data from the DMR experiment on COBE. The angular resolution of this map is about 7°. Contrast this map with the highly isotropic distribution measured at lower sensitivities, shown in Figure 5.3. (Edward L. Wright)

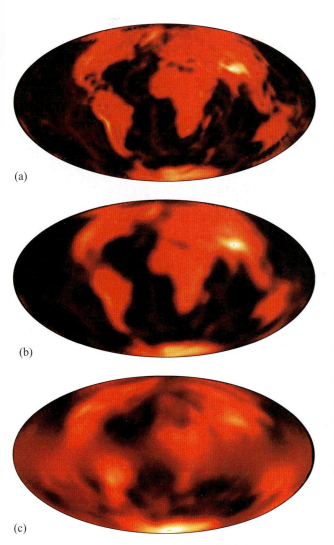

(a)

(b)

(c)

Figure 7.14 (a) A map of the Earth. (b) The same map viewed with the angular resolution that is inherent in the COBE anisotropy map. (c) The addition of noise makes the underlying pattern even harder to discern. (Courtesy of Ted Bunn)

The publication of the first anisotropy map in April 1992 was another very significant development in observational cosmology. Nonetheless, it is important to recognize the limitations of the COBE findings. In the first place, the data on which the map was based were very noisy. This meant that much of the visible structure in the map could have been illusory. A statistical analysis of the data justified the claim that anisotropies had been observed, but it did not guarantee that the coloured boundaries that appear in the map were accurately located. Moreover, the nature of the DMR's collecting horns meant that each of the paired regions being compared was about 7° across, so this was the minimum angular scale on which anisotropies might reliably be detected. To give some idea of the significance of this, Figure 7.14 shows a map of the Earth viewed with the same kind of resolution. As you can see, even the major continents are indistinct, while finer details, such as the existence of the British Isles, are completely lost. The features become even more obscure when 'noise' is deliberately added to the data to make the comparison with the DMR results more apposite. Clearly, despite COBE's success, much work remained to be done.

In the early 1990s, further measurements and analyses by the COBE Science Team confirmed and refined the original results for angular scales of about 7°. Since then other groups of researchers, mainly using ground-based or balloon-borne equipment, have proceeded to measure CMB anisotropies on smaller angular scales. One particularly notable effort was that of the BOOMERanG collaboration, an international team that used equipment suspended from a high-altitude balloon (Figure 7.15). BOOMERanG (which stands for 'balloon observations of millimetric extragalactic radiation and geophysics') mapped CMB anisotropies over only a limited region of sky, but did so with high sensitivity and an angular resolution of better than 1°. Many other degree-scale anisotropy observations have followed those of BOOMERanG, some from balloon experiments, others from ground-based detectors.

The next major mission was NASA's Wilkinson Microwave Anisotropy Probe (WMAP), which was launched in June 2001. This is the successor to COBE and, like COBE, it has produced an all-sky anisotropy map, although this time with an angular resolution of about 0.1°. The WMAP anisotropy map is shown in Figure 7.16 and is based on nine years of observations. As you can see, it is far more detailed than the COBE map that was produced roughly ten years earlier. These observations were further improved by ground-based telescopes, most notably the South Pole Telescope and the Atacama Cosmology Telescope. These did not map the whole sky but provided important statistical constraints on the smaller scale structures in the CMB. The most recent mapping of the CMB has been from the European Space Agency's Planck telescope, whose all-sky CMB map is compared to that of WMAP in Figure 7.17. Note the much finer angular resolution.

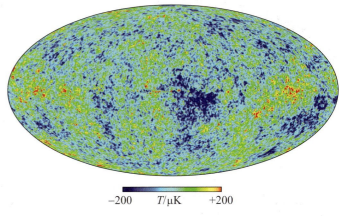

−200 $T/\mu K$ +200

Figure 7.15 The high-altitude balloon and instrument package used by the BOOMERanG team to measure CMB anisotropies from Antarctica. (BOOMERanG Collaboration)

Figure 7.16 An all-sky CMB anisotropy map, based on nine years of data obtained by the WMAP space probe. The angular resolution of this map is about 0.1°. Compare this map with the lower resolution COBE data shown in Figure 7.13. (Bennett *et al.*, 2003)

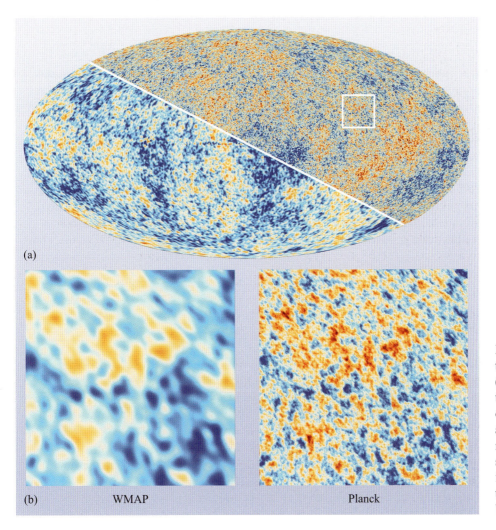

(a)

(b)   WMAP   Planck

Figure 7.17 (a) A comparison of the all-sky CMB map from Planck (upper right half of ellipse) with that of WMAP (lower left half of ellipse). Note the much higher angular resolution of the Planck map. (b) The region marked by the square in the all-sky map expanded for WMAP and Planck. (ESA and the Planck collaboration; NASA/WMAP Science Team)

The wealth of data from CMB experiments has been part of a revolution in quantitative cosmology. To see how, we will first need to understand how anisotropies in the CMB are characterized, which we will cover in the next section.

QUESTION 7.6

When comparing the predictions with the data in Figures 7.16 and 7.17, would you expect the big bang prediction to show precisely the same pattern of anisotropies as that observed by WMAP and Planck, or simply something 'similar' to the observations? Justify your answer.

7.4.4 Describing anisotropies in the CMB: a bluffer's guide to power spectra

The details of how structures in the CMB are measured would take us beyond the mathematical sophistication of this book. Fortunately, it's possible to gain an intuitive sense of how these structures are described without delving too far into the mathematics. The key to these insights is a surprising analogy with music. You won't need to be a musician or have a musical background to follow the analogy, but there will be some extra information in optional Box 7.1 or marginal notes to help musically inclined readers deepen their understanding. The only 'musical' experience you might find useful is turning down the 'treble' control of a stereo. This turns out to have a profound cosmological analogue in the early Universe!

The sine function is further described in the Appendix.

It will help to first understand a little about sound, and how musical instruments produce sound. The pitch of a note is determined by its frequency, so sounds that seem high-pitched are in fact high-frequency sound waves, while low-pitched notes are low-frequency sound waves. Figure 7.18 shows example high-pitched and low-pitched sound waves: note that the shape of these waves is described by what is called a sine function (they are sinusoidal waves). Note also that the low-frequency waves also have longer wavelengths. The frequency and the wavelength of a sound wave are related by the equation $c_s = \lambda \times f$, where λ is the wavelength, f is the frequency and c_s is the speed of sound. The speed of sound in air is (to all intents and purposes) constant, so large wavelengths always mean small frequencies, and vice versa.

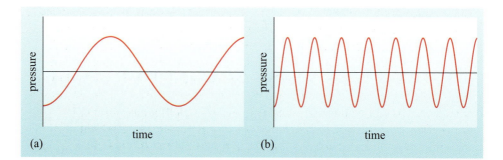

Figure 7.18 Schematic examples of (a) a deep-pitched, low-frequency sound wave and (b) a high-pitched, high-frequency sound wave. The air pressure measured at some particular location is plotted as a function of time, and as the sound waves pass through that location, the air pressure varies. Note that the low-frequency sound wave has a longer wavelength than the high-frequency sound wave.

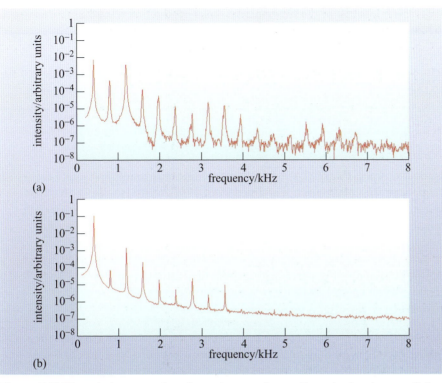

(a)

(b)

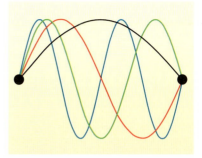

Figure 7.20 Possible oscillations of a string fixed at both ends at the points marked with black dots. The longest-wavelength oscillation possible is marked in black, the second-longest in red, the third-longest in green and the fourth-longest in blue. Each waveform is shown at a position of maximum amplitude. For example, the black waveform starts as a straight line connecting the two dots, stretches to the shape shown in the figure, returns to the straight line, then stretches in the opposite direction to make an upside-down version of the shape shown in the figure (i.e. U-shaped).

Figure 7.19 The relative strengths of sound waves from a flute playing the note G (upper panel), and a clarinet playing the same pitch (lower panel). Note the logarithmic spacing of the vertical axis marks. The flute and clarinet are ostensibly playing a note with the same pitch, but they are also making a wide range of other notes, particularly at the resonant frequencies (i.e. the spikes).

It may surprise you to know that when a musical instrument plays a particular note, the sound waves that come out are not purely that note. Sound waves of other notes are made as well. Figure 7.19 shows the strengths of the sound waves from two musical instruments as a function of frequency. Even though both instruments are ostensibly playing the same pitch, they are nonetheless both sending out sound waves with a wide range of frequencies. The instruments also differ in the strengths of these additional sound waves, which is partly what makes one instrument sound different to another.

What is particularly striking about Figure 7.19 is that these additional sound waves are not evenly spread in frequency, but instead concentrated into very distinct spikes. This comes about because of the concept of *resonance*, which as you'll see also has a deep cosmological application. Musical instruments in general are most efficient at producing sound waves where the wavelength fits neatly into the sound-producing parts. This is perhaps easiest to visualize in a stringed instrument. The string is fixed at both ends, so the only possible oscillations would have shapes much like Figure 7.20. The longest-wavelength oscillation (making the lowest-pitched sound) is called the *fundamental* or *first harmonic*, and the shorter-wavelength oscillations are called *overtones*, or sometimes *second harmonic*, *third harmonic*, and so on. The fundamental oscillation (or first harmonic) is done by fitting *half* a wavelength along the string, like the longest-wavelength wave in Figure 7.20. If λ is the wavelength of the first harmonic, then the higher harmonics have wavelengths of $\lambda/2$, $\lambda/3$, $\lambda/4$, and so on, because these are what fits neatly along the string. For a string of length L, the lowest-pitched oscillation has $\lambda = 2L$.

Wind instruments work slightly differently to strings, and the allowed resonances will depend on what holes are open on the instrument, but similar concepts still apply of fitting in wavelengths or fractions of wavelengths.

■ If f is the frequency of the fundamental note (i.e. the first harmonic), what are the frequencies of the higher harmonics?

☐ We know that $c_s = \lambda \times f$, with c_s constant. This means that halving λ means doubling f, and reducing λ by a third means increasing f by a factor of 3, and so on. Therefore the frequencies of the harmonics are f, $2f$, $3f$, $4f$, and so on.

From the question above, you should now be able to see why the spikes in Figure 7.19 are evenly spaced in frequency.

The link to cosmology and the cosmic microwave background starts with the mathematician Joseph Fourier, who lived from 1768–1830. He discovered that almost any periodic wave can be described as the sum of pure sine waves, with wavelengths of 1/2, 1/3, 1/4, and so on, of the longest wavelength in the sum. Similarly, the frequencies are 2×, 3×, 4× (and so on) the lowest-frequency wave in the sum. These are the same spacings of wavelengths and frequencies that you met in the harmonics of musical instruments. An example is shown in Figure 7.22. Note how adding shorter and shorter wavelengths progressively improves the approximation to the total 'target' waveform.

Musicians may be interested to know that an electric guitar played through an overdriven amplifier (the "distortion" effect) creates fairly square sound waves.

We are nearly at the stage where we have enough information to interpret the cosmological information in the CMB. A further aspect of sound waves is still needed: the effect of the treble control on the shape of a wave. Turning down the treble control on a stereo has the effect of suppressing the higher-frequency harmonics of the sound waves. In Figure 7.23, you will see that the effect on the waveform of turning down the 'treble' is to smooth out the waveform – and indeed

BOX 7.1 OCTAVES

A piano keyboard is shown in Figure 7.21. As you have read, sound waves have wavelengths that are related to their pitch. Notes that are one octave apart (shown in the figure) have wavelengths that are exactly a factor of two of each other. For example, in the song 'Somewhere over the rainbow', the pitch for 'Some' has a wavelength that's twice as long as the pitch for 'where', so it is one octave lower. On a stringed instrument such as a guitar or a violin, you can raise the pitch of a string by an octave by halving its length, which is done by pressing the string down halfway along its length with your finger. Other shortenings result in different notes.

■ Where is the octave in the song 'Singing in the rain'?

☐ From 'I'm' to 'singing'. Other songs with an octave somewhere in them (not necessarily at their starts) are 'Let it snow', 'There's no business like show business' and 'Money' by Pink Floyd. Can you think of others?

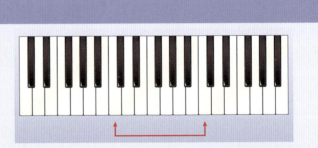

Figure 7.21 A piano keyboard. The red arrow connects two C notes, which are an octave apart. Any two notes separated by the distance of the red arrow are an octave apart. In the song 'Somewhere over the rainbow', the notes for 'Some' and 'where' are an octave apart.

The second harmonic is one octave above the first harmonic, because the wavelengths are exactly a factor of two of each other, e.g. if λ is the wavelength of the first harmonic and $\lambda/2$ the wavelength of the second, then their ratio is $\lambda/(\lambda/2) = 1/(1/2) = 2$. Similarly, the fourth harmonic is one octave above the second harmonic, because $(\lambda/2)/(\lambda/4) = (1/2)/(1/4) = 2$, and two octaves above the fundamental, because $\lambda/(\lambda/4) = 1/(1/4) = 4$. However, the third harmonic is *not* an octave above the second harmonic, nor the first harmonic.

the mathematical operation of smoothing turns out to be identical to attenuating (i.e. reducing the amplitude of) the higher-frequency harmonics.

Music players sometimes give the user more control than just a 'bass' and 'treble' control to amplify or attenuate the lower and higher harmonics respectively. Figure 7.24 shows a graphic equalizer, which controls the amplification or attenuation of harmonics in many specific frequency ranges. One can think of each slider as a volume control that affects only a particular range of frequencies or harmonics.

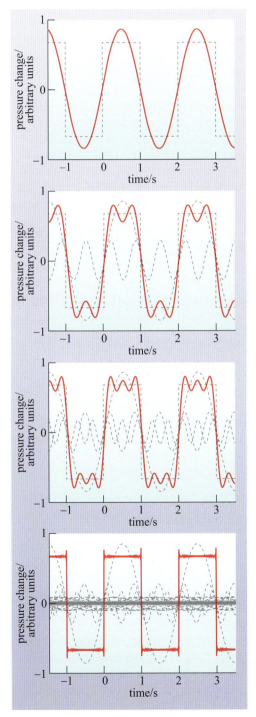

Figure 7.22 (Left) A square wave (dashed line), and its Fourier series approximations. A square wave is unusual in that it only has odd-numbered harmonics. As the number of harmonics added increases, the approximation to the 'target' square wave shape progressively improves, as follows. The top panel shows the simplest approximation with just the fundamental note in red. Clearly a single sine wave is not a good approximation to a square wave. Another harmonic has been added in the next panel down. The harmonics are shown separately as dashed lines, and the sum of the two harmonics (shown in red) is slightly closer to a square wave. A further harmonic has been added in the second from bottom panel, and again the total (in red) is slightly closer to the square wave shape. Finally, in the bottom panel, many more harmonics have been added, and the total waveform is now quite close to a square wave.

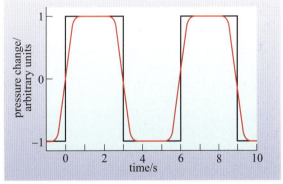

Figure 7.23 (Above) A square wave before (black) and after (red) turning down the 'treble'.

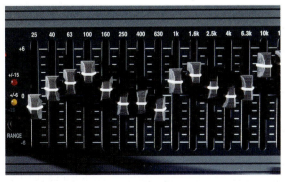

Figure 7.24 (Above) A graphic equalizer.

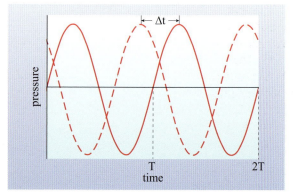

Figure 7.25 Two sine waves that differ only in their phases. As with Figure 7.18, this graph plots the pressure at a particular location against time. The period of both waves is T, but the dashed curve is offset by a time interval of Δt. The phase difference between the two waves is defined as $2\pi\Delta t/T$.

The reason this is all so useful for the cosmic microwave background is that the early Universe was awash with sound waves. You will see why in the next section. There are many similarities between the resonance effects you have just met and the sound waves in the primordial Universe, but there is one key difference: the primordial sound waves in the early Universe were random and incoherent. To see what this means for the sound waves themselves, you need to understand one final property of sound waves: their *phase*.

Figure 7.25 shows two sine waves that differ only in their phases. As you can see from this figure, the phase expresses where the waves start their upswings. Shifting the phase of a sine wave by one full cycle is the same as doing nothing at all to a wave. The harmonics from a musical instrument do not have random phases: for example, in Figure 7.22 all the phases are the same for all the harmonics, because all the harmonics have an amplitude of zero at time $t = 0$ and are all on an upswing at that point. However in the early Universe, each of the wave harmonics could have been produced with a downswing, or an upswing, or any intermediate random point. In other words, the primordial sound waves in the early Universe had random phases, which is what is meant by 'incoherent' waves.

Partly because the waves were incoherent, the convention is to measure the amount of 'hiss' or variance in the CMB at each harmonic, rather than the amplitudes and phases themselves. *Do not be tempted to draw the analogy with a musical instrument too far*: musical instruments resonate coherently, but the sound of the early Universe was an incoherent hiss or roar in a resonating cavity, and this cavity grew with time.

Roughly speaking, it is the imprint of the sound waves of this hiss onto the last scattering surface (Section 6.5) that gives rise to many of the observed anisotropies in the CMB.

We have been working towards an important diagram in cosmology (Figure 7.26), called the **angular power spectrum**, in which the variation in the CMB sky is plotted as a function of frequency. Sound waves such as those in Figure 7.18 are

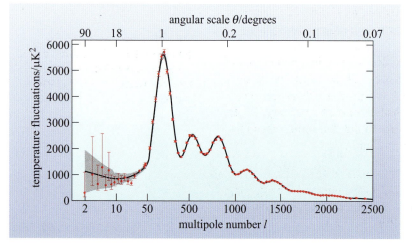

Figure 7.26 The angular power spectrum of the CMB as determined by the Planck satellite. The shaded band enclosing the best-fit line represents an effect known as **cosmic variance**. The cosmic variance is a consequence of the fact that we are compelled to estimate the angular power spectrum of the CMB from observations made at one cosmic location (i.e. on the basis of a single 'sample' of the CMB, made from within the Solar System). The smooth curve represents the 'best fit' to these data from a range of theoretical predictions based on various FRW models (more will be said about these predictions in the next section). (ESA and the Planck Collaboration)

one-dimensional waves, whereas the CMB image is two-dimensional (longitude and latitude on the sky), so there's been some averaging over different orientations on the sky to make Figure 7.26. The horizontal axis is similar to the frequencies in a graphic equalizer in Figure 7.24, or the frequency axes in Figure 7.19. The distances on the sky that these 'frequencies' refer to are given along the top of the plot, and note that long wavelengths are low frequencies, while short wavelengths are high frequencies, just as with musical sound waves. The equivalent term for frequency in this context is **multipole number**. The vertical axis measures the amount of hiss at each frequency, i.e. in each multipole. We have skipped over most of the technical details of how Figure 7.26 is made from the CMB image, for example in how the variance (hiss) is measured and how the curvature of the celestial sphere is accounted for. However, the essence of Figure 7.26 is that it measures the variation in the image as a function of some quantity analogous to frequency.

The relationship between distance on the sky and multipole number is quite simple. The angle θ may be associated with a multipole number l using the relation

$$l = 180°/\theta \tag{7.11}$$

implying that the multipole number l is the number of times the angle θ can be fitted into a 180° arc. Because of this relationship between multipole number and angle, the horizontal axis of an angular power spectrum is sometimes shown as an angular scale, but with the angle *decreasing* in unequal steps from left to right. (A scale of this kind is shown at the top of Figure 7.26.)

■ COBE was only able to detect anisotropies of angular size 7° or above. What is the corresponding range of l values?

☐ From Equation 7.11, $q = 7$ corresponds to an l value of about 26. Larger angles correspond to smaller values of l, so we can say that the COBE results were confined to the range $l = 0$ to $l = 26$ or thereabouts. Looking at Figure 7.26 you can see that this is a very small range compared with the Planck results; it doesn't even cover the first peak.

What you will notice in Figure 7.26 is that there are *bumps* in the CMB's power spectrum. These bumps have an astonishing origin: nearly the whole observable Universe at the time was acting like an acoustic resonating cavity, filled with incoherent hissy noise, and these bumps are the harmonics of the resonating cavity! It is the resonance of the whole Universe, working like a musical instrument, but in which the resonating cavity is filled with an incoherent roar of sound waves rather than a coherent musical oscillation. These bumps are known as the acoustic oscillations in the CMB and their fingerprint is still visible on the large-scale distribution of galaxies today. Measurements of these large-scale features in the galaxy distribution are what gave rise to the 'baryonic acoustic oscillations' constraint on the density parameters in Figure 7.11.

The astronomer Mark Whittle transposed the CMB sound waves up 50 or so octaves, to create an audible sound, now available on the web.

Another striking feature is that the amount of structure dies away at high frequencies. Something has 'turned down the treble' in the noise of the big bang. We will see in the next section what almost certainly caused this smoothing (i.e. high-frequency attenuation) of the sound waves of the big bang.

Figure 7.27 A carpet, with a 30 cm ruler for scale; for use in Question 7.8.

7.4.5 Predicting anisotropies in the CMB

The power spectrum of the CMB in Figure 7.26 is one of the most richly informative diagrams in cosmology, because by studying the overtones of the primordial sound waves, it's possible to infer the properties of the material that was oscillating in the early Universe, such as how much baryonic matter or dark matter it contained. This section will show how some of these inferences, and other inferences, are made.

There is one inference that can be made straight away. We know from Chapter 6 (Figure 6.18) that the size of the observable Universe at the time of recombination is estimated to appear only two degrees across on the CMB sky, so how is it that the CMB appears so uniform? Features in the CMB further apart than two degrees should not have been in communication with each other, as you saw in Section 6.5.5. The inevitable inference is that something else fundamental must be going on. This something else has to be new physics at much higher energies than probed by the largest particle accelerators, such as the Large Hadron Collider at CERN.

There is a proposed solution in the theory of inflation, which you met in Chapter 6: a small region of the Universe, smaller than the cosmic horizon size S at the time, was suddenly inflated to a diameter much bigger than S. This means that regions of the CMB that *appear* to have been out of contact with each other since the big bang were in fact in contact with each other prior to inflation. This also means that what we refer to in practice as the 'size of the Observable Universe' really means the 'size of the Observable Universe *since inflation ended*', and not since the big bang. This is arguably slightly sloppy terminology by astronomers. In practice though, any light signals from further away from before inflation would be so enormously red-shifted by inflation that it would not be possible to detect them, even neglecting the fact that the Universe was opaque at that time.

Inflation also provides a mechanism for supplying the initial fluctuations visible in the CMB. Quantum processes in the very early Universe first gave rise to density fluctuations, which were then amplified by inflation, and ultimately left their imprint on the CMB. One of the exciting things about modern observational cosmology is that it is now beginning to test inflation theories and some speculative proposals concerning inflation have already been ruled out by measurements of the CMB. However, whatever the role of inflation may have been, there is general agreement that the early Universe contained density fluctuations on essentially all size scales and with a range of strengths. Any detailed assumptions we make about the relative strengths of fluctuations of different sizes can be presented in the form of an assumed spectrum of fluctuations underlying the acoustic peaks in Figure 7.26. This underlying spectrum can be characterized mathematically by introducing some more cosmological parameters. The simplest credible description of CMB density fluctuations involves two new cosmological parameters: a *power spectrum normalization*, denoted A, and a *scalar spectral index*, denoted n_s. We shall not be much concerned with these two parameters, but we should add them to the list of cosmological parameters that observational cosmologists are trying to determine, and keep in mind the possibility that more parameters might be needed to fully describe the fluctuation spectrum.

■ What would the CMB power spectrum in Figure 7.26 look like if there were *no* underlying fluctuations? (This would happen with an underlying power spectrum normalization of $A = 0$.)

☐ If there were no underlying fluctuations, there would have been no sounds to resonate within the 'resonating cavity' that was the observable Universe at the time. There would therefore be no acoustic peaks in the CMB power spectrum in Figure 7.26. The CMB would be completely uniform and the data Figure 7.26 would be consistent with zero at all l values. (Then again, the underlying fluctuations later grow into galaxies and stars, so no underlying fluctuations means no Galaxy, no Earth, no astronomers, no observations and no Figure 7.26.)

An all-sky CMB anisotropy map, such as Figure 7.16, is essentially a snapshot of the spherical surface, centred on the Earth, at which the CMB radiation now reaching us was last scattered. That surface was discussed in Chapter 6, where it was referred to as the *last-scattering surface*. According to Chapter 6, the last scattering occurred at the time of recombination, when matter and radiation became decoupled. The time of this decoupling is determined from Planck observations to be $t_{dec} = 3.72 \times 10^5$ years, which we adopt from here on. So Figures 7.16 and 7.17 show that when the Universe was about 372 000 years old some parts of it were slightly warmer or slightly cooler than others. When making predictions of the CMB sky, no attempt is made to predict the particular pattern of anisotropies shown in the all-sky maps of Figures 7.16 and 7.17: it is the statistically significant angular power spectrum of the data in Figure 7.17 that is of interest, i.e. the power spectrum shown in Figure 7.26.

To work out how a spectrum of density fluctuations, characterized by A and n_s, influences the anisotropy of the CMB, we need to consider the interaction of matter and radiation in the early Universe, at least up to the time of decoupling ($t_{dec} = 3.72 \times 10^5$ years). This is best done by considering the various constituents of the Universe separately, and then considering the way they interact with each other.

Before decoupling, most of the matter in the Universe would have been non-baryonic dark matter, just as it is now, since the relative densities of baryonic and nonbaryonic matter are not expected to change with time. The density fluctuations would have been largely composed of dark matter, and can be roughly thought of as dark-matter 'halos', of various densities and sizes. These halos would have been expanding, like the rest of the Universe, but those that had a slightly greater than average density would have been growing a little less rapidly than average, and would have been increasing in 'strength' as a result.

The radiation in the Universe, prior to decoupling, would have been plentiful and energetic. As in the present-day Universe, the radiation would not have had much direct interaction with the dark matter. However, the radiation would have incessantly interacted with the electrically charged particles of baryonic matter (such as nuclei of hydrogen and helium). As a result, the baryons and the radiation would jointly form a sort of fluid, with a common temperature and pressure. This **photon–baryon fluid** would be subject to the gravitational influence of the dark-matter halos and would be drawn towards those haloes. However, as indicated in Chapter 6, the gravitational tendency to compress the photon–baryon fluid would have been resisted by the internal pressure of the fluid.

When the Universe had been expanding for a time t, effects due to the pressure in the photon–baryon fluid would not have been able to make themselves felt over distances greater than ct. (The speed of light represents an upper limit to the speed at which signals of any kind can travel.) Consequently, the photon–baryon fluid inside a dark-matter halo that is larger than ct at time t expands along with the halo, essentially undisturbed by the effects of pressure. However, as time passes, t increases and the scale of pressure effects grows. As ct comes to exceed the size of any particular dark-matter halo, the photon–baryon fluid collapsing into that halo will be subject to an increasing pressure that will halt the collapse, and then allow the photon–baryon fluid to 'spring back'. The photon–baryon fluid contained in some dark-matter halos can undergo several of these oscillations, allowing the density fluctuations to act as the generators of the **acoustic waves** (i.e. sound waves) in the photon–baryon fluid that you met in the previous section. An important point to remember is that at time t the greatest wavelength that any of these waves may have will be roughly ct. We make use of this fact later.

When the Universe gets to be about 372 000 years old, recombination occurs. This allows the photons to decouple from the baryons. It is the decoupled photons streaming away from the surface of last scattering at $t_{dec} \approx 372\,000$ years that eventually produce the CMB we now observe. The anisotropies that we see in the CMB are mainly caused by the unevenness of the last-scattering surface, and this is due to the density fluctuations and the acoustic waves that are present there. Now that we have identified the cause of the anisotropies we can account for the main features of their angular power spectrum.

On angular scales of a few degrees or more (i.e. multipole numbers of about 50 or less), the main source of anisotropy is called the **Sachs–Wolfe effect**. This is largely due to the general relativistic phenomenon of *gravitational redshift*, which causes photons coming from the denser parts of the last-scattering surface to be observed with slightly longer wavelengths than identical photons leaving other parts of the last-scattering surface. This causes the denser parts to appear slightly cooler and leads to the flat plateau seen on the left of the angular power spectrum in Figure 7.26.

On intermediate angular scales (i.e. multipole numbers between 50 and 1000, say), the angular power spectrum shows the effect of the acoustic waves at the time of decoupling. At that time there are some long wavelength waves that are just reaching their state of maximum compression for the first time. These compressions heat the photon–baryon fluid, causing the photons that escape from them to create temperature anisotropies. The moving charged particles associated with the waves will also cause the wavelengths of scattered photons to change (this is a Doppler effect), and this too will create temperature anisotropies. The size scale of the anisotropies would be comparable to the wavelength of the waves, and of the order of ct_{dec}.

■ Evaluate ct_{dec} in light-years.

☐ Since $t_{dec} \approx 372\,000$ years, it follows that $ct_{dec} \approx 372\,000$ light-years.

In a universe with a flat geometry (i.e. $k = 0$), a feature of this size on the last-scattering surface would span an angle of about two degrees. At these epochs, the speed of sound in the matter and radiation at the time (the photon–baryon fluid) was just over half the speed of light, and as a result the size of the 'resonating cavity' is at a scale of about one degree, rather than two. This accounts for the prominent peak in the angular power spectrum around $l = 220$, or about one degree.

The angular size of this feature depends sensitively on the geometry of the Universe. In a universe where the geometry is not flat (i.e. $k = 1$ or $k = -1$), the angular size of this feature is different. Figure 7.28 shows a detail of the CMB temperature anisotropies, observed with the BOOMERanG balloon-borne CMB telescope. This strong sensitivity to the geometry of the Universe is the

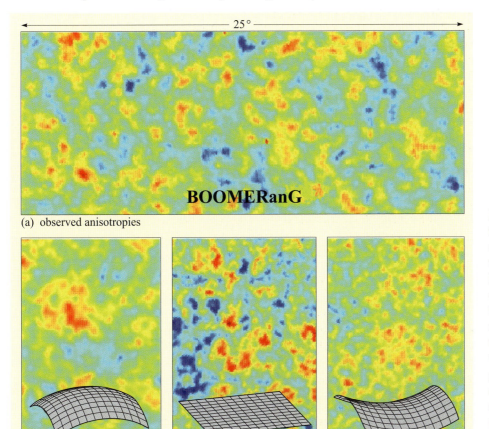

25°

BOOMERanG

(a) observed anisotropies

(b) $k = +1$ (c) $k = 0$ (d) $k = -1$

Figure 7.28 Evidence that the Universe has a flat ($k = 0$) spatial geometry. (a) When the anisotropies observed by BOOMERanG are compared with computer predictions (b, c and d) based on various cosmological models, it is the $k = 0$ predictions that provide the best agreement with the observations. The small grids printed at the bottom of the computer simulations provide a reminder of the geometric properties that correspond to the different values of k. (BOOMERanG Collaboration)

fundamental reason why in Figure 7.11 the constraints from the CMB are almost parallel to the line defining spatial flatness.

The other, lesser **acoustic peaks** in the angular power spectrum are the harmonics discussed in Section 7.4.4, and can be attributed to waves that have collapsed once, collapsed then rebounded once, fully collapsed twice, and so on. As you saw in Question 7.7, the second harmonic has a wavelength that is half that of the fundamental.

The relative ratios of these acoustic peaks depend very strongly on the nature of the fluid sloshing around, that is, on the composition of the photon–baryon fluid. To see how, first consider what forces are at play. If there is a region of higher than average density, matter around it will feel a net gravitational force toward that region. Once inside the region, the matter falling in from all sides would be compressed. This would increase the pressure in the photon–baryon fluid, which would send the fluid rebounding outwards again. The odd-numbered harmonics (first, third, etc.) are associated with how far down the matter falls into these regions, while the even-numbered peaks (second, fourth, etc.) are associated with how far the plasma rebounds.

Now consider what would happen if more baryonic matter were added in. The clump draws the matter further in, and the resulting oscillation becomes asymmetrical, with the photon–baryon fluid spending more time deep inside the clump. For this reason, adding more baryonic matter suppresses the even-numbered harmonics compared to the odd-numbered harmonics. (It also has a subtle effect on the frequency of the oscillations.) The comparison of the first and second harmonics therefore provides a very useful measure of the amount of baryonic matter in the Universe.

Figure 7.29 shows a set of predictions for the shape of the CMB angular power spectrum based on fixed values for all the main cosmological parameters apart from $\Omega_{b,0}$. The different curves in the figure show the effect of increasing the assumed current value of the baryon density parameter while keeping all the other parameters fixed. As you can see, the shape of the power spectrum changes significantly as $\Omega_{b,0}$ is increased, so in principle, it is possible to constrain the value of $\Omega_{b,0}$ from CMB observations.

The higher harmonics are particularly useful for measuring the amount of dark matter. The underlying reason is that the energy balance between radiation and

Musical readers may be interested to know that the CMB acoustic peaks are still fifty or so octaves below human hearing, even supposing humans could have survived those environments.

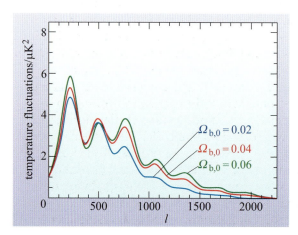

Figure 7.29 The angular power spectrum of CMB anisotropies, as predicted by big bang cosmology, for a range of values of the density parameter for baryonic matter, $\Omega_{b,0}$. The differently coloured traces correspond to values of $\Omega_{b,0}$ ranging from 0.02 to 0.06.

matter changes as the Universe expands. If the Universe expands by a factor of a, then a given volume will increase by a factor a^3. The energy density of matter (matter energy per unit volume) will therefore *decrease* by a factor of a^3. In other words, for an expansion factor $R(t)$, the energy density of matter will be proportional to $1/R^3(t)$. For photons though, the expansion stretches their wavelengths by the same factor a (as you saw in Section 6.2.2), which also reduces their energies by the same factor, so the net reduction in photon energy density is a factor of $a \times a^3 = a^4$. In other words, the photon energy density varies with the expansion factor as $1/R^4(t)$. Therefore, photon energy density drops more quickly as the Universe expands, compared to matter energy density. The energy budget of the Universe was therefore radiation-dominated at early epochs, and matter-dominated at later epochs. Now imagine a denser-than-average clump during the radiation-dominated era. The radiation-energy is what is dominating the density of the clump, and as material falls in, it becomes stabilized from the pressure of the photons. The density of the clump reduces as the Universe expands, and we reach a point where the gravitational binding is no longer sufficient to prevent the fluid rebounding out again, releasing its pressure. This strongly rebounding behaviour would not happen if the Universe is matter-dominated, which means there should be a difference between the modes that started oscillating during the radiation-dominated era and those that started in the matter-dominated era. The size of the observable Universe increased with time, which means that small-scale oscillations of the photon–baryon fluid started earlier than larger ones. For this reason, the smaller-scale multipoles above about $l > 500$ are a useful diagnostic of the relative energy density ratios of matter and radiation in the Universe.

As noted in Chapter 4, some astronomers have rejected the idea of dark matter, and explained the observations of galaxies and clusters of galaxies by modifying the law of gravity. However, there has yet to be any theoretical description of the CMB multipoles that does not invoke some quantity of dark matter, of some sort. For this reason, the CMB power spectrum turns out to be another powerful argument for the reality of dark matter.

The final feature of the CMB power spectrum (Figure 7.26) is suppression of the signal at high frequencies, i.e. high multipole numbers, in a manner that is analogous to turning down the 'treble' of a graphic equalizer (Section 7.4.4). The physical cause of this is the fact that recombination was not instantaneous. Before recombination the electrons are not bound to nuclei, and the photons scatter easily off these electrons. This rapid scattering of photons means that the Universe appears opaque, like a thick fog. In turn, this means that regions of high matter density are also regions of high photon density, and low matter density means low photon density. This is partly why the matter and radiation at the time is often referred to as a combined 'photon–baryon fluid'. After recombination the electrons are bound to protons and neutrons (making neutral atoms), and the photon scattering is very much reduced, so the Universe appeared much more transparent and photons can travel more or less unimpeded. However, *during* recombination, the photons begin to drift away from their locations, in random directions, in a statistically random process known as a **random walk**. This means the photons do not accurately trace where the matter is on small scales, and the effect is to smooth out the structures in the visible CMB on the smallest scales. If recombination lasted a time Δt, then structures smaller than the distance $c\Delta t$ that light can travel in this time would tend to be vulnerable to this smoothing out. This process is illustrated schematically in Figure 7.30, and is known as **Silk damping** after the

Figure 7.30 Illustration of Silk damping. Recombination is illustrated progressing from left to right. A harmonic of the density perturbations is illustrated as a single vertical wave, for simplicity. As recombination progresses and the Universe becomes progressively more transparent, the photons diffuse away from their initial positions. Photon paths are shown in yellow. This photon diffusion will tend to smooth out structures smaller than $c\Delta t$, where Δt is the duration of recombination. (W. Hu, University of Chicago)

astronomer Joe Silk who first proposed it. The duration of recombination depends on the age of the Universe at the time, so as a result the details of this smoothing depend on the cosmological parameters Ω_m, Ω_b, and so on. For this reason, the details of the smoothing provide astronomers with a very useful consistency check of their interpretation of the CMB acoustic peaks.

7.5 The concordance cosmology

The term precision cosmology was coined in 1996 by the American cosmologist Michael Turner. It expressed rather well the direction in which cosmology was then heading. Gone are the days when cosmologists were free to speculate about the origin and evolution of the Universe with hardly any observational data to guide or fetter them. Cosmologists, like most other scientists, are now fenced in by large amounts of data, some of them very precise. Hubble's original estimate of the constant that bears his name was wrong by a factor of about 10. The estimate based on Planck's CMB anisotropy observations is thought to be within about 2% of the true value. Many of the other cosmological parameters are now thought to be known with a similar level of precision.

It is not just the precision of the measurements that have made such a difference in modern cosmology, it is also the striking agreement between very different types of constraint, as you will see in this chapter. This broadly consensual view has come to be known as the **concordance cosmology**, though as you will see later in this chapter, this cosmological model is not without its anomalies.

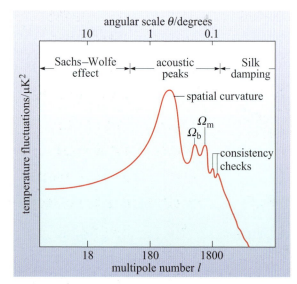

Figure 7.31 The causes of the main features in the angular power spectrum. The various effects named on the graph are discussed in the text. Note that in this case the multipole number has been plotted logarithmically, to allow the inclusion of large values of l.

7.5.1 The parameter inferences from the CMB

The key point to come from our discussion about analysis of fluctuations in the CMB is that the predicted shape of the angular power spectrum (Figure 7.29) of a particular cosmological model depends on the values of H_0, $\Omega_{m,0}$, $\Omega_{\Lambda,0}$, $\Omega_{b,0}$, A, n_s, etc. that are used in its calculation. This is an important point because it provides a method of deducing the values of all the cosmological parameters simultaneously. By comparing a large number of predicted power spectra with the available observational data regarding the power spectrum it is possible to identify a 'best-fit' spectrum, and hence to come to some conclusions about the most probable values for the various cosmological parameters. More detailed analysis even allows uncertainties in those parameters to be estimated. Figure 7.31 summarizes the nature of the constraints, and the underlying processes, in the CMB multipoles.

Description	Parameter	Best fit	Planck + WMAP 68% limits
Baryon density today $\times h^2$	$\Omega_{b,0}h^2$	0.022032	0.02205 ± 0.00028
Cold dark matter density today $\times h^2$	$\Omega_{c,0}h^2$	0.12038	0.1199 ± 0.0027
$100 \times$ approximation to r_*/D_A (CosmoMC)	$100\theta_{MC}$	1.04119	1.04131 ± 0.00063
Thomson scattering optical depth due to reionization	τ	0.0925	$0.089 ^{+0.012}_{-0.014}$
Scalar spectrum power-law index ($k_0 = 0.05$ Mpc^{-1})	n_s	0.9619	0.9603 ± 0.0073
Log power of the primordial curvature perturbations ($k_0 = 0.05$ Mpc^{-1})	$\ln(10^{10}A_s)$	3.0980	$3.089 ^{+0.024}_{-0.027}$
Dark energy density divided by the critical density today	$\Omega_{\Lambda,0}$	0.6817	$0.685 ^{+0.018}_{-0.016}$
Matter density (inc. massive neutrinos) today divided by the critical density	$\Omega_{m,0}$	0.3183	$0.315 ^{+0.016}_{-0.018}$
RMS matter fluctuations today in linear theory	σ_8	0.8347	0.829 ± 0.012
Redshift at which the Universe is half reionized	z_{re}	11.37	11.1 ± 1.1
Current expansion rate in km s^{-1} Mpc^{-1}	H_0	67.04	67.3 ± 1.2
$10^9 \times$ dimensionless curvature power spectrum at $k_0 = 0.05$ Mpc^{-1}	$10^9 A_s$	2.215	$2.196 ^{+0.051}_{-0.060}$
Total matter density today (inc. massive neutrinos) $\times h^2$	$\Omega_{m,0}h^2$	0.14305	0.1426 ± 0.0025
Total matter density today (inc. massive neutrinos) $\times h^3$	$\Omega_{m,0}h^3$	0.09591	0.09589 ± 0.00057
Fraction of baryonic mass in helium	Y_p	0.247695	0.24770 ± 0.00012
Current age of the Universe	Age/Gyr	13.8242	13.817 ± 0.048
Redshift for which the optical depth equals unity	z_*	1090.48	1090.43 ± 0.54
Comoving size of the sound horizon at $z = z_*$	r_*	144.58	144.71 ± 0.60
$100 \times$ angular size of sound horizon at $z = z_*$ (r_*/D_A)	$100\theta_*$	1.04136	1.04147 ± 0.00062
Redshift at which baryon-drag optical depth equals unity	z_{drag}	1059.25	1059.25 ± 0.58
Comoving size of the sound horizon at $z = z_{drag}$	r_{drag}	147.36	147.49 ± 0.59
Characteristic damping comoving wavenumber (Mpc^{-1})	k_D	0.14022	0.14009 ± 0.00063
$100 \times$ angular extent of photon diffusion at last scattering	$100\theta_D$	0.161375	0.16140 ± 0.00034
Redshift of matter-radiation equality (massless neutrinos)	z_{eq}	3403	3391 ± 60
$100 \times$ angular size of the comoving horizon at matter-radiation equality	$100\theta_{eq}$	0.8125	0.815 ± 0.011
BAO distance ratio at $z = 0.57$	$r_{drag}/D_v(0.57)$	0.07126	0.07147 ± 0.00091

Table 7.2 The values of the various cosmological parameters obtained by combining the Planck and WMAP observations. The first column gives the best-fit value. The second column gives the uncertainty range for each parameter. There is a 68% probability that the 'true' underlying value is within the uncertainty range for each parameter. Note that the uncertainties are sometimes asymmetrical. These values are taken from a paper by the Planck collaboration and draw on the results obtained by many different observational cosmologists. Some parameters are combined with h, which is the present-day Hubble parameter divided by 100 km s^{-1} Mpc^{-1}. (Note that some of these parameters have not been described in this book.)

This process of determining cosmological parameters from CMB anisotropy observations has already been carried out by a number of observational cosmology research groups. The latest values at the time of writing, those obtained by the Planck team, are shown in Table 7.2. Some of the parameters listed will be unfamiliar since they refer to aspects of cosmology that have not been discussed in this book. However, from the list it is possible to draw the following values for the key cosmological parameters:

$$H_0 = (67.3 \pm 1.2) \text{ km s}^{-1} \text{ Mpc}^{-1}$$

$$\Omega_{m,0} = 0.32 \pm 0.02$$

$$\Omega_{\Lambda,0} = 0.68 \pm 0.02$$

$$\Omega_{b,0} = 0.049 \pm 0.003$$

Also included in the list are the values of some 'secondary' parameters that depend on the key parameters. Among them is a value for the age of the Universe that provides a valuable cross-check on the reasonableness of the 'primary' results. In the case of the Planck findings, the favoured value for the age of the Universe is

$$t_0 = 13.82 \times 10^9 \text{ years}$$

Pleasingly, this does seem to be a realistic estimate for t_0, since it makes the Universe older than the oldest stars we know of, the Population II stars in globular clusters. In view of this and other consistency checks, the Planck results have been well received by cosmologists and are widely thought to represent the best numerical characterization of the Universe currently available.

7.5.2 Ω_m measurements besides the CMB

There are many methods of determining the current value of Ω_m apart from the CMB measurements. You have already met one type of constraint: measurements of high-redshift supernovae. Some of these non-CMB methods are of dubious reliability, and prior to 'precision cosmology' the results for $\Omega_{m,0}$ ranged from about 0.2 to 1.5, often with substantial uncertainties.

However, the situation has changed strikingly in the era of 'precision cosmology'. Perhaps the strongest constraint on $\Omega_{m,0}$ independent of the cosmic microwave background is 'baryonic acoustic oscillations' or simply 'baryon wiggles'. Primordial fluctuations (as seen in the CMB) were the initial seeds for the later formation of stars and galaxies. Material flowed from underdense regions into overdense regions, by the action of gravity, exacerbating the contrasts in density and ultimately giving rise to stars and galaxies. As a result, the galaxy distribution was also left with the imprint of the primordial acoustic fluctuations seen in the CMB. The wavelengths of these harmonics seen throughout the Universe give astronomers a 'standard rod', i.e. an object of a known size, which can be seen throughout much of the history of the Universe. One can make inferences about the expansion history of the Universe by measuring the apparent (angular) sizes on the sky of these large-scale features. This expansion history is strongly affected by the amount of matter in the Universe, which acts to slow down the expansion, and from this is derived a constraint on $\Omega_{m,0}$.

Table 7.3 shows the constraints on $\Omega_{m,0}$ from a selection of baryonic acoustic oscillation surveys, in isolation and in combination. The results are in good agreement with those from the CMB and supernovae, although there is still currently some discussion over the Hubble parameter, which we will discuss below.

Table 7.3 A selection of constraints on $\Omega_{m,0}$ and the Hubble constant from baryonic acoustic oscillation surveys, in isolation and in combination. Note that the uncertainties are often asymmetrical, so are given separately as positive and negative uncertainties. 6dF is the Six-Degree Field Galaxy Survey; SDSS is the Sloan Digital Sky Survey, while SDSS(R) is the SDSS data with a correction applied for galaxy motions relative to the rest-frame of the FRW metric; WiggleZ is another galaxy survey, carried out at the Anglo-Australian Telescope; BOSS is the Baryon Oscillation Spectroscopic Survey.

Sample	Ω_m	$H_0/\mathrm{km\ s^{-1}\ Mpc^{-1}}$
6dF	$0.305\,^{+0.032}_{-0.026}$	$68.3\,^{+3.2}_{-3.2}$
SDSS	$0.295\,^{+0.019}_{-0.017}$	$69.5\,^{+2.2}_{-2.1}$
SDSS(R)	$0.293\,^{+0.015}_{-0.013}$	$69.6\,^{+1.7}_{-1.5}$
WiggleZ	$0.309\,^{+0.041}_{-0.035}$	$67.8\,^{+4.1}_{-2.8}$
BOSS	$0.315\,^{+0.015}_{-0.015}$	$67.2\,^{+1.6}_{-1.5}$
6dF+SDSS+BOSS+WiggleZ	$0.307\,^{+0.010}_{-0.011}$	$68.1\,^{+1.1}_{-1.1}$
6dF+SDSS(R)+BOSS	$0.305\,^{+0.009}_{-0.010}$	$68.4\,^{+1.0}_{-1.0}$
6dF+SDSS(R)+BOSS+WiggleZ	$0.305\,^{+0.009}_{-0.008}$	$68.4\,^{+1.0}_{-1.0}$

Another independent method of measuring $\Omega_{m,0}$ is to use clusters of galaxies. As you saw in Chapter 4, clusters of galaxies contain large amounts of hot gas, and this gas interacts with the CMB photons that pass through it. This creates characteristic distortions in the CMB spectrum, called the Sunyaev–Zeldovich effect, which can be used to find clusters of galaxies. The formation of clusters of galaxies depends sensitively on the density of matter in the Universe, and as a consequence, the number of clusters that we can observe through the Sunyaev–Zeldovich effect provides a way to measure $\Omega_{m,0}$. The current best constraint is $\Omega_{m,0} = 0.29 \pm 0.02$, in very good agreement with the combined CMB and supernovae constraints. However, a subtlety is that the numbers of clusters of galaxies also depend on how strongly matter clusters in general, and measurements of this matter-clustering strength are slightly discrepant between CMB data and galaxy cluster data. This may just be a statistical fluke, or may be our first inkling of deep new physics, of which we will hear more later in this chapter.

An important part of the galaxy cluster constraints on $\Omega_{m,0}$ is that the Sunyaev–Zeldovich distortions in the CMB spectrum have to be converted into masses, so independent mass estimates are needed for at least some galaxy clusters in order to calibrate this conversion.

■ List and briefly describe the optical and X-ray techniques for determining cluster masses that were described in Chapter 4.

☐ *The virial mass method*, which is based on Doppler measurements of the velocity dispersion within the cluster and the assumption that the cluster is in dynamical equilibrium (i.e. relaxed or virialized).

The X-ray emission method, which is based on X-ray spectra and X-ray surface brightness measurements (to determine the temperature and density distribution of the intracluster gas) and the assumption that the gas is in hydrostatic equilibrium.

The gravitational lensing method, which is based on observations of distorted images of more distant galaxies and models of the mass distribution within the lensing cluster.

A further galaxy cluster-based method for measuring $\Omega_{m,0}$ is the method of mass-to-light ratios. The **mass-to-light ratio** of an astronomical system, such as a star or galaxy or even a cluster of galaxies, is the value obtained by dividing the mass M of the system by its luminosity L. The luminosity in this definition is normally restricted to some specified range of wavelengths, usually the optical range. So, in the case of the Sun where the mass is $M_\odot$ and the (optical) luminosity is $L_\odot$, the mass-to-light ratio is given by $M_\odot/L_\odot$.

The mass-to-light ratio of the Sun is not of much direct interest in cosmology, but if observers could determine the mass-to-light ratio of a representative portion of the Universe then they could multiply that quantity by the observable luminosity density of the Universe, j_{Univ}, to obtain the average matter density ρ and hence the density parameter for matter Ω_m. In terms of symbols

$$\Omega_m = \rho/\rho_{crit} = [(M/L)_{Univ} \times j_{Univ}]/\rho_{crit} \tag{7.12}$$

where $(M/L)_{Univ}$ represents the mass-to-light ratio of the Universe.

■ What would be appropriate SI units of $(M/L)_{Univ}$ and j_{Univ}?

□ $(M/L)_{Univ}$ could be expressed in terms of kg W^{-1} (i.e. kilograms per watt) and j_{Univ} could be measured in W m^{-3} (i.e. watts per cubic metre). In practice, astronomers are quite likely to use $M_\odot/L_\odot$ as a unit of mass-to-light ratio, and to express the luminosity density in terms of solar luminosities per cubic parsec ($L_\odot$, pc^{-3}), but these are not SI units.

Any attempt to use mass-to-light ratios to determine $\Omega_{m,0}$ must, of course, take account of many details. For instance, the same range of wavelengths must be used in determining $(M/L)_{Univ}$ and j_{Univ}, and the range over which all quantities are measured, although large enough to be cosmologically representative, must also be small enough to ensure that it is the *current* values of $(M/L)_{Univ}$ and j_{Univ} that are being determined. (Presumably, the values of both $(M/L)_{Univ}$ and j_{Univ} have changed over time.) Nonetheless, the basic challenges of this method are clear: determine $(M/L)_{Univ}$ and measure j_{Univ}.

Over the years, these challenges have been confronted by a number of astronomers, starting with Fritz Zwicky (see Section 4.3.2) in the 1950s. It has been clear for a long time that mass-to-light ratios of astronomical systems tend to increase as the size of the system becomes larger, at least up to scales of hundreds of kiloparsecs. When dealing with galaxies and clusters of galaxies, this growth is usually interpreted as signifying the presence of large amounts of dark matter, since dark matter tends to increase the mass of a system without producing any corresponding increase in luminosity. In fact, the growth of mass-to-light ratios with the size scale on which they are measured is often quoted as evidence of the existence of dark matter. However, there is also evidence that, as the size of the systems considered increases to a megaparsec and beyond, so the growth in the mass-to-light ratio ceases and M/L attains a constant value. It is this constant limiting value that is taken to represent the average 'cosmic' value of the mass-to-light ratio.

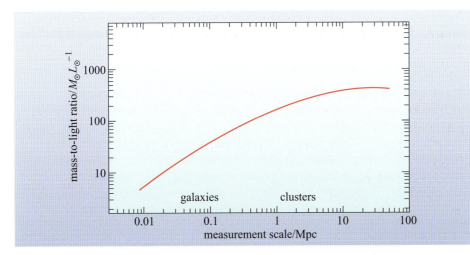

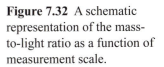

Figure 7.32 A schematic representation of the mass-to-light ratio as a function of measurement scale.

Figure 7.32 provides a highly schematic representation of the way that observed mass-to-light ratios increase with the scale on which they are measured. As you can see, for the visible parts of galaxies (corresponding to measurement scales of about 10 kpc) the mass-to-light ratio is about five times $M_\odot/L_\odot$, while on the larger scale of entire galaxies (a scale of 100 kpc or so, which would include the galaxy's dark halo) the mass-to-light ratio is about 20 times $M_\odot/L_\odot$, and on the even larger scale of groups and clusters (scales around 1 Mpc) the mass-to-light ratio is more like 240 times $M_\odot/L_\odot$. The belief that the mass-to-light ratio becomes roughly constant on the largest of these scales is supported by various measurements, including the finding that the mass-to-light ratio of the supercluster MS302 is essentially equal to the mass-to-light ratios of the clusters of which it is composed. (In this latter case, all the masses were determined using the gravitational lensing method.)

It appears that mass and light do share the same overall distribution on the large scale (i.e. the mass-to-light ratio does become constant) but there is still a great need for caution when making measurements because relatively dense regions such as rich clusters of galaxies exhibit higher mass-to-light ratios than do lower-density regions. This so-called 'bias' effect is attributed to the relatively more highly evolved state of the high-density regions, where the emission of blue light from stars has declined more than in the relatively less evolved, lower-density regions. Taking this bias into account, and the effect of the slightly different clustering strength of luminous matter compared to matter as a whole, the conclusion is that

$$\Omega_{m,0} = 0.26 \pm 0.02 \qquad (7.13)$$

This is slightly lower than the widely favoured value obtained from supernova cosmology ($\Omega_{m,0} \approx 0.3$), but not vastly different given the difficulty of the measurements.

7.5.3 Determinations of $\Omega_{b,0}$

The methods of density determination that have been discussed in the last section have been sensitive to dark matter, and have indicated its presence in substantial amounts. On the basis of mass-to-light ratios, it is widely assumed that much of this dark matter is contained in dark-matter halos of galaxies. However, these measurements give little indication of the nature of the dark matter. One obvious possibility is that it might be some form of ordinary baryonic matter. One way to investigate this possibility is to determine the density parameter for baryonic matter $\Omega_b(t)$; if the current value of this quantity, $\Omega_{b,0}$, is significantly less than the current value of the density parameter for all kinds of matter, $\Omega_{m,0}$, then much of the dark matter must be non-baryonic.

This section presents two arguments for the belief that the current average density of baryonic matter is much less than the current total density of matter, independently of the evidence from the CMB. The first of these arguments is based on the requirement that the relative abundances of light elements predicted by the theoretical account of *primordial nucleosynthesis* (as discussed in Chapter 6) should agree with the observational evidence regarding those abundances. The second argument is based on direct observational assessments of the density of baryonic matter in various parts of the Universe; that is to say a *baryon inventory*.

Primordial nucleosynthesis

According to Chapter 6, the abundances of light elements (essentially hydrogen, helium and lithium) predicted by primordial nucleosynthesis are related to the current value of the density parameter for baryonic matter. This link exists because the total number of baryons in the Universe is a *conserved* quantity that is not expected to change with time, according to the standard model of particle physics. So the current mean density of baryonic matter is related to the density of baryonic matter at the time of cosmic nucleosynthesis, and this primordial density played an important role in determining the extent to which light nuclei were synthesized between about three and 30 minutes after the start of cosmic expansion.

One of the great successes of big bang cosmology is its ability to simultaneously predict primordial abundances of deuterium, helium and lithium that are consistent with the abundances currently observed in stars and gas clouds (once allowances have been made for more recent astronomical processes such as stellar nucleosynthesis). In order to achieve this consistency, however, it is necessary that the current value of the density parameter for baryonic matter should be in the range

$$0.02 \leq \Omega_{b,0} \leq 0.05 \tag{7.14}$$

This is much lower than the estimates of $\Omega_{m,0}$ based on CMB and supernova cosmology which, as you saw earlier, are around 0.3. So it seems that baryons account, very roughly, for only somewhere between about a fifteenth and a sixth of the total density of matter in the Universe.

A baryon inventory

Where are the baryons (protons, neutrons, etc.) in the Universe? Well, some are in you and me. We each contain about 2 to 4×10^{28} protons and neutrons. But there are many more in the Earth, far more in the Sun, and enormously more in the Milky Way and other galaxies. Perhaps surprisingly, stars and their remnants are thought to account for less than 30% of the baryonic matter in the Universe. The majority of the baryons are believed to reside in the various forms of ionized gas that exist within and between the galaxies. Astronomers have attempted to add up all these contributions to make an independent measure of $\Omega_{b,0}$.

A baryon inventory of this kind is open to the criticism that many of the entries, including some of the most important, are hard to estimate. It is also the case that some repositories of baryons may have been overlooked. The general conclusion, after taking account of all known repositories of baryons and making reasonable estimates wherever data are lacking, is that $\Omega_{b,0}$ is in the range

$$0.007 \leq \Omega_{b,0} \leq 0.041 \tag{7.15}$$

with a 'best guess' value of

$$\Omega_{b,0} \approx 0.021 \tag{7.16}$$

Because of the uncertainties this 'best guess' should not be taken too seriously. However, the upper part of the range given in Equation 7.14 is consistent with the requirements of primordial nucleosynthesis, and gives further evidence that the dark matter cannot be entirely, or even mainly, baryonic matter.

QUESTION 7.9

Justify the rough claim made earlier that 'baryons account for only somewhere between about a fifteenth and a sixth of the total density of matter in the Universe'.

7.5.4 Venturing into the unknown

The overall picture from the previous sections is that some very diverse experimental constraints are all zeroing in on very similar values for cosmological parameters. From the CMB to the clustering of galaxies to the abundances of elements in the present-day Universe to the properties of galaxy clusters to measurements of distant supernovae, everything seems to be almost exactly consistent. This is a great triumph in our understanding of the cosmos.

But that does not mean that the picture is *entirely* self-consistent yet. One of the significant surprises from the Planck satellite is that it appears to support slightly lower values for H_0 than other determinations. Figure 7.33 compares the Hubble parameter from various sources. Might there be a problem in the derivation of H_0 from the Planck CMB maps? As you saw in Figure 7.11, the value of one parameter can be closely related to that of another, so perhaps the values assumed for the other cosmological parameters are incorrect. The subtlety of CMB parameter estimation is that the constraints are always inevitably model-dependent. Alternatively, might the local determinations of H_0 have uncorrected systematic uncertainties? Some astronomers have argued that this might be the case, but it can't entirely remove the anomaly, although it might *just* be enough to be able to write off the remaining differences as flukes. There are also more exotic possible explanations. One exotic option is to suppose the existence of new species of fundamental particle called a *sterile neutrino* that interacts mainly through gravity alone. The addition of these particles changes the CMB models in such a way as to resolve these differences. Another exotic possibility is that dark energy is not quite the cosmological constant originally envisaged by Einstein.

This is not the only anomaly present in the concordance cosmology. We have already mentioned the discrepancy in the clustering strength of matter on the scale of massive galaxy clusters. It is not clear what the underlying cause of this

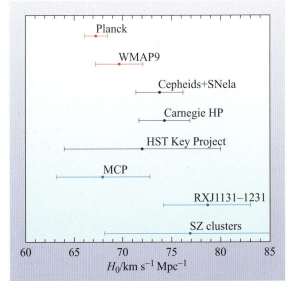

Figure 7.33 Comparison of H_0 measurements, with 68% confidence regions for each. The data points are marked with their associated particular classes of objects, single objects, or studies, that resulted in the measurement. The Carnegie HP project uses a recalibration of the HST Key Project data using the Spitzer space telescope.

discrepancy is, but a 'sterile neutrino' is again an exotic option that could explain this discrepancy. All the constraints are highly model-dependent, so at the time of writing it seems far too early to rewrite the textbooks, but the results are intriguing nonetheless.

There are stranger anomalies still, and the proposed explanations are even more exotic. A big surprise from the Planck data is the detection of large-scale anomalies in the CMB sky, which have been accentuated in Figure 7.34. The observed structures on scales of 6 degrees to 90 degrees are about 10% weaker than models predict. Also, both Planck and WMAP found that one hemisphere of the CMB appears to be slightly more structured than the other. These strange effects are yet to have an agreed explanation. There is also a mysterious 'cold spot' in the CMB, first detected by WMAP. It had been thought that this spot and the hemispheric asymmetry could both be instrumental artefacts in the WMAP data, but their confirmation by Planck appears to rule out this possibility. One explanation of the cold spot is a giant underdense region in the Universe called a 'supervoid'. If we live in such a supervoid, it could be enough to explain the acceleration of the expansion of the Universe, because the underdense region would expand more quickly. We would however need to be located very close to its centre. Could we really be living right in the middle of a vast and mysterious hole in the distribution of galaxies?

Cosmology is a fast-moving field, and it remains possible that the addition of new data, or a more careful reassessment of existing data, could remove some of these anomalies. Alternatively, if these anomalies are not themselves just statistical flukes, they may be our first indications of deep new revelations about the underlying fundamental processes of our Universe.

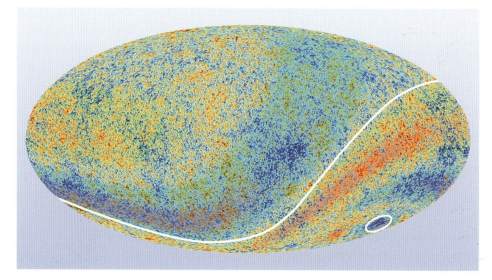

Figure 7.34 The CMB sky seen by Planck, with anomalous large-scale features accentuated with red and blue shading to make them more visible. The regions separated by the long white line have slightly different average temperatures. The 'cold spot' is marked with a white ellipse. (ESA and the Planck Collaboration)

7.6 Summary of Chapter 7

Measuring the Hubble constant, H_0

- H_0 measures the current rate of expansion of the Universe.

- H_0 is traditionally determined by means of the Hubble diagram, a plot of redshift against distance for distant galaxies. When making such a plot, the redshift and the distance must be determined independently. The Hubble constant is obtained from the gradient of the plotted line.

- The HST Key Project team used Cepheid variables to calibrate five other methods of distance measurement. Using these together with independent measurements of galaxy redshifts, they concluded that $H_0 = 72 \pm 8$ km s^{-1} Mpc^{-1}.

- Other methods of determining H_0 include those based on gravitational lensing (via time delays between fluctuations in different images of the same lensed galaxy) and those based on observations of anisotropies in the cosmic microwave background radiation (CMB). The current best evidence from the Planck CMB satellite is that $H_0 = 67.3 \pm 1.2$ km s^{-1} Mpc^{-1}. This is slightly discrepant with other current non-CMB determinations and these differences are currently still a matter of discussion among astronomers.

Measuring the current value of the deceleration parameter, q_0

- q_0 measures the current rate of change of cosmic expansion.

- q_0 may be determined from the Hubble diagram by observing the curvature of the plotted line at $z > 0.2$, but early attempts to do this were highly inconsistent.

- Results obtained using Type Ia supernovae as distance indicators suggest that q_0 is negative, implying that the expansion of the Universe is speeding up. It has become traditional to express the value of q_0 in terms of $\Omega_{m,0}$ and $\Omega_{\Lambda,0}$, using the relation

$$q_0 = \frac{\Omega_{m,0}}{2} - \Omega_{\Lambda,0}$$

Measuring the current values of the density constants, $\Omega_{\Lambda,0}$, $\Omega_{m,0}$ and $\Omega_{b,0}$

- The current values of the density parameters $\Omega_{\Lambda,0}$, $\Omega_{m,0}$ and $\Omega_{b,0}$ measure the densities associated with the cosmological constant (dark energy), matter of all kinds and baryonic matter, relative to the critical density $\rho_{crit} = 3H_0^2/8\pi G$.

- Results based on observations of anisotropies in the CMB strongly favour $\Omega_{\Lambda,0} + \Omega_{m,0} = 1$, implying a flat Universe (i.e. $k = 0$).

- When these are combined with the results of observations of Type Ia supernovae, favoured values of $\Omega_{m,0}$ and $\Omega_{\Lambda,0}$ are typically $\Omega_{m,0} = 0.32 \pm 0.02$ and $\Omega_{\Lambda,0} = 0.68 \pm 0.02$.

- The current value of the density constant for baryonic matter can be determined in a number of ways. This quantity is constrained by primordial nucleosynthesis calculations that, if they are to agree with observations, require $0.02 \leq \Omega_{b,0} \leq 0.05$.

- The acoustic oscillations in the CMB require $\Omega_{b,0}$ to lie in the range $0.046 \leq \Omega_{b,0} \leq 0.052$.

- Direct assessments of $\Omega_{b,0}$, based on baryon inventories, generally favour lower values: $0.007 \leq \Omega_{b,0} \leq 0.041$. However these assessments are beset by many uncertainties, in particular in being potentially incomplete.

Anisotropies in the CMB and precision cosmology

- Although highly isotropic, the CMB exhibits anisotropies in intensity at the level of a few parts in 100 000 over a range of angular scales. These can be mapped, and are usually shown as variations in the temperature of the CMB.

- The angular power spectrum of an anisotropy map shows the level of variation that is present on any specified angular scale (or, equivalently, the angular power at multipole number l).

- Values of cosmological parameters may be extracted from anisotropy measurements by comparing the observed angular power spectrum with that predicted by big bang cosmology. Recent results from the Planck space probe imply a Universe dominated by dark energy, and in which most of the matter is non-baryonic dark matter.

- Planck measurements also indicate that the age of the Universe is $t_0 = (13.82 \pm 0.05) \times 10^9$ years.

- We have now entered an era of precision cosmology in which cosmological speculations are tightly constrained by measurements, and quantities that were previously very uncertain are now accurately determined by several different methods. Nevertheless, despite these successes there are still subtle anomalies between several data sets that could indicate systematic uncertainties not fully recognized, or could be tantalizing hints of new fundamental physical processes in the Universe.

Questions

QUESTION 7.10

Using a variety of measurements (not just CMB), outline the observational basis of the claim that

(a) the Universe is dominated by dark energy

(b) most of the matter is dark

(c) most of the dark matter is non-baryonic.

QUESTION 7.11

On the basis of the very incomplete account given in this chapter, outline the role that space technology has played in observational cosmology and the role that it is expected to play in the future development of the subject.

QUESTION 7.12

Quote some examples to show the importance of terrestrial (as opposed to space-based) observations in cosmology.

CHAPTER 8
QUESTIONING COSMOLOGY –
OUTSTANDING PROBLEMS ABOUT THE UNIVERSE

8.1 Introduction

'Don't let me catch anyone talking about the Universe in my department.'

Lord Rutherford

Rutherford – the discoverer of the atomic nucleus – was a practical man. Perhaps one of the greatest experimental physicists of all time, he was profoundly sceptical of notions that were not grounded in hard experimental evidence. We can guess that he might not have been happy with some of the ideas of modern cosmology! But, however challenging the concepts of cosmology – and ideas about space being curved or expanding and the vacuum possessing energy are certainly challenging – there can be no doubt that Rutherford would have been deeply impressed and intrigued by the vast amount of observational evidence that cosmologists have now acquired. His main interest, however, might well have been in those areas of cosmology where there are still clear gaps in our knowledge; the areas where work remains to be done and where new insights can be expected to arise.

The last few chapters have been largely concerned with the development and testing of models of the Universe. By a model we mean some simplified representation of the real world that helps us to understand reality by focusing on some specific aspects of it. The model should be simple, but not so simple that the phenomena of interest are inadequately represented. The Earth's orbit around the Sun, for example, can be modelled by a circle. This is an adequate model for explaining the occurrence of certain annual events, but a better model, such as an ellipse, is required to explain finer details such as the precise timing of those events. Neither model fully represents reality, but both are useful within their own ranges of validity, and the greater precision of the elliptical model is bought at the price of greater mathematical complexity.

■ In what important way did the FRW cosmological models of Chapter 5 simplify reality?

☐ They treated the contents of the Universe as a simple uniform fluid, the properties of which could be specified at any time by a density $\rho(t)$ and a pressure $p(t)$. Small-scale departures from uniformity, such as stars and galaxies, or even whole clusters of galaxies, were ignored.

■ In what important way did the big bang model of Chapter 6 improve on this?

☐ In Chapter 6 the matter in the Universe was treated more realistically by taking account of the variety of interacting particles that it contains, and by acknowledging the presence of density fluctuations.

Despite its conceptual difficulty, the big bang is now widely accepted as the best available model of how the Universe evolved into its present state. However, it is quite clear that the model, as it has been presented in this book, and insofar as it is accepted by the majority of cosmologists, is still inadequate in several important respects. There are a number of major questions about our Universe that the standard big bang model does not address. This does not mean the model is wrong, but it does indicate deficiencies in certain areas and the model may need to be extended in those areas. This chapter concerns some of these outstanding problems, and considers the ways in which the standard big bang model might be extended to provide a more adequate account of reality.

The questions we shall consider are these:

Problem 1: What is dark matter? (Section 8.2)

Problem 2: What is dark energy? (Section 8.3)

Problem 3: Why is the Universe so uniform? (Section 8.4.1)

Problem 4: Why does the Universe have a flat ($k = 0$) geometry? (Section 8.4.2)

Problem 5: Where did structure come from? (Section 8.5)

Problem 6: Why is there more matter than antimatter? (Section 8.6)

Problem 7: Did inflation happen? (Section 8.7)

Problem 8: Why is the Universe the way it is? (Section 8.8)

8.2 The nature of dark matter

As you have seen many times in this book, the visible matter in the Universe – the stuff of stars and nebulae – accounts for only a small fraction of the whole. By recent estimates about 85% of the matter in the Universe is dark matter, and its nature is still a mystery. This is hardly satisfactory! So far you have been asked to accept the existence of dark matter without really enquiring what it is, but the time has now come to address that question – Problem 1 in our list – head on.

To prepare for that, we will review some of what you have already learned about dark matter.

■ What do you understand by baryonic and non-baryonic dark matter?

□ Baryonic dark matter is non-luminous matter in which most of the mass is attributable to baryons, most probably neutrons and protons. Non-baryonic dark matter is non-luminous matter made from something else.

In Chapter 6 you saw that the physics of the early Universe, especially the nucleosynthesis of the elements, puts constraints on the density of baryonic matter. No more than about 15% of the matter in the Universe can be baryonic. To further explore this issue, surveys have been conducted to determine what proportion of dark matter may be tied up in astronomical bodies of very low luminosity, such as stellar remnants or substellar masses (and referred to by the term **MACHOs**, for *massive astrophysical compact halo objects*). These surveys, which were based on a technique called gravitational microlensing, found that less than 20% of the dark matter in the halos of galaxies could be attributed to such objects. Such results add further support to the view that the dominant component of dark matter is non-baryonic in nature. Figure 8.1 summarizes our current understanding of the composition of the Universe.

It is widely assumed that non-baryonic dark matter is in the form of an, as yet, undiscovered subatomic particle. The known fundamental particles of nature are ruled out because they do not have the properties that are required. In particular, the hypothesized dark-matter particle cannot interact with other particles by the electromagnetic or strong interaction, although it may be subject to the weak interaction (and, of course, it has to interact gravitationally). Also, as noted in Chapter 2, in order to explain the formation of galaxies and other structure in the Universe, dark matter must move at relatively low speeds compared to the speed of light – this is what is meant by the term 'cold dark matter' (CDM). It is for this reason that the known types of neutrino (which satisfy the requirements of not interacting by the strong or electromagnetic interaction) are not good candidates for dark matter.

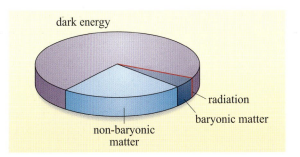

Figure 8.1 The contributions to the total (energy) density of the Universe from various sources. According to recent estimates, about 68% of the density of the Universe is currently due to dark energy, and about 32% is due to matter of all types. Only 4% of the total density is due to baryonic matter (i.e. roughly 13% of the matter). The contribution from radiation is only about 0.005%.

8.2.1 WIMPs

There are several types of hypothetical particle that could fit the bill, but the one that has received most attention is the WIMP (weakly interacting massive particle), which was mentioned in Chapter 6 in relation to the growth of structure under the influence of gravity in the early Universe. In that context, the justification for introducing the idea of a particle that responds only to gravity and the weak interaction was that density fluctuations composed of such particles would be able to grow prior to recombination. It should be stressed that at the time of writing no one knows what the WIMPs are, but several candidates have been proposed.

Here we shall consider only one WIMP candidate called the **neutralino**. This particle is associated with a proposed new symmetry of nature known as **supersymmetry**. The notion of symmetry plays an important part in the standard model of elementary particles, since it implies various relationships between the fundamental particles and between the laws that govern them. Supersymmetry – first proposed in the 1970s but still unconfirmed – would, if it existed, extend the known symmetries of nature and imply the existence of many kinds of particles that have not yet been observed in nature. On theoretical grounds it is expected that there should be a stable neutralino with a relatively high mass (m_W) – roughly in the range 10 to 1000 times the mass of the proton. Like neutrinos, neutralinos are predicted to interact with other particles only through the weak interaction and through gravity. If the theory of supersymmetry is correct, then some of the neutralinos created in the early moments of the big bang will still be present in the Universe today and these might be the WIMPs mainly responsible for cold dark matter.

Whether WIMPs are neutralinos or other types of particles, there are essentially three experimental approaches that can be adopted to demonstrate their existence. Firstly, we may seek to directly detect dark-matter particles as they pass through an Earth-bound laboratory. Secondly, we could attempt to create candidate particles using high-energy collider experiments. Finally, we could use an indirect technique that depends on detecting the products of an interaction between a WIMP and its antiparticle.

Direct detection

Although dark matter in the Milky Way is believed to form a widely dispersed dark-matter halo, the visible parts of the Galaxy lie within this halo. Thus the dark-matter particles are expected to permeate the entire Galaxy, and we may expect to find WIMPs in our neighbourhood. They interact only weakly with ordinary matter, so they will pass freely through the Earth. The predicted number density of these particles is such that more than a hundred million of them will pass through your head as you read this sentence.

This opens up the attractive possibility of detecting WIMPs in a laboratory experiment. Many such experiments have been developed. They all work on the principle that if presented with a suitable target, a small proportion of the WIMPs passing through the Earth might interact with that target and deposit a measurable amount of energy (up to about 100 keV per collision). It is expected that such experiments may be able to detect about one interaction a day, and the biggest problem is to distinguish between genuine WIMPs and various kinds of background radiation. As a consequence, such experiments are often placed deep underground in order to reduce the signal from the cosmic rays that are present at the surface of the Earth.

Particle colliders

By simulating the high-energy conditions that were present in the early stages of the big bang, it should be possible to create WIMPs. Such extreme conditions are routinely achieved in particle accelerators by making particles collide with

Figure 8.2 The Large Hadron Collider (LHC) at CERN near Geneva occupies a circular tunnel 27 km in circumference. At the energies available to the LHC, it may be possible to create neutralinos, one of the candidate particles of cold dark matter. (CERN)

each other at high energies. The interaction energies available at the Large Hadron Collider (LHC) at the European Laboratory for Particle Physics (CERN, Figure 8.2) should be sufficient to investigate whether such particles exist. A complicating factor is that the LHC is not suitable for creating WIMPs directly. However, it is expected that the LHC could create other particles predicted by supersymmetry, which may then decay into WIMPs. A further difficulty is that WIMPs, by their nature, are only weakly interacting and cannot be detected directly within the LHC. The presence of dark-matter particles would have to be inferred from decays that imply the creation of an unseen massive particle. If candidate dark-matter particles could be created, then their properties would need to be thoroughly investigated to determine whether they provide a detailed match to the requirements of dark matter.

Indirect detection

As mentioned above, WIMPs are expected to be created in particle–antiparticle pairs in the early Universe. The condition for this is that the typical interaction energy exceeds $2m_W c^2$. After the interaction energy dropped below this energy, no further WIMPs would have been created. However, unlike particles such as the electron and its antiparticle the positron, WIMPs are not expected to have undergone a process of widespread annihilation. This is because the rate of annihilations is very low, and drops even further as the Universe expands. This situation has continued ever since, but note that WIMP particle–antiparticle annihilation continues to happen, albeit at a rate that has a negligible effect on the density parameter of dark matter. This provides another way to search for dark matter – by looking for the signature of these annihilation events in regions of the highest dark-matter density. Although high-energy photons, in the form of γ-rays are one possible outcome of annihilations, other possibilities are the production of electron–positron pairs or high-energy neutrinos.

Specialized instruments to detect astrophysical sources of high-energy γ-rays have been developed, such as the Fermi Large Area Telescope. In principle, such instruments can be used to look for dark-matter annihilation in locations where such a signal might be detectable, such as the halo of our Galaxy. This might be expected to be seen as a diffuse glow in γ-rays all around us. Another promising location for such a search is the region near the centre of the Galaxy, where the dark-matter density is highest. A major problem with this approach is that there are many other potential astrophysical sources of γ-rays that all have to be understood and accounted for in order to isolate any signal arising from dark matter.

Turning now to consider the production of electron–positron pairs, it is expected that these particles will have energies roughly corresponding to the rest-mass energy of the WIMP. Thus the range of expected energies of electrons and positrons arising from WIMP annihilation is expected to be about 10 to 1000 GeV, and it is usually assumed that electron and positron energies are typically ~100 GeV.

- ■ Imagine that a space-based cosmic ray detector found evidence for electrons with energies of about 100 GeV. Why might this, on its own, not be sufficient evidence for the dark-matter annihilation?

- ☐ Electrons are ubiquitous in the local Universe, and there are other energetic astrophysical processes (such as supernovae) that may have accelerated them to such high energies.

Hence electrons are not a suitable particle to detect dark-matter annihilation. There are far fewer astrophysical processes that might result in the generation of high-energy positrons, and it is for this reason that they have been the subject of much experimental effort. Cosmic rays, which are very high-energy particles coming from space, have long been detected and studied. The presence of very high-energy (~100 GeV) positrons in the background of cosmic rays would be a likely indicator of dark-matter annihilation, and several space-borne experiments, such a PAMELA (Payload for Antimatter/Matter Exploration and Light-nuclei Astrophysics) and AMS (Alpha Magnetic Spectrometer experiment), have been designed specifically to detect such particles. While there have been some results that suggest the presence of such a signal, like the γ-ray measurements, all astrophysical sources need to be properly accounted for before any firm claim to a genuine dark-matter signal can be made.

WIMP annihilation that results in high-energy neutrinos may seem a much more challenging prospect for detection, since neutrinos too are weakly interacting particles. However, it is predicted that a small proportion of WIMPs passing through the Earth or the Sun collide with atomic nuclei, lose energy and become gravitationally trapped at the core of the respective body. These WIMPs then form a reservoir in which the WIMP annihilation rate becomes much greater, and if neutrinos are produced, then Earth-bound neutrino detectors, such as Super-Kamiokande or IceCube, may be able to measure them.

8.2.2 Other dark-matter candidates

WIMPs have received much attention as dark-matter candidates primarily because the idea of supersymmetry requires only a minimal extension to the standard model of particle physics. However, many other candidate dark-matter particles have been suggested from more speculative theories in particle physics. Here we shall concentrate only on other hypothetical dark-matter candidates that offer a solution to a possible astrophysical problem. One of the features of the ΛCDM scenario is that it predicts that the very central regions of galactic halos should be highly condensed. As techniques for mapping masses in galaxies have improved, it has become apparent that these so-called *halos cusps* are not as pronouced as predicted by numerical simulations.

One solution to this problem could be provided if dark matter moves just fast enough to smooth out some of the small-scale structure. Such so-called *warm dark matter* may be due to the **sterile neutrino** (mentioned in Chapter 7) that has a much greater mass than the three known types of neutrino. This hypothetical neutrino is called 'sterile' because, unlike the known types of neutrino, it does not interact via the weak interaction. It is expected that such a neutrino may undergo a decay in which about one half of its mass energy is released as a photon. Such decays would produce an emission line that is expected to be in the X-ray part of the spectrum. If such a spectral line were to be observed, for instance in a rich cluster of galaxies, this may then provide evidence for dark matter being due to sterile neutrinos.

Another solution to the halo cusp problem could be provided if the dark matter is self-interacting in the sense that it is subject to interactions that are analogous to, but hidden from, the interactions of the standard model. A feature of this **hidden dark matter** is that in addition to the dark-matter particle, there is also an interaction that behaves like electromagnetism but only acts on and between these particles – there is no interaction with the particles of the standard model of particle physics. The idea of dark matter with its own set of hidden interactions may seem an over-complicated solution to the dark-matter problem, but some aspects of this idea are amenable to observation. The degree to which dark matter is self-interacting is constrained by situations such as the collision seen in the Bullet Cluster that you met in Chapter 4, where dark-matter mapping places observational limits on how much self-interaction of dark matter could have occurred.

However, the solution to the halo cusp problem may not require any modification to the simplest dark-matter scenario. Because baryons and dark matter interact gravitationally, an alternative suggestion is that the lack of halo cusps is a result of the feedback processes that you met in the context of galaxy formation – strong outflows of baryonic material driven by star formation or AGN that essentially drag dark matter outwards from the very central regions of galaxies. Again, this highlights the combined effort needed from astrophysicists and particle physicists in order to make progress on one of the biggest questions in cosmology.

QUESTION 8.1

Table 8.1 presents a number of candidates for dark matter. Classify each candidate by placing ticks in the appropriate columns.

Table 8.1 For use with Question 8.1

Dark matter candidate	Baryonic	Non-baryonic	Cold	Warm	Hot
neutron stars					
neutrinos					
neutralinos					
sterile neutrinos					

8.3 The nature of dark energy

While the prospect of a Universe filled with exotic new matter has itself challenged our understanding of the physical world, an even more startling cosmological discovery is that the rate of cosmic expansion is accelerating. As we saw in Chapters 5 and 7, this is the conclusion from studies of remote Type Ia supernovae, and measurements of CMB anisotropies. The accelerating expansion can be accounted for by attributing about 73% of the energy density of the Universe to dark energy, the effect of which is to oppose the deceleration of cosmic expansion. The presence of dark energy allows the Universe to maintain the critical density, even though the density of matter is low and decreases with time.

■ What is the connection between dark energy and dark matter?

☐ None that we know of. They refer to completely different and seemingly unconnected phenomena. In particular, dark energy is *not* the energy equivalent of dark matter.

The dark energy acts to oppose the gravitational attraction of the matter in the Universe. But what does that mean physically? In general relativity, gravity is a consequence of space–time curvature, and that is determined by the distribution of energy and momentum, not just by the presence of massive bodies. Even the pressure in a fluid influences its gravitational effect since, as the fluid expands or contracts, the pressure will affect the internal energy of the fluid, and this energy will influence the curvature of space. (This is one way in which general relativity differs from Newtonian gravity.) A normal fluid uniformly filling the Universe, as envisaged in the Friedmann–Robertson–Walker models, would exert a positive pressure at every point, and this, like the density of the fluid, would have the gravitational effect of decelerating the cosmic expansion. The accelerating effect of dark energy indicates that it, in contrast to a normal fluid, is a source of negative pressure. The *density* of dark energy actually tends to retard cosmic expansion, but the negative pressure more than compensates for this, so the overall effect of dark energy is to accelerate the expansion. The negative pressure, or rather the associated gravitational effect, can be thought of as driving the cosmic acceleration. (Incidentally, if you think that *positive* pressure is normally responsible for pushing things apart, you are probably thinking of the effect of *differences* in pressure between one region and another. FRW cosmology is concerned with uniform universes, where there are no pressure differences, but where uniform pressure, like uniform density, does have a gravitational effect.)

Negative pressure may be unfamiliar but it is not unphysical. A phenomenon known as the **Casimir effect** (see Figure 8.3) shows that the presence of two narrowly separated, parallel metal plates modifies the electrical properties of the vacuum between them, producing a negative pressure in that region. In this case there *is* a pressure difference, and it creates an effective attraction between the plates that can be demonstrated and measured experimentally. This is not a gravitational effect, but it is a demonstration of negative pressure. Don't worry if you find the idea of negative pressure hard to grasp. Accept for now that a uniform distribution of dark energy with negative pressure will have a repulsive gravitational effect, and such a distribution of dark energy will tend to accelerate the Hubble expansion.

What is the nature of dark energy? This is our Problem 2, and although cosmologists have no definitive answers, there are some lines of attack that may help in addressing this problem. It is possible that the dark energy simply represents the effect of a cosmological constant Λ, and has no deeper explanation. Unlike matter and radiation, the energy associated with the cosmological constant would not be diluted by the expansion of the Universe. It would stay the same, exerting a constant negative pressure throughout the expansion of the Universe. The value of the constant, Λ, would then be a new constant of nature, much like the gravitational constant, G. We would not be able to explain why it had the value it had, any more than we can explain why G has the value it has. This view of Λ has some merit in a scenario where there are 'many universes', in each of which the physical constants have random and arbitrary values (we shall return to this idea in Section 8.7.2).

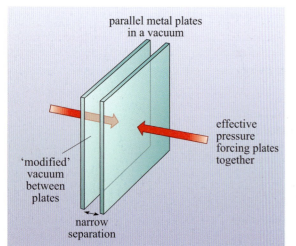

parallel metal plates
in a vacuum

effective
pressure
forcing plates
together

'modified'
vacuum
between
plates

narrow
separation

Figure 8.3 The Casimir effect. Two narrowly separated, uncharged conducting plates located in a vacuum will be attracted towards one another. The attraction arises from the influence that the plates have on the region between them and the negative pressure that this produces in that region.

To many cosmologists however, the fact that Λ quantifies the energy density of empty space suggests that it should be possible to calculate its value. The area of physics that has to be applied to calculating the dark energy density comes not from general relativity but from quantum physics. One of the central features of quantum theory is **Heisenberg's uncertainty principle**, which can be expressed in several ways, including the simple formula

$$\Delta E \Delta t > h/2\pi \qquad\qquad (8.1)$$

where h is the Planck constant. Its usual interpretation is that we cannot know both the precise energy of a particle and the precise time we measure that energy. If we wish to know the energy of a particle to an uncertainty ΔE, then the time we take to measure it must be at least Δt. We cannot know both the energy and the time to greater precision.

The implications of this are profound and not a little disturbing when applied to empty space. Let's do a thought experiment. Suppose we take a small box, as small as we wish, and clear it of all particles. We also shield it from the outside world to make sure no fields are present within it. It's as empty as we can get it.

■ What would be the energy density inside this box?

☐ Common sense tells is that if the box is empty the energy density inside must be zero. But the uncertainty principle tells us otherwise!

If space were devoid of particles we would be able to say 'the energy in this part of space at this time is zero'. But we would then know the energy exactly and that would violate the uncertainty principle! The uncertainty principle forces us to recognize that even in 'empty' space, particles of energy ΔE could exist for a time Δt. It also implies that more massive particles (with greater mass energy) will be shorter lived than less massive particles. These **virtual particles**, which always appear as pairs of particles and antiparticles, are created and destroyed before they can be observed. (The process is similar to the process of pair production that was discussed in Chapter 6.) Space, far from being empty, is teeming with particles continually popping in and out of existence (Figure 8.4). The collective energy of these particles is known as **vacuum energy**.

There is no doubt that vacuum energy exists. It provides the explanation of the Casimir effect. The presence of the parallel conducting plates limits the kinds of virtual particles that can form in the region between the plates, and it is this that causes that region of 'empty space' to have different properties from the surrounding 'empty space'. The fact that vacuum energy is a property of the vacuum itself ensures that it will not be 'diluted' by the expansion of the Universe, and that it will have the required negative pressure.

With this interpretation of the vacuum energy, it is possible to calculate its density. The result is unsettling, for it comes out to about a factor of 10^{120} – 120 orders of magnitude – higher than the measured density of dark energy! This seems bizarre. Can there really be so much energy in empty space? Wouldn't we notice it? The repulsive gravitational effect of so much vacuum energy would be so great that the Universe would expand explosively (much like inflation in fact, where the vacuum energy is also implicated, as you saw in Chapter 6).

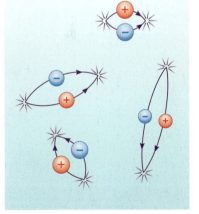

Figure 8.4 Virtual particles. According to quantum physics pairs of particles and antiparticles are continually being created and destroyed in empty space. The more massive the particle, the shorter its life.

The huge discrepancy between the observed and calculated energy densities (Nobel laureate Steven Weinberg has called it 'the worst failure of an order-of-magnitude estimate in the history of science') is a barrier to making any progress using this line of attack.

A different approach to the problem is to set up a framework to describe the behaviour of dark energy based on very general theoretical considerations, with the expectation that this formulation will have some measurable parameters that can be used to constrain the properties of dark energy. It turns out that a very useful way to characterize dark energy is by the relationship that energy density (ρc^2) has to pressure (p); this is done by defining the **equation-of-state parameter** for dark energy

$$w = p/\rho c^2.$$

If dark energy were accurately described by a cosmological constant, then $w = -1$ for all cosmic times. More generally, a framework that is sometimes referred to as **quintessence** allows w to take on other values. Quintessence is based on the idea that there is a quantity, ϕ, at every point in space, which determines the dark energy density. This quantity is what is known as a **scalar field** and it may vary with time, or indeed location (although for homogeneous cosmological models, this is assumed not to be the case). The physical nature of the scalar field for quintessence is not known. For every value of ϕ, there is assumed to be an associated potential energy that contributes to the dark energy. It is expected that the potential energy of the scalar field should evolve towards a minimum state (as shown schematically in Figure 8.5), but the rate at which this may happen is unknown. If the potential energy of the scalar field changes with time, then there will be a change of the dark energy density. The effect of this is that w need not necessarily have the value of -1, and indeed w itself may change with time.

Since time is related to redshift z, cosmologists find it useful to express w in terms of a constant part and a part that varies with z,

$$w = w_0 + w_a z/(1 + z).$$

Progress may then be made by finding what are essentially two new cosmological parameters, w_0 and w_a. It is important to stress that these parameters serve to help explore the phenomenon of dark energy and the underlying scalar field, rather than being part of a fully developed physical model.

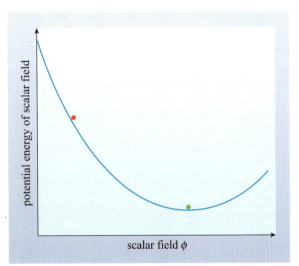

Figure 8.5 The potential energy of a scalar field ϕ, which applies to all points in space. The scalar field may change with time: the value of the potential energy at a given time is indicated by the red dot, and as time progresses it would be expected to evolve towards the minimum (green dot). Any change in the potential energy of the scalar field causes a change in the dark energy density. (Adapted from Frieman *et al.*, 2008)

■ What values of w_0 and w_a would you expect if the dark energy density is given by a cosmological constant Λ?

☐ In the case of the dark energy density being given by the cosmological constant, $w = -1$ for all times and hence all redshifts. So $w_0 = -1$, $w_a = 0$.

Current observational programmes to investigate dark energy, in essence, seek to determine w_0 and w_a. So far, there is no evidence that the behaviour of dark energy is inconsistent with $w_0 = -1$

and $w_a = 0$. Figure 8.6 shows the results of one such study. As observational programmes develop, they may reveal values that differ from the expectations of a simple cosmological constant, and this would provide insights into the underlying physical theory for this mysterious component of our Universe.

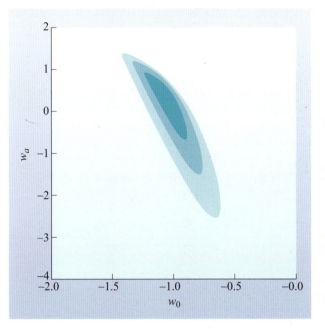

Figure 8.6 An example of a study to constrain the values of the dark energy parameters w_0 and w_a using data from observations of Type Ia supernovae, baryonic acoustic oscillations and fluctuations in the cosmic microwave background. The different shading represents 68.3%, 95.4%, and 99.7% confidence limits. (Amanullah *et al.*, 2010)

8.4 The horizon and flatness problems

Two observed properties of the Universe that should be accounted for by cosmological models relate to its uniformity and the flatness of its three-dimensional spatial geometry. Neither of these properties is a natural outcome of the hot big bang model. Consequently the issues of explaining the uniformity and spatial flatness of the Universe have been labelled as 'problems' called, respectively, the horizon and flatness problems.

We start by considering the problem of explaining the uniformity of the Universe (Problem 3 in our list).

8.4.1 The horizon problem

You first came across the horizon problem in Chapter 6, when we considered the uniformity of the cosmic background radiation. This problem arises from the observation that the temperature of the cosmic background radiation is uniform to a few parts in 10^5 across the sky, yet points on the sky more than about two degrees apart are separated by a distance that is greater than the horizon distance at the time of last scattering (Figure 8.7). Remember that the horizon distance at a given time represents the maximum distance that a physical signal could propagate through space in the time elapsed since the very first instant of the big bang.

Although we have concentrated on the horizon distance at the time of last scattering, it is possible to consider the horizon distance and uniformity of

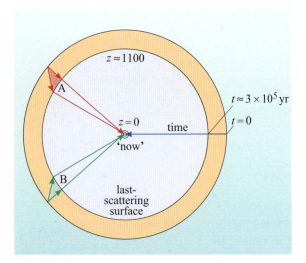

Figure 8.7 Regions A and B individually have an extent equal to the horizon distance at the time of last scattering. Both A and B subtend an angle of about 2° on the last-scattering surface (note that these angles are exaggerated in this diagram). However, A and B are separated by an angle greater than 2° and so lie outside each other's horizon at the time of last scattering, and hence no physical communication could have occurred between these two regions. The fact that the CMB appears to have the same temperature at A and B, despite this apparent lack of communication, is an example of the horizon problem.

the Universe at later times, and to arrive at a similar conclusion about the existence of the horizon problem. The advantage of discussing horizon distances in terms of the last-scattering surface is that it is readily observable through observations of the CMB. Thus, according to the standard account of the big bang, there is no reason to expect the CMB to be uniform on scales greater than about 2°.

A solution to the horizon problem has already been mentioned in Chapter 7. Despite the arguments we have advanced to suggest that regions of the last-scattering surface could not have been in communication with each other, nevertheless, they *have* colluded together. In the very early stages of the Universe's development, they *were* in contact with each other and reached a state of thermal equilibrium. What is required then is that these regions were separated. A key idea here is that such a separation would create large-scale uniformity by the expansion *of space* rather than the propagation of a physical signal *through space*. The expansion would have to be by an enormous factor so that two regions that were once in contact, would later appear to be separated by many horizon distances.

■ What process in the early Universe could have had this effect?

☐ The process of inflation – a brief period of very rapid expansion could have caused regions that were once in contact to become separated by more than the horizon distance.

To recap, the key idea of inflation is that close to the time that the grand unified era came to an end, the Universe underwent a brief but rapid phase of exponential expansion. The net result was that by the time that inflation was over, the scale factor had increased by a huge amount. As noted in Chapter 6, the growth in the scale factor during inflation is very uncertain – but it is thought that an increase by a factor of as much as 10^{27} or more could have occurred during this process. Thus a region of the Universe, having first had a chance to become homogeneous and to reach thermal equilibrium before inflation, might have expanded to a size that is much greater than the horizon distance.

So inflation can solve the horizon problem by enlarging and sweeping apart regions of the Universe that had already become homogenized and leaving them separated by vast distances. When the cosmic background radiation became decoupled from matter at recombination (when the age of the Universe was about 372 000 years), it was necessarily isotropic because it was emitted by matter that was already homogeneous.

It is important to appreciate that inflation should be considered as an addition to the standard hot big bang model. Whereas the predictions of the standard hot big bang model from times of about 10^{-9} s onwards are widely accepted and stand up to observational scrutiny, the inflationary hypothesis is rather more speculative.

The appealing aspect of inflation is that it offers a single solution to several cosmological problems; not only the horizon problem that we have just considered, but also the problems of structure formation and spatial flatness. It is to the latter problem that we now turn.

8.4.2 The flatness problem

The next outstanding problem on our list, Problem 4, relates to the observation that the average density of the Universe is almost equal to the critical density, i.e. $\Omega = 1$. In the context of the Friedmann–Robertson–Walker models this causes the curvature parameter, k, to be zero (see Question 5.9) and implies that three-dimensional space will have a flat geometry. Why should this be a problem?

In order for the total density parameter, $\Omega(t)$, to be close to 1 today, it had to be even closer to 1 in the past. This is because the extent to which $\Omega(t)$ differs from 1 is predicted to grow with time. If $\Omega(t)$ were *exactly* equal to 1 today, at time t_0, then at any earlier time, it would also have been exactly equal to 1. If, however, the current value of the density parameter turned out to be somewhat less than 1, $\Omega(t_0) = 0.90$ say, what would its value have been at some earlier time when the age of the Universe was only a fraction of t_0? The answer depends on the details of the FRW model used to represent the Universe, but in one case, for example, the difference grows at a rate proportional to $t^{2/3}$. In this particular case, a current difference of 0.1 implies that when the Universe was a thirtieth of its present age the difference between $\Omega(t)$ and 1 would have been smaller by a factor $30^{2/3} \approx 10$. Thus if $\Omega(t_0) = 0.90$, then $\Omega(t_0/30) \approx 0.99$. The fact that the Universe is now about 10^{17} seconds old, and has a density parameter that is still close to 1 means that at very early times, $t < 10^{-6}$ s say, $\Omega(t)$ must have been *extremely* close to 1. Explaining why $\Omega(t)$ should be so close to 1 at very early times is the crux of the flatness problem.

Now, of course, it is possible that the total density of the Universe just happens to have the critical value, in which case $k = 0$ and $\Omega(t)$ will always be equal to one.

However, this would be another of those 'just so' explanations that, although possible, are never much trusted by cosmologists. Their preference is always for 'mechanisms' or 'processes' that force the cosmological parameters to take on their observed values. One of the motivations for proposing that there might have been an era of inflation in the very early Universe is that this can solve the flatness problem just as neatly as it solves the horizon problem.

The most direct result of inflation is an enormous increase in the cosmic scale factor $R(t)$, perhaps by a factor of 10^{27} or more. As suggested by Figure 8.8, this will have the effect of reducing the curvature of space, which depends on the quantity k/R^2. A sufficient amount of inflation will result in such a small value of k/R^2 that the effective value of k is zero, and space is geometrically 'flat', irrespective of the true value of k prior to inflation.

Now a geometrical argument of this kind might seem quite convincing at first sight, but you might still wonder how such an argument can have any bearing

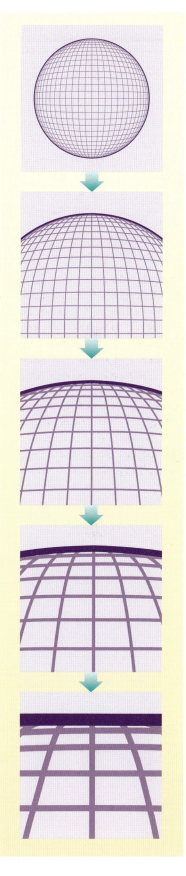

Figure 8.8 A spherical balloon analogue for illustrating how the inflation of the Universe at an early epoch would have resulted in a flat spatial geometry for the observable Universe regardless of the curvature prior to inflation.

on the density of the Universe. Figure 8.8 might suggest that, in effect, $k = 0$, but how can it account for the fact that $\Omega(t)$ is correspondingly close to 1? To understand this you have to recognize that the link between the effective curvature and the density comes directly from the Friedmann equation and remains true no matter what the source of the cosmic energy density may be. Thus, near the end of inflation, when space is effectively flat, the Universe might have become quite cold and most of its energy might take the form of some exotic kind of vacuum energy, but the density of that vacuum energy will be just what is required to produce an effectively flat geometry. The usual assumption is that as the Universe ceases to inflate, some of this vacuum energy is converted into more conventional forms of matter and radiation, and that the Universe is reheated to a temperature similar to that it would have had in the absence of inflation. This has the interesting effect of causing all the matter and radiation in the observable Universe to be a direct consequence of inflation (any pre-existing matter or radiation will have been so diluted by inflation as to be unobservable), but it will not alter the fact that when all forms of matter and radiation are taken into account, as well as any dark energy, the total density of the Universe will be very close to the critical density and $\Omega(t)$ will be correspondingly close to 1.

Although inflation leads in a natural way to a Universe that is close to having a critical density, it does not convert the Universe into one that has *exactly* the critical density. If the density before the onset of inflation was greater (or less) than critical, it will still be greater (or less) afterwards, though only barely so.

■ Supposing that prior to inflation the density were *less* than critical, what would be the analogue of Figure 8.8 for a Universe undergoing inflation?

☐ It would be a saddle-shaped rubber sheet that, after inflation, had been almost completely flattened over the tiny part constituting the 'observable' Universe.

As you have seen in Chapter 7, measurements in fluctuations in the CMB show that the Universe is very close to having a flat spatial geometry and hence the sum of the density parameters is very close to 1. You have also seen that the contribution from the dark energy is about 68% of the critical density with matter at 32%. Hence the observation that two apparently unconnected components of the Universe, matter and dark energy, add up to the critical density is simply explained by inflation. Otherwise this coincidence seems very hard to understand.

At a stroke, inflation solves both the horizon *and* flatness problems that afflict the standard model of the big bang. Indeed, inflation also provides a mechanism to explain the slight inhomogeneities that do occur – those responsible for triggering the formation of galaxies and those that manifest themselves as the ripples in the microwave background radiation. We shall learn more about this in the next section.

8.5 The origin of structure

In Chapter 2 you were introduced to the idea that galaxies formed from primordial fluctuations in the density of matter in the early Universe. Then, in Chapters 6 and 7 you learned how these fluctuations left their imprint on the cosmic background radiation.

Now it is time to address our Problem 5: where did those primordial fluctuations come from? In a sense this is the inverse of the uniformity problem. Earlier we asked why the Universe is so uniform, and now we ask why it is not perfectly uniform. From where did all this diverse structure come?

Inflation enlarges a tiny region of space by a huge factor. Solving the horizon problem may require space to expand by many orders of magnitude. This implies that current cosmological scales, tens to hundreds of megaparsecs, corresponded to subatomic scales during inflation. This intense magnification means that large-scale structures in the present-day Universe may have been rooted in subatomic irregularities in the pre-inflationary Universe. But where could such irregularities have come from in the first place?

As you have seen in the discussion of vacuum energy, empty space is seething with virtual particles that flit in and out of existence as a consequence of the Heisenberg uncertainty principle. Similar **quantum fluctuations** occurring during inflation would have been enlarged by the very rapid expansion of the Universe and could have been the cause of macroscopic variations in density. These, after much subsequent evolution, might have been the source of the large-scale structure we see around us today.

Although there is no agreement about exactly what caused inflation, there are a number of proposed models that make quite detailed predictions about the quantum fluctuations that might have occurred. Generally speaking these fluctuations would have been caught up in inflation and would have grown in size until they exceeded the horizon scale of the then-observable Universe (Figure 8.9). Once this happened, information could no longer travel from one side of a fluctuation to the other, so the fluctuation would have been unable to smooth itself out and would have become 'frozen in', expanding along with space as the scale factor $R(t)$ continued to grow.

As earlier fluctuations were stretched beyond the horizon scale, newer quantum fluctuations were generated, creating more small-scale density variations that were stretched in their turn. The precise outcome varies from one model of inflation to another, but the general result is a range of density fluctuations with roughly the same 'strength' on a wide range of size scales, i.e. variations in density that are much the same on large scales as on small scales.

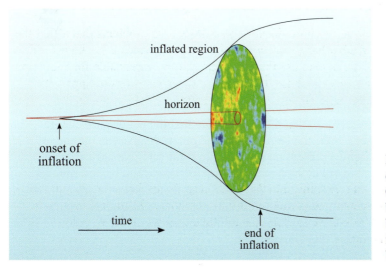

Figure 8.9 A schematic illustration of the expansion of a small region of the Universe due to inflation. Quantum fluctuations within this region are expanded beyond the horizon distance, at which point they become 'frozen in' and form the primordial density fluctuations from which subsequent structure in the Universe develops.

Following inflation, the Universe continued to expand, but the rate of expansion was so much reduced that the growing cosmic horizon would have encompassed more and more of the matter in the Universe as time passed. As the horizon expanded at the speed of light, material that had been swept over the horizon during inflation re-entered the horizon bringing the frozen-in density variations with it. The last density fluctuations to be inflated beyond the horizon would have been the first to re-enter, but these would have been followed by other density variations on larger and larger scales. Once back inside the horizon, the variations in density ceased to be 'frozen-in' and were able to become stronger or weaker depending on the prevailing conditions at the time they re-entered the horizon. It has been known for some time that the pattern of density variations predicted by inflation (with the same amplitude on all scales) is just what is required to give rise to the observed range of superclusters, clusters, etc. Hence inflation provides a possible explanation for the origin of all of the structure we see in the Universe – it might all have come from quantum fluctuations. Intriguingly, according to the inflationary hypothesis, the largest macroscopic structures might have had their origin in the microscopic quantum world.

8.6 The matter of antimatter

Inflation can explain many things, but it cannot account for the fact – Problem 6 – that the Universe contains far more matter than antimatter. Why should this be so?

In the account of inflation given in Chapter 6, it was noted that the end of inflation would have been accompanied by the release of a vast amount of energy in the form of particle–antiparticle pairs. At a later stage the antiparticles annihilated with the particles to create photons, so by now we might expect the Universe to be devoid of matter but full of radiation. In numerical terms this is very nearly the case: there are about a billion photons (mainly in the cosmic microwave background) for each proton or electron in the Universe. Even so, the fact that there are any matter particles at all, implies that rather more matter than antimatter must have been formed.

Another way of stating this problem is in terms of the baryon number of the Universe. If, for every baryon in the Universe, there was a corresponding antibaryon, the baryon number of the Universe would be zero. However, we know that in the real Universe there is a surplus of baryons over antibaryons – so the baryon number of the Universe is a positive number. The conservation of baryon number in nuclear reactions is an important principle of physics, and we might naïvely expect that because the baryon number of the Universe currently has a positive value, this must always have been the case. However, physicists are reluctant to accept that the Universe must have started out with a positive baryon number. It seems more natural for the baryon number of the Universe to have originally been zero – in much the same way that the net electric charge of the Universe is zero. So if the Universe started out with a baryon number of zero – how did it reach its present non-zero value? One possible answer is that there may have been an era in the history of the Universe when baryon number was not conserved, i.e. reactions may have occurred that violate the principle of conservation of baryon number. Such reactions could have caused the baryon number of the Universe to change from an initial value of zero to the positive value we observe in the present-day Universe.

In Chapter 6 you saw that the unification of the strong and electroweak forces (grand unification) is believed to occur at energies of around 10^{15} GeV. The speculative grand unified theories that describe reactions at these enormous energies predict that baryon number need not always be conserved. Particle interactions at such energies could therefore give rise to the slight excess of matter that we see in the present-day Universe. Note that such interactions must have occurred *after* the process of inflation. According to the inflationary model, the vast majority of particles in the Universe were created from the energy released at the end of inflation. The imbalance between matter and antimatter could only have developed after these particles were created, and so inflation must have occurred at, or before, the end of the grand unified era.

Reactions in which baryon number is not conserved occur at such high energies that they are far beyond anything we can reproduce in laboratories today. However, it might still be possible to find experimental evidence in support of grand unification theories. The fundamental processes that allowed an excess of baryons to form in the very early Universe might also, very rarely, allow protons to decay in the present-day Universe. So far, no proton has ever been observed to decay, despite many attempts to detect such a process. However, if the proton has a mean life of the order of 10^{33} years or more, the decay rate would result in a signal below current experimental limits. So proton decay cannot be ruled out, but if it does occur, it does so on a very long timescale.

Before we leave this subject, we need to address an assumption we have made, namely that there are no significant amounts of antimatter in today's Universe. How do we know this? If matter and antimatter could somehow have become segregated in the early Universe, then there may be regions of space in which antimatter dominates. Antiprotons, antineutrons and antielectrons (i.e. positrons) could have come together to form anti-atoms, which in turn could have formed anti-molecules.

QUESTION 8.2

Suppose you suspected that a newly discovered galaxy was made of antimatter.

(a) Could you tell from its emitted radiation whether the galaxy was made of matter or antimatter?

(b) Given that matter and antimatter will annihilate each other to form γ-rays, are there any other observations you could make?

8.7 The very early Universe

According to the classical Friedmann–Robertson–Walker models, the Universe started expanding from a condition in which the scale factor was zero, implying a state of infinite density, often referred to as the **initial singularity**. If we naïvely extrapolate the big bang model back towards $t = 0$, many of its physical properties (the energy density of matter and radiation, the pressure and temperature, and the curvature of space–time) approach infinity, i.e. they diverge. When a model predicts infinite values we can take it as a warning that we have probably pushed the model beyond its limits of validity.

This should not come as a great surprise. You saw in Chapter 6 that quantum physics sets a natural limit – represented by the Planck time, about 10^{-43} seconds – on the earliest moment at which we can have any confidence in the big bang model. As we approach the Planck time quantum effects become as important as general relativistic effects and behaviour cannot be understood within the framework of existing physical theory. Neither general relativity nor conventional quantum physics are much help to us here, and a new theory is needed if we are to explore back beyond the Planck time.

Before doing so however, we should critically consider another aspect of the early Universe, and question whether inflation actually happened – this is our Problem 7.

8.7.1 Is there evidence for inflation?

We have seen that a process of inflation solves several cosmological problems. This in itself is a strong argument that some process of very rapid cosmological expansion occurred in the early Universe. Despite this, cosmologists need to investigate whether there is any observational evidence to support the idea that inflation occurred. It is important to appreciate that inflation is not a unique physical theory, and that the mechanism by which it may have occurred is not known. Rather, it should be viewed as a framework within which various theoretical models can be developed. This may sound similar to the approach taken to dark energy that was described in Section 8.3, and with good reason. Like the problem of dark energy, cosmologists can attribute inflation to a scalar field in the early Universe, without knowing in detail what the physical origin of that field is.

The term 'inflaton' refers to a hypothetical particle associated with the scalar field that gives rise to inflation.

Rather remarkably, despite the unknown nature of this so-called *inflaton* field, inflation has observable consequences. In particular, we have already noted that inflation amplifies quantum fluctuations to the macroscopic scale, and these fluctuations seem to be in accord with what is required for structure to develop. The amplification of fluctuations during inflation would have another effect in that it results in the generation of a background of gravitational waves. These waves, which are predicted by Einstein's general theory of relativity, are propagating disturbances in the curvature of space–time. Such waves would leave a distinctive imprint called *tensor modes* on the cosmic microwave background. The level of such a signal would be small, and can easily be mimicked by astrophysical effects, so a great deal of care is needed to properly claim that tensor modes due to gravitational waves have been detected. At the time of writing (2014), detection of such tensor modes was claimed by the BICEP-2 experiment, but subsequent analysis by the Planck team and others found that much of this signal could arise from dust in our Galaxy. Further analysis is underway.

8.7.2 Inflation and the multiverse

We have noted that inflation solves many cosmological problems, but it does not answer the question of why inflation happened in the first place, or why it is that the physical constants such as G or the charge of an electron $(-e)$ have the values they do. An explanation has been suggested as an extension to inflation, by the cosmologist Andrei Linde. In a scenario called **chaotic inflation**, the initial condition of the Universe is random. However, in this model the Universe is much larger than we can see, and is partitioned into *domains* with differing laws of physics (see Figure 8.10). We live in a region where the right conditions for inflation have arisen by chance. This region has inflated and has provided the right conditions for life, but other domains may not be capable of this.

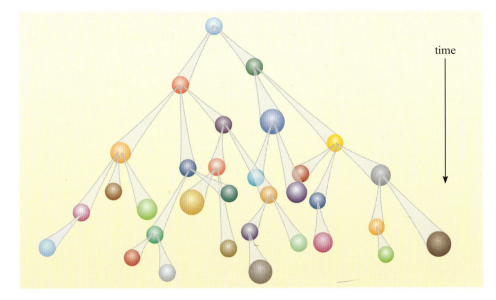

time

Figure 8.10 Chaotic inflation suggests that the Universe may consist of numerous inflationary domains or 'universes' each with different laws of physics (different colours in this diagram). Our observable Universe is just a tiny part of one of these domains.

This concept of many separate domains or 'universes' is often referred to as the **multiverse**. A problem with this idea is that we cannot probe these other domains unless they impinge on our domain. This model is therefore not testable. One day, chaotic inflation may be found to be part of testable theory that explains how the right conditions could arise. At present, however, it is not clear if this will ever be the case.

8.7.3 M-theory

We have noted that the Planck time represents the current limit of physical theory, and that theoretical physicists are seeking a theory that will correctly describe gravity and quantum theory – a so-called 'theory of everything' that describes all the known interactions of nature. One promising development is **M-theory**, which subsumes the earlier ideas of *superstring theory*. In M-theory the fundamental objects are not particles, as usually assumed, but strings or even sheets. These strings and sheets can vibrate, and the different vibrations describe the different particles and their masses. One of the attractive features of these models is that they predict a massless particle that has the expected properties of the graviton, the quantum particle of the gravitational field. Hence M-theory looks promising as a way forward in creating a quantum theory of gravity. However, despite these properties M-theory is far from being fully developed, and even further from being tested by experiment.

One of the stranger features of M-theory is that it predicts more than four dimensions. In fact it requires 11 dimensions to work at all. Usually it is assumed that the extra dimensions are 'rolled up' and hidden away to leave the familiar three dimensions of space and one of time – the extra dimensions would only become apparent at very high energies. However, it seems that some of the extra dimensions need not be rolled up. One line of development regards our Universe as residing on a sheet (known as a 'brane' since it is derived from the term membrane) that itself resides in a five-dimensional space called the 'bulk'.

This picture has allowed cosmologists to develop new models of the early Universe, including one that does not require inflation to have occurred at all. According to this **ekpyrotic model** (the name means 'out of fire'), a key event in

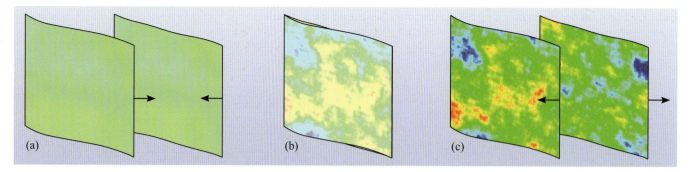

Figure 8.11 The ekpyrotic model, which invokes colliding sheets or 'branes' (representing our Universe and a parallel universe). (a) Two branes come together, resulting in (b) a collision that creates (c) the primordial fluctuations that are required to give rise to the observed large-scale structure and to the observed anisotropies in the cosmic microwave background.

cosmic history was a collision between the brane on which our Universe resides and some other 'parallel' brane (Figure 8.11). This collision would have been the cause of the fluctuations that led to large-scale structures being formed. In an extension of this model, the sheets collide over and over again, allowing an infinite number of past big bangs in a 'cyclic' universe. Just before the collision, the forces on the sheets cause an accelerated expansion, which would look like the currently observed acceleration, and which flattens and smoothes out the Universe, thus dispensing with the flatness and horizon problems.

In this theory, dark energy is a manifestation of the energy that controls the separation of the colliding branes, so it plays a vital role in the evolution of the cosmos rather than being a sort of 'uninvited guest'. The model can also be tested against observation – one generic prediction of all variants of the ekpyrotic model is an absence of primordial gravitational waves, implying no detectable tensor modes in the CMB (Section 8.7.1).

While the ekpyrotic model is not widely supported, it acts as a reminder to cosmologists that there may be alternatives to inflation that should be explored and tested.

8.8 The anthropic Universe

There seem to be a multitude of possible universes, but only one Universe. Why is our Universe the way it is? (This was the last question posed at the beginning – Problem 8.) Why do the fundamental constants of nature have the values they do? Could they have been different? In particular, if the Universe had started out with different conditions would we, or someone like us, still be here to puzzle about it?

The notion that the Universe must be able to give rise to life (and cosmologists!), has been given the status of a principle, the **anthropic principle**. It comes in two basic forms, the weak and the strong, and there are various versions of each.

The *weak anthropic principle* holds that the initial conditions of the big bang were such as to allow the eventual emergence of carbon-based life. Universes with different initial conditions – perhaps different values of the speed of light, the Planck constant, the gravitational constant, and so on – may have been possible, but if those conditions were not favourable to life then we would not be here. We would not, therefore, expect to find anything in the Universe incompatible with our own existence. So far the anthropic principle is not saying anything remarkable, beyond being able to make broad inferences about the Universe from the fact that it contains life. Before the age of precision cosmology, this was one route to placing rough numerical constraints on cosmological parameters.

But the *strong anthropic principle* goes further, and holds that the Universe *necessarily* had the initial conditions that eventually allowed carbon-based life to emerge. Proponents of the strong principle argue that of all possible universes our Universe is so extraordinarily improbable that it cannot be an accident. Some go further and hold that the Universe was in some sense destined to develop not only life but self-awareness. They point to numerous coincidences that seem to imply that the Universe is finely tuned to favour the emergence of life.

The classic example, which actually pre-dates the anthropic principle, is the nucleosynthesis of carbon. Carbon is produced in post main sequence stars by the triple-alpha reaction. Two helium nuclei fuse to form an unstable nucleus of beryllium. Before the beryllium can fall apart again it is hit by a third helium nucleus to form a stable carbon nucleus. The trouble is, such a triple collision is so improbable that it's hard to explain the amount of carbon seen in stars today. The puzzle was solved in 1953 by the cosmologist Fred Hoyle, who predicted that carbon must possess a 'resonant' state such that at a certain collision energy a carbon nucleus is formed much more readily than expected. Without Hoyle's resonant state – which was later confirmed by experiment – there would be no carbon and no cosmologists. The energy of the resonant state is so finely tuned that if it were a little higher, the carbon would be rapidly converted into oxygen and there would again be no cosmologists. Hoyle himself was deeply affected by his discovery.

There are many other supposed coincidences of this nature, although – unlike Hoyle's prediction – they have all been recognized in hindsight. Stable orbits are only possible in a universe with three spatial dimensions. Any more than three and there would be no planets for life to make its home on and no atoms either. The initial expansion of the big bang must have been just right – any faster and galaxies and stars would not have been able to form, any slower and the Universe would have collapsed again before life could emerge. Gravity must also be the right strength. If G is too big, only massive stars will form and burn out before life can take hold. If G is too small, stars will not get hot enough to start nuclear reactions.

Many people are unimpressed by such arguments, pointing out that no one should be surprised that the Universe permits us to exist. For them the anthropic principle begins and ends with its weak form and is not saying anything very profound. Douglas Adams, the author of the *Hitchhiker's Guide to the Galaxy*, liked to tell the story of a puddle of water lying in the road. Suppose the puddle suddenly becomes conscious and starts to contemplate its situation. It starts to sense its surroundings, probing the surface of the road beneath it, and notes that the depression in which it lies is the same shape as its own body. It comes to the conclusion not only that the Universe is perfectly adjusted to the emergence of puddles but that its own existence was somehow predestined. Can it be a coincidence?

You win the National Lottery against odds of 14 million to one. What an extraordinarily improbable coincidence! Yet someone had to win, and your win has to be seen in the context of millions of ticket holders who did not win. On this view, our Universe may be one of countless possible past and future universes – perhaps arising from chaotic inflation – the difference being that ours holds the winning ticket.

Others argue that aside from the meaninglessness of hypothetical and unobservable 'other universes', such critics are missing the point. Russell Stannard, a physicist at the Open University, cites a counter-example of a prisoner sentenced to be executed by firing squad. At the crucial moment all ten marksmen miss their target and the prisoner is reprieved. Asked to explain his amazing deliverance the prisoner is unimpressed. 'Of course they missed, or else I wouldn't be here.' Such an explanation is not wholly satisfactory, since it fails to address why all ten skilled marksmen so improbably missed their target. Likewise, some people seek an explanation for why the Universe is set up the way it is beyond the explanation that it just has to be that way or else we would not be here to ask the question.

QUESTION 8.3

What do you think about the anthropic principle? Do you find it trivial, like Douglas Adams, or profound, like Russell Stannard? Can you find flaws in either of their stories? Can you reconcile the two views? How much is your opinion coloured by your own philosophical or religious beliefs? These are questions you will not find answered at the back of the book!

8.9 Epilogue

This chapter started with a quotation so it will finish with one too. Albert Einstein, whose insights into the nature of space and time made modern cosmology possible, once remarked that 'the most incomprehensible thing about the Universe is that it is comprehensible'. Perhaps after reading about multidimensional sheets colliding in 11-dimensional space, you may be inclined to think that Einstein, on this occasion, got it wrong!

But stay with us. Einstein was saying two things. First, he was expressing a faith that underpins all science, not just cosmology, namely that we will be able to understand the Universe and find it makes sense. The world is not chaotic and that makes science possible. But he was also saying something else, namely that we really have no right to expect the Universe to be that way. Why *should* we find the Universe comprehensible?

This could be a cue for another excursion into the anthropic principle, but we shall not do that. The models we have been discussing in the last few chapters may stretch your imagination to the limit (and beyond!) but they are the cosmologist's way of making the Universe comprehensible. So far our models have been able to keep up with new surprises sprung on us by the Universe. One day our luck and our imagination may run out and we may then have to admit that the Universe makes no sense after all. Until that happens, cosmology will continue to be one of the most exciting and mind-stretching of all the sciences.

QUESTION 8.4

Make brief notes *in your own words* to answer each of the questions posed at the beginning of this chapter. How satisfied are you with the answers? Check them against the summary at the end of this chapter.

8.10 Summary of Chapter 8

The models devised by cosmologists are simplified representations of the Universe. Like all models, they are only partial analogies to reality and break down outside their limits of validity. The big bang model is successful as far as it goes, but there are several problems it cannot answer.

- *Problem 1: What is dark matter?* Dark matter makes up about 23% of the Universe. A little of it is baryonic, in the form of MACHOs, which are simply familiar objects that are too faint to see. Some of them can be revealed by gravitational microlensing. About 85% of the dark matter has to be non-baryonic but apart from a very small proportion of neutrinos its nature is largely unknown. The best candidate is the neutralino, a form of WIMP, which may soon be discovered in laboratory experiments or detected from an astrophysical signal of its annihilation.

- *Problem 2: What is dark energy?* Dark energy is a source of negative pressure that fills the Universe and drives the accelerating expansion. Its nature is still a mystery, but the leading contenders are Einstein's cosmological constant (a source of 'repulsive' gravity arising from general relativity), quantum vacuum energy (a consequence of Heisenberg's uncertainty principle) or 'quintessence' (arising from an underlying scalar field).

- *Problem 3: Why is the Universe so uniform?* This is the horizon problem, which asks why widely separated regions have the same temperature and density, even though each has been beyond the horizon of the other throughout the history of the Universe. Inflation provides a possible answer. A small region of the Universe that had become homogeneous might have expanded so rapidly and by such an enormous factor that the whole of the currently observable part of the Universe (and perhaps more) is contained within the inflated homogeneous region.

- *Problem 4: Why does the Universe have a flat (k = 0) geometry?* Again, inflation may make it so. During the inflationary period large amounts of matter and energy were released into the Universe from the vacuum energy, leaving its density very close to the critical density, which corresponds to a flat geometry. Equivalently, whatever curvature the early Universe may have had would have been smoothed out by inflation, leaving the spatial geometry of the observable Universe indistinguishable from that of a 'flat' space.

- *Problem 5: Where did structure come from?* Clusters of galaxies were formed from density fluctuations in the early Universe that have left their imprint on the cosmic background radiation. Those fluctuations in turn may have arisen from tiny quantum fluctuations that were stretched by inflation from the microscopic scale up to and beyond the size of the then-observable Universe. At that point they would have become 'frozen in' as large-scale primordial fluctuations from which galaxies could condense.

- *Problem 6: Why is there more matter than antimatter?* Although one might expect equal numbers of particles and antiparticles to have been created in the early Universe, grand unified theories of physics allow a slight imbalance of matter over antimatter of 1 part in 10^9. The matter now in the Universe is that left over when the bulk of the matter and antimatter annihilated.

- *Problem 7: Did inflation happen?* The process of inflation solves several important cosmological problems and that, in itself, is evidence of an early stage of rapid cosmological expansion. Strong supporting evidence for inflation would come from the detection of a particular signal in the CMB called tensor modes, which are expected to have arisen from the gravitational wave background caused by inflation.

- *Problem 8: Why is the Universe the way it is?* According to the anthropic principle, because we are here to ask the question!

ANSWERS AND COMMENTS

Since $v = d/t$, the time it takes the Sun to travel a distance d around its orbit at speed v is given by $t = d/v$.

The distance d the Sun travels in one orbit is the circumference of a circle of radius $R = 8.5$ kpc, so $d = 2\pi R = 2 \times 3.14 \times 8.5$ kpc $= 53.4$ kpc. As the speed is given in km s^{-1} rather than kpc s^{-1}, it is convenient to convert this distance to units of km. Since 1 pc $= 3.09 \times 10^{13}$ km, $d = 53.4 \times 10^3$ pc $\times 3.09 \times 10^{13}$ km pc$^{-1} = 1.65 \times 10^{18}$ km.

We can then calculate the time taken to complete one orbit of the Galactic centre as

$$t = d/v = 1.65 \times 10^{18} \text{ km}/200 \text{ km s}^{-1} = 8.25 \times 10^{15} \text{ s}$$

The number of seconds in one year is $60 \times 60 \times 24 \times 365.25 \approx 3.16 \times 10^7$. Therefore

$$t = 8.25 \times 10^{15} \text{ s}/3.16 \times 10^7 \text{ s yr}^{-1} = 2.61 \times 10^8 \text{ yr}$$

Note that since R is given to only two significant figures, we should round the final results to the same accuracy: $t = 8.3 \times 10^{15}$ s (in SI units) and $t = 2.6 \times 10^8$ yr (in years).

The locations to which stars travel within the Galaxy are determined by their motions. For example, stars with higher speeds are more likely to be able to travel to great distances away from the Galactic centre. Hence location is also an indicator of motion.

(a) The circumference of a circle of radius R is $d = 2\pi R$, so the distance the Earth travels in each orbit is $d = 2 \times \pi \times 150 \times 10^6 \times 10^3$ m $= 9.425 \times 10^{11}$ m. (This should be rounded to three significant figures in the final answer.)

(b) The speed of the Earth, v_E, can be calculated from the known distance, d, around the orbit, and the time taken, $T = 365.25 \times 24 \times 60 \times 60$ s $\approx 3.156 \times 10^7$ s

i.e. $\quad v_E = 9.425 \times 10^{11}$ m$/3.156 \times 10^7$ s $= 2.986 \times 10^4$ m s^{-1} (about 30 km s^{-1})

(c) The formula of the rotation curve gives

$$M_\odot = v_E^2 r/G = (2.986 \times 10^4 \text{ m s}^{-1})^2 \times 150 \times 10^9 \text{ m}/(6.673 \times 10^{-11} \text{ N m}^2 \text{ kg}^{-2})$$

$$= 2.004 \times 10^{30} \text{ m s}^{-2} \text{ N}^{-1} \text{ kg}^2$$

Since 1 N $= 1$ kg m s^{-2}, we can write

$$M_\odot = 2.004 \times 10^{30} \text{ m s}^{-2} \text{ (kg m s}^{-2})^{-1} \text{ kg}^2 = 2.004 \times 10^{30} \text{ kg}$$

(d) The distance from the Earth to the Sun was given to three significant figures, whereas all other values are known to higher precision, so the final answer is also known to three significant figures, 2.00×10^{30} kg.

QUESTION 1.4

(a) To use the equation $M(r) = v^2 r / G$, we need to know the speed the Sun moves, v, the radius of its orbit, r, and G the universal gravitational constant.

$$M_{MW}(8.5\ \text{kpc}) = v^2 r / G$$

$$= (200 \times 10^3\ \text{m s}^{-1})^2 \times 8.5 \times 10^3\ \text{pc} \times 3.09 \times 10^{16}\ \text{m pc}^{-1}/(6.673 \times 10^{-11}\ \text{N m}^2\ \text{kg}^{-2})$$

$$= 1.574 \times 10^{41}\ \text{m}^2\ \text{s}^{-2}\ \text{m N}^{-1}\ \text{m}^{-2}\ \text{kg}^2$$

$$= 1.574 \times 10^{41}\ \text{m}^3\ \text{s}^{-2}\ (\text{kg m s}^{-2})^{-1}\ \text{m}^{-2}\ \text{kg}^2$$

$$= 1.574 \times 10^{41}\ \text{kg}$$

This should be quoted to at most two significant figures, 1.6×10^{41} kg.

Since $M_\odot = 1.99 \times 10^{30}$ kg, we can write

$$M_{MW}(8.5\ \text{kpc}) = 1.574 \times 10^{41}\ \text{kg}/1.99 \times 10^{30}\ \text{kg}\ M_\odot^{-1}$$

$$= 7.9 \times 10^{10} M_\odot$$

(b) As the distance of the Sun from the Galactic centre is given to two significant figures, we could quote the result to this many figures. However, you might wonder whether the result really has that much accuracy because we know the Galaxy does not perfectly meet one of the assumptions of the method – that the mass in the Galaxy is distributed in a way that is spherically symmetric. You might justifiably wonder whether only the first digit is really significant, in which case it would be appropriate to quote the result as $M_{MW}(8.5\ \text{kpc}) = 8 \times 10^{10} M_\odot$. In fact, given the uncertainties in the method, the mass is probably best quoted as an order of magnitude estimate, yielding a value of $10^{11} M_\odot$.

QUESTION 1.5

The rotation curve is a plot of speed against distance from the centre, so the more useful form of the rotation-curve equation is $v(r) = (GM(r)/r)^{1/2}$. To sketch the rotation curve, we need to know how v varies with r.

(a) $M(r) = kr$ so $v(r) = (GM(r)/r)^{1/2}$ becomes $v(r) = (Gkr/r)^{1/2} = (Gk)^{1/2}$ = constant. That is, the speed has the same value, irrespective of the distance from the centre, and hence the rotation curve is a horizontal straight line (it is flat) (see Figure 1.36a).

(b) $M(r) = \rho \times \frac{4}{3} \pi r^3$ so $v(r) = (GM(r)/r)^{1/2}$ becomes $v(r) = (G\rho \frac{4}{3} \pi r^3/r)^{1/2} = (G\rho \frac{4}{3} \pi \times r^2)^{1/2}$ = const $\times\ r$. That is, the speed rises in proportion to the distance from the centre, and hence the rotation curve is a straight line passing through the origin (see Figure 1.36b).

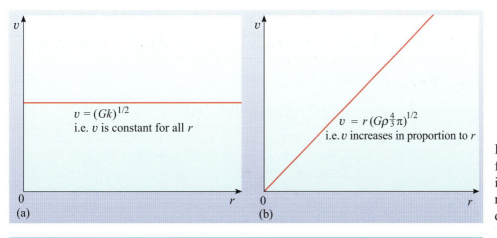

Figure 1.36 Rotation curves for (a) a mass distribution increasing linearly with radius, and (b) a uniform density sphere.

QUESTION 1.6

(a) The time, t, to complete one orbit is given by the distance travelled, d, divided by the speed, v. For a circular orbit, $d = 2\pi r$, where r is the radius of the orbit.

Hence

$$t = d/v = 2\pi r/v = 2\pi \times 4 \times 10^3 \text{ pc} \times 3.09 \times 10^{16} \text{ m pc}^{-1}/(200 \times 10^3 \text{ m s}^{-1})$$
$$= 3.88 \times 10^{15} \text{ s}$$

Since the conversion factor from years to seconds is $365.25 \times 24 \times 60 \times 60$ s yr^{-1} $\approx 3.16 \times 10^7$ s yr^{-1}, we can write $t = 3.88 \times 10^{15}$ s$/31.6 \times 10^6$ s $\text{yr}^{-1} = 1.23 \times 10^8$ yr. Therefore, over 4.5×10^9 yr, at 4 kpc the arm would make 4.5×10^9 yr$/1.23 \times 10^8$ yr $= 37$ rotations.

(b) Similarly, at 10 kpc an arm would complete one revolution in

$$t = 2\pi \times 1.0 \times 10^4 \text{ pc} \times 3.09 \times 10^{16} \text{ m pc}^{-1}/(200 \times 10^3 \text{ m s}^{-1}) = 9.70 \times 10^{15} \text{ s}$$

or 9.70×10^{15} s$/3.16 \times 10^7$ s $\text{yr}^{-1} = 3.07 \times 10^8$ yr

Over 4.5×10^9 yr, at 10 kpc the arm would make 4.5×10^9 yr$/3.07 \times 10^8$ yr $= 15$ rotations.

(c) Over this length of time, the parts at 4 kpc would have made $37 - 15 = 22$ more rotations than the 10 kpc part, which means the spiral arm would be wound around the Galaxy 22 times. However, images of other galaxies (e.g. Figure 1.2), and even the limited maps of our own Galaxy (e.g. Figure 1.21), suggest that this does not happen. Spiral arms can usually be traced for *at most* a few circuits of a galaxy.

QUESTION 1.7

If the co-rotation distance is at 15 kpc, then the spiral-arm speed and star speed must be the same at this distance. From Figure 1.11c, the speed here is about 200 km s^{-1}. Since the arms rotate rigidly, the speed must increase linearly with distance from the Galactic centre, so the graph of the rotation curve must resemble Figure 1.37.

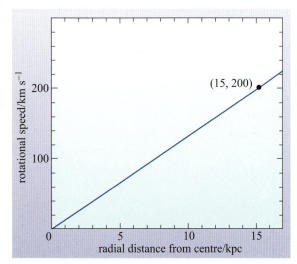

Figure 1.37 Rotation curve for a rigidly rotating spiral pattern that rotates at 200 km s^{-1} at 15 kpc from the Galactic centre. Note that although this rotation 'curve' passes through the centre of the Galaxy, the spiral arms do not extend to radial distances smaller than about 2 kpc from the centre.

The speed of the spiral pattern at 8.5 kpc is therefore (8.5/15) × 200 km s^{-1} = 113 km s^{-1}. Since the Sun is travelling at 200 km s^{-1}, it will approach the wave at 200 km s^{-1} − 113 km s^{-1} = 87 km s^{-1}.

QUESTION 1.8

(a) At the centre of a globular cluster, the number density of stars is 10^4 pc^{-3}, so each star can occupy a volume $1/(10^4$ pc$^{-3})$, which equals 10^{-4} pc^3. To ensure we leave no gaps between adjacent volumes, we must consider cubic spaces rather than spherical ones. Since a cube with sides of length s has a volume s^3, we can write $s^3 = 10^{-4}$ pc^3, so $s = 10^{-4/3}$ pc = 0.046 pc. So the average separation between stars at the centre of a globular cluster is 0.05 pc (to 1 significant figure).

(b) 0.046 pc/1.3 pc ≈ 0.035, so the stellar separation at the centre of a globular cluster is typically 0.035 (i.e. about 1/28) times the distance to Alpha Centauri. Clearly stars are packed together much more tightly in globular clusters than they are in the region of the Galaxy near the Sun.

QUESTION 1.9

The given equation

$$m_V - M_V = 5 \log_{10} (d/\text{pc}) - 5$$

can be rearranged to give the logarithm of the distance in terms of the other quantities:

$$\log_{10}(d/\text{pc}) = (m_V - M_V + 5)/5$$

This can be evaluated using the magnitudes provided ($m_V \approx +20.5$, $M_V \approx +0.5$)

$$\log_{10}(d/\text{pc}) = (20.5 - 0.5 + 5)/5 = 5$$

This implies (from the definition of the $\log_{10}$ function) that $d = 10^5$ pc. That is, RR Lyrae stars having $m_V \approx 20.5$ could be seen to a distance of 100 kpc, well out into the most distant parts of the stellar halo.

QUESTION 1.10

Initially, losses to stellar remnants will continue, so the total amount of ISM will decrease below the current 10% of the stellar mass, and for a time, the metallicity will continue to increase due to mass loss from stars. However, as there will be fewer new stars forming in the Galaxy in future because it will have less gas, there will be fewer high-mass stars to enrich the ISM. The ISM will then be replenished only slowly, by the evolution of long lived, low-mass stars. Ultimately the intergalactic medium will be the main means of replenishment, assuming (boldly) that this source is unlimited. Then, the metallicity of the ISM will begin to reflect that of the infalling intergalactic gas.

QUESTION 1.11

As massive stars evolve, they convert hydrogen into helium, helium into carbon and oxygen, and carbon and oxygen into still heavier elements. Some of these freshly synthesized elements are ejected into the interstellar medium when stars reach the end of their lives, so the metallicity of the interstellar medium increases with time. Surviving Pop. II stars are very old and hence formed from material with a low metallicity, whereas Pop. I stars are much younger and hence formed more recently from more metal-rich gas.

QUESTION 1.12

While this view is understandable, it is flawed for the following reason. During their main sequence lifetimes, stars convert H into He, but do not produce additional heavier elements. Not until helium burning begins during the giant phase do they produce carbon and oxygen, and even then it occurs only in the core of the star, and is not observed at the surface until late in the evolution of the star.

QUESTION 1.13

High-velocity stars are not part of the disc population (Population I), they really belong to the halo population (Population II). As members of this older population they formed from material that had not yet been enriched in heavy elements by nucleosynthesis and mass loss from stars and supernovae. Therefore they are expected to have, on average, lower metallicity than the Sun.

QUESTION 1.14

According to Figure 1.11c, a star 8.5 kpc from the Galactic centre will have a rotation speed of 200 km s^{-1}. The circumference of a circular orbit of radius 8.5 kpc is $d = 2\pi r = 2\pi \times 8.5 \times 10^3$ pc $\times 3.09 \times 10^{16}$ m pc^{-1} = 1.65×10^{21} m.

Thus the time required for a star, such as the Sun, to execute such an orbit is

$$t = \frac{d}{v} = \frac{1.65 \times 10^{21} \text{ m}}{2.0 \times 10^5 \text{ m s}^{-1}} = 8.25 \times 10^{15} \text{ s}$$

There are 3.16×10^7 s in one year. So the time required for one complete orbit by the Sun is

$$\frac{8.25 \times 10^{15} \text{ s}}{3.16 \times 10^7 \text{ s yr}^{-1}} = 2.61 \times 10^8 \text{ yr}$$

Since the Sun has existed for 4.5×10^9 yr, it follows that the number of orbits is

$$\frac{4.5 \times 10^9 \text{ yr}}{2.61 \times 10^8 \text{ yr}} = 17.2$$

Thus there will have been 17 orbits. (Of course, the two-figure 'precision' in this calculation is largely spurious, given the uncertainties that arise in such a calculation.)

QUESTION 1.15

The disc has a radius of about 15 kpc. Thus the area of the disc is $\pi(15\text{ kpc})^2$ and, since it is ≈ 1 kpc thick, its volume is $\pi(15)^2$ kpc^3. By similar reasoning, the optically observable volume of the disc is $\pi(5)^2$ kpc$^2 \times 1$ kpc $= \pi(5)^2$ kpc^3. Thus the fraction of the disc's volume that can be observed is

$$\frac{\pi(5)^2}{\pi(15)^2} = \frac{25\pi}{225\pi} = \frac{1}{9}$$

This limitation is mainly the result of dust in the plane of the Galaxy.

QUESTION 1.16

The Sun is about 4.5×10^9 years old. This is older than all but a very few of the longest-lived open clusters. Thus, even if the Sun was originally part of an open cluster, it would have long since escaped from the cluster. Possible causes of the escape are gravitational disruption (possibly through an encounter with a giant molecular cloud complex), the 'evaporation' of the cluster due to stars occasionally exceeding the escape speed, or simply the dispersive effect of differential rotation over a long period of time.

QUESTION 1.17

In the density wave theory, the spiral pattern moves around rigidly with an unchanging shape, and does not wind up. Matter in the Milky Way revolves differentially, with a longer orbital period for matter at a greater distance from the Galactic centre. Such matter passes into the spiral arms and then out again. Thus the matter highlighting the spiral arms at any time is not permanently present within the arms and thus the arms have no tendency to wind up.

QUESTION 1.18

Tracers of spiral arms include:

- open clusters
- OB associations
- bright HII regions
- dense molecular clouds
- clouds of neutral hydrogen.

QUESTION 1.19

Orange. The brightest stars in a globular cluster will be those at the highest point on the H–R diagram of a cluster of *old* stars. In the globular cluster H–R diagrams the stars in this position are cool red giants, i.e. orange in colour.

QUESTION 1.20

The evidence that the galaxy continues to evolve can be summarized by the following points:

- Star formation is still occurring;
- Enriched gas is returned to the ISM via stellar winds, planetary nebulae and supernovae;

- Infall of intergalactic gas is inferred from gas recycling and high-velocity clouds;
- Some gas from the ISM becomes locked away in the cores of stellar remnants;
- Young stars have higher metallicities than older stars;
- The Sagittarius dwarf galaxy is currently merging with the Milky Way;
- Open star clusters are disrupted by differential rotation of the Galaxy long before most of their stars die;
- Many high-velocity clouds have large velocities towards the Milky Way.

QUESTION 2.1

High-mass main sequence stars, open clusters, HII regions and an abundance of Population I stars (relative to Population II stars), are all indicators of continuing star formation. Since new stars are unlikely to be formed in the absence of cool gas (the raw material needed to make them) it is to be expected that each of these types of object will increase or become more significant in going from ellipticals (which have little cool gas) to spirals, which are actively forming stars in their discs.

QUESTION 2.2

The completed Table 2.1 is shown below.

Property	Ellipticals	Spirals	Irregulars
approximate proportion of all galaxies	$\geq$60%	$\leq$30%	$\leq$15%
mass of molecular and atomic gas as % of mass of stars	small, 1% say	5–15%	15–25%
stellar populations	Population II	Populations I and II	Populations I and II
approximate mass range	$\sim10^5 M_\odot$ to $\sim10^{13} M_\odot$	$\sim10^9 M_\odot$ to a few times $10^{12} M_\odot$	$\sim10^7 M_\odot$ to $10^{10} M_\odot$
approximate luminosity range	a few times $10^5 L_\odot$ to $\sim10^{11} L_\odot$	$\sim10^9 L_\odot$ to a few times $10^{11} L_\odot$	$\sim10^7 L_\odot$ to $10^{10} L_\odot$
approximate diameter range ^{a}	$(0.01-5)d_{MW}$	$(0.02-1.5)d_{MW}$	$(0.05-0.25)d_{MW}$
angular momentum per unit mass	low	high	low

$^a d_{MW}$, diameter of Milky Way.

It is important to realize that many of the properties in the table are difficult to determine and that approximate figures are often poorly determined.

QUESTION 2.3

(a) The diameter of the ring $2a$ can be found using Equation 2.2

$$2a = \frac{c\Delta t}{\sqrt{\left(1 - \left(\frac{b}{a}\right)^2\right)}}$$

(2.2)

The time delay is 340 days, so

$$c\Delta t = (3.00 \times 10^8 \text{ m s}^{-1}) \times (340 \times 24 \times 60 \times 60 \text{ s}) = 8.81 \times 10^{15} \text{ m}$$

The ratio (b/a) can be measured from Figure 2.21 as the ratio between the short and long axes of the ellipse (which is ($2b/2a$) = (b/a))

($2b/2a$) = (short axis of ellipse)/(long axis of ellipse) = (49 mm)/(69 mm) = 0.710

Substituting these values into Equation 2.2

$$2a = \frac{8.81 \times 10^{15} \text{ m}}{\sqrt{\left(1 - (0.710)^2\right)}} = \frac{8.81 \times 10^{15} \text{ m}}{0.704} = 1.25 \times 10^{16} \text{ m}$$

So the diameter of the ring around SN 1987A is 1.3×10^{16} m.

(b) To find the distance to a feature of length l that subtends an angle θ as viewed from the Earth, we use the relation $l = d \times (\theta \text{ /radians})$ given in Section 2.4.1. Thus

$$d = \frac{l}{(\theta / \text{radians})}$$

From part (a), the diameter of the ring is 1.25×10^{16} m. The angular diameter of the ring is given as 1.66 arcsec, and this needs to be expressed in radians

$$(\theta / \text{radians}) = (\theta / \text{arcsec}) \times (1/57.3) \times (1/3600) = 8.05 \times 10^{-6}$$

It follows that the distance to SN 1987A is

$$d = \frac{1.25 \times 10^{16} \text{ m}}{8.05 \times 10^{-6}} = 1.55 \times 10^{21} \text{ m} = \frac{1.55 \times 10^{21} \text{ m}}{3.09 \times 10^{16} \text{ m pc}^{-1}} = 5.02 \times 10^4 \text{ pc}$$

So the distance to SN 1987A using this method is found to be 50 kpc. (This is consistent with the value of 52 ± 3 kpc that is given in the text.)

QUESTION 2.4

(a) Absorption will reduce the flux received from an object in comparison with the flux that would be measured in the absence of absorption. Thus a distance estimate that is based on applying Equation 2.3 to the flux received when there is absorption will be greater than the value obtained if there were no absorption. So if the effects of absorption are simply ignored, distances will be overestimated.

(b) If the flux is measured over a narrow range of wavelengths, the standard candle method can still be used provided that the luminosity used is that which is emitted over an identical range of wavelengths. Thus, in the visual band (V-band), we could write Equation 2.3 in the form

$$d = [L_V/(4\pi F_V)]^{1/2}$$

Where L_V and F_V are the luminosity and flux in the V-band respectively. (This assumes that radiation does not undergo any significant shift in wavelength between the source and the observer.)

QUESTION 2.5

For a Cepheid with a period of 10 days, Figure 2.25 shows that the average absolute visual magnitude M_V is -4.2.

QUESTION 2.6

The following items of information are needed.

(i) The observed flux density from each supernova at peak brightness. (In practice this would be limited to particular wavebands.)

(ii) A value for the distance, d, to each host galaxy. (In principle this might be based on observations of Cepheid variable periods, but in practice the distances used in these particular cases were based on other bright star observations.)

(iii) An estimate of the amount of radiation absorbed or scattered between the supernova and the flux detector. (Again, in practice this would be limited to a particular waveband.)

The observed flux density should be increased by the amount that was lost due to scattering and absorption, and the resulting total, F, used in conjunction with the distance, d, to find the luminosity, L, where:

$$L = 4\pi d^2 F$$

(If F had been limited to some particular band of wavelengths then L would be limited in the same way. In practice, the calibration of the Type Ia supernova method uses other information besides the three nearby examples mentioned in the question.)

QUESTION 2.7

Let quantities relating to the two galaxies be denoted by subscripts A and B. The relationship between the velocity dispersions of galaxies A and B can be expressed as

$$(\Delta v)_A = 1.2\,(\Delta v)_B$$

(a) The luminosities are related to the velocity dispersion according to the Faber–Jackson relation (Equation 2.5)

$$L_A \propto (\Delta v)_A^4$$

$$L_B \propto (\Delta v)_B^4$$

Hence $(L_A / L_B) = ((\Delta v)_A / (\Delta v)_B)^4 = (1.2)^4 = 2.07$

Thus the luminosity of galaxy A is 2.1 times greater than that of galaxy B.

(b) The velocity dispersion, mass M and length scale R are related according to the relation quoted in Section 2.3.2

$$(\Delta v) \propto (M/R)^{1/2}$$

Which can be rearranged to give

$$M \propto (\Delta v)^2 R$$

Thus

$$M_A \propto (\Delta v)_A^2 R_A$$

and

$$M_B \propto (\Delta v)_B^2 R_B$$

These relations imply that

$$(M_A / M_B) = ((\Delta v)_A / (\Delta v)_B)^2 (R_A / R_B)$$

But the radii are identical and hence the length scale R is the same for both galaxies, i.e. $R_A = R_B$,

so $(M_A / M_B) = ((\Delta v)_A / (\Delta v)_B)^2 = (1.2)^2 = 1.44$

So the mass of galaxy A is a factor of 1.4 times greater than that of galaxy B.

QUESTION 2.8

The first stage is to rearrange Equation 2.7

$$d = \frac{cz}{H_0}$$

Using the measured redshift of $z = 0.048$, the assumed value of $H_0 = 72 \text{ km s}^{-1} \text{ Mpc}^{-1}$, and ensuring that c is expressed in units of km s^{-1}, the distance is

$$d = \frac{(3.00 \times 10^5 \text{ km s}^{-1}) \times 0.048}{72 \text{ km s}^{-1} \text{ Mpc}^{-1}} = 200 \text{ Mpc}$$

So, using Hubble's law, the distance of this galaxy is 200 Mpc.

QUESTION 2.9

To express the Hubble constant in SI units, speed should be expressed in m s^{-1} and distance in terms of m.

$$H_0 = 72 \text{ km s}^{-1} \text{ Mpc}^{-1}$$

$$H_0 = 7.2 \times 10^4 \text{ m s}^{-1}/(10^6 \times 3.09 \times 10^{16} \text{ m})$$

$$H_0 = 2.33 \times 10^{-18} \text{ s}^{-1}$$

So a value of H_0 of 72 km s^{-1} Mpc^{-1} is equivalent to 2.3×10^{-18} s^{-1}.

QUESTION 2.10

(a) The redshift due to a random motion is a Doppler shift. The equation that relates Doppler shift and radial velocity is Equation 1.7

$$v = c(\lambda_{obs} - \lambda_{em})/ \lambda_{em}$$

But $\quad z = (\lambda_{obs} - \lambda_{em})/ \lambda_{em}$

So, $\quad v = cz$

Rearranging

$$z = v/c$$

So if $v = 300$ km s^{-1},

$$z = (300 \times 10^3 \text{ m s}^{-1})/(3.00 \times 10^8 \text{ m s}^{-1})$$

i.e. $z = 1.0 \times 10^{-3}$

The typical random velocities of galaxies will give rise to redshifts typically of the order of 0.001. Since the motions are random they are equally likely to be towards us as away from us, and so the redshifts may be negative or positive.

(b) The distance at which the redshifts due to Hubble's law will be a factor of ten greater than the redshifts caused by random motion is found using Hubble's law. From part (a) we know that random motions cause a redshift of 0.001, thus we need to find the distance at which Hubble's law predicts $z = 10 \times 0.001 = 0.01$. Using

$$d = \frac{cz}{H_0}$$

$$d = \frac{(3.00 \times 10^5 \text{ km s}^{-1}) \times 0.01}{72 \text{ km s}^{-1} \text{ Mpc}^{-1}} = 42 \text{ Mpc}$$

So the redshifts predicted by Hubble's law will be a factor of ten greater than the redshifts due to random motion for distances greater than about 40 Mpc.

QUESTION 2.11

The flattening factor for an elliptical galaxy is $f = (a - b)/a$. For the ellipse shown in Figure 2.3, a is 23 mm and b is 13 mm, so

$$f = \frac{23 \text{ mm} - 13 \text{ mm}}{23 \text{ mm}} = 0.43$$

In assigning a Hubble type to an elliptical galaxy, the number that follows the E is the nearest integer to $10 \times f$. So in this case the appropriate Hubble type would be E4.

QUESTION 2.12

(a) NGC 7479 has wide-flung arms, and there is a bar across its centre; it is an SBc galaxy.

(b) M101 also has wide-flung arms and a relatively small bulge; it is a spiral galaxy of type Sc.

(c) NGC 4449 has no obvious symmetry; it is an irregular galaxy (Irr).

QUESTION 2.13

The ellipsoid is the only three-dimensional shape that presents an elliptical outline to all observers, irrespective of the direction from which it is observed. Oblate and prolate spheroids (and triaxial ellipsoids) are special cases of the general ellipsoid.

QUESTION 2.14

Shortcomings of the standard candle methods include the following.

(i) The difficulty of selecting classes of objects or bodies that have a uniform luminosity (i.e. standard candles).

(ii) The difficulty of determining the luminosity of those standard candles (i.e. the calibration problem).

(iii) The likelihood that the standard candles, whatever they may be, will simply be too faint to be seen at all in the more distant galaxies.

(iv) The problems associated with the absorption and/or scattering of radiation along the path between the source and the detector. These effects generally reduce the flux density received from the source and make it seem further away than it really is.

(v) The possibility, particularly in the case of spirals, that it may be necessary to take into account the orientation of the galaxy relative to the observer (for example, when using the Tully–Fisher method).

(It is also possible that standard candles observed at great distances (and hence at earlier times, because of the finite speed of light) may not be the same as those relatively nearby objects used for calibration due to evolutionary effects.)

QUESTION 2.15

See Figure 2.45.

QUESTION 2.16

The rotation curve method depends on the stars in a galaxy moving in near circular orbits (Box 1.2). Stars in elliptical galaxies are moving randomly, unlike the fairly orderly rotation of matter in spirals, so there is little net rotation. Consequently the rotation curve analysis cannot be applied to matter in elliptical galaxies.

QUESTION 2.17

(a) M31 is a spiral galaxy, like the Milky Way. Therefore it is to be expected that the central bulge will mainly consist of Population II stars whereas the disc will be dominated by Population I stars. Since these populations are significantly different, it makes sense to model them separately.

(b) In an E2 galaxy, where there is little or no active star formation, the stars will be mainly long-lived types of the sort common in Population II. Thus, lower main sequence stars (i) and red giants (ii) should be well represented, whereas upper main sequence stars (iii) and Cepheid variables (iv) will be rare.

QUESTION 2.18

The correct sequence is (c), (a), (b). In Figure 2.44c, all the stars have formed and the main sequence is well populated, although massive blue stars are much less common than low-mass red stars. A few million years later (Figure 2.44a) the very massive blue stars have burnt themselves out and some of the slightly less massive stars have already left the main sequences and started to become

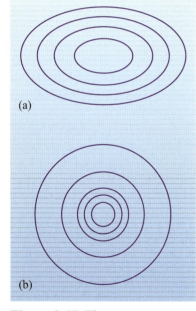

Figure 2.45 The answer to Question 2.15: (a) an E4 galaxy; (b) a face-on S0 galaxy.

cooler, although no less luminous. Overall, by this stage there has been some reduction in luminosity and a definite change towards a yellower integrated spectrum. After billions of years (Figure 2.44b) even intermediate-mass stars have started to leave the main sequence before entering their giant phase. Overall, owing to the exhaustion of the more massive stars, there has been a further lowering of luminosity and a movement towards a redder spectrum. (Note that the chronological sequence of these H–R diagrams is similar to that shown for the three clusters of stars shown in Figure 1.28.)

QUESTION 3.1

For the Sun's photosphere we have $T = 6000$ K and $m = m_H = 1.67 \times 10^{-27}$ kg. So the velocity dispersion is given by

$$\Delta v \approx \left(\frac{2kT}{m}\right)^{1/2} = \left(\frac{2 \times 1.38 \times 10^{-23} \text{ J K}^{-1} \times 6000 \text{ K}}{1.67 \times 10^{-27} \text{ kg}}\right)^{1/2} = 9960 \text{ m s}^{-1}$$

So hydrogen atoms in the Sun's atmosphere are moving at around 10 km s^{-1}.

Rearranging Equation 3.1 we have

$$\Delta\lambda = \frac{\lambda \Delta v}{c} = \frac{656.3 \text{ nm} \times 9960 \text{ m s}^{-1}}{3.0 \times 10^8 \text{ m s}^{-1}} = 0.022 \text{ nm}$$

so the Doppler broadening of the solar Hα line is 0.02 nm (to 1 significant figure). (This is a tiny broadening, about 1 part in 30 000, and rather difficult to observe.)

QUESTION 3.2

The rotation curve shows that the Galaxy is rotating at between roughly 200 and 250 km s^{-1}. Edge-on, this is the approach speed at one extremity and the recession speed at the other. So the line-width that would be observed if the Galaxy were viewed edge-on is 400–500 km s^{-1}.

QUESTION 3.3

The Hβ line has a wavelength of about 485 nm and a width of roughly 5 nm. So the velocity dispersion is

$$\Delta v = \frac{c \Delta\lambda}{\lambda} = \frac{3.0 \times 10^5 \text{ km s}^{-1} \times 5 \text{ nm}}{485 \text{ nm}} \approx 3100 \text{ km s}^{-1}$$

Rearranging Equation 3.2 and putting $m = m_H$ we have

$$T = \frac{m_H (\Delta v)^2}{2k} = \frac{1.67 \times 10^{-27} \text{ kg} \times (3.1 \times 10^6 \text{ m s}^{-1})^2}{2 \times 1.38 \times 10^{-23} \text{ J K}^{-1}} = 6 \times 10^8 \text{ K}$$

In view of the difficulty of measuring the width of the line, it would be appropriate to give the temperature as approximately 10^9 K. (As is explained in the text following this question, the Hβ emitting region does *not* have such a high temperature.)

QUESTION 3.4

In the optical region ($\lambda = 0.5$ μm), galaxy A has $\lambda F_\lambda = 0.5 \times 10^{-29}$ W m^{-2}. For galaxy B, $\lambda F_\lambda = 0.5 \times 10^{-30}$ W m^{-2}. So galaxy A is 10 times brighter in the optical.

In the far-infrared ($\lambda = 100$ μm), the upper limit to λF_λ is 10^{-30} W m^{-2} whereas galaxy B has $\lambda F_\lambda = 10^{-28}$ W m^{-2}. The far-infrared flux density of galaxy B is not only greater than that of galaxy A at this wavelength, but also exceeds the flux density at optical wavelengths of both galaxies. On the basis of these (very sparse!) data, it is concluded that galaxy B is the more luminous galaxy.

QUESTION 3.5

The spectrum shows two distinct peaks, one at the red end of the optical (similar to a normal galaxy) and one far into the infrared, near 100 μm. The far-infrared peak is at a similar wavelength to the small peak in a normal spiral galaxy, but it is higher than the optical peak, suggesting that this galaxy emits most of its energy in the far-infrared. There is no significant emission in the UV or X-ray region.

This is not a normal galaxy and you might have guessed that it is an active galaxy. In fact, it is a starburst galaxy. The infrared radiation is coming from dust heated by the continuing star formation and is another distinguishing characteristic of a starburst galaxy, in addition to the strong narrow optical emission lines that you encountered earlier.

QUESTION 3.6

There are several things you may have thought of. Table 3.2 summarizes many of the characteristics and includes some pieces of new information as well. What all active galaxies have in common is a powerful, compact nucleus which appears to be the source of their energy.

Table 3.2 Features of active galaxies compared to those of normal galaxies.

Characteristic		Active galaxies			
	Normal	Seyfert	Quasar	Radio galaxy	Blazar
Narrow emission lines	weak	yes	yes	yes	some cases
Broad emission lines	no	some cases	yes	some cases	some cases
X-rays	weak	some cases	some cases	some cases	yes
UV excess	no	some cases	yes	some cases	yes
Far-infrared excess	no	yes	yes	yes	yes
Strong radio emission	no	no	some cases	yes	yes
Jets and lobes	no	no	some cases	yes	no
Variability	no	yes	yes	yes	yes

QUESTION 3.7

(a) An angular size limit of 0.1 arcsec corresponds to an angle in radians of

$$(\theta/\text{rad}) = 0.1 \times (1/3600) \times (1/57.3) = 4.8 \times 10^{-7}$$

Multiplying this by the distance shows that the upper limit on the size is
$$= (50 \times 10^6 \text{ pc}) \times (4.8 \times 10^{-7} \text{ rad}) = 24 \text{ pc}.$$

(b) A week is 7 days which is $7 \times 24 \times 60 \times 60$ s. The upper limit from the variability is

$$R \sim c\Delta t = (3.0 \times 10^8 \text{ m s}^{-1}) \times (7 \times 24 \times 60 \times 60 \text{ s}) = 1.8 \times 10^{14} \text{ m} = 0.006 \text{ pc}$$

(Thus variability constraints provide a much lower value for the upper limit to the size of the AGN than does the optical imaging observation.)

QUESTION 3.8

The relationship between flux density F, luminosity L and distance d is given by Equation 2.3 which can be rearranged to give

$$F = \frac{L}{4\pi d^2}$$

Using this relationship it can be seen that if the AGN is at twice the distance but appears as bright as the normal galaxy in the optical, then it must be emitting four times the optical light of the normal galaxy like our own. If only one-fifth of the AGN's energy is emitted in the optical, then its luminosity is $4 \times 5 = 20$ times that of the normal galaxy like our own, assuming that (as usual) the normal galaxy emits mostly at optical wavelengths. The AGN luminosity is thus about $20 \times 2 \times 10^{10} L_\odot = 4 \times 10^{11} L_\odot$.

QUESTION 3.9

A mass m has a rest energy of mc^2.

(a) If 1 kg of hydrogen were to undergo nuclear fusion to produce helium, the energy liberated would be 0.007 (i.e. 0.7%) of its rest energy:

$$E = 0.007 mc^2 = 0.007 \times 1 \text{ kg} \times (3 \times 10^8 \text{ m s}^{-1})^2 \text{ J} = 6 \times 10^{14} \text{ J}$$

(b) If 1 kg of hydrogen were to fall into a black hole, the energy liberated would be approximately $0.1 mc^2 = 0.1 \times 1 \times (3 \times 10^8 \text{ m s}^{-1})^2 \text{ J} = 9 \times 10^{15} \text{ J}$.

You would expect much *less* energy from the chemical reaction.

QUESTION 3.10

For the Seyfert nucleus, $L = 4 \times 10^{10} L_\odot = 1.6 \times 10^{37}$ W. By Equation 3.5, $Q = L/(0.1c^2)$. Substituting for L,

$$Q = \frac{1.6 \times 10^{37}}{(0.1 \times 9 \times 10^{16})} \text{ kg s}^{-1} \approx 2 \times 10^{21} \text{ kg s}^{-1}$$

This can be converted into solar masses per year, by using 1 year $\approx 3 \times 10^7$ s, and $M_\odot \approx 2 \times 10^{30}$ kg, giving

$$Q = \frac{(2 \times 10^{21} \times 3 \times 10^7)}{2 \times 10^{30}} M_\odot \text{ yr}^{-1} = 0.03 M_\odot \text{ yr}^{-1}$$

The Eddington limit places an upper limit on the luminosity for a black hole of given mass.

QUESTION 3.11

Wien's displacement law relates the temperature of a black body to the wavelength at which the spectral flux density has its maximum value. In this case, the dust grains on the inner edge of the torus will be at

$$(\lambda_{max} / \text{m}) = \frac{2.9 \times 10^{-3}}{(T/\text{K})} = \frac{2.9 \times 10^{-3}}{2000} \approx 1.5 \times 10^{-6}$$

So, λ_{max} is about 1.5 μm.

Grains further from the engine will be cooler, and their emission will peak at longer wavelengths, so the torus can be expected to radiate in the infrared at wavelengths of 1.5 μm or longer. (Note that although the spectrum emitted by dust grains is *not* a black-body spectrum, it is similar enough for the above argument to remain valid.)

QUESTION 3.12

From Equation 3.7 we have

$$r = \left(\frac{L}{16\pi\sigma T^4}\right)^{1/2} = \left[\frac{1 \times 10^{38} \text{ W}}{16\pi \times (5.67 \times 10^{-8} \text{ W m}^{-2} \text{ K}^{-4}) \times (2000 \text{ K})^4}\right]^{1/2}$$

$$= 1.48 \times 10^{15} \text{ m}$$

$$= \frac{1.48 \times 10^{15} \text{ m}}{3.09 \times 10^{16} \text{ m pc}^{-1}} = 4.8 \times 10^{-2} \text{ pc}$$

Thus, according to this calculation, the radius of the inner edge of the dust torus is 1.5×10^{15} m or 0.05 pc. (A more rigorous calculation, which takes account of the efficiency of dust grains in absorbing and emitting radiation, gives a radius of 0.2 pc.)

QUESTION 3.13

The NLR is illuminated by radiation from the central engine. As the engine is partly hidden by the dust torus, radiation can only reach the NLR through the openings along the axis of the torus. Any gas near the plane of the torus lies in its shadow and will not be illuminated. The visible NLR would take the form of a double cone of light corresponding to the conical beams of radiation emerging from either side of the torus.

The best view would be from near the plane of the torus, where a wedge-shaped glow would be visible on either side of the dark torus.

QUESTION 3.14

(a) If the mass of the 'seed' black hole is M_{seed}, then the mass of the black hole (M_{bh}) after one doubling will be

$$M_{bh} = M_{seed} \times 2$$

After two doublings it is

$$M_{bh} = M_{seed} \times 2 \times 2 = M_{seed} \times 2^2$$

And after n doublings it is

$$M_{bh} = M_{seed} \times 2^n$$

(b) Making the assumptions that black hole growth is limited by the Eddington luminosity and that 10% of the rest mass of accreting material is radiated away, we know that black holes can double in mass every 3.4×10^7 years, i.e. 34 million years. The time available for black hole growth is 750 million years. So the number of times the mass doubles is:

$$n = (750 \text{ million years}) / (34 \text{ million years}) = 22.1$$

(c) The mass of the black hole after n doublings is given by part (a)

$$M_{bh} = M_{seed} \times 2^n$$

Using the value of $n = 22.1$ obtained in part (b) gives a final mass of

$$M_{bh} = 10 \, M_\odot \times 2^{22.1}$$

$$M_{bh} = 4.5 \times 10^7 \, M_\odot$$

It appears that based on the assumptions made here, that the black hole could just about grow to a size which is typical for a supermassive black hole in an active galaxy.

Comment: The calculation of the number of doublings in part (b) returned a non-integer value ($n = 22.1$). Since the growth of the black hole is a continuous process, it is valid to calculate the mass at a given time using a value of n which is not an integer. On the other hand, given the approximate nature of the calculation, it is also justified to use the value n rounded to two significant figures ($n = 22$) here. Both approaches are valid and lead to the same conclusion in part (c).

QUESTION 3.15

For the gas motion use Equation 3.1, $\Delta\lambda/\lambda = \Delta v/c$, where Δv is the velocity dispersion. Then $\Delta\lambda/\lambda = 2.0 \text{ nm}/654.3 \text{ nm} \approx 0.0030$. Thus the overall spread of internal speeds is $\Delta v \approx 0.0030 \times c \approx 900 \text{ km s}^{-1}$, which is too large for a normal galaxy.

QUESTION 3.16

In the radio wave region, $\lambda = 10^5$ μm so

$$\lambda F_\lambda = 10^5 \text{ μm} \times 10^{-28} \text{ W m}^{-2} \text{ μm}^{-1} = 10^{-23} \text{ W m}^{-2}$$

In the far-infrared region $\lambda = 100$ μm so that

$$\lambda F_\lambda = 100 \text{ μm} \times 10^{-23} \text{ W m}^{-2} \text{ μm}^{-1} = 10^{-21} \text{ W m}^{-2}$$

In the X-ray region, $\lambda = 10^{-4}$ μm so

$$\lambda F_\lambda = 10^{-4} \text{ μm} \times 10^{-20} \text{ W m}^{-2} \text{ μm}^{-1} = 10^{-24} \text{ W m}^{-2}$$

The largest of these λF_λ values is $10^{-21} \text{ W m}^{-2}$, so we conclude that the far-infrared emission dominates.

QUESTION 3.17

The wavelengths λ are 0.5 μm, 5 μm and 50 μm, therefore the λF_λ values are 5×10^{-28} W m^{-2}, 5×10^{-28} W m^{-2} and 5×10^{-27} W m^{-2}, respectively. The largest of these values is 5×10^{-27} W m^{-2}, so the dominant flux is at 50 μm, which is in the far-infrared. The object is likely to be either a starburst galaxy or an active galaxy.

QUESTION 3.18

If the galaxy were active, one would expect to see strong emission lines in the optical and spectral excesses at non-optical wavelengths.

QUESTION 4.1

(a) First, we must convert the angular diameter θ into radians:

$$\theta = 1.9° \times 1/57.3 = 0.0332 \text{ rad}$$

The cluster diameter ($2R$) is given by $d \times \theta$ where d is the distance to the cluster:

$$2R = d \times \theta = 120 \text{ Mpc} \times 0.0332 \text{ radians} = 3.98 \text{ Mpc}$$

So the diameter of the cluster is 4.0 Mpc (to 2 significant figures).

(This is a typical cluster with a radius equal to the Abell radius.)

(b) The cluster is now viewed from distance $d = 420$ Mpc, and we know that the diameter ($2R$) of the cluster is 3.98 Mpc

$$\theta = 2R/d$$

$$\theta = 3.98 \text{ Mpc}/420 \text{ Mpc} = 9.48 \times 10^{-3} \text{ rad} = 9.48 \times 10^{-3} \times 57.3° = 0.543°$$

So as seen from a distance of 420 Mpc, the angular diameter of the cluster would be 0.54°.

QUESTION 4.2

Using Equation 2.7:

$$z = \frac{H_0}{c} d$$

Gives $\quad d = \dfrac{cz}{H_0} = \dfrac{3 \times 10^5 \text{ km s}^{-1} \times 0.25}{72 \text{ km s}^{-1} \text{ Mpc}^{-1}} = 1042 \text{ Mpc}$

So the distance to is 1.0×10^3 Mpc (to 2 significant figures).

(Note that the simple linear relationship between redshift and distance (Equation 2.7) only holds for redshifts less than 0.2. Consequently the result of the answer to this question should be regarded as an *upper limit* to the distance. The deviations from Equation 2.7 for a redshift of 0.25 will be small, so it is reasonable to say that the distance to the furthest cluster in the Abell survey is *approximately* 1000 Mpc.)

QUESTION 4.3

We start by converting the Abell radius of 2 Mpc into metres

$$R_A = 2 \times 10^6 \times (3.09 \times 10^{16}) \text{ m}$$

$$R_A = 6.18 \times 10^{22} \text{ m}$$

Using Equation 4.1,

$$M = \frac{R_A (\Delta v)^2}{G}$$

$$= \frac{6.18 \times 10^{22} \text{ m}}{6.67 \times 10^{-11} \text{ m}^3 \text{ kg}^{-1} \text{ s}^{-2}} \left(5.5 \times 10^5 \text{ m s}^{-1}\right)^2 = 2.80 \times 10^{44} \text{ kg}$$

$$= \frac{2.80 \times 10^{44} \text{ kg}}{1.99 \times 10^{30} \text{ kg}} M_\odot = 1.41 \times 10^{14} M_\odot$$

So, using the virial mass method, the mass of the Virgo cluster is found to be $1 \times 10^{14} M_\odot$ (to 1 significant figure).

QUESTION 4.4

Equation 4.2 is an expression for the angular radius of the Einstein ring

$$\theta_E = \sqrt{\frac{4GM}{c^2} \frac{D_{LS}}{D_L D_S}} \tag{4.2}$$

This needs to be rearranged to give an expression for the mass M of the cluster. So both sides of Equation 4.2 are squared

$$\theta_E^2 = \frac{4GM}{c^2} \frac{D_{LS}}{D_L D_S}$$

and this is then rearranged to give

$$M = \frac{\theta_E^2 c^2 D_L D_S}{4G D_{LS}} \tag{i}$$

The question states that we can assume that the cluster is mid-way between Earth and the background galaxies, and so $D_{LS} = D_L$. Substituting this value for D_{LS} into Equation (i) gives

$$M = \frac{\theta_E^2 c^2 D_L D_S}{4G D_L} = \frac{\theta_E^2 c^2 D_S}{4G} \tag{ii}$$

D_S is the distance to the background *source* galaxy, which is twice as far away as Abell 2218 itself, so $D_S = 1400$ Mpc.

We must also convert the angle θ_E into radians

$$\theta_E = (1.0/60) \times (1/57.3) \text{ rad} = 2.91 \times 10^{-4} \text{ rad}$$

Substituting these values into Equation (ii) gives

$$M = \frac{(2.91 \times 10^{-1})^2 \times (3 \times 10^8 \text{ m s}^{-1})^2 \times 1400 \times 3.09 \times 10^{22} \text{ m}}{4 \times 6.67 \times 10^{-11} \text{ N m}^2 \text{ kg}^{-2}}$$

$$= 1.24 \times 10^{45} \text{ kg}$$

$$M = \frac{1.24 \times 10^{45} \text{ kg}}{1.99 \times 10^{30} \text{ kg}} \; M_\odot = 6.23 \times 10^{14} \; M_\odot$$

So the mass of the cluster is $6.2 \times 10^{14} M_\odot$ (to 2 significant figures). (As mentioned in the text, cluster masses typically lie between 10^{14} and $10^{15} M_\odot$, so this estimate is towards the upper end of the range of cluster masses. This is consistent with the fact that Abell 2218 is one of the richest clusters in the Abell catalogue.)

QUESTION 4.5

The completed table of distances and lengths is given as Table 4.3.

Table 4.3 The scales of different types of cosmic structures.

Feature	Distance or length/Mpc
Milky Way (diameter)	0.03
Distance to Large Magellanic Cloud	0.05
Distance to the Andromeda Galaxy	0.8
Extent of the Local Group	~2
Typical diameter of a cluster	~4
Distance to nearest rich cluster (Virgo)	20
Extent of a typical supercluster	30–50
Extent of voids	~60
Scale on which the Universe appears uniform	~200

QUESTION 4.6

Redshift is given by Equation 2.6

$$z = \frac{\lambda_{obs} - \lambda_{em}}{\lambda_{em}}$$

The original wavelength λ_{em} of the Lyman α line is 121 nm. (Note that although the original wavelength is called λ_{em}, this is of course the wavelength of an *absorption* line due to atomic hydrogen in the intergalactic medium.)

A red-shifted wavelength of $\lambda_{obs} = 372$ nm gives

$$z = \frac{372 \text{ nm} - 121 \text{ nm}}{121 \text{ nm}} = 2.07$$

From Figure 4.21 a redshift of 2.07 corresponds to a distance of approximately 5.4×10^3 Mpc.

QUESTION 4.7

The Virgo cluster is at a distance of about 20 Mpc. One parsec equals 3.26 light-years, so light will take 3.26×20 million years, or about 65 million years to travel 20 Mpc. Therefore we are seeing the Virgo cluster as it was about 65 million years ago. Expressed as a fraction of the age of the Universe this is $(65 \times 10^6 \text{ years})/(1.38 \times 10^{10} \text{ years}) = 4.7 \times 10^{-3}$, or about 0.5% of the age of the Universe.

QUESTION 4.8

The three methods for determining the total mass of a cluster of galaxies are: (i) using the velocity dispersion of galaxies within the cluster, (ii) from X-ray observations of the intracluster gas, and (iii) from the effect that the cluster has as a gravitational lens.

For the velocity dispersion method it is assumed that the cluster is relaxed or virialized. The determination of mass from X-ray measurements assumes that the intracluster gas is in a state of hydrostatic equilibrium such that the gradient in gas pressure is balanced by gravity. The method of using gravitational lensing to determine the cluster mass does not rely on making any assumptions about the physical state of the cluster.

QUESTION 5.1

In a Universe where the baryonic matter is 75% hydrogen and 25% helium (by mass), it has already been shown that there will be 12 hydrogen nuclei for each helium nucleus. Assuming that all the helium is helium-4, it follows that for the two neutrons in each helium-4 nucleus there will be 14 protons (two from the helium-4 nucleus and 12 from the 12 hydrogen nuclei).

Thus, for every 10 neutrons there will be 70 protons. In arriving at this conclusion we have assumed that all hydrogen nuclei consist of a single proton, that all helium nuclei are helium-4, that neutrons and protons have identical mass, and that the universal mix of 75% hydrogen and 25% helium (by mass) is accurate.

QUESTION 5.2

The uniformity of the CMB argues (but does not prove) that the CMB is cosmic. In the case of sunlight, for example, it is clear that there is a nearby source since part of each day is dark when the Earth is between us and the Sun. Most other sources of radiation give similar indications of their local origin by being non-uniform.

QUESTION 5.3

In Section 5.2.3 it was argued that the matter and radiation in the Universe are uniformly distributed at any time. On this basis it is to be expected that the energy and momentum associated with that matter and radiation are also uniformly distributed.

QUESTION 5.4

In the Einstein model the scale factor R is constant. This means that the graph of R against t will be a horizontal line, as shown in Figure 5.34.

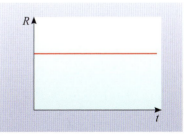

Figure 5.34 R against t graph for the scale factor of the Einstein model.

QUESTION 5.5

(a) There are no such ranges, all the FRW models are consistent with the cosmological principle which demands homogeneity and isotropy. (Questions have been raised about the possibility that some models, such as the Eddington–Lemaître model, might evolve into an inhomogeneous state, but these are beyond the scope of this chapter.)

(b) All ranges of k and Λ allow a big bang, but $k = +1$ models with $0 < \Lambda < \Lambda_E$ allow the possibility of universes that began without a big bang. The case $k = +1$, $\Lambda = \Lambda_E$ allows the possibility that the universe might be static (hence no big bang) or that there might not have been a big bang in a non-static universe. Among the limiting cases (like the de Sitter model) that arise as the density approaches zero, there are cases in which the big bang happened an infinitely long time ago.

(c) This is possible for $k = +1$ and $0 < \Lambda \leq \Lambda_E$. (The symbol $\leq$ means 'less than or equal to'.)

(d) There are no ranges that allow the big bang to be associated with a unique point in space. Such an association would violate the cosmological principle. Take good note of this point since it is a widespread misconception to suppose that the big bang was the 'explosion' of a dense primeval 'atom' located at some particular point in space. Rather than thinking of the big bang as an event in space you should think of it as giving rise to space (or rather space–time).

(e) This is true in any model with $k = 0$.

(f) This is true in any model with $k = +1$.

(g) This is true in all models with $k = 0$ or -1. Of course, due to the finite speed of light we can have no direct observational knowledge of those parts of the Universe that are so distant that light emitted from them has not yet reached us.

QUESTION 5.6

The line is described by Hubble's law, $z = (H_0/c)d$, so the gradient of the line represents H_0/c, i.e. the Hubble constant divided by the speed of light in a vacuum. The gradient of the graph is found by dividing the vertical 'rise' Δz of the line by the corresponding horizontal 'run' Δd. In the case of Figure 5.27, this implies

$$\frac{H_0}{c} = \frac{\Delta z}{\Delta d} = \frac{0.15}{0.64 \times 10^3 \text{ Mpc}} = 2.34 \times 10^{-4} \text{ Mpc}^{-1}$$

It follows that

$$H_0 = 3.00 \times 10^5 \text{ km s}^{-1} \times 2.34 \times 10^{-4} \text{ Mpc}^{-1}$$

i.e. $H_0 = 70 \text{ km s}^{-1} \text{ Mpc}^{-1}$

Note that this question is based on artificial data, not real measurements.

QUESTION 5.7

In the Einstein model R does not change with time, so $\dot{R} = 0$ at all times. This implies that for the Einstein model, the Hubble parameter will also be zero at all times. So, $H(t) = 0$. (Einstein produced this model before Hubble's law was discovered, so he did not appreciate the need for an expanding Universe.)

QUESTION 5.8

Figure 5.23 indicates that all the accelerating FRW models correspond to $\Lambda > 0$. So, if q_0 is indeed negative, as recent observations indicate, then we should expect that $\Lambda > 0$ if the real Universe is well described by an FRW model.

QUESTION 5.9

According to the Friedmann equation

$$\left(\dot{R}\right)^2 = \frac{8\pi G R^2}{3}\left[\rho + \frac{\Lambda c^2}{8\pi G}\right] - kc^2$$

Now $H(t) = \dot{R}(t)/R(t)$, so to get H^2 on the left-hand side of the Friedmann equation, divide both sides by R^2:

$$\left(\frac{\dot{R}}{R}\right)^2 = \frac{8\pi G}{3}\left(\rho + \frac{\Lambda c^2}{8\pi G}\right) - \frac{kc^2}{R^2}$$

i.e. $$H^2 = \frac{8\pi G}{3}\left[\rho + \frac{\Lambda c^2}{8\pi G}\right] - \frac{kc^2}{R^2}$$

as required. Multiplying both sides of this equation by $3/8\pi G$ gives

$$\frac{3H^2}{8\pi G} = \rho + \frac{\Lambda c^2}{8\pi G} - \frac{3kc^2}{8\pi G R^2}$$

i.e. $$\rho_{\text{crit}} = \rho + \rho_\Lambda - \frac{3H^2}{8\pi G}\frac{kc^2}{H^2 R^2}$$

Dividing both sides by ρ_{crit} gives

$$1 = \frac{\rho}{\rho_{\text{crit}}} + \frac{\rho_\Lambda}{\rho_{\text{crit}}} - \frac{kc^2}{H^2 R^2}$$

i.e. $$1 = \Omega_{\text{m}} + \Omega_\Lambda - \frac{kc^2}{H^2 R^2}$$

Letting $\Omega = \Omega_{\text{m}} + \Omega_\Lambda$ gives

$$\Omega - 1 = \frac{kc^2}{H^2 R^2}$$

again, as required.

If $\Omega_{\text{m}} + \Omega_\Lambda = 1$ then $\Omega = 1$, so $\Omega - 1 = 0$

Therefore $\dfrac{kc^2}{H^2 R^2} = 0$, so $k = 0$.

QUESTION 5.10

Multiplying both sides of the given equation by $\dfrac{-R}{2\dot{R}^3}$ gives

$$\frac{-R\ddot{R}}{\dot{R}^2} = \frac{8\pi G}{3}\frac{R^2}{\dot{R}^2}\left(\frac{\rho}{2} - \rho_\Lambda\right)$$

But $\quad \dfrac{R^2}{\dot{R}^2} = \dfrac{1}{H^2}\quad$ and $\quad \dfrac{-R\ddot{R}}{\dot{R}^2} = q$

So, $\quad q = \dfrac{8\pi G}{3H^2}\left(\dfrac{\rho}{2} - \rho_\Lambda\right)$

Now, $\quad \dfrac{8\pi G}{3H^2} = \dfrac{1}{\rho_{crit}}$

So, $\quad q = \dfrac{1}{2}\dfrac{\rho}{\rho_{crit}} - \dfrac{\rho_\Lambda}{\rho_{crit}}$

i.e. $\quad q = \dfrac{\Omega_m}{2} - \Omega_\Lambda$

Substituting the currently favoured values $\Omega_{m,0} = 0.3$ and $\Omega_{\Lambda,0} = 0.7$ gives

$$q_0 = 0.15 - 0.7$$

i.e. $\quad q_0 = -0.55$

QUESTION 5.11

1916	Einstein's theory of general relativity published. This included the field equations in their original form.
1917	Einstein modified the field equations to include the cosmological constant and published the first (static) cosmological model. The de Sitter model was also published within a year.
1919	Eddington led the eclipse expedition that confirmed the general relativistic prediction that starlight passing close to the edge of the Sun would be bent.
1922 and 1924	Friedmann published papers on the behaviour of the scale factor in homogeneous and isotropic relativistic cosmologies.
1925	Lemaître introduced the model (later championed by Eddington) that allowed cosmic expansion to follow an indefinitely long period of effectively static behaviour.
1931	Lemaître formulated his theory of the primeval atom, a crude precursor of the modern big bang theory.
1936	Walker published his independent work on the improvement and generalization of Friedmann's investigations.
1965	Cosmic microwave background radiation discovered.
1997	Discovery that cosmic expansion is currently accelerating.

QUESTION 5.12

The assumptions underpinning the FRW models are that:

- space and time behave in accordance with general relativity; and
- energy and momentum are distributed homogeneously and isotropically on the large scale. (This is the cosmological principle.)

An additional assumption underpinning the Friedmann equation is that:

- the Universe is uniformly filled with a gas of density ρ (that may depend on time).

There are of course other unstated assumptions behind these, such as the belief that space and time are three-dimensional and one-dimensional respectively.

QUESTION 5.13

Since the question concerns FRW models, which are homogeneous and isotropic, it follows that the curvature must be uniform (i.e. the same everywhere and in all directions) at any given time. In a three-dimensional space of (uniform) positive curvature, space will have a finite total volume, straight lines will close back upon themselves, pairs of nearby parallel lines will converge and may meet, any plane triangle will have interior angles that sum to more than 180°, and the circumference of any circle will be less than 2π times its radius.

QUESTION 5.14

In making the step from Equation 5.17 to Equation 5.18 it is stated that

$$\Delta R(t) = \Delta t \times \dot{R}(t)$$

This would be exactly true if the rate of change of R was constant, and it *is* approximately true because we have limited the discussion to cases where Δt is small. However it is not *exactly* true if $\dot{R}$ is changing, i.e. if there is acceleration or deceleration.

QUESTION 5.15

$$H_0 = (72 \pm 8) \text{ km s}^{-1} \text{ Mpc}^{-1}$$

and $\quad H_0 = (2.3 \pm 0.3) \times 10^{-18} \text{ s}^{-1}$

$$\Omega_{m,0} \approx 0.3$$

$$\Omega_{\Lambda,0} \approx 0.7$$

$$q_0 \approx -0.55 \text{ (see answer to Question 5.10)}$$

The age of the Universe t_0 is quoted as 13.8×10^9 years.

To show that the two values of H_0 are equivalent, note that

$$1 \text{ Mpc} = 3.09 \times 10^{22} \text{ m} = 3.09 \times 10^{19} \text{ km}$$

Dividing both sides by 1 Mpc shows that

$$1 = 3.09 \times 10^{19} \text{ km Mpc}^{-1}$$

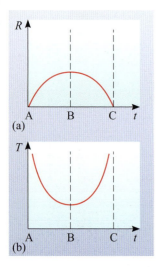

Figure 6.25 (a) Scale factor, and (b) temperature as functions of time for a Friedmann– Robertson– Walker model with $k = 0, \Lambda < 0$. A, B and C are times that are referred to in Table 6.5.

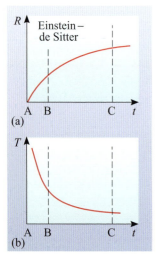

Figure 6.26 (a) Scale factor, and (b) temperature as functions of time for a Friedmann– Robertson–Walker model with $k = 0, \Lambda = 0$ (the Einstein–de Sitter model).

The units make the right-hand side of this equation a complicated way of writing 1.

Dividing the first quoted value of H_0 by this conversion factor gives

$$H_0 = \frac{(72 \pm 8)\,\text{km}\,\text{s}^{-1}\,\text{Mpc}^{-1}}{3.09 \times 10^{19}\,\text{km}\,\text{Mpc}^{-1}} = (2.3 \pm 0.3) \times 10^{-18}\,\text{s}^{-1}$$

QUESTION 6.1

The approach to drawing sketches of how the temperature T varies with time t for all Friedmann–Robertson–Walker models with $k = 0$ shown in Figure 5.23, is similar to that adopted in Example 6.1. We shall look at each model in turn.

$k = 0, \Lambda < 0$ model

The curve showing $R(t)$ is shown in Figure 6.25a. The behaviour of the scale factor R at times A, B and C, and the inferred behaviour of the temperature T at these times is summarized in Table 6.5. The sketch of $T(t)$ is shown in Figure 6.25b.

Table 6.5 The behaviour of the scale factor R at various times indicated on Figure 6.25a and the inferred behaviour of the temperature T at those times.

Time	Behaviour of R at this time	Behaviour of T at this time
A	$R = 0$	$T = 1/R = \infty$
B	R has increased to a maximum value and now does not vary much with time	T must decrease to some minimum value and also only change slowly with time
C	$R = 0$	$T = 1/R = \infty$

$k = 0, \Lambda = 0$ model (the Einstein–de Sitter model)

The curve showing $R(t)$ is shown in Figure 6.26a. The behaviour of the scale factor R at times A, B and C, and the inferred behaviour of the temperature T at these times is summarized in Table 6.6. The sketch of $T(t)$ is shown in Figure 6.26b.

Table 6.6 The behaviour of the scale factor R at various times indicated on Figure 6.25a and the inferred behaviour of the temperature T at those times.

Time	Behaviour of R at this time	Behaviour of T at this time
A	$R = 0$	$T = 1/R = \infty$
B	R is increasing rapidly	T must decrease rapidly
C	R is increasing slowly	T must decrease slowly

$k = 0, \Lambda = \Lambda_E$ model

The curve showing $R(t)$ is shown in Figure 6.27a. The behaviour the scale factor R at times A, B, C and D and the inferred behaviour of the temperature T at these times is summarized in Table 6.7. The sketch of $T(t)$ is shown in Figure 6.27b.

Table 6.7 The behaviour of the scale factor R at various times indicated on Figure 6.27a and the inferred behaviour of the temperature T at those times.

Time	Behaviour of R at this time	Behaviour of T at this time
A	$R = 0$	$T = 1/R = \infty$
B	R is increasing rapidly	T must decrease rapidly
C	R is increasing slowly	T must decrease slowly
D	R is increasing to very high values	T must decrease to very small values

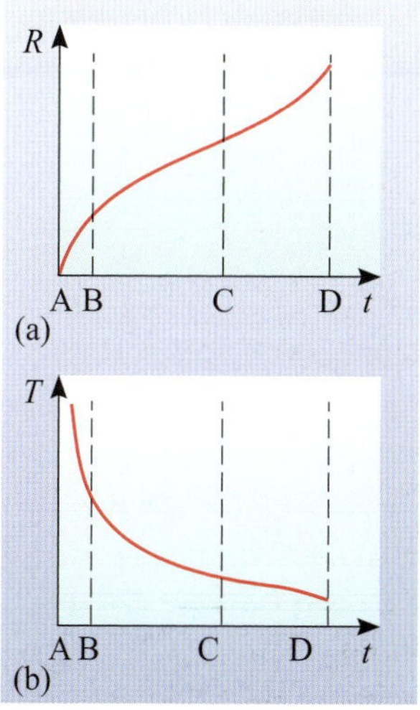

Figure 6.27 (a) Scale factor, and (b) temperature as functions of time for a Friedmann–Robertson–Walker model with $k = 0$, $\Lambda = \Lambda_E$.

QUESTION 6.2

The question states that the current average mass density of luminous and dark matter, $\rho_m \approx 3 \times 10^{-27}$ kg m^{-3}.

The definition of mass density is that if a volume V contains a mass m, then $\rho_m = m/V$.

The energy density u is defined in an analogous way; if a volume of space V contains an energy E, then $u = E/V$.

To find the energy density due to matter u_m we need to calculate the energy equivalent of the mass m that is contained within the volume V. To do this we use the mass-energy equivalence relation $E = mc^2$,

$$u = E/V = mc^2/V = (m/V)c^2$$

but $(m/V) = \rho_m$, so

$$u = \rho_m c^2$$

Using the value of ρ_m given in the question,

$$u = 3 \times 10^{-27} \text{ kg m}^{-3} \times (3.00 \times 10^8 \text{ m s}^{-1})^2$$

i.e. $u = 2.7 \times 10^{-10}$ J m^{-3}

Thus the average energy density due to matter is currently 3×10^{-10} J m^{-3} (to 1 significant figure).

QUESTION 6.3

The temperature T at time t is given by Equation 6.19

$$(T/\text{K}) \approx 1.5 \times 10^{10}(t/\text{s})^{-1/2}$$

This can be rearranged to give

$$(t/\text{s})^{1/2} \approx 1.5 \times 10^{10}/(T/\text{K})$$

Squaring both sides gives

$$(t/s) \approx (1.5 \times 10^{10}/(T/K))^2 \qquad \text{(i)}$$

When $T = 10^6$ K,

$$(t/s) \approx (1.5 \times 10^{10}/(10^6 \text{ K/K}))^2 = (1.5 \times 10^4)^2 = 2.25 \times 10^8$$

$$t = 2.25 \times 10^8 \text{ s} = 2.25 \times 10^8 \text{ s}/(365 \times 24 \times 60 \times 60 \text{ s yr}^{-1}) = 7.13 \text{ yr}$$

So when the temperature was 10^6 K, the age of the Universe was 7 years (to 1 significant figure).

QUESTION 6.4

It is stated in the text that grand unification occurs at an interaction energy of about 10^{15} GeV. Thus the strong and electroweak interaction became distinct when interaction energies dropped below this value. The corresponding temperature can be found using Equation 6.20, which can be rearranged to give

$$T \sim E/k$$

An interaction energy of 10^{15} GeV therefore corresponds to a temperature of

$$T \sim (10^{15} \times 10^9 \text{ eV} \times 1.60 \times 10^{-19} \text{ J (eV)}^{-1})/(1.38 \times 10^{-23} \text{ J K}^{-1})$$

$$T \sim 1.16 \times 10^{28} \text{ K}$$

So, grand unification occurs when the temperature exceeds 10^{28} K.

In order to calculate the time at which the temperature was 10^{28} K, we can use Equation (i) from the answer to Question 6.3,

$$(t/s) \sim (1.5 \times 10^{10}/(10^{28} \text{ K/K}))^2 = 2.25 \times 10^{-36}$$

So, expressing this to the nearest power of ten, the time at which grand unification occurred was $t \sim 10^{-36}$ s.

QUESTION 6.5

(a) (i) Before the decay, the only particle is a single neutron. This has a baryon number of +1. (ii) The baryon numbers of the products of the decay are +1 (proton), 0 (electron) and 0 (electron antineutrino). The baryon number before and after the decay is thus +1 and so baryon number is conserved.

(b) (i) The single neutron has a lepton number of 0. (ii) The lepton numbers of the products are 0 (proton), +1 (electron) and −1 (electron antineutrino). The lepton number before and after the decay is thus 0 and so lepton number is conserved.

(c) (i) A neutron comprises one up and two down quarks (u d d). (ii) A proton comprises two up and one down quark (u u d). The β^--decay reaction thus involves a down quark being transformed into an up quark. Hence β^--decay can be expressed in terms of quarks and leptons as

$$d \rightarrow u + e^- + \overline{v}_e$$

QUESTION 6.6

Electron–positron pair production requires an amount of energy given by

$$E = 2m_e c^2$$

(note that the mass of the positron is equal to the mass of the electron m_e). The value of $m_e c^2$ is given in Table 6.3 as being 0.511 MeV, thus

$$E = 2 \times 0.511 \text{ MeV} = 1.02 \text{ MeV} = 1.02 \times 10^6 \text{ eV}$$

So the interaction energy required for electron–positron pair production is 1.02×10^6 eV.

The temperature can be found using Equation 6.20

$$E \sim kT$$

Which can be rearranged as

$$T \sim E/k = (1.02 \times 10^6 \text{ eV} \times 1.60 \times 10^{-19} \text{ J})/(1.38 \times 10^{-23} \text{J K}^{-1})$$
$$= 1.18 \times 10^{10} \text{ K}$$

So the temperature for electron–positron pair production is about 1×10^{10} K. (This is called the *threshold temperature* for electron–positron pairs.)

QUESTION 6.7

The temperature T at time t is given by Equation (i) of the answer to Question 6.3.

(a) When $T = 10^9$ K, then

$$(t/s) \approx (1.5 \times 10^{10}/(10^9 \text{ K/K}))^2 = 15^2 = 225$$

So the temperature was 10^9 K when $t = 2 \times 10^2$ s (to 1 significant figure).

(b) When $T = 5 \times 10^8$ K, then

$$(t/s) \approx (1.5 \times 10^{10}/(5 \times 10^8 \text{ K/K}))^2 = 30^2 = 900$$

So the temperature was 5×10^8 K when $t = 9 \times 10^2$ s (to 1 significant figure).

QUESTION 6.8

At around $t \approx 1$ s the ratio n_n/n_p had a value of 0.22. After this time, the dominant reaction affecting free neutrons was β^--decay. If we consider a region of the Universe that contained 100 neutrons at $t \approx 1$ s, then the number of protons in this region would have been $N_p = 100/0.22 = 455$. After this time, the neutrons decayed according to the curve shown in Figure 6.12. The time at which deuterium was first formed was $t = 225$ s. From Figure 6.12 it can be seen that a fraction of 0.78 of the sample of neutrons would remain at this time. Thus the sample would have contained 78 neutrons ($N_n = 78$). However, due to β^--decay, the number of protons would have *increased* by $(100 - 78) = 22$, so the total number of protons (N_p) in the sample would have been $455 + 22 = 477$. So

$$N_n/N_p = 78/477 = 0.164$$

The ratio N_n/N_p is the same as the ratio of number densities of neutrons and protons n_n/n_p. So at the time that deuterium started to form, $n_n/n_p = 0.16$ (to 2 significant figures).

QUESTION 6.9

The mass fraction in helium (Y) is given by Equation 6.33b. The value of n_n/n_p that was obtained in Question 6.8 is 0.16, so

$$Y = 2\left[\frac{1}{1 + (n_p/n_n)}\right] = 2\left[\frac{1}{1 + (1/0.16)}\right] = 0.276$$

So the mass fraction in helium is 0.28 (to 2 significant figures).

QUESTION 6.10

The metallicity Z of a sample of material is defined as the mass of the sample that is in metals divided by the total mass of the sample. Since lithium is the only metal (i.e. element with mass number greater than 4) produced in the big bang, the metallicity will be

Z = (mass of lithium)/(mass of sample)

The mass of the sample can be found from the definition of the hydrogen mass fraction

X = (mass of hydrogen)/(mass of sample)

(mass of sample) = (mass of hydrogen)/X

So the metallicity can be expressed as

$Z = X \times$ (mass of lithium)/(mass of hydrogen)

This is a useful expression because Figure 6.13 gives the abundance of lithium as the ratio of the mass of lithium to the mass of hydrogen within a sample.

Using the approximation that $X \sim 0.75$,

$Z = 0.75 \times$ (mass of lithium)/(mass of hydrogen)

For the purposes of this order of magnitude calculation it is reasonable to equate the metallicity to the relative abundance of lithium shown in Figure 6.13. Since the maximum value of the lithium abundance is about 10^{-8}, the maximum metallicity of material formed in the big bang would be of order of magnitude $Z \sim 10^{-8}$.

The oldest observed stars have measured values of $Z \sim 10^{-6}$ (Chapter 1). Thus the metallicity of material formed in the big bang is expected to be at least a factor of 10^2 smaller than the metallicities observed even in the least chemically enriched stars. So the oldest stars cannot be formed from material that has not been subject to some enrichment after the era of primordial nucleosynthesis.

QUESTION 6.11

In all three cases, A, B and C, the value of $\Omega_{b,0}$ can be found by identifying the locations on the curves in Figure 6.13 that correspond to the given abundances. The values of $\Omega_{b,0}$ are shown in Table 6.8.

In cases A and B the lithium abundances correspond to three different values of $\Omega_{b,0}$. In case C, the lithium abundance corresponds to a range of values of $\Omega_{b,0}$. Thus, in all cases, the lithium abundances on their own do *not* allow $\Omega_{b,0}$ to be determined uniquely.

Case	$\Omega_{b,0}$
A	0.006
B	0.07
C	0.02

Table 6.8 The values of $\Omega_{b,0}$ determined from the abundances in Question 6.11.

QUESTION 6.12

The energy required for the ionization of hydrogen is 13.6 eV. Because the number of photons exceeds the number of protons by a factor of 10^9, by analogy with the case of the photodisintegration of deuterium, recombination will only occur once the mean photon energy is a factor of 9.6 lower than the ionization energy (see Section 6.4.1).

So the mean photon energy

$$\varepsilon_{mean} = (13.6 \text{ eV})/9.6 = 1.42 \text{ eV} = 1.42 \text{ eV} \times 1.60 \times 10^{-19} \text{ J eV}^{-1}$$

$$= 2.27 \times 10^{-19} \text{ J}$$

However, ε_{mean} is related to the absolute temperature T by Equation 6.27

$$\varepsilon_{mean} = 2.7kT$$

Thus $\quad T = \dfrac{\varepsilon_{mean}}{2.7k} = \dfrac{2.27 \times 10^{-19} \text{ J}}{2.7 \times 1.38 \times 10^{-23} \text{ J K}^{-1}} = 6092 \text{ K}$

So recombination occurred at a temperature of 6.1×10^3 K.

(This is, in fact, something of an overestimate. The ionization of hydrogen may occur, not only by an atom in the ground state absorbing a photon with energy greater than 13.6 eV, but also by an atom absorbing a photon such that it is in an excited state, and then absorbing another photon. The effect of this is to lower the temperature at which recombination starts to occur – to about 4500 K.)

QUESTION 6.13

The relationship between temperature and scale factor is given by Equation 6.6,

$$T \propto \dfrac{1}{R(t)}$$

Thus the relationship between the temperature of the background radiation at the present time T_0 and that at the time of last scattering T_{last} is

$$\dfrac{T_{last}}{T_0} = \dfrac{R(t_0)}{R(t_{last})}$$

where t_{last} is the time at which the last scattering of photons occurred.

The relationship between redshift and scale factor is given by Equation 5.13. In this case, the time at which the photon is observed is t_0 and the time at which the photon was emitted is t_{last}, so Equation 5.13 can be written as

$$z = \frac{R(t_0)}{R(t_{last})} - 1$$

So

$$z = \frac{T_{last}}{T_0} - 1 \qquad\qquad\qquad\qquad \text{(ii)}$$

The question states that $T_{last} = 3.0 \times 10^3$ K, and $T_0 = 2.7$ K, so

$$z = \frac{3.0 \times 10^3 \text{ K}}{2.7 \text{ K}} - 1 = 1110$$

So the redshift at which the last scattering of cosmic background photons occurred is 1.1×10^3 (to 2 significant figures).

QUESTION 6.14

A 'time-line' for the history of the Universe is shown in Figure 6.28.

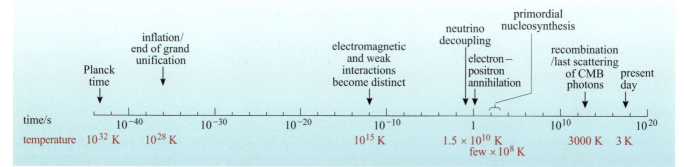

Figure 6.28 The major events in the evolution of the Universe from the Planck time ($t \sim 10^{-43}$ s) to the present day.

QUESTION 6.15

A 'theory of everything' is a theory that accounts for all four of the fundamental interactions of nature in a unified manner. In particular, such a theory would provide a consistent way to describe processes that are currently described by two separate and mutually incompatible theories: the standard model of elementary particles and general relativity (which describes gravitational interactions). Such a theory is needed to describe processes in the very early Universe; before the Planck time (10^{-43} s) the strength of all four fundamental interactions would have been similar, and gravity would have played a role in particle interactions.

QUESTION 6.16

The relationship between scale factor and temperature is given by Equation 6.6, which shows that temperature is inversely proportional to scale factor. Thus a dramatic *increase* in scale factor would cause a dramatic *fall* in temperature.

An increase in the scale factor by, for example, a factor of 10^{27}, would cause the temperature to fall by the same factor. Even if the temperature prior to inflation was about 10^{28} K, it would plummet to 10^{28} K$/10^{27} \sim 10$ K at the end of inflation. So the temperature at the end of inflation would be expected to be extremely low.

(This might seem at odds with the whole idea of a hot big bang model. How could the Universe that is just above absolute zero at the end of inflation have proceeded to cause, for example, primordial nucleosynthesis? It turns out that the energy that is released at the end of inflation re-heats the Universe, and it does so as if the cooling due to inflation had never occurred!)

QUESTION 6.17

If the photodisintegration of deuterium required photons of a much higher energy than 2.23 MeV, then, in comparison to the real Universe, a significant amount of deuterium would have been formed at earlier times when the temperature was higher. Thus deuterium would have formed at an earlier time, when the ratio of neutron to proton number densities (n_n/n_p) would have been higher. Most of the neutrons that are present at the time that deuterium production starts, end up in helium nuclei. A higher value of (n_n/n_p) at the start of deuterium production would therefore result in a higher value of the helium mass fraction.

QUESTION 6.18

Following the same reasoning as is used to arrive at Equation (ii) in the answer to Question 6.13, the redshift z is related to the temperature T_{em} of the background radiation at that redshift by

$$z = \frac{T_{em}}{T_0} - 1$$

This can be rearranged to give

$$T_{em} = T_0(z + 1)$$

Inserting the given values leads to

$$T_{em} = (2.73 \text{ K}) \times (2.5 + 1) = 9.56 \text{ K}$$

So, their measurement of the temperature of the cosmic microwave background would be 9.6 K (to 2 significant figures).

(Although we don't have any communication with astronomers anywhere else in the Universe, a similar principle applies to a real observational technique: it is possible to measure the temperature of the cosmic background radiation as experienced by Lyα clouds at redshifts of $z \approx 2$. Such measurements, which are based on detailed analysis of spectral lines, show that the temperature of the cosmic background does increase with redshift in this way.)

QUESTION 7.1

(a) Systematic. This uncertainty would influence the modelling of the lensing galaxy and hence the calculation of the time-of-passage lag, but it would influence repeated measurements in the same way, and would therefore be a systematic effect rather than a random one.

(b) Systematic. Repeated measurements will be affected in the same way each time they are performed.

(c) Random. Performing similar observations several times will lead to different results due to this source of uncertainty.

QUESTION 7.2

Using Hubble's law $d = cz/H_0$, with $z = 1$ and $H_0 = 72$ km s^{-1} Mpc^{-1} (the HST value), gives

$$d = (3.0 \times 10^5 \text{ km s}^{-1})/(72 \text{ km s}^{-1} \text{ Mpc}^{-1})$$

i.e. $d = 4.2 \times 10^3$ Mpc

The simple form of Hubble's law used above should not be trusted as a reliable indicator of distances at such large redshift, since it ignores the effect of any acceleration or deceleration in the rate of cosmic expansion. For instance, from the more accurate Equation 7.4, it can be seen that if $q_0 = 0$, then d would increase to 150% of the quoted value. In either case, this would be very much larger than the diameter of the local supercluster, which was described in Chapter 4 as being 20 to 50 Mpc across.

QUESTION 7.3

In the supernova cosmology measurements the emphasis was on determining the magnitude of distant Type Ia supernovae relative to nearer ones. For this purpose it is only necessary to measure the apparent magnitudes of the supernovae rather than the absolute magnitudes. To measure the Hubble constant we would also need to know the absolute magnitudes, in order to determine the distances.

QUESTION 7.4

From Figure 7.11, the lowest point on the 99.7% confidence contour corresponds to $\Omega_{\Lambda,0} = 0.23$, while the highest point on that contour corresponds to $\Omega_{\Lambda,0} = 1.11$. In contrast, the points at which the same 99.7% contour crosses the 'flat geometry' ($k = 0$) line correspond to $\Omega_{\Lambda,0} = 0.64$ and $\Omega_{\Lambda,0} = 0.80$.

QUESTION 7.5

The dipole anisotropy is a large-scale non-uniformity in the CMB, as observed from Earth. It causes the direction of maximum intensity (or effective temperature) to be diametrically opposite the direction of minimum intensity (or temperature), with the intensity increasing progressively from the minimum to the maximum. The dipole anisotropy is caused by the motion of the Earth relative to the frame of reference in which the CMB shows large-scale isotropy. (In other words the dipole anisotropy is a result of the Doppler effect.)

The existence of the dipole anisotropy is implicitly acknowledged in Point 1 by saying that the radiation (CMB) is 'intrinsically' uniform to better than one part in 10 000. It is more explicitly recognized in the next sentence by the inclusion of the phrase 'after correcting for effects due to the motion of the detector'.

In Point 2, the word 'intrinsic' is again used to indicate 'after correcting for effects due to the motion of the detector'.

Point 5 also includes the word 'intrinsic' with the same implication.

QUESTION 7.6

The predictions are based on the general principles of big bang cosmology and specific assumptions about the values of the cosmological parameters. Since the predictions don't relate to specific features such as the exact region of sky being observed, or the detailed history of that region, the predictions cannot be expected to show anything more than 'general' agreement even when the appropriate values for the cosmological parameters have been chosen.

QUESTION 7.7

From Figure 7.26, the second and third peaks are located at $l \approx 500$ and $l \approx 800$, respectively. From Equation 7.11, these values correspond to angular scales $\theta = 0.4°$ and $\theta = 0.2°$, respectively. The maximum angular power associated with these peaks is roughly 2500 $(\mu K)^2$ in each case.

The angular scales of these two peaks are broadly consistent with being harmonics of a fundamental note with a wavelength just less than a degree.

QUESTION 7.8

The carpet would look uniform from a distance, so the power spectrum would be close to zero on large scales. As you zoom in to the ~0.2–0.5 cm scale you would see clumps of shading, so the carpet would look highly structured on that scale, so the power spectrum would peak. As you zoom in closer still e.g. on an individual fibre on scales of less than a millimetre, the carpet would look more uniform again, so the power spectrum would drop to a low value.

QUESTION 7.9

The claim that 'baryons only account for somewhere between about a fifteenth and a sixth of the total density of matter in the Universe' was based on Equation 7.13, according to which

$$0.02 \leq \Omega_{b,0} \leq 0.05$$

The widely favoured value for $\Omega_{m,0}$, based on supernova cosmology and measurements of the cosmic microwave background radiation is

$$\Omega_{m,0} \approx 0.3$$

This suggests that

$$0.067 \leq \frac{\Omega_{b,0}}{\Omega_{m,0}} \leq 0.167$$

i.e. between a fifteenth and a sixth.

QUESTION 7.10

(a) The CMB anisotropy measurements indicate that $k = 0$ and hence $\Omega_{\Lambda,0} + \Omega_{m,0} = 1$.

However, Type Ia supernovae, mass-to-light ratios and CMB measurements all indicate that $\Omega_{m,0}$ is much less than 1, probably about 0.3. This implies that $\Omega_{\Lambda,0} \approx 0.7$. Thus the energy associated with the cosmological constant (dark energy) dominates the density of the Universe. The nature of this energy is not well understood at the time of writing.

(b) Among other results discussed in earlier chapters, the growth of mass-to-light ratios with the scale of observations indicates that most of the matter in the Universe is dark matter.

(c) Although observations indicate that $\Omega_{m,0} \approx 0.3$, estimates of $\Omega_{b,0}$ based on primordial nucleosynthesis constraints, baryon inventories and CMB measurements, indicate that $\Omega_{b,0} \approx 0.04$. Hence most of the (dark) matter must be non-baryonic.

QUESTION 7.11

Even from the incomplete account given in this chapter it is clear that space technology has profoundly influenced observational cosmology. This is clear from:

- the role of the HST in facilitating the determination of the Hubble constant from Cepheid variable observations;

- the use of the HST to study some of the faintest (and therefore most distant) Type Ia supernova candidates used in the determination of the current value of the deceleration parameter;

- the role of space-based X-ray observations in the determination of cluster masses that are an essential ingredient in measurements of the mass-to-light ratio on large scales;

- the additional role of X-ray and other space-based observations in compiling a baryon inventory;

- the importance of COBE in confirming the black-body nature of the CMB spectrum, determining the mean temperature of the radiation and revealing the existence of anisotropies in that temperature; and

- the role of the WMAP and Planck missions in measuring the CMB angular power spectrum with great precision so that the key cosmological parameters can be precisely determined.

QUESTION 7.12

Terrestrial observations, including those performed from planes and balloons, are still of enormous importance in cosmology, despite the many advances that can be attributed to space technology. Examples of important ground-based or air-based experiments include:

- observations of gravitational lensing using optical and radio telescopes (although important results have also come from the HST);

- studies of supernovae carried out by large telescopes at sites such as Mount Palomar and, more recently, Cerro Tololo and Hawaii (Keck);

- observations of various kinds used in establishing mass-to-light ratios, baryon inventories and the relative abundances of light elements that are needed to check primordial nucleosynthesis predictions;

- a range of CMB observations including the initial breakthrough by Penzias and Wilson, the discovery of the dipole anisotropy and the anisotropy measurements carried out by BOOMERanG and various other projects; and

- the large-scale galaxy surveys that are helping to refine our knowledge of the galaxy power spectrum.

QUESTION 8.1

The completed Table 8.1 is shown below.

Dark matter candidate	Baryonic	Non-baryonic	Cold[a]	Warm[a]	Hot[a]
neutron stars	✓				
neutrinos		✓			✓
neutralinos		✓	✓		
sterile neutrinos		✓		✓	

[a] Note that the terms hot, warm and cold only apply to candidate dark matter particles, since they refer to the speed of particles at the time of decoupling and not to temperature in the conventional sense.

QUESTION 8.2

(a) An antihydrogen atom (made from an antiproton and a positron) would have the same energy levels as normal hydrogen and would absorb and emit photons in the same way. The photons themselves would be identical to photons produced in our own Galaxy. The same applies to all other atoms. So the spectrum of an antigalaxy would look much the same as an ordinary galaxy.

(b) If the galaxy really is an antigalaxy, then somewhere between the galaxy and our own Milky Way there must be a boundary between matter and antimatter. We would expect annihilations to be taking place at the boundary, perhaps in the intergalactic medium, and the boundary could be revealed by a search for γ-rays. In fact, the absence of such γ-radiation is strong evidence that there are no significant amounts of antimatter in our part of the Universe.

QUESTION 8.3

You didn't really expect to find an answer to this one, did you?

QUESTION 8.4

The end-of-chapter summary provides an overview of the current status of the cosmological problems that were outlined at the beginning of the chapter.

APPENDIX

Useful quantities

Quantity	Symbol	Value[a]
Physical constants		
speed of light in a vacuum	c	$3.00 \times 10^8 \, \mathrm{m\,s^{-1}}$
Planck constant	h	$6.63 \times 10^{-34} \, \mathrm{J\,s}$
Boltzmann constant	k	$1.38 \times 10^{-23} \, \mathrm{J\,K^{-1}}$
gravitational constant	G	$6.67 \times 10^{-11} \, \mathrm{N\,m^2\,kg^{-2}}$
Stefan–Boltzmann constant	σ	$5.67 \times 10^{-8} \, \mathrm{W\,m^{-2}\,K^{-4}}$
charge of electron	$-e$	$-1.60 \times 10^{-19} \, \mathrm{C}$
mass of hydrogen atom	m_H	$1.67 \times 10^{-27} \, \mathrm{kg}$
mass of electron	m_e	$9.11 \times 10^{-31} \, \mathrm{kg}$
Astronomical data		
mass of the Earth	M_E	$5.98 \times 10^{24} \, \mathrm{kg}$
radius (equatorial) of the Earth	R_E	$6.38 \times 10^6 \, \mathrm{m}$
mass of the Sun	$M_\odot$	$1.99 \times 10^{30} \, \mathrm{kg}$
radius of the Sun	$R_\odot$	$6.96 \times 10^8 \, \mathrm{m}$
luminosity of the Sun	$L_\odot$	$3.84 \times 10^{26} \, \mathrm{W}$
Hubble constant[b]	H_0	$72 \pm 8 \, \mathrm{km\,s^{-1}\,Mpc^{-1}}$

[a]Values are given to 3 significant figures. Many of these are known more accurately.
[b]Freedman, W. L. *et al. Astrophysical Journal,* **533**, 47-72, (2001).

Units

Quantity	SI Unit	Other units	In SI	Alternative SI Units
length	metre, m	Astronomical unit, AU	$1.50 \times 10^{11} \, \mathrm{m}$	
		parsec, pc	$3.09 \times 10^{16} \, \mathrm{m}$	
time	second, s	year, yr	$3.16 \times 10^7 \, \mathrm{s}$	
frequency	hertz, Hz			$\mathrm{s^{-1}}$
force	newton, N			$\mathrm{kg\,m\,s^{-2}}$
pressure	pascal, Pa			$\mathrm{kg\,m^{-1}\,s^{-2}}, \mathrm{N\,m^{-2}}$
temperature	kelvin, K	°C	(kelvin − 273)	
energy	joule, J	electronvolt, eV	$1.60 \times 10^{-19} \, \mathrm{J}$	$\mathrm{kg\,m^2\,s^{-2}}$
power	watt, W			$\mathrm{kg\,m^2\,s^{-3}}, \mathrm{J\,s^{-1}}$
angle	radian, rad	degree, °	$1/57.3$ rad	
		$1° = 60$ arcmin $= 3600$ arcsec		
		arcsec, "	$1/206265$ rad	

Useful mathematical functions

Any mathematical function relates a given value of a variable, x say, to some other value that can be denoted $f(x)$. Here we provide some details on two functions that are used in the text – the exponential and the sine functions.

The exponential function

The exponential function makes use of an important mathematical constant, usually denoted e, which has the value 2.718 to four significant figures. In the case of the exponential function, this value is given by $f(x) = e^x$ for any value of x. (This last statement is one of many ways of defining the exponential function.) So, if $x = 1$, the exponential function has the value $f(1) = e^1 = 2.718$ (to four significant figures); if $x = 2$ then $f(2) = e^2 = 7.389$. Of course, x may be negative, for instance if $x = 1$ then $f(-1) = e^{-1} = 1/e = 0.3679$; and so on.

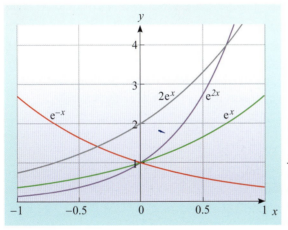

Figure A.1 Exponential curves described by the equation $y = y_0\, e^{bx}$, for various values of y_0 and b. The value of y_0 is the value of y when $x = 0$, and the value of b determine how rapidly y increases or decreases as x changes

The exponential function is of great importance throughout physics and astronomy because many natural processes exhibit the phenomenon of exponential growth or exponential decay. Such processes are described by an equation of the form

$$y = y_0\, e^{bx}$$

where y_0 and b are independent parameters, i.e. two fixed values that may be chosen to fit any particular case. As indicated in Figure A.1, the chosen value of y_0 will be the value of y when $x = 0$, and the chosen value of b will determine how rapidly y increases or decreases as x changes. Note that positive values of b imply exponential growth, and negative values of b imply exponential decay. Examples of exponential decay relevant to astronomy include the rate of decay of the radioactive nuclei ^{56}Ni and ^{56}Co which maintain the light output of supernovae, and the absorption of starlight as it passes through an absorbing gas cloud.

The sine function

The sine function is a periodic (or repeating) function that arises from trigonometry (i.e. from the consideration of the properties triangles, and especially of right-angled triangles). We do not describe the origins of the function here, but note that in the sine function $f(x) = \sin(x)$ the variable x is interpreted as an angle, and consequently would naturally be expressed in radians. The values of $\sin(x)$ lie between −1 (at $x = \pi/2$) and +1 (at $x = \pi/2$), and the function has a characteristic wave-like profile (called a sinusoid). The function is periodic in that if x is changed by a whole number multiple of 2π radians, the value of $f(x)$ remains the same. The green curve in Figure A.2, shows the behaviour of the function $\sin(x)$ for $-2\pi \le x \le 2\pi$.

The sine function is of great use in physics and astronomy because it represents the way in which many periodically varying systems behave. Such systems can be described using an equation of the form,

$$y = y_0 \sin(bx + \phi)$$

where y is the periodically varying quantity. The quantity y_0, which is termed the amplitude, is a positive constant that is equal to the maximum value of y over a full cycle. The relationship between the variable x and the periodically varying quantity is determined by the constant b. The quantity y will undergo one complete cycle as the quantity bx changes by a value of 2π radians (e.g. from 0 to 2π). A higher value of b will result in a greater number of cycles within a given range of x as illustrated by the blue curve in Figure A.2 (in which $b = 2$). Conversely, if b is made smaller, the number of cycles within a given range of x also becomes smaller (e.g., the red curve in Figure A.2, in which $b = 0.5$). Note that if the periodic variation is an oscillation, i.e. x represents time, then the period T, which is the time taken to undergo one full oscillation is $T = 2\pi/b$. If the periodic variation is with distance, then the distance in which the wave repeats itself is the wavelength $\lambda = 2\pi/b$.

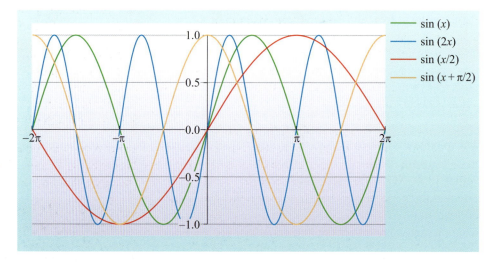

Figure A.2 Examples of sinusoidal curves of the form $y = y_0 \sin(bx + \phi)$, for various values of b and ϕ. Note that $y_0 = 1$ throughout, and where no value of ϕ is shown then $\phi = 0$. The value of b determines the distance between successive peaks, while non-zero values of ϕ cause the wave pattern to be shifted horizontally.

The constant ϕ, which is known as the *phase* of the oscillation, can be visualised as quantifying a shift of the wave pattern relative to the x-axis in Figure A.2. When $\phi = 0$, the oscillating quantity is zero (but rising) at $x = 0$. For a phase value of $\phi = \pi/2$, the oscillating quantity reaches this zero point at $x = -\pi/2$ (or equivalently $x = 3\pi/2$), see the yellow curve in Figure A.2. Finally, it is worth noting that the sine function with a phase shift of $\pi/2$ radians (the yellow curve in Figure A.2) is equivalent to another trigonometrical function – the cosine (usually denoted by $f(x) = \cos(x)$).

GLOSSARY

21 centimetre radiation Electromagnetic radiation with a wavelength of 21 cm, in the radio part of the spectrum, which is emitted or absorbed by a hydrogen atom when it undergoes a spin–flip transition. The spin–flip transition of a hydrogen atom may be described in terms of classical (i.e. non-quantum) physics by saying that the proton and electron which comprise the atom, change from having their spins aligned parallel with one another to being antiparallel (180° out of alignment), or vice versa.

Abell radius The typical radius of a cluster of galaxies. It is now known to be about 2 Mpc.

absolute visual magnitude An intrinsic property of a star, equal to the apparent visual magnitude the star would have if observed from a standard distance of 10 parsecs, in the absence of interstellar absorption. The absolute visual magnitude provides a measure of the star's luminosity.

accelerating model A cosmological model, belonging to the class of Friedmann–Robertson–Walker models, in which the rate of change of the scale factor is itself changing at a positive rate (i.e. $\ddot{R} > 0$). Although any FRW model with a sufficiently large cosmological constant Λ may exhibit accelerated expansion at late times in its development, the term 'accelerating model' is often used to refer specifically to the model with positive cosmological constant ($\Lambda > 0$) and zero curvature parameter ($k = 0$).

accretion disc A disc of gas which forms around a massive object such as the accreting star in an interacting binary system, or around the massive black hole in the engine of an AGN. Material spirals inwards within the disc and falls onto the central object from the inner edge of the disc.

acoustic peaks A set of peaks, the largest of which is called the Doppler peak, seen in the angular power spectrum of the cosmic background radiation for multipole numbers in the range from about $l = 50$ to $l = 1000$ (i.e. on angular scales between about 0.1 degrees and a few degrees). The phenomenon is partly due to the localized heating of the last-scattering surface caused by the acoustic waves in the photon–baryon fluid that is present there, and partly due to the effect that the moving charged particles, associated with those waves, have on the wavelengths of scattered photons.

acoustic waves Waves in a fluid arising from the tendency of localized concentrations of pressure or density to return to some preferred equilibrium value by increasing the pressure or density in neighbouring regions. Acoustic waves are also known as sound waves.

active galactic nucleus (AGN) The bright, point-like object at the centre of an active galaxy. AGN typically have high luminosities and often exhibit rapid variability.

active galaxy A galaxy that typically exhibits an unusually high and varying luminosity, and which may additionally show signs of energetic processes connected with its central regions. The term embraces: Seyfert galaxies, quasars, radio galaxies and blazars.

age–metallicity relation The tendency for older stars to have lower metallicities.

AGN *See* active galactic nucleus.

angular power The quantity, usually measured in units of $(\mu K)^2$, that is plotted on the vertical axis of an angular power spectrum, and which measures the amount of variation present in the corresponding (anisotropy) map on the angular scale $\theta = 180°/l$.

angular power spectrum A mathematical entity, often presented as a plot of angular power against multipole number, that is used to describe the statistically important data contained in a two-dimensional map of temperature anisotropies in the cosmic microwave background.

anisotropies (in the cosmic microwave background) Variations in the intensity of the cosmic microwave background radiation over the celestial sphere.

annihilation The interaction between a particle and its corresponding antiparticle in which both the particle and antiparticle are destroyed and energy is released. This is the opposite process to pair-creation.

anthropic principle An assertion that the existence of intelligent life in the Universe is of cosmological significance. In its weak form, the anthropic principle states that conditions in the Universe are those which allowed the emergence of intelligent life. In its strong form, the principle states that the Universe necessarily had conditions that led to the emergence of intelligent life.

antibaryon An elementary particle that comprises three antiquarks. An example of an antibaryon is the antiproton.

apparent surface brightness The quantity that describes the apparent brightness at any point on an extended object, such as a galaxy. The apparent surface brightness at any chosen point is the amount of radiant flux that would reach $1 \, m^2$ at Earth from a small, uniformly bright, square region, of angular area $1 \, (arcsec)^2$, surrounding the chosen point. An acceptable SI unit of apparent surface brightness is the $W \, m^{-2} \, arcsec^{-2}$.

apparent visual magnitude A quantity that describes the apparent brightness of a body. For a star, it is a measure of the flux density received in the V band, i.e. a band that approximates the wavelength response of human vision.

atomic hydrogen Hydrogen in the electrically neutral state in which it contains a single proton bound to a single electron. This form may be distinguished from other common forms of hydrogen such as molecular hydrogen and ionized hydrogen.

band-shifting The effect whereby electromagnetic radiation from a highly red-shifted object will be observed in a different wavelength band to that in which it was emitted.

bar instability The process by which a system of stars orbiting their common centre of gravity in a flattened (disc-like) configuration, tends to develop an elongated (bar-shaped) rather than circular distribution.

barred galaxy Any spiral, lenticular or irregular galaxy with a central bar-like feature.

barred spiral galaxy Any spiral galaxy whose central stars form an elongated (bar-shaped) distribution, with the long axis lying in the plane of the disc.

baryon An elementary particle that comprises three quarks. Protons and neutrons are examples of particles that are baryons.

baryon number A quantity that is conserved in all particle interactions (with the exception of some speculative interactions predicted by grand unified theories). The baryon number of any baryon is +1 and that of any antibaryon is −1 respectively (or equivalently, the baryon number of any quark is +1/3, whereas that of any antiquark is −1/3). The baryon number of any other particle is zero.

baryonic dark matter Dark matter consisting of baryons that do not produce any detectable radiation. It is generally believed that most dark matter is non-baryonic.

big bang The early part of the expansion of the Universe, as described by those Friedmann–Robertson–Walker models that start with the scale factor R equal to zero at time $t = 0$ and with the rate of change of the scale factor being positive ($\dot{R} > 0$).

big crunch The late part of the contraction of the Universe, as described by those Friedmann–Robertson–Walker models that end with the scale factor R equal to zero and with the rate of change of the scale factor being negative ($\dot{R} < 0$).

BL Lac object (BL Lacertae object) A subclass of blazar. In contrast to flat spectrum radio quasars, the spectra of BL Lac objects show no emission lines.

black hole A region of space from which, according to general relativity, signals (including particles of matter and electromagnetic radiation) are unable to escape due to the action of gravity. Such regions are bounded by an event horizon, and may be created by the catastrophic collapse of massive stars.

black-body spectrum The spectrum of an ideal thermal source of radiation (i.e. a black body). This is a continuous spectrum with a characteristic 'humped' shape, the peak wavelength depending on the temperature of the source, in accord with Wien's displacement law. A characteristic of sources that produce spectra that are close to the black-body form is that there is a high degree of interaction between electromagnetic radiation and the material that makes up the source. (This leads to the formal definition of a black-body source as one that has the property of absorbing perfectly any electromagnetic radiation that is incident on it, and emits a black-body spectrum.)

blazar A kind of active galaxy, characterized by a point-like appearance, a relatively featureless optical spectrum, and rapid variability across the electromagnetic spectrum. Blazars may be subclassified as BL Lac objects or flat spectrum radio quasars.

bottom-up scenario *See* hierarchical scenario.

broadband spectrum A spectrum covering a wide range of wavelengths or frequencies, which indicates the energy distribution of a source. It does not generally show narrow features such as absorption lines.

broad-line region (BLR) The body of gas responsible for emitting the broad emission lines in an AGN. The typical width of broad emission lines corresponds to a radial velocity dispersion of about 5000 km s^{-1}.

bulge The region around the centre of a spiral galaxy, where the galaxy is thicker and brighter and the concentration of matter is greater than elsewhere. Its outer parts are dominated by the light of old stars, but towards the centre it may contain material associated with the inner parts of the disc including sites of star formation.

calibration problem The difficulty of establishing the relationship between various methods for measuring relative distances and of deducing absolute distances from relative distances, arising largely from difficulties in determining the luminosity of standard candles.

Casimir effect The effect whereby two parallel, uncharged, metal plates will experience a mutual attraction, of electromagnetic origin, when narrowly separated in a vacuum. The effect arises because the plates modify the physical properties of the intervening vacuum. In particular, the pressure of the vacuum between the metal plates is lower than the pressure of the surrounding vacuum, giving rise to the forces that act on the plates.

cD galaxy A supergiant elliptical galaxy with a large diffuse envelope.

Cepheid A type of giant/supergiant star which pulsates regularly with a period in the range from about a day to about 100 days. The changes in radius, temperature, and hence luminosity, arise from instabilities in the envelopes of such evolved giant or supergiant stars. Classical Cepheids, which are Population I stars, can be distinguished from another category of stars with similar but nevertheless distinct properties, the Population II Cepheids.

Cepheid variable method The use of a (classical) Cepheid as a standard candle for the purposes of distance determination. The method is based on the fact that for classical Cepheids the absolute magnitude is related to the period of brightness variation.

chaotic inflation A hypothesis that the entire Universe is partitioned into domains, and that the laws of physics may differ from one domain to another. In the chaotic inflation scenario, the physical laws that lead to inflation occur by chance in some of these domains. Such a scenario avoids the need to explain why conditions in a single Universe were 'just right' to produce inflation.

chimneys Regions of hot, low-density gas in the disc of the galaxy where supernova explosions have heated the local interstellar medium and caused it to break out from the disc. Such structures are believed to provide channels whereby gas from the disc can flow into the tenuous halo.

closed model A cosmological model, belonging to the class of Friedmann-Robertson-Walker models, that starts with a big bang, expands to maximum value of the scale factor, and recollapses in a big crunch. The closed model is characterized by a positive curvature parameter ($k = +1$).

cluster (of galaxies) A concentration of galaxies in a region of space, of order 4 Mpc across.

cold dark matter (CDM) Dark matter that is comprised of particles whose speeds are low in comparison to the speed of light.

collisional excitation The process in which an ion, atom or molecule is raised to a higher energy state as a result of its collision with another particle.

colour index The quantity that describes the colour of a star, obtained by subtracting its apparent magnitude in one wavelength band (e.g. its apparent visual magnitude) from its apparent visual magnitude in a different band (e.g. in blue light). In this example, the colour index would be denoted $B - V$. The colour index of a star depends primarily on its temperature.

comoving A term used to indicate a state of expansion or contraction matched to that of the Universe as a whole. The term is typically applied to 'comoving coordinates' which allow points moving with the Hubble flow to be described by fixed coordinate values despite their increasing physical separation.

comoving volume Any volume of space whose boundary is fixed in comoving coordinates.

concordance cosmology The current consensus of precise cosmological parameter measurements from a wide variety of different experiments.

confidence level A numerical quantity, usually expressed as a percentage, describing the likelihood that the true value of a measured quantity lies within some specified range of values.

Copernican principle The principle that the Earth does not occupy a privileged position in the Universe.

co-rotation radius The distance, measured from the centre of a spiral galaxy, at which the orbital speed of the stars about the galaxy and the pattern speed of the spiral arms is the same.

cosmic background radiation The electromagnetic radiation that pervades the Universe. At the present time, the peak of the spectral energy distribution of the cosmic background radiation occurs at microwave wavelengths and is observed as the cosmic microwave background.

cosmic microwave background (CMB) The contribution to the observed astronomical 'background radiation' that has no identifiable stellar or galactic source and which occupies the wavelength range from about 0.1 mm to 0.1 m. The cosmic microwave background is of great importance in modern cosmology and is often represented by the abbreviation CMB. In terms of total energy content, the CMB represents the dominant form of radiation in the Universe. It is characterized by a black-body spectrum corresponding to a temperature of (2.725 ± 0.002) K, and has a highly isotropic distribution with intrinsic temperature anisotropies of no more than a few parts in 10^5.

cosmic recycling The cycling of gas through various forms in the Galaxy, from the interstellar medium into stars and then back into the interstellar medium (towards the end of the star's life). In this process the metallicity of the gas is increased by the production of heavier elements in stars.

cosmic shear The effect whereby distant galaxies appear to be distorted due to the gravitational deflection of the light from those galaxies as it encounters non-uniformities in the large-scale distribution of matter. An analysis of the consequences of cosmic shear provides a way of mapping the distribution of matter in the Universe. *See also* weak gravitational lensing.

cosmic variance A source of uncertainty in the determination of the angular power spectrum of the cosmic microwave background, arising from the fact that the temperature anisotropies on which the determination is based are being measured from just one location in the Universe (i.e. the spectrum of a cosmic phenomenon is being estimated on the basis of a single sample of data).

cosmic web A description of the structure of the distribution of matter on scales exceeding about 100 Mpc that alludes to its interconnected filaments and sheets.

cosmological constant A constant, usually denoted Λ, that appears in the Einstein field equations of general relativity, and through them plays a role in many relativistic cosmological models. A positive cosmological constant causes an effective repulsion between distantly separated points in space, and may result in an eventual acceleration in the rate of cosmic expansion. *See also* dark energy.

cosmological model A mathematical model of the Universe as a whole, usually involving equations and parameters. Typically, a cosmological model describes the large-scale geometry of space and time, the contents of space and time, and the evolution of the parameters that describe the geometry and contents of space and time. *See also* Friedmann–Robertson–Walker models.

cosmological principle The principle (essentially an assumption based on increasingly good observational evidence) that on sufficiently large size scales, the Universe is homogeneous and isotropic. In this context, the phrase 'sufficiently large size scales' is usually taken to mean a few hundred megaparsecs or more.

cosmological redshift The contribution to the redshift (i.e. the fractional increase in wavelength $z = (\lambda_{obs} - \lambda_{em}) / \lambda_{em}$) of radiation emitted from a distant source which arises from the large-scale expansion of the Universe. Note that the observed redshift of a distant galaxy is usually the sum of a cosmological redshift and another contribution arising from the peculiar motion of the galaxy relative to the large-scale expansion.

cosmology The branch of science concerned with the Universe as a whole, including its origin, structure, composition, evolution and eventual fate.

critical density The value of the cosmic density, $\rho_{crit}(t) = 3[H(t)]^2/(8\pi G)$, that would cause a Friedmann–Robertson–Walker model with Hubble parameter $H(t)$ at time t and zero cosmological constant ($\Lambda = 0$) to be a critical model. The critical density provides a useful reference value in discussions of the cosmic density and is used in defining the density parameter for matter and the density parameter for the cosmological constant. (In a FRW model where $H(t_0) = 72$ km s^{-1} Mpc^{-1} at some time t_0, the critical density at that time is $\rho_{crit}(t_0) \approx 1 \times 10^{-26}$ kg m^{-3}.)

critical model A cosmological model, belonging to the class of Friedmann–Robertson–Walker models, that starts with a big bang and expands continuously but in such a way that the rate of change of the scale factor approaches zero as the time t approaches infinity. The critical model is characterized by zero cosmological constant ($\Lambda = 0$) and zero curvature parameter ($k = 0$). In such a model the cosmic density is always equal to the critical density. The critical model is also referred to as the Einstein–de Sitter model.

curvature A geometric property of space, or of space–time, that may be used to describe departures from 'flat' geometry.

curvature parameter A parameter (i.e. a quantity that may vary from case to case, but which takes a constant value in any given case) that appears in the Robertson–Walker metric, and which helps to characterize the curvature of space or space–time, and which may take the value +1, 0 or −1.

damped Lyman α system A relatively dense cloud of un-ionized gas that is detectable from the very strong Lyman α absorption that it causes in the spectrum of a background quasar. It is speculated that such clouds may be galaxies that are in the process of forming.

dark energy The energy, whatever its nature, that may be associated with an effective cosmological constant Λ via the relation $\rho_\Lambda = \Lambda c^2/(8\pi G)$, where $\rho_\Lambda c^2$ represents the (uniform) energy density of the dark energy.

dark energy density The energy density $\rho_\Lambda c^2$ associated with dark energy.

dark matter Matter that can be detected through its gravitational attraction, but which appears neither to emit nor absorb electromagnetic radiation, and hence gives few clues as to its nature. Some fraction of the dark matter is made up of baryons (baryonic dark matter), but most is believed to be composed of something else (non-baryonic dark matter).

dark-matter halo An approximately spherical volume surrounding the luminous parts of a galaxy where a large quantity of dark matter resides. The luminous parts of galaxies probably occupy the highest density part of the dark-matter halo and are held in place by the gravity of the dark matter.

de Sitter model A cosmological model describing a universe in which there is a negligible amount of matter and the cosmological constant Λ is positive. In the de Sitter model, space is infinite and in a state of perpetual

expansion as described by the scale factor $R \propto e^{Ht}$, where $H = (\Lambda c^2/3)^{1/2}$. The de Sitter model is a limiting case of the Friedmann–Robertson–Walker model with $k = 0$ and $\Lambda > 0$.

deceleration parameter The time-dependent quantity $q(t) = -R\ddot{R}/[\dot{R}]^2$ that arises in any Friedmann–Robertson–Walker model in which the scale factor at time t is $R(t)$, its rate of change at the time t is $\dot{R}(t)$ and the rate of change of $\dot{R}(t)$ at that time is $\ddot{R}(t)$. To the extent that such a model describes the real Universe, the current value of the deceleration parameter $q(t_0)$ should equal the quantity q_0 that quantifies departures from Hubble's law in the formula $H_0 d = cz[1 + (1 - q_0)z/2]$.

deep survey An astronomical survey that is performed with sufficient sensitivity to detect very faint sources. Typically, deep surveys require long observation times and are consequently restricted to small areas of the sky.

dense cloud One of the coldest and densest kinds of cloud to be found in the interstellar medium, usually rich in molecules. Dense clouds give birth to stars, mainly in the form of open clusters.

density parameter for dark energy An alternative way to refer to the *density parameter for the cosmological constant*. This usage emphasizes the point that dark energy need not have its physical origin in a cosmological constant.

density parameter for matter The time-dependent quantity $\Omega_m(t) = \rho(t)/\rho_{crit}(t)$ that arises in any Friedmann–Robertson–Walker model where the density of matter at time t is $\rho(t)$ and the critical density at that time is $\rho_{crit}(t) = 3[H(t)]^2/(8\pi G)$.

density parameter for the cosmological constant The time-dependent quantity $\Omega_\Lambda(t) = \rho_\Lambda/\rho_{crit}(t)$, that arises in any Friedmann–Robertson–Walker model where $\rho_\Lambda = \Lambda c^2/(8\pi G)$ is the 'density' associated with the cosmological constant Λ and $\rho_{crit}(t) = 3[H(t)]^2/(8\pi G)$ is the critical density at time t. Note that $\rho_\Lambda c^2$ is sometimes referred to as the density of dark energy and that $\Omega_\Lambda(t)$ may accordingly be referred to as the density parameter for dark energy.

density wave theory An explanation of the formation and maintenance of the density enhancements thought to be responsible for spiral arms. The density wave sweeps around the galaxy, compressing the material it traverses and triggering star formation.

deuteron A nucleus of deuterium. It comprises one proton and one neutron.

differential rotation A pattern of rotation in which the rotation period of one part of the rotating system may differ from that of another. In the case of the Milky Way, for example, the rotation period of different parts of the disc varies with their distance from the centre. *See* rotation curve.

dipole anisotropy The large-scale variation in the intensity of the cosmic microwave background due to the Earth's motion with respect to the Hubble flow.

disc A major structural component of spiral galaxies, containing most of the visible matter of the galaxy in a highly flattened distribution.

distance ladder A synthesis of techniques for measuring astronomical distances. The distance ladder is based on using one method of distance determination to calibrate another method that is appropriate to measurements of larger distances – which can then be used to calibrate a method that is used over yet larger distances and so on.

distance modulus The difference between the apparent and absolute magnitudes of an astronomical body. The distance modulus provides a measure of the distance d to the body, since (using the V-band magnitudes) $m_V - M_V = 5 \log_{10}(d/\text{pc}) - 5$.

Doppler broadening The effect whereby the width of a spectral line is increased as a result of movements within the region where the line originates. *See* Doppler effect; Doppler shift.

Doppler shift The difference, arising from the relative motion of an observer and a source of radiation, between the observed wavelength (or frequency) of the radiation and the wavelength (or frequency) of that radiation at its point of emission.

dust Small solid particles, around 10^{-7} or 10^{-6} m across, found mixed with interstellar gas. Dust grains are predominantly composed of carbonaceous material and silicates, but may be surrounded by an icy mantle. Dust is very effective at absorbing and scattering ultraviolet and visible light.

dwarf elliptical Any small, intrinsically faint, elliptical galaxy, typically of type E0 and with a mass of about $10^6 M_\odot$.

Eddington limit The limiting luminosity of an accreting body such as an accreting massive black hole which is set by the outward radiation pressure on infalling material. This limit is proportional to the mass of the accreting body.

Eddington–Lemaître model A cosmological model belonging to the class of Friedmann–Robertson–Walker models, in which the rate of change of the scale factor is positive at any positive time t (i.e. $\dot{R} > 0$ for all $t > 0$) implying perpetual expansion, but in which $\dot{R}$ approaches zero as t approaches zero indicating a long period of quasi-static behaviour (similar to the behaviour of the Einstein model) before the expansion really takes hold. The Eddington–Lemaître model is characterized by a positive cosmological constant ($\Lambda = 4\pi G\rho/c^2$) and a positive curvature parameter ($k = +1$).

Einstein–de Sitter model An alternative name for the critical model.

Einstein field equations The key equations of general relativity that relate the geometric properties of space–time (such as curvature) to the distribution of energy and momentum. In applying general relativity to problems in cosmology, Einstein argued for the inclusion of a term involving the cosmological constant, Λ, that was absent from his original formulation of the field equations. The significance of this modification has remained controversial since its introduction.

Einstein model A cosmological model describing a static universe in which space is finite but unbounded and 'straight' lines close back upon themselves. The Einstein model was the first relativistic cosmological model and is now regarded as a special case in the family of Friedmann–Robertson–Walker models characterized by a uniform distribution of matter with density ρ, a positive cosmological constant $\Lambda_E = 4\pi G\rho/c^2$ and a curvature parameter $k = +1$.

Einstein ring The circular image produced when a point source of light lies directly behind a symmetrical gravitational lens.

ekpyrotic model A speculative model for the early Universe which does not invoke the process of inflation. In the ekpyrotic model, our Universe corresponds to a sheet or 'brane' that moves through a higher dimensional space (the 'bulk'). The collision of the brane on which our Universe resides, with the brane of another 'Universe' may give rise to the effects that are commonly attributed to inflation.

ellipsoid A three-dimensional shape whose cross-section is always elliptical. It has three principal axes.

elliptical (galaxy) Any member of the Hubble class of galaxies characterized by an overall elliptical shape and central concentration of brightness. Membership of this class is indicated by the letter E, followed by a number that denotes the flattening factor of the galaxy.

energy A property of systems (such as arrangements of particles of matter or distributions of radiation) that measures their ability to do work. According to general relativity, the distribution of energy throughout a region of space–time is one of the factors that plays a role in determining the curvature of that region of space–time. The SI unit of energy is the joule (J).

engine The power source within an AGN. It is generally believed that the engine is an accreting supermassive black hole.

epoch of reionization The stage in the evolution of the Universe at which the neutral gas that had been present since the time of recombination first became ionized, possibly due to the intensity of ultraviolet radiation from newly formed stars or AGN. It is believed that reionization occurred when the age of the Universe was less than 10% of its current value.

equation-of-state parameter A quantity used to characterize the physical behaviour of dark energy from the relationship between its density $\rho_A c^2$ and pressure p. The equation of state parameter is defined by $w = \rho_A c^2 / p$. It is convenient to express w in terms of two further parameters that can be constrained by observation: a constant term w_0, and the coefficient w_a of the part of w that varies with $z / (1 + z)$ (where z is redshift); i.e. $w = w_0 + w_a\, z / (1 + z)$.

evaporation The process by which there is a gradual loss of stars from an open cluster due to their acquiring sufficient kinetic energy to escape. The energy to escape is provided by gravitational forces exerted by other stars.

event horizon The bounding surface of a black hole, at which the escape speed is equal to the speed of light in a vacuum. According to general relativity, the event horizon encloses a region of space from which signals (including particles of matter and electromagnetic radiation) cannot escape.

exponential function A mathematical function of the form $y = y_0 e^{ax}$, where $e \approx 2.718$, a is a parameter which may be positive or negative, and y_0 is the value of y when $x = 0$. Many natural processes can be described quantitatively by the exponential function.

Faber–Jackson relation A relationship between the luminosity and the velocity dispersion of elliptical galaxies.

feedback In the context of galaxy formation, feedback refers to physical processes that arise following gravitational collapse that have an effect on subsequent star formation rates. Examples of feedback processes are stellar winds arising from massive star formation and outflows from accreting supermassive black holes.

field galaxy Any galaxy that is not a member of a cluster of galaxies.

finite A property of certain cosmological models (or more specifically of certain space–times) implying that the total volume of space is of limited extent.

flat spectrum radio quasar A subclass of blazar. In contrast to BL Lac objects, the spectra of flat spectrum radio quasars do show strong broad and narrow emission lines.

flux density (F) A quantity describing the rate at which energy transferred by radiation is received from a source, per unit area facing the source. The SI unit of flux density is the watt per square metre (W m^{-2}).

forbidden line A spectral line that can only be produced in a very low density gas. Forbidden lines cannot normally be produced in the laboratory.

Friedmann equation An equation, arising in the context of the Friedmann–Robertson–Walker models, that relates the value of the scale factor $R(t)$ and its rate of change $\dot{R}(t)$ to the curvature parameter k, the cosmic density ρ and the cosmological constant A. Given the value of the parameters k, ρ and A, the process of 'solving' the Friedmann equation leads to an expression for R as a function of the time t that may be presented as a graph of R against t. Such an expression (or graph) substantially determines the evolution of the cosmological model.

Friedmann–Robertson–Walker models A class of cosmological models based on general relativity and the assumption that the Universe is homogeneous and isotropic (i.e. the cosmological principle). The geometric properties of space–time in these models are described by the Robertson–Walker metric, which includes a curvature parameter k, and a scale factor $R(t)$ that satisfies the Friedmann equation.

FRW models A common abbreviation of Friedmann–Robertson–Walker models.

Galactic coordinates A coordinate system on the sky, whose two elements are Galactic longitude l, and Galactic latitude b, resembling longitude and latitude on the Earth. The orientation of the coordinate system is defined to make it useful for describing the locations of objects in the Galaxy from the viewpoint of the Sun. The Galactic equator is chosen to coincide more-or-less with the Galactic plane, the direction $(l, b) = (0°, 0°)$ is roughly in the direction of the Galactic centre, $(l, b) = (90°, 0°)$ is roughly in the direction of motion of the Sun around the Galactic centre, and $b = 90°$ is the direction in the sky perpendicular to the disc, that lies above the northern hemisphere of the Earth.

Galactic disc A major structural component of the Galaxy, containing most of the visible matter, which lies in a highly flattened distribution. The Sun is located in the disc.

Galactic equator The directions in the sky, close to the Galactic plane, whose Galactic latitude is $b = 0°$. *See* Galactic coordinates.

Galactic fountain A flow of hot gas away from the disc of the Galaxy due to heating by supernova explosions. The gas is believed to cool and then be attracted back to the disc by gravity.

Galactic latitude The Galactic coordinate, denoted b, that measures the angular position of an object relative to the plane marked out by the Galactic equator.

Galactic longitude The Galactic coordinate, denoted l, that measures the angular distance of an object around the Galactic equator, from a reference point $l = 0°$ roughly in the direction of the Galactic centre.

Galactic plane The plane defined by the distribution of stars in the flattened disc of the Galaxy, with equal amounts of material on either side. The Sun is located close to, but not exactly in the plane. It may be distinguished from the Galactic equator, which is the plane of a coordinate system defined for convenience to pass through the Sun, tied only approximately to the true distribution of matter in the disc.

Galactic spheroid A structural component of the Milky Way, with the shape of a spheroid, consisting of the halo and nuclear bulge. It extends several tens of kiloparsecs from the Galactic centre. The disc lies within the spheroid, but is not considered to be part of the spheroid.

galaxies Collections of luminous stars, non-luminous dark matter, and in the case of spiral and irregular galaxies some amount of gas and dust, that are gravitationally bound to one another and separated from other similar structures usually by distances of tens of kiloparsecs or more. Various categories of galaxies may be defined based on their appearance, such as spiral galaxies (barred or normal), elliptical galaxies, lenticular galaxies and irregular galaxies.

gaseous corona The body of very hot tenuous gas in the halo of a spiral galaxy.

general relativity A theory of gravity, proposed by Albert Einstein in 1916, according to which gravitational phenomena are a consequence of the geometric distortion of space and time (described mathematically by the curvature of space–time). Formally, the theory is based on the Einstein field equations, but it is often summarized by the somewhat overly simple statement 'matter tells space how to curve; space tells matter how to move'.

geometrical methods (of distance measurement) Any method of distance measurement based on measuring the angular diameter of a feature of known linear diameter. In practice, geometrical methods of distance measurement are not commonly used because there are few astronomical bodies with known linear diameters.

geometry The branch of mathematics concerned with the study of points, lines, surfaces and volumes in space or space–time and the relationships between them.

globular clusters Clusters of 10^5 to 10^6 very old stars tightly bound by gravity into a spherical region of space less than about 50 pc in diameter. The 150 or so globular clusters associated with the Milky Way are found in a spherical distribution about the centre of our Galaxy. Similar distributions are seen in other galaxies.

grand unified theory (GUT) A physical theory which, it is supposed, should describe the strong, weak and electromagnetic interactions as different manifestations of a single type of interaction. Several candidate grand unified theories exist, but all are speculative and difficult to test experimentally.

gravitational instability The process by which a region of enhanced density becomes more pronounced as a result of its own enhanced gravitational attraction.

gravitational lens An object that, by virtue of its gravitational field, forms an image (or images) of a background source of electromagnetic radiation.

gravitationally bound (system) A system of bodies whose gravitational field and distribution of velocities is such that members of the system cannot escape from the system (except by the relatively slow process of evaporation).

group A collection of galaxies that contains fewer than about 50 bright members. They are believed to be gravitationally bound systems.

Gunn–Peterson effect The effect, expected to be seen in the spectra of sufficiently distant quasars, whereby electromagnetic radiation at wavelengths shorter than the Lyman α line (121 nm) should be absorbed by smoothly distributed neutral hydrogen in the intergalactic medium. This effect has been observed at high resolution but is not seen in the spectra of quasars with redshifts up to about 5, indicating that the reionization of the intergalactic medium occurred when the Universe was less than about 10% of its present age.

hadron An elementary particle that consists of a cluster of three quarks (or three antiquarks) or of a quark–antiquark pair.

halo A major component of spiral galaxies, spheroidal in shape, and extending several tens of kiloparsecs from the centre of the galaxy. The disc lies within the halo, but is not considered part of the halo. The halo contains mainly Population II stars, with some tenuous gas (see gaseous corona). The halo is sometimes referred to as the 'stellar halo' in order to distinguish it from the dark-matter halo.

Heisenberg's uncertainty principle A fundamental principle of quantum physics which, in one form, states that it is not possible to determine to an arbitrarily high precision both the energy of a system and the time at which the measurement is made. Mathematically, the relationship between the uncertainties in the energy of the system (ΔE) and in the time of measurement (Δt) can be expressed as ($\Delta E \times \Delta t$) > $h/2\pi$ (where h is the Planck constant).

Hertzsprung–Russell diagram A diagram showing the luminosity and temperature of stars, which is useful for comparing large numbers of stars and for tracking their evolution. Photospheric temperature is shown along the horizontal axis (increasing to the left), and luminosity is shown along the vertical axis. A star appears as a point on the diagram, corresponding to its observed temperature and luminosity.

hidden dark matter A hypothetical type of matter that interacts gravitationally but does not undergo the interactions of the standard model of elementary particles. It does however have its own set of interactions.

hierarchical scenario A proposed process for the formation of structure in the Universe which proceeds by the merging of relatively low-mass structures to form more massive structures. It is also referred to as the bottom-up scenario.

high-velocity clouds Clouds of atomic hydrogen well away from the Galactic disc, that are moving rapidly relative to the Sun. Their distances are almost impossible to judge, and there is uncertainty whether they are located within or beyond the Galactic halo.

high-velocity stars Stars, typical of Population II but seen as their orbits carry them through the Galactic disc, whose velocities consequently are abnormally high relative to the disc stars that surround them.

HII region Hot, luminous region of the interstellar medium, comprising ionized hydrogen gas that is made visible by the presence of a hot, young star or stars. Strong ultraviolet radiation from hot stars ionizes the hydrogen, and the occasional recombination of an electron and proton to form a neutral hydrogen atom results in the emission of light, before the hydrogen is reionized.

homogeneous A term meaning 'the same everywhere'.

horizon distance At any instant in the history of the Universe, the maximum distance that a physical signal could have travelled in the time that had elapsed up to that instant.

horizon mass The mass contained within a sphere with a radius equal to the horizon distance.

horizontal branch A region on the Hertzsprung–Russell diagram occupied by stars of low mass and low metallicity after they have left the red giant branch during helium core burning. It is often seen in H–R diagrams of globular clusters, where many stars have similar luminosity, but a wide range of surface temperatures and hence lie in an approximately horizontal strip.

host galaxy (of an AGN) The galaxy in which an AGN is found.

hot big bang A theory of cosmic evolution, according to which the current state of the Universe results from the expansion and cooling of a hot, dense and highly uniform initial state.

hot dark matter (HDM) Dark matter which is comprised of particles that are moving at speeds close to the speed of light.

H–R diagram *See* Hertzsprung–Russell diagram.

Hubble classes The four major classes of galaxy: elliptical (E), irregular (Irr), lenticular (S0) and spiral (S). The lenticular and spiral galaxies can be further classified as barred or non-barred.

Hubble classification scheme A classification scheme for galaxies based on their observed shape and structure. *See* Hubble classes; Hubble types.

Hubble constant *See* Hubble's law.

Hubble diagram A plot of redshift against distance for a sample of distant objects such as galaxies or clusters of galaxies. Such a plot may employ linear or logarithmic axes; if linear axes are used, then the gradient (i.e. slope) of the straight line drawn through the plotted data should equal the Hubble constant divided by the speed of light in a vacuum.

Hubble flow A term used to describe the smooth overall expansion of the Universe that is described by Hubble's law. Distant galaxies provide observable tracers of this expansion, but only imperfectly since each individual galaxy will have its own 'peculiar motion' relative to the Hubble flow.

Hubble parameter The time-dependent quantity $H(t) = \dot{R}(t)/R(t)$ that arises in any Friedmann–Robertson–Walker model in which the scale factor at time t is $R(t)$ and the rate of change of the scale factor at that time is $\dot{R}(t)$. To the extent that such a model represents the real Universe, the current value of the Hubble parameter, $H(t_0)$, should equal the observed Hubble constant H_0.

Hubble time The time $t = 1/H_0$, where H_0 is the Hubble constant, that provides a useful reference value in discussions of cosmic age.

Hubble types The subdivisions of the Hubble classes in the Hubble classification scheme for galaxies. The galaxies belonging to the elliptical class may be typed as E0, E1, E2 … E7, according to their observed shape. Spiral (and barred spiral) galaxies can be typed as Sa, Sb, Sc (and SBa, SBb, SBc), according to the openness of the spiral arms and the size of the galactic nucleus relative to the disc of the galaxy.

Hubble's law The observationally based law, discovered by Edwin Hubble, according to which the distance (d) and redshift (z) of (moderately) distant galaxies are approximately related by

$$z = \frac{H_0}{c} d$$

where H_0 is the Hubble constant and c is the speed of light in a vacuum.

ICM *See* intracluster medium.

inflation An episode of rapid and accelerating expansion in the early Universe. Such a process would have occurred if the effective value of the cosmological constant temporarily became very large. It is speculated that this may have happened immediately before the end of grand unification, as a result of the development of a 'false vacuum' with a high density of vacuum energy. However, there is no accepted theoretical explanation of why such conditions might have occurred. Despite the lack of a mechanism for inflation, it is an attractive hypothesis because it provides a natural solution to the horizon and flatness problems, as well as offering an explanation of the origin of cosmic structure.

infrared dark cloud A cloud in the interstellar medium that is sufficiently dense to significantly absorb infrared radiation at a wavelength of about 10 μm. Infrared dark clouds usually correspond to compact dense regions within giant molecular clouds.

initial singularity The state, in some cosmological models, in which some physical quantities (such as density) have implied values that are infinite at the time $t = 0$.

instability strip A roughly vertical region on the Hertzsprung–Russell diagram where the structure of stars is unstable. Any star in this region pulsates and therefore shows variability. Amongst the stars found in this region are classical Cepheids and RR Lyrae variables.

integrated spectrum The overall spectrum of electromagnetic radiation from an entire galaxy, or from a large region of a galaxy, made up from the spectra of stars and other luminous matter.

interacting galaxies Two (or more) galaxies, interacting with each other in a manner that wreaks profound internal changes in both.

interaction energy The typical amount of energy available in a particle interaction. In a system that is in thermal equilibrium at a temperature T, the interaction energy has a value of approximately kT (where k is the Boltzmann constant).

intercloud medium A component of the interstellar medium characterized by very low density, within which the other components of the interstellar medium, such as dense clouds and HII regions, are embedded.

interstellar medium (ISM) The matter that thinly fills interstellar space in the Galaxy. It consists of gas (mainly hydrogen), with a trace of dust, and occurs as several varied types of region, such as dense clouds, HII regions and the intercloud medium.

intracluster medium (ICM) The gas that lies between the galaxies within a cluster of galaxies. Typically such gas is very hot and ionized, and has a very low density.

ionized hydrogen Hydrogen in a state where the single proton and single electron that form atomic hydrogen have acquired sufficient energy (13.6 eV or more) that they are no longer bound to one another. The energy may come from collisions with other fast-moving particles, or from electromagnetic radiation of sufficiently high energy/short wavelength.

irregular (galaxy) Any member of the Hubble class of galaxies characterized by having no overall symmetry or regularity. Membership of this class is denoted by the symbol Irr.

ISM *See* interstellar medium.

isochrone A curve in the Hertzsprung–Russell diagram showing the theoretically expected locations (i.e. temperatures and luminosities) of stars of different masses, but the same age (and the same initial metallicity).

isophote A curve linking points of equal apparent surface brightness.

isotropic A term meaning 'the same in all directions'.

Jeans mass The minimum mass that a uniform, spherical, non-rotating cloud must have if it is to collapse under its own gravitation. In the context of evolutionary cosmology, the Jeans mass at any time (based on the mean cosmic density and temperature at that time) determines which of two evolutionary pathways an over-dense region will follow. A region that exceeds the Jeans mass will contract under the influence of gravity.

A region that has a mass lower than the Jeans mass will be supported by its internal pressure and will be stable against gravitational collapse.

Keplerian orbit An orbit arising when the mass of a gravitating system is dominated by a single body. This applies in the case of the Solar System where the Sun dominates, but not in the case of the disc of a galaxy where large fractions of the mass lie away from the centre.

kiloparsec A unit of distance, equal to one thousand parsecs and usually denoted 1 kpc, that is convenient for measuring distances on the scale of a galaxy. 1 kpc = 1000 pc = 3.09×10^{19} m.

λF_λ (lambda-eff-lambda) The product of multiplying the spectral flux density F_λ at a wavelength λ by that wavelength. When plotted against wavelength (to form a spectral energy distribution) this quantity provides a measure of the contribution to the total luminosity of a source that arises from different parts of the electromagnetic spectrum.

large-scale structure A generic term for the distribution of matter in the Universe on or exceeding the scales of superclusters, i.e. on linear scales exceeding several tens of megaparsecs.

last-scattering surface The surface defined by the locations at which photons in the cosmic microwave background last underwent significant interaction with matter (with the exception of gravitational effects). This interaction was due primarily to electron scattering, and so last-scattering occurred at about the time of recombination.

Lemaître model A cosmological model, belonging to the class of Friedmann–Robertson–Walker models, which starts with a big bang and which expands perpetually, but in such a way that there is a 'coasting' or 'pseudo-static' phase at intermediate times during which the rate of change of the scale factor approaches zero so that the model behaves like the Einstein model. The Lemaître model is characterized by a positive cosmological constant ($\Lambda > 4\pi G\rho/c^2$) and a positive curvature parameter ($k = +1$).

lenticular (galaxy) Any member of the Hubble class of galaxies characterized by having a disc but no spiral arms, possibly related to spiral galaxies. Membership of this class is denoted by the symbol S0, or SB0 in the case of a barred lenticular galaxy.

lepton One of a family of six elementary particles that includes the electron and the three types of neutrino.

lepton number A quantity that is conserved in all particle interactions (with the exception of some speculative interactions predicted by grand unified theories). The lepton number of any lepton is +1 and that of any antilepton is −1 respectively. The lepton number of any other particle is zero.

light curve A diagram showing the variation of brightness (e.g. magnitude, flux density or luminosity) with time, for a celestial object such as a variable star or supernova.

Local Group A sparse cluster of over 100 galaxies within about 1 Mpc of the Milky Way, and including the Milky Way.

Local Supercluster The supercluster of clusters of galaxies to which the Local Group belongs. It is 25–50 Mpc across, and contains 1000 or so bright galaxies.

lookback time The time interval between the emission and observation of electromagnetic radiation from an astronomical source. For distant galaxies, the lookback time can be a significant fraction of the age of the Universe.

luminosity A quantity describing the rate at which energy is carried away from a luminous object by electromagnetic radiation. The SI unit of luminosity is the watt (W), where $1\ \mathrm{W} = 1\ \mathrm{J\ s^{-1}}$.

Lyman α (line) The spectral line that arises from the electronic transitions in the hydrogen atom from $n = 1$ to $n = 2$ (absorption) or from $n = 2$ to $n = 1$ (emission).

Lyman α forest A set of absorption lines (which are predominantly due to Lyman α absorption) appearing in the spectrum of a quasar. The absorption lines are due to clouds of neutral intergalactic gas that lie along the line of sight to that quasar.

Lyman series The series of electronic transitions in the hydrogen atom that involve a change to or from the $n = 1$ state.

MACHO (massive astrophysical compact halo object) An astronomical body with a moderate mass but a low luminosity (such as a stellar remnant or a body of substellar mass) that might exist undetected in the halo of a galaxy. A large population of such objects might account for a significant amount of (baryonic) dark matter.

main sequence turn-off The point on the main sequence of the Hertzsprung–Russell diagram of a star cluster above which no stars are present. It corresponds to stars that are just reaching the end of their time on the main sequence, and is therefore an indication of the age of the cluster.

major merger A collision between two galaxies of roughly similar mass. Major mergers typically result in dramatic changes in galaxy morphology.

mass accretion rate The rate at which material is transferred to an astronomical body.

mass-to-light ratio The value obtained by dividing the mass M of a system by its luminosity L.

mathematical model A mathematical representation of some process or system that captures certain essential features of its subject but does not attempt to recreate every detail. A mathematical model is usually based on one or more equations and many involve one or more parameters that might have to be determined by observation.

merger tree A schematic representation of the history of a galaxy in terms of the merger events that have led to the formation of that galaxy.

metallicity (Z) A numerical measure of the proportion of heavy elements in a sample of material, obtained by dividing the mass of heavy elements (i.e. 'metals' to an astronomer) in the sample by the total (baryonic) mass of the sample. In the Sun, $Z \approx 0.02$.

metals To an astronomer, all elements except hydrogen and helium.

Milky Way The name given to the Galaxy of which the Sun is a member. Also the name given to the diffuse band of light seen when an observer on the Earth looks in a direction near the plane of the Galaxy, where uncountable numbers of unresolved stars produce a background glow.

minor merger A collision between two galaxies of very different masses. Typically the morphology of the more massive galaxy does not change, while the less massive galaxy is likely to be completely disrupted.

molecular clouds Any cloud-like region in the interstellar medium in which hydrogen is predominantly in the form of molecular hydrogen. Dense clouds are found in this type of region.

molecular hydrogen Hydrogen in a state where many pairs of hydrogen atoms have become bound to one another to form hydrogen molecules (H_2). This is only possible in cold, dense clouds where neither collisions with fast-moving particles, nor ultraviolet radiation is likely to break apart (dissociate) the molecules.

momentum A property of systems (such as arrangements of particles of matter or distributions of radiation) that measures their ability to impart an impulse. According to general relativity, the distribution of momentum throughout a region of space–time is one of the factors that plays a role in determining the curvature of that region of space–time. The SI unit of momentum is the kilogram metre per second ($kg\ m\ s^{-1}$).

monolithic collapse A scenario for galaxy formation in which the gravitational collapse of a single over-dense region gives rise to a single galaxy.

morphology (of a galaxy) The observed shape and large-scale structure of a galaxy, used as the basis of the Hubble classification.

M-theory A speculative physical theory that unifies gravitation with the strong, weak and electromagnetic interactions. One feature of M-theory is that it requires 11 dimensions, rather than the four dimensions of space–time with which we are familiar.

multipole number The numerical quantity, usually denoted l, that is plotted on the horizontal axis of an angular power spectrum, and which indicates an angular scale (on an anisotropy map or elsewhere) of $\theta = 180°/l$.

multiverse A hypothetical set of domains in which physical constants, and hence laws of nature, are different. If it is assumed that physical constants are set at random, then only some of these domains will lead to inflation and a Universe in which life can evolve.

narrow-line region (NLR) The body of gas responsible for emitting the narrow emission lines in an AGN. For narrow emission lines, the width corresponds to a radial velocity dispersion in the range 200 to 900 km s^{-1}.

neutralino An uncharged elementary particle that is predicted by some supersymmetric extensions to the standard model. The existence of such particles has not been established. The neutralino is a candidate weakly interacting massive particle (WIMP).

neutrino decoupling A process in which changing physical conditions prevent the frequent interaction between neutrinos and other types of elementary particles. The term neutrino decoupling is often used to refer to the postulated episode in cosmic history in which the declining density and temperature of matter caused cosmic neutrinos to cease their frequent interactions with every other kind of particle (except for the effects of gravity). Neutrino decoupling is believed to have occurred when the age of the Universe was about 0.7 s.

non-baryonic dark matter A component of dark matter which can be shown not to be made of baryons.

normal galaxy A galaxy which has an approximately constant luminosity that can largely be accounted for in terms of the stars and gas that the galaxy contains.

number density The quantity used to describe the number per unit volume of particles or bodies of some specified type (e.g. electrons or stars). The SI unit of number density is the per cubic metre (m^{-3}).

OB association A group of young stars containing many stars of spectral types O and B.

oblate spheroid An ellipsoid having the shape of a flattened sphere, i.e. with two principal axes of equal length and a shorter third axis.

observational cosmology The branch of science concerned with measuring the parameters that characterize the Universe. These parameters include the Hubble constant, the current value of the deceleration parameter and the current values of the density parameter for matter and the density parameter for the cosmological constant.

open cluster A cluster of up to a few hundred stars, formed from a cloudlet that has fragmented from a larger dense cloud. The stars are only loosely bound together by gravity, hence the name 'open', in contrast to the much stronger binding of stars in a globular cluster.

open model A cosmological model, belonging to the class of Friedman–Robertston–Walker models, that starts with a big bang and expands continuously without limit, so the rate of change of the scale factor is always positive (i.e. $\dot{R} > 0$ at all times). The open model is characterized by a zero cosmological constant ($\Lambda = 0$) and a negative curvature parameter $(k = -1)$. In such a model the cosmic density is always less than the critical density.

Orion Spur Also called the Orion–Cygnus arm. A strip-shaped region of the Galaxy near the Sun occupied by astronomically young objects, either a spur of a spiral arm, or an arm in its own right. The Sun is located in the Orion Spur.

pair-creation The physical process in which, given sufficient energy, a particle and its antiparticle can spontaneously form. This is the opposite process to annihilation.

parallax The quantity that describes the change in direction to a celestial body (relative to a background of far more distant bodies) resulting from a given change in position of the observer perpendicular to the direction of the body. The term parallax is often used to refer specifically to stellar parallax, p, where the change in position of the observer is one astronomical unit. This quantity is important in the determination of the distance of nearby stars.

parsec The distance to a celestial body that has a parallax of one arc second. 1 pc $\approx$ 3.26 light-years $\approx 3.09 \times 10^{16}$ m.

peculiar galaxy A galaxy of more-or-less readily apparent Hubble type, but with some abnormal feature (such as a jet); denoted by 'p' after the Hubble type.

peculiar motion (of a galaxy) The component of motion of a galaxy as a whole that is additional to that arising from its participation in the Hubble flow.

period–luminosity relationship A correlation between period and luminosity; in particular the relationship between period and luminosity of Cepheid variables that enables these stars to be used as standard candles. (Absolute visual magnitude, M_V, is generally used in place of luminosity when displaying this relationship.)

Perseus Arm A strip-shaped region of the Galaxy, slightly further from the Galactic centre than the Sun, occupied by astronomically young objects. It is one of the local spiral arms of the Galaxy.

photodisintegration The process in which a nucleus is split apart by the absorption of a gamma-ray photon. This type of reaction plays an important role in the later stages of stellar nucleosynthesis.

photon The particle of electromagnetic radiation in the photon model of light. The photon energy ε is proportional to the frequency f of the associated radiation; $\varepsilon = hf$ where h is the Planck constant.

photon–baryon fluid The fluid-like system in the early Universe, formed by the incessant interaction of radiation and charged baryons, that is characterized in any sufficiently localized region by a specific temperature and pressure. The photon–baryon fluid is subject to the gravitational influence of the dark matter that is also present in the early Universe.

photon energy distribution function A quantity that describes the relative numbers of photons at different energies. Specifically, at some specified energy, the photon energy distribution function is the proportion of photons whose energies lie in a narrow energy range around that energy.

Planck era The period in the history of the Universe prior to the Planck time.

Planck time A time determined by a combination of physical constants $((Gh/2\pi c^5)^{1/2} = 5.38 \times 10^{-44}$ s) that represents the earliest time in cosmic history at which currently established physical theory might be used to study the nature and evolution of the Universe. Prior to the Planck time, gravity might have played a significant role in particle interactions.

Population I Stars found in the discs of spiral galaxies, generally less than 10^{10} yr old, and have a metallicity similar to that of the Sun ($Z \sim 0.01$ to 0.04).

Population II Stars found in the Galactic spheroid of spiral galaxies and in elliptical galaxies, generally more than 10^{10} yr old. Population II stars in the halo have metallicities $Z < 0.002$, but Population II stars in the bulge have metallicities similar to Population I stars ($Z \sim 0.01$ to 0.04).

population synthesis A method of investigating the stellar content of a galaxy by modelling the expected galaxy spectrum as stars evolve. The overall model uses models of stellar evolution and stellar spectra to obtain a time-dependent integrated spectrum of a galaxy that can be compared with observations.

precision cosmology A term used to describe recent developments in cosmology whereby the values of a range of key cosmological parameters have been determined with high precision (and possibly good accuracy).

primordial nucleosynthesis The nuclear processes that were responsible for the initial formation of the nuclei of light elements (such as helium and lithium) in the early Universe. It is generally believed that primordial nucleosynthesis began when the age of the Universe was about 3 minutes, and that it continued for about thirty minutes.

prolate spheroid An ellipsoid with the shape of an elongated sphere, i.e. with two principal axes of equal length and a longer third axis.

Pythagoras's theorem A theorem of geometry according to which the square of the length of the longest side of a right-angled triangle is equal to the sum of the squares of the lengths of the other two sides ($c^2 = a^2 + b^2$).

quantum fluctuations Variations in energy density of the vacuum that occur on a microscopic scale due to the presence of virtual particles.

quantum theory A wide-ranging theory that describes, amongst other things, the structure and behaviour of atoms and their interaction with electromagnetic radiation. It accounts for the phenomena that are embraced by the photon model of light, and implies the existence of energy levels in atoms.

quark–hadron phase transition A process in which changing conditions cause a 'gas' of free quarks and antiquarks to transform itself into a gas of hadrons and antihadrons. The term is also used to describe the episode in cosmic history in which this process is believed to have affected the baryonic matter in the Universe. It is believed that the quark–hadron phase transition occurred when the age of the Universe was about 10^{-5} s.

quasar/QSO A kind of active galaxy, typically characterized by a point-like appearance and a very large redshift. Quasars provide very distant and very bright examples of the effect of an AGN in a galaxy where the rest of the galaxy is so faint that it can only be discerned with difficulty, if at all.

quintessence A hypothetical and exotic form of matter, probably better thought of as a scalar field filling the Universe, that would exert a negative pressure. The energy associated with this field would constitute the dark energy. In contrast to some of the other proposed explanations of dark energy, the energy density of quintessence might vary with time and spatial position.

radiation-dominated era The period of the history of the Universe when the energy density of radiation exceeded that of matter and that of dark energy.

radiation pressure A pressure exerted by photons on any object that absorbs or scatters them.

radio galaxy A kind of active elliptical galaxy which shows (usually) two regions of diffuse radio emission from either side of the galaxy – radio lobes. An AGN is needed to power the radio lobes, and can be seen in the centre of the parent galaxy.

random uncertainties Uncertainties in the measured value of a quantity that cause repeated measurements of that quantity to vary about some mean value.

random walk A random process in which each step is in a random direction, and that direction does not depend on previous steps.

recombination A process in which an electron and an ion combine, i.e. the opposite of ionization. The electron is typically captured into a high-energy orbit and then cascades downward through the atom's energy levels emitting photons as it does so. The term recombination is also used to refer to the postulated episode in cosmic history in which the baryonic matter of the Universe made the transition from being predominantly plasma to predominantly neutral atoms. Recombination is believed to have occurred when the age of the Universe was about 3.7×10^5 years.

redshift The numerical quantity used to measure the shift in wavelength of a spectral line. If a spectral line is emitted at a wavelength λ_{em} and observed at a wavelength λ_{obs}, the redshift is defined by

$$z = \frac{\lambda_{obs} - \lambda_{em}}{\lambda_{em}}$$

reionization A process in which a formerly ionized medium that had undergone recombination to become neutral is again ionized. The term reionization is often used to refer to the postulated episode in cosmic history in which much of the neutral hydrogen became ionized, perhaps due to an increase in the intensity of ultraviolet light from hot stars or from AGN. Reionization is believed to have occurred when the age of the Universe was less than 10% of its current value.

relative density fluctuation The numerical quantity used to describe the extent to which the density of a given region of the Universe departs from the mean density of the Universe by expressing the difference between the density of the region and the mean density as a fraction of the mean density.

retrograde Motion around the Galaxy in the opposite direction to that of disc stars.

reverberation mapping A method of determining the distance of the broad-line region from the active galactic nucleus by comparing variations in the broad lines and the continuous spectrum, measuring, for example, the time delay between a brightening of the continuous spectrum and the corresponding brightening of the emission lines.

richness A quantity used in the classification of clusters of galaxies that describes the number of galaxies (above a certain threshold luminosity) present within a cluster. Rich clusters contain relatively high numbers of galaxies.

rigid body rotation A pattern of rotation in which all parts of the rotating system have the same period of rotation, irrespective of their distance from the centre of rotation. This is the kind of rotation exhibited by a solid planet and should be contrasted with the differential rotation that characterizes objects such as the Sun and the Milky Way.

Robertson–Walker metric An expression, applying to any space–time that is homogenous and isotropic, that relates the physical separation ds of two narrowly separated events to the coordinate differences (typically dx, dy, dz and dt) that may be used to describe the relative locations of those events. The Robertson–Walker metric involves a curvature parameter k and a scale factor $R(t)$ that respectively characterize the curvature of space and its expansion (or contraction) with time.

rotation curve A plot of rotation speed against radial distance for a rotating system.

RR Lyrae stars A type of regular variable star. RR Lyrae stars are found on the horizontal branch of the H–R diagram, and within the instability strip and hence their envelopes pulsate. Pulsation periods are around 12 hours. As all horizontal branch stars have very similar absolute magnitudes, RR Lyraes are good standard candles for measuring distances within the Galaxy. They are Population II stars.

Sachs–Wolfe effect The dominant source of angular power in the angular power spectrum of the cosmic microwave background for multipole numbers of 50 or less (i.e. on angular scales of a few degrees or more). The effect is largely due to the gravitational redshift of radiation coming from the denser parts of the last-scattering surface.

Sagittarius A* A very compact, strong radio source that lies at the very centre of the Galaxy. The motions of stars around Sgr A* suggest it is a black hole with a mass about 4.3×10^6 times greater than the mass of the Sun.

Sagittarius Arm A strip-shaped region of the Galaxy, slightly closer to the Galactic centre than the Sun, occupied by astronomically young objects. It is one of the local spiral arms of the Galaxy.

Sagittarius dwarf galaxy A dwarf galaxy (discovered in 1994) that is in the process of merging with the Milky Way.

scalar field A physical quantity which is defined by a single value at every point in space. The potential energy associated with the value of such a scalar field contributes to the energy density of space. Variation in (different) hypothetical scalar fields is invoked in both quintessence and inflation.

scale factor A time-dependent quantity, usually denoted $R(t)$, that appears in the Robertson–Walker metric where it describes the expansion or contraction of space in a homogeneous and isotropic Universe. The scale factor plays a crucial role in relating the coordinates of points to the physical distance between these points. In an expanding Universe, two points that have a fixed comoving coordinate separation r will be separated by a growing physical distance l that is proportional to $R(t)$, and which may be, for example, $l = R(t) \times r$.

scale height The distance, measured from the Galactic plane, over which the number density of disc stars decreases to $1/e$ times the density in the Galactic plane. (The value $e \approx 2.718$ is the basis of the exponential function.)

Schwarzschild radius The radial distance from the centre of a black hole at which the escape speed equals the speed of light.

SED *See* spectral energy distribution.

semimajor axis Half the longest diameter of an ellipse.

semiminor axis Half the shortest diameter of an ellipse.

Seyfert galaxy A kind of active spiral galaxy which has an AGN that appears as a central, point-like source in optical images.

SFR *See* star formation rate.

Silk damping The dominant source of angular power in the angular power spectrum of the cosmic background radiation for multipole numbers in excess of $l = 1000$ (or angular scales of less than 0.1 degree or so). The effect arises from the suppression of acoustic waves of very short wavelength due to the free movement of photons between encounters with charged particles.

space The aspect of space–time that consists of all the possible positions that a particle might occupy according to some observer. Space is three-dimensional and possesses a range of geometrical properties.

space–time The four-dimensional entity that unites space and time. It consists of all the possible events at which a particle might be present, and is characterized by geometrical properties such as curvature. When making measurements, any observer will divide space–time into space and time, but the way in which that division is made by two different observers will generally be different and depends on the relative motion of those two observers.

spectral energy distribution (SED) A form of the broadband spectrum of an astronomical source that shows the quantity λF_λ against wavelength λ. It shows the relative contributions to the total luminosity that are emitted in different wavelength ranges.

spectral excess The feature in the spectrum of a galaxy that represents the excess of emission in a certain wavelength band over that which would be expected due to emission from stars alone. Spectral excesses in the infrared are characteristics of starburst and active galaxies.

spheroid A three-dimensional shape that may be pictured as a sphere that has been flattened or stretched in one direction. This is the *shape* of both the (stellar) halo and the dark-matter halo of the Galaxy, and it is the *name* given to the region of the Galaxy associated with the halo and sometimes also the bulge.

spiral arms The regions of a spiral galaxy traced out by bright stars, HII regions and other astronomically young objects. These mark out a fragmented, roughly spiral pattern within the disc of the galaxy, extending outwards from near its centre.

spiral-arm tracers Those young, short-lived astronomical objects associated with recent star formation that map out the spiral arms of a spiral galaxy. They include HII regions, O and B stars, classical Cepheid variable stars, and T Tauri stars.

spiral density wave A long-lived, self-consistent pattern of density enhancement that may arise in a disc of stars and gas, thought possibly to account for the pattern of star formation that gives rise to spiral arms in spiral galaxies.

spiral galaxy Any member of the Hubble class of galaxies characterized by having a disc and spiral arms. Membership of this class is indicated by the letter S, or SB in the case of a barred spiral galaxy.

standard candle Any type of object whose luminosity is directly indicated by its observable properties, thus allowing its distance to be inferred from the difference between its apparent brightness and its true brightness.

standard candle methods Any method of distance determination based on the use of a standard candle, such as a Cepheid variable of known period, or a Type Ia supernova with a known rate of decline in brightness.

standard model (of elementary particles) A theory which describes all known elementary particles and their interactions. No discrepancy between the predictions of the standard model and experimental results has yet been discovered.

star formation rate (SFR) The rate at which stars are forming, usually quoted as the number of solar masses per year, in some specified volume.

starburst galaxy A galaxy in which a recent episode of star formation is believed to have occurred, leading to optical emission lines and infrared radiation being emitted from an extended region of the galaxy.

static A property of certain cosmological models (or more specifically of certain space–times) implying that space is neither expanding nor contracting.

stellar halo A term sometimes used to distinguish the halo from the dark-matter halo. *See* halo.

stellar population A grouping of stars characterized by their age, composition and location or motion. *See* Population I; Population II.

sterile neutrino A hypothetical type of neutrino that, in contrast to the known neutrinos, does not interact by the weak interaction.

strong gravitational lensing The effect whereby the paths of rays of light (or other electromagnetic radiation) from a distant source are bent by the gravitational effect of an intervening mass such that multiple images of the source are produced. The images are heavily distorted, and typically form arcs (or an Einstein ring) around the lensing mass.

sublimate The process in which a solid material (e.g. dust) is transformed, on heating, into the gas phase without going through a liquid phase.

sublimation radius The distance from the engine of an AGN at which the temperature is just sufficient to cause dust particles to sublimate.

superbubble A hot, bubble-like region of the interstellar medium where the gas has been heated greatly by large numbers of supernova explosions from stars in that region.

supercluster An association of galaxies of order 25 to 50 Mpc across.

supermassive black hole A black hole with a mass in excess of about $10^6 M_\odot$.

supersymmetry A type of symmetry which has been suggested, but not proven, to apply to particle interactions. Supersymmetric theories predict the existence of particles that have not been observed in nature. Some of these particles (such as the neutralino) are of cosmological significance because they are candidates for WIMPs.

surface brightness profile A plot of the apparent surface brightness of a galaxy as a function of radial distance from its centre.

systematic uncertainties Uncertainties in the measured value of a quantity that cause repeated measurements of that quantity to always differ from the true value in the same way.

thermal bremsstrahlung A process by which X-rays are generated in a plasma. Electrons pass close to ions, but without being captured. In doing so their paths are deflected and X-ray photons are emitted.

thermal equilibrium A state in which there is a high level of interaction between matter and radiation, leading to the radiation having a black-body spectrum with a characteristic temperature that is the same as that of the matter.

thick disc A component of the disc of the Galaxy, a few times thicker than the thin disc – the scale height of the thick disc is around 1000 parsecs – and containing fewer stars.

thin disc The main component of the disc of the Galaxy, with a scale height around 300 parsecs. Thin-disc stars belong to Population I.

time The aspect of space–time that consists of all the possible instants at which a particle might exist according to some observer. Time is one-dimensional.

top-down scenario A proposed process for the formation of structure in the Universe which proceeds by the fragmentation of relatively high-mass structures to form less massive structures.

triaxial ellipsoid An ellipsoid with three unequal principal axes.

Tully–Fisher relation The relationship between the luminosity of a spiral galaxy and the width of its 21cm hydrogen emission line. This relation is used to determine the luminosities and hence distances of spiral galaxies.

Type Ia supernovae A class of supernova, used as a standard candle.

Ultra-luminous infrared galaxy (ULIRG) A galaxy which has a far-infrared luminosity $> 10^{12} L_\odot$. The high far-infrared luminosity is believed to arise from very high star formation rates in such galaxies.

unbarred galaxy *See* barred (spiral) galaxy.

unbounded A property of certain cosmological models (or more specifically of certain space–times) implying that any 'straight' line may be extended infinitely without ever encountering any boundary or edge. Note, however, that in an unbounded space–time there is no guarantee that a straight line may not close back upon itself, as shown by the finite but unbounded space of the Einstein model.

vacuum energy The energy possessed by a region of space that contains no radiation or real particles. This energy is not zero because of the existence of virtual particle–antiparticle pairs, which are continuously being created and annihilated in the vacuum. Vacuum energy provides a possible explanation of the nature of dark energy, though attempts to provide a quantitative basis for such an explanation have so far been unsuccessful.

velocity dispersion A quantity that describes the range of velocities found within a collection of moving objects (*see* virial theorem). The SI unit of velocity dispersion is the metre per second ($m\ s^{-1}$).

virial theorem A theorem relating the total kinetic energy to the total gravitational potential energy for a system of gravitationally interacting particles that has settled into a state of equilibrium. A statistical treatment of such systems, leads to the result that the kinetic energy of the system is equal to $-1/2$ times the gravitational potential energy of the system. A consequence of this is that the velocity dispersion of the components of the system is expected to depend on the size of the region they occupy and on their total mass. This result is used to determine the masses of elliptical galaxies and of clusters of galaxies.

virialized The equilibrium state of a gravitationally bound system. The virial theorem applies only to systems in this state.

virtual particles Particle–antiparticle pairs that have a fleeting existence in the vacuum. The rest energy required for particle–antiparticle pair creation is related to the lifetime of the pair by the Heisenberg uncertainty relation.

voids Large (~60 Mpc) regions of the Universe in which the number density of galaxies is very low.

weak gravitational lensing The effect whereby the paths of rays of light (or other electromagnetic radiation) from a distant source are bent by the gravitational effect of an intervening mass such that the resulting image is subject to a small distortion. In contrast to strong gravitational lensing, multiple images are not produced. *See also* cosmic shear.

weakly interacting massive particle *See* WIMP.

WIMP (weakly interacting massive particle) A hypothetical elementary particle that has a relatively high mass and which only interacts by the weak interaction and gravity. A large population of such particles might account for a significant amount of (non-baryonic) dark matter.

winding dilemma The observation that if a galaxy's spiral arms consisted of an unchanging population of stars then differential rotation would cause the arms to smear out within a time that is short compared with the age of that galaxy.

FURTHER READING

General works at a level comparable to, or somewhat below this Book

Morison, I. (2008) *Introduction to Astronomy and Cosmology*, John Wiley & Sons Inc, Chichester.

Seeds, M.A. (2006) *Foundations of Astronomy*, Brooks/Cole, Pacific Grove.

Zeilik, M. (2002) *Astronomy: The Evolving Universe*, Cambridge University Press, Cambridge.

More specialized works at a level comparable to this book

Green, S.F. and Jones, M.H. (2015) *An Introduction to the Sun and Stars,* Cambridge University Press, Cambridge.

Guth, A.H. (1997) *The Inflationary Universe*, Jonathan Cape, London.

Silk, J. (2005) *On the Shores of the Unknown: A Short History of the Universe*, Cambridge University Press, Cambridge.

Tayler, R.J. (1993) *Galaxies: Structure and Evolution*, Cambridge University Press, Cambridge.

Works at a more advanced level

Binney, J. and Merrifield, M. (1998) *Galactic Astronomy*, Princeton University Press, Princeton.

Lambourne, R.J.A. (2010) *Relativity, Gravitation and Cosmology*, Cambridge University Press, Cambridge.

Liddle, A.R. (2003) *An Introduction to Modern Cosmology*, John Wiley & Sons Inc, Chichester.

Netzer, H. (2013) *The Physics and Evolution of Active Galactic Nuclei*, Cambridge University Press, Cambridge.

Roos, M. (2003) *Introduction to Cosmology*, John Wiley & Sons Inc, Chichester.

Rowan-Robinson, M. (2003) *Cosmology*, Oxford University Press, Oxford.

Serjeant, S. (2010) *Observational Cosmology*, Cambridge University Press, Cambridge.

Sparke, L.S. and Gallagher, J.S. (2007) *Galaxies in the Universe: An Introduction*, Cambridge University Press, Cambridge.

ACKNOWLEDGEMENTS

The production of this book involved a number of Open University staff, to whom we owe a considerable debt of thanks for their commitment and the high professional standards of their contributions. The following people have been instrumental in the production of this revised edition: Angela Russell, Chris Hough, Tara Hawes, Sharni Hirschy, Elliot O'Reilly, Claire Judd, Michael Hackett, Lisa Mills, Tracey Woodcraft, and especially Gill Knight. We are also grateful to Gill Gowans (Open University) and Vince Higgs (Cambridge University Press) for their support and help with co-publication.

We acknowledge the contributions made by individuals in preparing the earlier edition of this book. The original text was prepared by the following academic authors: David J. Adams, Alan Cayless, Anthony W. Jones, Barrie W. Jones, Lesley I. Onuora, Sean G. Ryan, Elizabeth Swinbank and Andrew N. Taylor. Other staff involved in the production of the original book were: Christopher Edwards, Valerie Cliff, Debbie Crouch, Pam Owen, Jane Henley, and especially Rebecca Graham. We also thank the following people who have acted as reviewers of the original or revised editions: Malcolm Longair (University of Cambridge), Michael Merrifield (University of Nottingham), the anonymous referees appointed by Cambridge University Press, Carole Haswell, Anthony W. Jones Fiona Vincent, Simon Green, Dr Heinz Andernach and students of the Open University.

Many other individuals and organizations furnished and/or granted permission for us to use their diagrams or photographs and to them we also express our gratitude.

Cover

Background image Spiral Galaxy M83, NGC5236, Southern Pinwheel, Hydra constellation: NASA, ESA and the Hubble Heritage Team (STSci/AURA); *Thumbnail images* Arp 148, Mayall's Object, interacting galaxies, constellation Ursa Major: NASA, ESA, the Hubble Heritage (STScI/AURA)-ESA/Hubble Collaboration, and A. Evans (University of Virginia, Charlottesville/NRAO/Stony Brook University); Abell 520 Cluster of galaxies (optical / X-ray / DM): NASA, ESA, CFHT, CXO, M.J. Jee (University of California, Davis), and A. Mahdavi (San Francisco State University); RXJ1131 Gravitational lens (of AGN) (optical / X-ray): NASA / CXC / Univ of Michigan / R.C.Reis et al; *An all-sky map of the fluctuations in the cosmic microwave background as measured by the European Sp*: ESA and the Planck Collaboration; NASA / WMAP Science Team;

Figures

Figure 1.1 Dennis di Cicco, Sky Publishing Corp.; *Figure 1.2* 1999–2002, Anglo-Australian Observatory, photograph by S. Lee, C. Tinney and D. Malin; *Figure 1.3* Atlas Image courtesy of 2MASS/UMass/IPAC-Caltech/NASA/NSF; *Figure 1.5* J. C. Barentine and G. A. Esquerdo (PSI), Kitt Peak, NOAO; *Figure 1.6* NASA/ESA. Image courtesy of Space Telescope Science Institution, Baltimore; *Figure 1.8* Astronomical Society of Australia 1999; *Figure 1.10* NASA, ESA, M. Robberto (STSci), Hubble Space Telescope Orion Treasury Project Team. This file is licensed under the Creative Commons Attribution Licence http://creativecommons.org/licenses/by/3.0/; *Figure 1.11c* Sofue, Y. *et al.* (2009) 'Unified Rotation Curve of the Galaxy – Decomposition into de Vaucouleurs Bulge, Disk, Dark Halo and

photograph by David Malin; *Figure 2.8* © 1987–2002, Anglo-Australian Observatory, photograph by David Malin; *Figure 2.10a* NASA and John Biretta (STScI/JHU); *Figure 2.10b* NASA and Hubble Heritage Team (STScI); *Figure 2.10c* NASA and S. Gallagher (Pennsylvania State University); *Figure 2.10d* European Southern Observatory; *Figure 2.10e* NASA, H. Ford (JHU); *Figure 2.11* ESA / Hubble & NASA; *Figure 2.13* The compressed files of the 'Palomar Observatory–Space Telescope Science Institute Digital Sky Survey' of the northern sky, based on scans of the Second Palomar Sky Survey are copyright © 1993–2000 by the California Institute of Technology and are distributed herein by agreement. All Rights Reserved; *Figure 2.14* Gadotti, D.A. (2008) 'Image decomposition of barred galaxies and AGN hosts', Monthly Notices of the Royal Astronomical Society, vol. 384, Issue 1, February 2008, pp. 420–439, Oxford Journals; *Figure 2.16* Westerbork Synthesis Radio Telescope, Netherlands; *Figure 2.17* D. Mihalas and J. Binney (1981), Galactic Astronomy, copyright © 1981 W. H. Freeman and Co.; *Figure 2.18* from Astrophysics and Space Research Group, The University of Birmingham, www.sr.bham.ac.uk; *Figure 2.19a* Han, Z. *et al.* (2007) 'A binary model for the UV-upturn of elliptical galaxies', Monthly Notices of the Royal Astronomical Society, vol. 380, Issue 3, September 2007, pp. 1098–1118, Oxford Journals; *Figure 2.20a and 2.20b* Anglo-Australian Observatory/David Malin Images; *Figure 2.21* NASA and ESA; *Figure 2.24* Bart J. Bok and Priscilla F. Bok (1974), reprinted by permission of the publishers from The Milky Way, Cambridge, Mas., Harvard University Press, copyright © 1941, 1945, 1957, 1974 by the President and Fellows of Harvard College; *Figure 2.26* R. Berendzen, R. Hart and D. Seeley (1976) Man Discovers the Galaxies, New York, Science History Publications; *Figure 2.27* UCL / University of London Observatory / Steve Fossey / Ben Cooke / Guy Pollack / Matthew Wilde / Thomas Wright. This file is licensed under the Creative Commons Attribution Licence http://creativecommons.org/licenses/by/2.0/; *Figure 2.31* © 2014 The Illustris Collaboration; *Figure 2.32* Michael Merrifield; *Figure 2.33* © 2014 The Illustris Collaboration; *Figure 2.34* Kennicutt, Jr, R. C. (1998) Star formation in galaxies along the Hubble sequence. Annual Review of Astronomy and Astrophysics 1998, vol. 36, Annual Reviews; *Figure 2.35a* Royal Observatory Edinburgh/Anglo Australian Telescope Board. Photograph by David Malin; *Figure 2.35b* from Alar Toomre (MIT) and Juri Toomre (University of Colorado); *Figure 2.39* The Sauron Project; *Figure 2.40* R. Williams and the HDF Team (STScI) and NASA; *Figure 2.40* NASA, ESA, G. Illingworth, D. Magee, and P. Oesch (University of California, Santa Cruz), R. Bouwens (Leiden University), and the HUDF09 Team; *Figure 2.41* Buitrago, F. *et al.* (2013) 'Early-type galaxies have been the predominant morphological class for massive galaxies since only z ~ 1', Monthly Notices of the Royal Astronomical Society, vol. 428, Issue 2, pp. 1460–1478, 11 January 2013, Oxford Journals; *Figure 2.42* Madau, P. and Dickinson, M. (2014) 'Cosmic Star-Formation History', Annual Review of Astronomy and Astrophysics, vol. 52, pp. 415–486, August 2014, Annual Reviews; *Figure 2.43a, 2.43b and 2.43c* NOAO;

Figure 3.4 R. C. Kennicutt, Astrophysical Journal, vol. 388, p. 310, © 1992 The American Astronomical Society; *Figure 3.10* NASA/IPAC Extragalactic Database NED; *Figure 3.11* NASA/IPAC Extragalactic Database NED; *Figure 3.12* NASA/ IPAC Extragalactic Database (NED), Jet Propulsion Laboratory, California Institute of Technology, the National Aeronautics and Space Administration; *Figure 3.14* Courtesy of Barry Poteete 2000 ©; *Figure 3.15* Kaufmann III, W. J. Galaxies and Quasars, 1979. W. H. Freeman and Company

Lucent Technologies) and NASA; *Figure 4.13* NASA, Andrew Fruchter and the ERO Team [Sylvia Baggett (STScI), Richard Hook (ST-ECF), Zoltan Levay (STScI)] (STScI); *Figure 4.15* L. J. King (U. Manchester), NICMOS, HST, NASA; *Figure 4.17a* Von der Linden, A. *et al.* (2014) 'Weighing the Giants – 1. Weak-lensing masses for 51 massive galaxy clusters: project overview, data analysis methods and cluster images', Monthly Notices of the Royal Astronomical Society (MNRAS), vol. 439, Issue 1, pp. 2–27, 21 March 2014, Oxford Journals; *Figure 4.17b* Von der Linden, A. *et al.* (2014) 'Weighing the Giants – 1. Weak-lensing masses for 51 massive galaxy clusters: project overview, data analysis methods and cluster images', Monthly Notices of the Royal Astronomical Society (MNRAS), vol. 439, Issue 1, pp. 2–27, 21 March 2014, Oxford Journals; *Figure 4.18* a Clowe, D. *et al.* (2006) 'A Direct Empirical Proof of the Existence of Dark Matter', The Astrophysical Journal, vol. 648, No. 2, pp. 109–113, 10 September 2006, IOP Publishing; *Figure 4.18b* Clowe, D. *et al.* (2006) 'A Direct Empirical Proof of the Existence of Dark Matter', The Astrophysical Journal, vol. 648, no. 2, pp. 109–113, 10 September 2006, IOP Publishing; *Figure 4.19* Courtesy of R. Brent Tully/Sky & Telescope; *Figure 4.20* © Copyright Maddox, S. *et al.*; *Figure 4.22a* Margaret, J. Geller and John, P. Huchra, Smithsonian Astrophysical Observatory; *Figure 4.22b* Image by Luis Teodoro, University of Durham, taken from www-astro.physics.ox.ac.uk; *Figure 4.23a* Mount Stromlo Observatory; *Figure 4.23b* Michael R. Blanton, Department of Physics, New York University / SDSS; *Figure 4.24a, 4.24b and 4.24c* Images sourced from http//www.astro.princeton.edu/~refreg/ weak_lensing/pictures.html, Alexandre Refregier, University of Cambridge; *Figure 4.28* Van Waerbeke, L. *et al.* (2013) 'CFHTLenS: mapping the large-scale structure with gravitational lensing', Monthly Notices of the Royal Astronomical Society (MNRAS), vol. 433, Issue 4, pp. 3373–3388, Oxford Journals; *Figure 4.29* Springel, V. *et al.* (2005) 'Simulations of the formation, evolution and clustering of galaxies and quasars', Nature, vol. 435, Issue 7042, p. 629–636, 2 June 2005, Nature Publishing Group;

Figure 5.1 AURA/STScI/NASA; *Figure 5.2* Sparke, L. S. and Gallagher, J. S. Galaxies in the Universe: An Introduction (2000) Cambridge University Press; *Figure 5.3* Douglas Scott; *Figure 5.4* Robert Smith, The 2df Quasar Redshift Survey; *Figure 5.6a* Adapted from Kirshner, R. (2004) 'Hubble's diagram and cosmic expansion', Proceedings of the National Academy of Sciences of the United States of America (PNAS), vol. 101, No. 106, January 2004, pp. 8–13; *Figure 5.6b* (Saul Perlmutter): © University of California, Lawrence Berkeley National Laboratory–Roy Kaltschmidt, photographer; (Brian Schmidt): SIPA USA / REX; (Adam Riess): ill Kirk / homewoodphoto.jhu.edu; *Figure 5.7, 5.17, 5.20 and 5.24* Science Photo Library; *Figure 5.14* Kittel, C., Knight, W. D. and Ruderman, M. A. Mechanics-Berkeley Physics Course – Volume 1 1965. Education Development Center, Inc; *Figure 5.15* Raine, D. J. The Isotropic Universe. 1981 Adam Hilger Ltd; *Figure 5.21* © Howard Percy Robertson; *Figure 5.22* © Arthur Geoffrey Walker; *Figure 5.23* Adapted from P.T. Landsberg and D.A Evans, Mathematical Cosmology, Oxford University Press, 1977; *Figure 5.25* The Royal Astronomical Society Library; *Figure 5.26* Sky Publishing Corp.; *Figure 5.33* adapted from Hannu Kurki-Suonio http://www.helsinki.fi/~hkurkisu/cosmology/Cosmo4.pdf;

Figure 6.13 T.P. Walker. *et al.* 'Primordial Nucleosynthesis Redux', Astrophysical Journal (1991) vol. 376, p. 51 © American Astronomical Society; *Figure 6.17* Image courtesy of Bell Laboratories, AT & T; *Figure 6.19* NASA / WMAP Science

FIGURE REFERENCES

Abell, G.O., Corwin, H.G. and Olowin, R.P. (1989) A catalog of rich clusters of galaxies, *Astrophysical Journal Supplement Series*, **70**, p. 1.

Ade, P.A.R. *et al.* (2013) Planck 2013 results. XVI. Cosmological parameters, arXiv:1303.5076

Amanullah, R. *et al.* (2010) Spectra and Hubble Space Telescope Light Curves of Six Type Ia Supernovae at $0.511 < z < 1.12$ and the Union2 Compilation, *The Astrophysical Journal*, **716**, p. 712.

Arp, H.C. (1958) *Handbook of Physics*, **51**, p. 75.

Belokurov, V. *et al.* (2006) The Field of Streams: Sagittarius and its Sibling*s*. *Astrophysical Journal*, **642**, p. 137.

Bennett, C.L. *et al.* (2003) First-Year Wilkinson Microwave Anisotropy Probe (WMAP) Observations: Preliminary Maps and Basic Results, *Astrophysical Journal Supplement Series*, **148**, p. 1.

Berendzen, R., Hart, R. and Seeley, D. (1976) *Man Discovers the Galaxies*, Science History Publications, New York.

Binney, J. and Merrifield, M. (1998) *Galactic Astronomy*, Princeton University Press, Princeton.

Blitz, L. and Spergel, D.N. (1991) Direct evidence for a bar at the Galactic center, *Astrophysical Journal*, **379**, p. 631.

Bok, B.J. and Bok, P.F. (1974) *The Milky Way*, Harvard University Press, Cambridge, Massachusetts.

Buitrago, F. *et al.* (2013) Early-type galaxies have been the predominant morphological class for massive galaxies since only $z \sim 1$, *Monthly Notices of the Royal Astronomical Society*, **428**, p. 1460.

Burton, W.B. (1985) Leiden-Green Bank survey of atomic hydrogen in the galactic disk. I. l, v and b, v maps, *Astronomy and Astrophysics Supplement Series*, **62**, p. 365.

Castelli, F. and Kurucz, R.L. (2004) New Grids of ATLAS9 Model Atmospheres, arXiv:astro-ph/0405087.

Clowe, D. *et al.* (2006) A Direct Empirical Proof of the Existence of Dark Matter, *Astrophysical Journal*, **648**, p. 109.

Compiegne , M. *et al.* (2011) The global dust SED: tracing the nature and evolution of dust with DustEM, *Astronomy & Astrophysics*, **525**, A103.

Condon, J.J., Helou, G., Sanders, D.B. and Soifer, B.T. (1996) A 1.425 GHz Atlas of the IRAS Bright Galaxy Sample, Part II, *Astrophysical Journal Supplement Series*, **103**, p. 81.

Davies, R.L. *et al*. (2001) Galaxy Mapping with the SAURON Integral-Field Spectrograph: The Star Formation History of NGC 4365, *Astrophysical Journal*, **548**, L33.

D'Onghia , E. *et al.* (2013) Self-Perpetuating Spiral Arms in Disk Galaxies, *Astrophysical Journal,* **766**, p. 34.

Elvis, M. *et al*. (1994) Atlas of quasar energy distributions, *Astrophysical Journal Supplement Series*, **95**, p. 1.

Finkbeiner, A.K. (1998) Cosmic yardsticks: supernovae and the fate of the Universe, *Sky and Telescope*, September, p. 43.

Francis, P.J. *et al*. (1991) A high signal-to-noise ratio composite quasar spectrum, *Astrophysical Journal*, **373**, p. 465.

Freedman, W.L. *et al*. (2001) Final Results from the Hubble Space Telescope Key Project to Measure the Hubble Constant, *Astrophysical Journal*, **553**, p. 47.

Frieman, J.A. *et al*. (2008) Dark Energy and the Accelerating Universe, *Annual Review of Astronomy and Astrophysics*, **46**, p. 385.

Gadotti, D.A. (2008) Image decomposition of barred galaxies and AGN hosts, *Monthly Notices of the Royal Astronomical Society,* **384**, p. 420.

Genzel, R. and Cesarsky, C. (2000) Extragalactic Results from the Infrared Space Observatory, *Annual Review of Astronomy and Astrophysics*, **38**, p. 761.

Gillessen, S. *et al*. (2009a) Monitoring Stellar Orbits around the Massive Black Hole in the Galactic Center, *Astrophysical Journal,* **692**, p. 1075.

Gillessen , S. *et al*. (2009b) The Orbit of the Star S2 Around SGR A* from Very Large Telescope and Keck Data, *Astrophysical Journal*, **707**, L114.

Han, Z. *et al*. (2007) A binary model for the UV-upturn of elliptical galaxies, *Monthly Notices of the Royal Astronomical Society*, **380**, p. 1098.

Hartmann, D. and Burton, W.B. (1997) *Atlas of Galactic Neutral Hydrogen*, Cambridge University Press, Cambridge.

Henbest, N. and Marten, M. (1983) *The New Astronomy, Cambridge University* Press, Cambridge.

Hesser, J., Harris, W. and VandenBerg, D. (1987) Inferences from a color magnitude diagram for 47-Tucanae, *Publications of the Astronomical Society of the Pacific,* **99**, p. 1148.

Hibbard, J.E., Guhathakurta, P., van Gorkom, J.H. and Schweizer, F. (1994) Cold, warm, and hot gas in the late-stage merger NGC 7252, *Astronomical Journal*, **107**, p. 67.

Jenkins, A. *et al*. (The Virgo Consortium) (1998) Evolution of Structure in Cold Dark Matter Universes, *Astrophysical Journal*, **499**, p. 20.

Kalberla , P.M.W. and Kerp, J. (2009) The HI Distribution of the Milky Way, *Annual Review of Astronomy and Astrophysics*, **47**, p. 27.

Kaufmann III, W.J. (1979) *Galaxies and Quasars*, W.H. Freeman and Company Publishers, San Francisco.

Kennicutt, R.C. (1992) The integrated spectra of nearby galaxies – *General properties and emission-line spectra, Astrophysical Journal*, **388**, p. 310.

Kennicutt, R.C. (1998) Star Formation in Galaxies along the Hubble Sequence, *Annual Review of Astronomy and Astrophysics*, **36**, p. 189.

Kerr, F.J., Bowers, P.F., Jackson, P.D. and Kerr, M. (1986) Fully sampled neutral hydrogen survey of the southern Milky Way, *Astronomy and Astrophysics Supplement Series*, **66**, p. 373.

Kirshner, R. (2004) Hubble's diagram and cosmic expansion, *Proceedings of the National Academy of Sciences of the United States of America*, **101**, p. 8.

Kittel, C., Knight, W.D. and Ruderman, M.A. (1965) Mechanics-Berkeley Physics Course – Volume 1, Education Development Center, Inc.

Kolb, E.W. and Turner, M.S. (1990) *The Early Universe*, Addison-Wesley, Reading, Massachusetts.

Landsberg, P.T. and Evans, D.A. (1977) *Mathematical Cosmology*, Oxford University Press, Oxford.

Lee, J.C. *et al.* (2002) The shape of the relativistic iron Kα line from MCG – 6–30–15 measured with the Chandra High Energy Transmission Grating Spectrometer and the Rossi X-ray Timing Explorer, *Astrophysical Journal*, **570,** L47.

Longair, M.S. (1998) *Galaxy Formation*, Springer, Berlin/ London.

Madau , P. and Dickinson, M. (2014) Cosmic Star-Formation History, *Annual Review of Astronomy and Astrophysics*, **52**, p. 415.

Majewski , S.R. *et al.* (2003) A Two Micron All Sky Survey View of the Sagittarius Dwarf Galaxy. I. Morphology of the Sagittarius Core and Tidal Arms, *Astrophysical Journal*, **599**, p. 1082.

Mihalas, D. and Binney, J. (1981) *Galactic Astronomy*, Freeman and Co., Oxford.

Mirabel, I.F. *et al.* (1998) The dark side of star formation in the Antennae galaxies, *Astronomy and Astrophysics*, **333**, L1.

Morse, J.A., Cecil, G., Wilson, A.S. and Tsvetanov, Z.I. (1998) Inclined gas disks in the lenticular Seyfert Galaxy NGC 5252, *Astrophysical Journal*, **505**, p. 159.

Nicolson, I. (1982) The Sun, Mitchell Beazley, London.

Osterbrock, D.E. Koski, A.T. and Phillips, M.M. (1976) The optical spectra of 3C 227 and other broad-line radio galaxies, *Astrophysical Journal*, **206**, p. 898.

Ouyed, R. *et al.* (2013) Quark-Novae Ia in the Hubble diagram: Implications For Dark Energy, *Research in Astronomy and Astrophysics*, arXiv:1310.4535.

Perley, R.A., Dreher, J.W. and Cowan, J.J. (1984) The jet and filaments in Cygnus A, *Astrophysical Journal*, **285**, L35.

Peterson, B.M. (1997) *An Introduction to Active Galactic Nuclei*, Cambridge University Press, Cambridge.

Raine, D.J. (1981) *The Isotropic Universe*, Adam Hilger Ltd., Bristol.

Roth, J. (1997) Dating the cosmos: a progress report, *Sky and Telescope*, October, p. 46.

Ryder, S.D., Walsh, W. and Malin, D. (1999) HI study of the NGC 6744 system, *Publications of the Astronomical Society of Australia*, **16**, p. 84.

Sandage A., and Bedke, J. (1994) *The Carnegie Atlas of Galaxies*, Volume I, Carnegie Institution of Washington.

Saunders, W., Saunders, W., Sutherland, W.J., Maddox, S.J., Keeble, O., S.J., Rowan-Robinson, M., McMahon, R.G., Efstathiou, G.P., Tadros, H., White, S.D.M., Frenk, C.S., Carraminana, A. and Hawkins, M.R.S. (2000) The PSCz catalogue, *Monthly Notices of the Royal Astronomical Society*, **317**, p. 55.

Scarpa, R., Urry, C.M., Falomo, R., Pesce, J.E. and Treves, A. (2000) The Hubble Space Telescope Survey of BL Lacertae Objects. I. Surface Brightness Profiles, Magnitudes, and Radii of Host Galaxies, *Astrophysical Journal*, **532**, p. 740.

Schild, R. and Thomson, D. (1997) The Q0957+561 time delay from optical data, *Astronomical Journal*, **113**, p. 130.

Seeds, M.A. (1988) *Foundations of Astronomy*, (2nd edn) Wadsworth Publishing, Belmont, California.

Silva, D.R. and Cornell, M.E. (1992) A new library of stellar optical spectra, *Astrophysical Journal Supplement Series*, **81**, p. 865.

Sofue, Y. *et al.* (2009) Unified Rotation Curve of the Galaxy – Decomposition into de Vaucouleurs Bulge, Disk, Dark Halo and the 9-kpc Rotation Dip, *Publications of the Astronomical Society of Japan* **61**, p. 227.

Sparke, L.S. and Gallagher, J.S. (2000) *Galaxies in the Universe: An Introduction*, Cambridge University Press, Cambridge.

Springel, V. *et al.* (2005) Simulations of the formation, evolution and clustering of galaxies and quasars, *Nature*, **435**, p. 629.

Stern, D. *et al.* (2000) Discovery of a Color-selected Quasar at $z = 5.50$, *Astrophysical Journal*, **533**, L75.

Sukyoung Yi. *et al.* (2001) Toward Better Age Estimates for Stellar Populations: The Y 2 Isochrones for Solar Mixture, *Astrophysical Journal*, **136**, p. 417.

Suzuki, N. *et al* (2011) The Hubble Space Telescope Cluster Supernova Survey..., The Supernova Cosmology Project, arXiv:1105.3470.

Suzuki , N. *et al.* (2012) The Hubble Space Telescope Cluster Supernova Survey. V. Improving the Dark-energy Constraints above $z > 1$ and Building an Early-type-hosted Supernova Sample, *Astrophysical Journal*, **746**, p. 85.

Tully, R.B. *et al.* (1996) The Ursa Major Cluster of Galaxies. I. Cluster Definition and Photometric Data, *Astronomical Journal*, **112**, p. 2471.

Urquhart, J.S. *et al.* (2013) The RMS survey: galactic distribution of massive star formation, *Monthly Notices of the Royal Astronomical Society*, **437**, p. 1791.

Van Waerbeke, L. *et al.* (2013) CFHTLenS: mapping the large-scale structure with gravitational lensing, *Monthly Notices of the Royal Astronomical Society*, **433**, p. 3373.

Von der Linden, A. *et al.* (2014) Weighing the Giants - 1. Weak-lensing masses for 51 massive galaxy clusters: project overview, data analysis methods and cluster images, *Monthly Notices of the Royal Astronomical Society*, **439**, p. 2.

Walker, T.P., Steigman, G., Kang, Ho-Shik, Schramm, D.M. and Olive, K.A. (1991) Primordial nucleosynthesis redux, *Astrophysical Journal*, **376**, p. 51.

Weinberg, M.D. (1992) Detection of a large-scale stellar bar in the Milky Way, *Astrophysical Journal*, **384**, p. 81.

Wilson, B.A. *et al.* (2005) A uniform CO survey of the molecular clouds in Orion and Monoceros, *Astronomy & Astrophysics*, **430**, p. 523.

Wolfe, A.M., Turnshek, D.A., Lanzetta, K.M. and Lu, L. (1993) Damped Lyman alpha absorption by disk galaxies with large redshifts. IV – More intermediate resolution spectroscopy, *Astrophysical Journal*, **404**, p. 480.

Zinn, R. (1985) The globular-cluster system of the Galaxy IV. The halo and disk subsystems, *Astrophysical Journal*, **293**, p. 424.

INDEX

Entries in bold refer to glossary terms and the page they appear in **bold**. Page numbers in *italics* refer to entries in figures or tables.

C